96.50 80E

Advances in

Heterocyclic Chemistry

Volume 38

Editorial Advisory Board

R. A. Abramovitch, *Clemson, South Carolina*
A. Albert, *Canberra, Australia*
A. T. Balaban, *Bucharest, Romania*
A. J. Boulton, *Norwich, England*
H. Dorn, *Berlin, G.D.R.*
S. Gronowitz, *Lund, Sweden*
T. Kametani, *Tokyo, Japan*
C. W. Rees FRS, *London, England*
Yu. N. Sheinker, *Moscow, U.S.S.R.*
E. C. Taylor, *Princeton, New Jersey*
M. Tišler, *Ljubljana, Yugoslavia*
J. A. Zoltewicz, *Gainesville, Florida*

Advances in

HETEROCYCLIC CHEMISTRY

Edited by

ALAN R. KATRITZKY FRS

Kenan Professor of Chemistry
Department of Chemistry
University of Florida
Gainesville, Florida

1985

Volume 38

ACADEMIC PRESS, INC.
(Harcourt Brace Jovanovich, Publishers)

Orlando San Diego New York London
Toronto Montreal Sydney Tokyo

COPYRIGHT © 1985, BY ACADEMIC PRESS, INC.
ALL RIGHTS RESERVED.
NO PART OF THIS PUBLICATION MAY BE REPRODUCED OR
TRANSMITTED IN ANY FORM OR BY ANY MEANS, ELECTRONIC
OR MECHANICAL, INCLUDING PHOTOCOPY, RECORDING, OR
ANY INFORMATION STORAGE AND RETRIEVAL SYSTEM, WITHOUT
PERMISSION IN WRITING FROM THE PUBLISHER.

ACADEMIC PRESS, INC.
Orlando, Florida 32887

United Kingdom Edition published by
ACADEMIC PRESS INC. (LONDON) LTD.
24-28 Oval Road, London NW1 7DX

LIBRARY OF CONGRESS CATALOG CARD NUMBER: 62-13037

ISBN: 0-12-020638-2

PRINTED IN THE UNITED STATES OF AMERICA

85 86 87 88   9 8 7 6 5 4 3 2 1

Contents

CONTRIBUTORS	vii
PREFACE	ix

Recent Advances in the Chemistry of Dihydroazines
ALEXANDER L. WEIS

I.	Introduction	3
II.	Scope and Limitations	6
III.	General Overview of Dihydroazine Chemistry	6
IV.	Dihydropyridazines	23
V.	Dihydropyrimidines	45
VI.	Dihydro-1,2,3-triazines	82
VII.	Dihydro-1,2,4-triazines	83
VIII.	Dihydro-1,3,5-triazines	91
IX.	Conclusions	102

Recent Advances in the Chemistry of Benzisothiazoles and Other Polycyclic Isothiazoles
MICHAEL DAVIS

I.	Introduction	106
II.	1,2-Benzisothiazoles	107
III.	1,2-Benzisothiazolin-3(2H)-one 1,1-Dioxide (Saccharin) and Derivatives	116
IV.	2,1-Benzisothiazoles	122
V.	Other Isothiazole Systems	130
VI.	Selenium and Tellurium Analogues	132

1,4-Benzothiazines, Dihydro-1,4-benzothiazines, and Related Compounds
C. BROWN AND R. M. DAVIDSON

I.	Introduction	135
II.	Natural Occurrence	138
III.	Pharmacology of 1,4-Benzothiazines and Their Derivatives	142
IV.	Agricultural Uses	145
V.	Industrial Uses	146
VI.	Synthesis	146
VII.	Reactions of 1,4-Benzothiazines	167

The Chemistry of Hydantoins
C. AVENDAÑO LÓPEZ AND G. GONZÁLEZ TRIGO

I. Introduction	178
II. Methods of Synthesis	178
III. Physicochemical Studies on Hydantoins	203
IV. Reactivity of Hydantoins and Their Derivatives	209
V. Uses and Applications	224

Recent Progress in Barbituric Acid Chemistry
JACEK T. BOJARSKI, JERZY L. MOKROSZ, HENRYK J. BARTÓN, AND MARIA H. PALUCHOWSKA

I. Introduction	229
II. Physicochemical Properties of Barbiturates	231
III. Reactivity of Barbiturates	263
IV. Analytical Methods for Barbiturates	288
V. Correlation Analysis in Barbituric Acid Chemistry	292
VI. Closing Remarks	295
VII. Addendum	297

Heterocyclic β-Enamino Esters, Versatile Synthons in Heterocyclic Synthesis
HEINRICH WAMHOFF

I. Introduction	300
II. Synthetic Approaches	301
III. Structure and Spectral Properties	309
IV. General Considerations of Chemical Reactivity	316
V. Electrophilic Attacks on the 2-Amino Group (or the Carbon-3 Atom). Synthesis of Polynuclear Heterocyclic Systems	318
VI. Nucleophilic Attacks of Amines and Diamines	343
VII. Cycloadditions and Reactions with Acetylenic Esters	347
VIII. 2-(Triphenylphosphorylidenamino) Esters. The Cycloaddition–Ring Enlargement Sequence	351
IX. Photochemistry	355
X. Heterocyclic β-Enamino Nitriles	357
XI. Conclusion and Outlook	367

CUMULATIVE INDEX OF TITLES	369

Contributors

Numbers in parentheses indicate the pages on which the authors' contributions begin.

HENRYK J. BARTOŃ, *Department of Organic Chemistry, Nicolaus Copernicus Academy of Medicine, 30–048 Kraków, Poland* (229)

JACEK T. BOJARSKI, *Department of Organic Chemistry, Nicolaus Copernicus Academy of Medicine, 30–048 Kraków, Poland* (229)

C. BROWN, *The Chemical Laboratory, University of Kent at Canterbury, Canterbury, Kent CT2 7NH, England* (135)

R. M. DAVIDSON, *The Chemical Laboratory, University of Kent at Canterbury, Canterbury, Kent CT2 7NH, England* (135)

MICHAEL DAVIS, *Department of Organic Chemistry, La Trobe University, Bundoora, Victoria 3083, Australia* (105)

C. AVENDAÑO LÓPEZ, *Departamento de Química Orgánica y Farmacéutica, Facultad de Farmacia, Universidad Complutense, Madrid-3, Spain* (177)

JERZY L. MOKROSZ, *Department of Organic Chemistry, Nicolaus Copernicus Academy of Medicine, 30–048 Kraków, Poland* (229)

MARIA H. PALUCHOWSKA, *Department of Organic Chemistry, Nicolaus Copernicus Academy of Medicine, 30–048 Kraków, Poland* (229)

G. GONZÁLEZ TRIGO, *Departamento de Química Orgánica y Farmacéutica, Facultad de Farmacia, Universidad Complutense, Madrid-3, Spain* (177)

HEINRICH WAMHOFF, *Institut für Organische Chemie und Biochemie der Universität Bonn, D–5300 Bonn 1, Federal Republic of Germany* (299)

ALEXANDER L. WEIS, *Department of Organic Chemistry, The Weizmann Institute of Science, 76 100 Rehovot, Israel* (1)

Preface

Volume 38 of *Advances in Heterocyclic Chemistry* is composed of six chapters which deal with six diverse classes of heterocyclic compounds.

A. L. Weis of the Weizmann Institute gives the first authoritative account of the chemistry of dihydroazines, a group of compounds that until recently has been rather neglected.

Two chapters deal with sulfur-containing compounds. The benzisothiazoles and other polycyclic isothiazoles are covered by M. Davis, in a sequel to earlier reviews in *Advances in Heterocyclic Chemistry*, Volumes 14 and 15. 1,4-Benzothiazines and related compounds are dealt with by C. Brown and R. M. Davidson; these compounds have not been previously comprehensively reviewed.

C. Avendaño and G. G. Trigo review the chemistry of the hydantoins, a subject that has not been fully treated for nearly 30 years. Another biologically important group of compounds, the barbituric acids, of which the last definitive account appeared more than a quarter of a century ago, are brought up to date by J. T. Bojarski *et al.*

Finally, H. Wamhoff discusses heterocyclic β-enamino esters, which are versatile synthetic building blocks.

ALAN R. KATRITZKY, FRS

Recent Advances in the Chemistry of Dihydroazines

ALEXANDER L. WEIS

Department of Organic Chemistry, The Weizmann Institute of Science, Rehovot, Israel

I. Introduction	3
II. Scope and Limitations	6
III. General Overview of Dihydroazine Chemistry	6
A. Nomenclature	6
B. Structure	9
1. Constitution and Stability	9
2. Molecular Geometry and Conformation	12
C. Preparation of Dihydroazines	15
D. Tautomerism and Rearrangements	18
1. Amidinic Tautomerism	19
2. Imine–Enamine Tautomerism	21
E. Physicochemical Properties and Reactivity	22
IV. Dihydropyridazines	23
A. General Comments	23
B. Synthesis	23
1. Cyclization Methods	23
2. Cycloaddition Reactions	27
3. From Pyridazine and its Derivatives	32
a. Addition of Organometallic Reagents	32
b. From 6-Oxo-1,4,5,6-tetrahydropyridazines	35
c. Reduction with Complex Hydrides	37
d. Electrochemical Reduction	37
4. Miscellaneous	39
C. Tautomerism of Dihydropyridazines	39
D. Physicochemical Properties	40
1. Infrared Spectra	40
2. Ultraviolet Spectra	41
3. Nuclear Magnetic Resonance Spectra	41
E. Reactivity	44
V. Dihydropyrimidines	45
A. General Comments	45
B. Synthesis	45
1. Cyclization Methods	45
a. From α,β-Unsaturated Carbonyls	45
b. From β-Dicarbonyl Compounds	53
c. Other Cyclizations	54

- 2. From Pyrimidine Derivatives 55
  - a. Addition of Organometallic Reagents 55
  - b. Reduction with Complex Metal Hydrides 58
  - c. From 4-Halopyrimidines 59
  - d. Electrophilic Aromatic Substitution with Protonated Pyrimidines . . . 60
  - e. Desulfurization of Pyrimidine-2(1H)-thiones 60
  - f. Electrochemical Reduction 61
- 3. Miscellaneous 62
- C. Tautomerism of Dihydropyrimidines 63
  - 1. Tautomerism of 1,4- and 1,6-Dihydropyrimidines 63
    - a. Structures of the Tautomers 63
    - b. External Parameters Affecting the Rate of Tautomerism 65
    - c. Description of NMR Spectra 66
    - d. Kinetic Measurements 70
    - e. Intra- and Intermolecular Mechanisms of Proton Transfer in Amidines . 73
  - 2. Tautomerism of 1,2-Dihydropyrimidines 75
- D. Physicochemical Properties 77
  - 1. Infrared Spectra 77
  - 2. Ultraviolet Spectra 78
  - 3. Nuclear Magnetic Resonance Spectra 78
  - 4. Crystal Structure 79
- E. Reactivity 80
- VI. Dihydro-1,2,3-triazines 82
- VII. Dihydro-1,2,4-triazines 83
  - A. General Comments 83
  - B. Synthesis 84
    - 1. Cyclization Methods 84
      - a. From N-Acylamino Ketones 84
      - b. Other Cyclizations 85
    - 2. From 1,2,4-Triazine Derivatives 85
      - a. Reduction of 1,2,4-Triazines 85
      - b. From 1,2,4-Triazines or 1,2,4-Triazinones and Grignard Reagents . . . 86
      - c. Electrochemical Reduction 86
      - d. Other Methods 87
    - 3. From Other Heterocycles 87
  - C. Physicochemical Properties 88
    - 1. Infrared Spectra 88
    - 2. Ultraviolet Spectra 89
    - 3. Nuclear Magnetic Resonance Spectra 89
  - D. Reactivity 90
- VIII. Dihydro-1,3,5-triazines 91
  - A. General Comments 91
  - B. Synthesis 91
    - 1. From Nitriles 91
    - 2. From Amidines 93
    - 3. From 1,3,5-Triazines 96
    - 4. From 1,3,5-Oxadiazines 97
  - C. Tautomerism in Dihydro-1,3,5-triazines 98
  - D. Physicochemical Properties 99
    - 1. Infrared Spectra 99

| 2. Ultraviolet Spectra and Photochromism 100
| 3. Nuclear Magnetic Resonance Spectra 100
| E. Reactivity. 101
| IX. Conclusions 102

I. Introduction

Aromatic, six-membered, nitrogen-containing heterocyclic compounds (e.g., pyridines, diazines) have long been known, and their preparation, chemical and physical properties, and occurrence and function in biological systems have been extensively studied. At the same time, despite the large number of known *reduced* azines, their chemistry remains largely unexplored. Among the partially or totally saturated azines, particular attention has lately been given to the dihydroazines (DHA) which, as their name indicates, contain two additional hydrogens on the azine ring. They comprise a group of compounds that had previously been regarded primarily as unstable intermediates formed during azine synthesis. This lively interest in dihydroazines not only reflects the general renaissance of heterocyclic chemistry, but it is also the result of recent advances in theoretical, bio-, and medicinal chemistry.

Of relevance to theoreticians, for example, is the ability to predict the structure, binding properties, chemical reactivity, etc. of dihydro compounds from a knowledge of the number and positions of the nitrogen atoms along with the disposition of double bonds within the DHA ring. New models for determining the relative stabilities of different DHA isomers open completely new approaches for predicting possible spontaneous isomerization of a derivative as well as its redox properties—a fundamental question of no less importance. Such quantum mechanical calculations also enable an evaluation of the degree of aromatic character in potential "homoaromatic" and "antiaromatic" isomers.

From the biochemical point of view, dihydroazines and particularly those containing the 1,4-dihydropyridine moiety[1-3] are of intense interest because of the presence of this group at the active site of "hydrogen transferring coenzyme" NADH (1) (reduced nicotinamide adenine dinucleotide). This nucleotide is a central participant in metabolic processes in living organisms, where it participates in the reduction of various unsaturated functionalities. In order to understand better the energetics and mechanism of NADH derivatives—both chiral and achiral—for use as NADH analogs.

[1] U. Eisner and J. Kuthan, *J. Chem. Rev.* **72**, 1 (1972).
[2] J. Kuthan and A. Kurfürst, *Ind. Chem. Eng. Prod. Res. Dev.* **21**, 191 (1982).
[3] D. M. Stout and A. I. Meyers, *J. Chem. Rev.* **82**, 223 (1982).

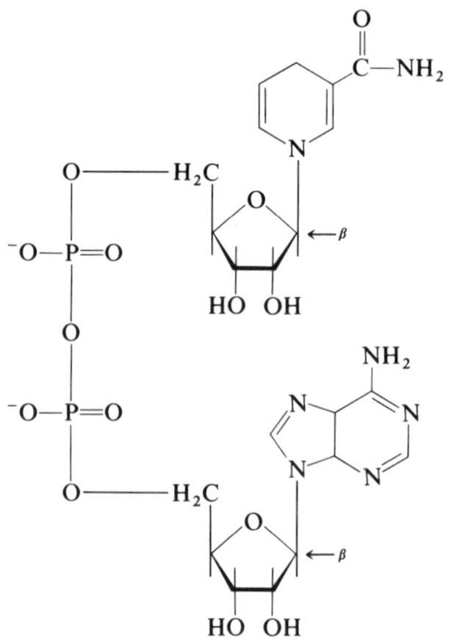

(1)

In addition, biochemists are also interested in the 1,4-dihydroazine[4,5] structure, present in several families of naturally occurring compounds, including the 1,5-dihydroflavins (2), which play a part in many biochemical processes, and the luciferins (3), which endow some interesting organisms with the property of bioluminescence.

Pharmaceutical chemists, too, have evinced a keen interest in DHA chemistry, particularly the 4-aryldihydropyridines, which exhibit powerful vasodilating activities via the blocking of calcium channels and modifying movement of Ca^{2+} into and within the cell. The explosion in activity in this area of heterocyclic synthesis has produced an exponential growth in patent applications and papers and has led to the marketing of a new drug, nifedipine (4).[6] Other active 4-aryldihydropyridine derivatives, such as nimodipine and nicardipine (cerebral vasodilators) and nitrendipine (antihypertensive), are presently under clinical trials.[7] An excellent list of other practical applications of dihydropyridines has been collected by Kuthan and Kurfürst.[2]

The author originally intended this work as an overall survey of dihydroazine chemistry. However, the appearance in 1982 of excellent reviews of

[4] R. R. Schmidt, *Angew. Chem., Int. Ed. Engl.* **14**, 581 (1975).
[5] W. Kaim, *Angew. Chem., Int. Ed. Engl.* **22**, 171 (1983).
[6] W. Vater, G. Kronenberg, F. Hoffmeister, H. Kaller, A. Meng, A. Oberdorf, W. Puls, K. Schlossmann, and K. Stoepel, *Arzneim.-Forsch.* **22**, 1 (1972).
[7] F. Bossert, H. Meyer, and E. Wehinger, *Angew. Chem., Int. Ed. Engl.* **20**, 762 (1981).

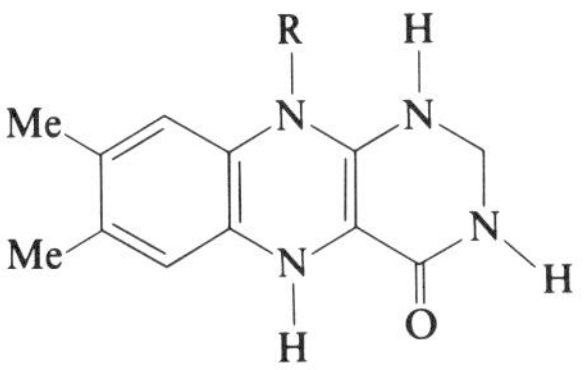

(2)

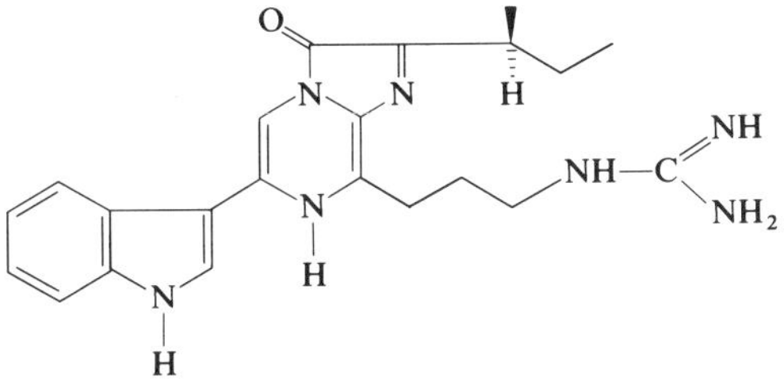

(3)

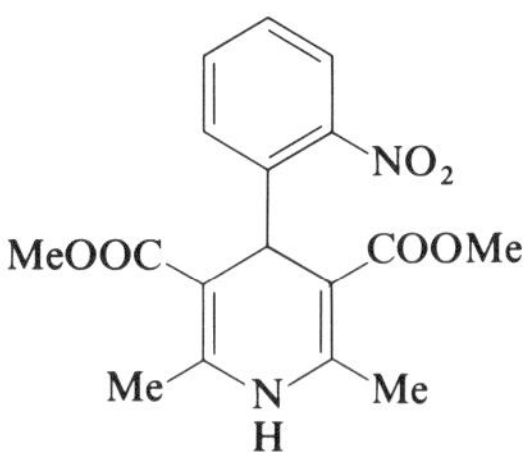

(4)

dihydropyridine[2,3] and dihydropyrazine[8] chemistry has required some modification in order to avoid unnecessary duplication.

This work, therefore, will not be an exhaustive treatise on the subject. It will summarize present knowledge and concentrate on the presentation of some important new trends and achievements in dihydroazine chemistry.

We hope, in particular, to illustrate the generality and scope of many of the synthetic processes discussed. It is also our aim to point out how similarities in structure of various dihydroazines lead to similarities in their physical and chemical properties and how such common structural features can be utilized to predict properties of compounds that have not yet been obtained or studied experimentally.

All readily accessible literature has been examined, including references which have appeared up to, and including, Volume 97 of *Chemical Abstracts* (December 1982), and some unpublished data (mainly from the author's laboratory) have been included. Because of the subjectivity inherent in choosing material for inclusion in this survey, it is possible that some important contributions have been unintentionally overlooked, for which the author expresses his sincere apologies.

---

[8] G. B. Barlin, in "The Chemistry of Heterocyclic Compounds" (A Weissberger, and E. C. Taylor, eds.), Vol. 41, p. 344. Wiley, Chichester, 1982.

II. Scope and Limitations

Following the approach taken by Eisner and Kuthan in their review of dihydropyridines,[1] this article will incorporate only isolable or spectroscopically identified monocyclic dihydrodiazines and dihydrotriazines. We specifically exclude compounds containing exocyclic double bonds, i.e., particularly azine methenes, as well as oxo-, thio- or iminodihydroazines (**5a**), which can be considered as tautomeric azines (**5b**), and hydroxy, mercapto, and amino DHA (**6a**), which can exist in the tautomeric tetrahydroazine form (**6b**).

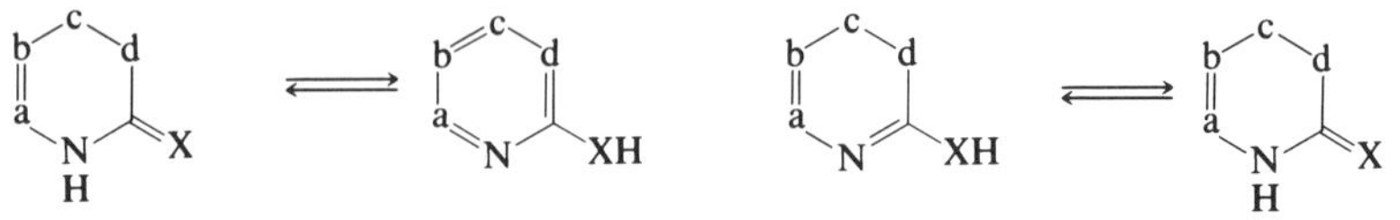

(**5a**)　　　(**5b**)　　　(**6a**)　　　(**6b**)

X = CH_2, O, S, NH　　　　　a, b, c, d = C or N

Particular attention will be given to those dihydroazines bearing no substituents on the ring nitrogen atom(s), which can undergo oxidation to the corresponding azines.

III. General Overview of Dihydroazine Chemistry

A. NOMENCLATURE

During the preparation of this review, minor inconsistencies in the nomenclature of dihydroazines were often encountered in the literature. Therefore, the author would like to present some systematic rules for numerating hydrogenated heterocycles, a convention which will be followed in this survey. (1) When the same heteroatom occurs more than once in a ring, the numbering is chosen to give lowest locants to the heteroatoms. (2) If there is still an ambiguity in numbering the dihydroazine, the lowest number is assigned to the heteroatom with the higher degree of saturation. (3) If there is still an ambiguity, numbering will be assigned such that the extra hydrogens will have the lowest locants.

According to these rules, the following names will be afforded to the possible isomeric dihydroazines.

1. Dihydropyridines:

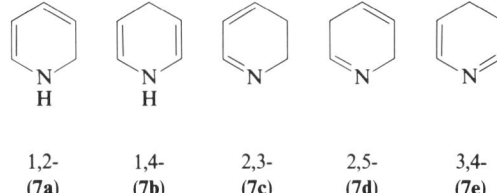

| 1,2- | 1,4- | 2,3- | 2,5- | 3,4- |
| (7a) | (7b) | (7c) | (7d) | (7e) |

2. Dihydropyridazines:

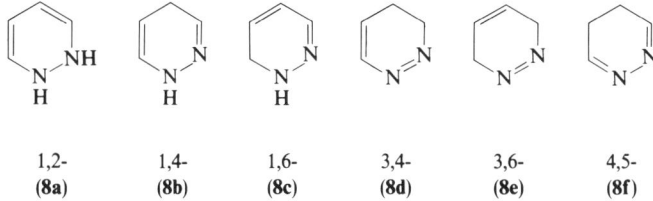

| 1,2- | 1,4- | 1,6- | 3,4- | 3,6- | 4,5- |
| (8a) | (8b) | (8c) | (8d) | (8e) | (8f) |

3. Dihydropyridimines:

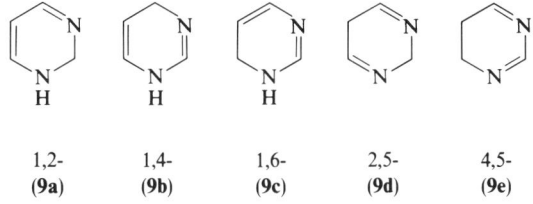

| 1,2- | 1,4- | 1,6- | 2,5- | 4,5- |
| (9a) | (9b) | (9c) | (9d) | (9e) |

4. Dihydropyrazines:

| 1,2- | 1,4- | 2,3- | 2,5- |
| (10a) | (10b) | (10c) | (10d) |

5. Dihydro-1,2,3-triazines:

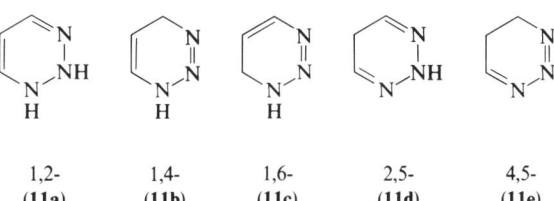

| 1,2- | 1,4- | 1,6- | 2,5- | 4,5- |
| (11a) | (11b) | (11c) | (11d) | (11e) |

6. Dihydro-1,2,4-triazines:

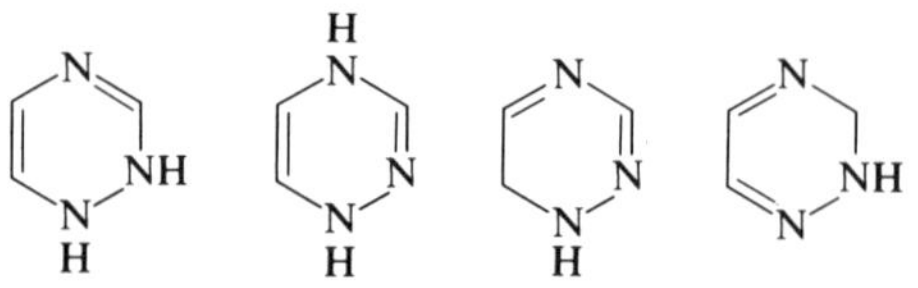

1,2-	1,4-	1,6-	2,3-
(12a)	(12b)	(12c)	(12d)

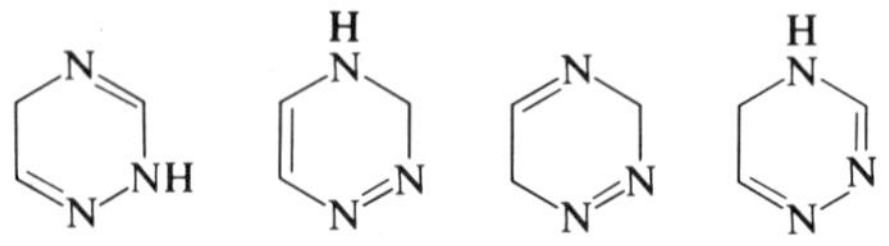

2,5-	3,4-	3,6-	4,5-
(12e)	(12f)	(12g)	(12h)

7. Dihydro-1,3,5-triazines:

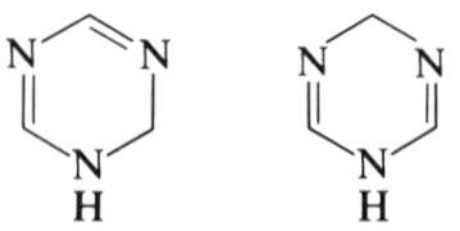

1,2-	1,4-
(13a)	(13b)

8. Dihydro-1,2,3,4-tetrazines:

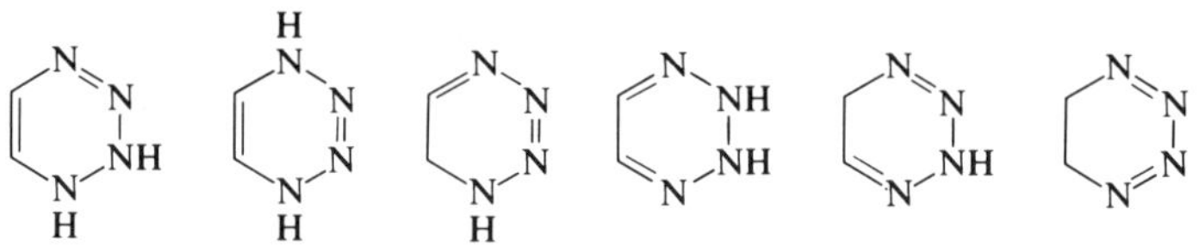

1,2-	1,4-	1,6-	2,3-	2,5-	5,6-
(14a)	(14b)	(14c)	(14d)	(14e)	(14f)

9. Dihydro-1,2,3,5-tetrazines:

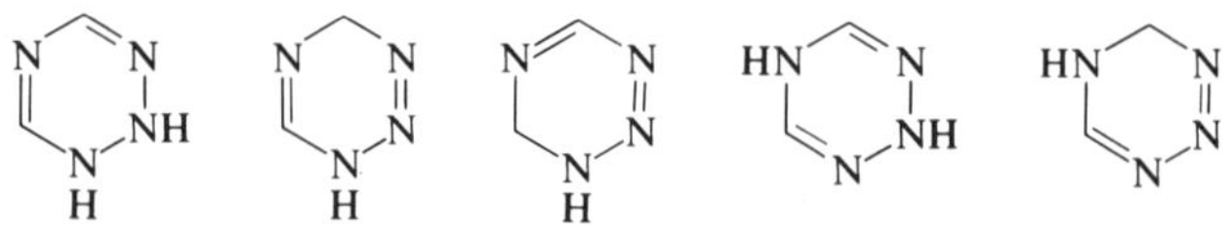

3,2-	1,4-	1,6-	2,5-	4,5-
(15a)	(15b)	(15c)	(15d)	(15e)

10. Dihydro-1,2,4,5-tetrazines:

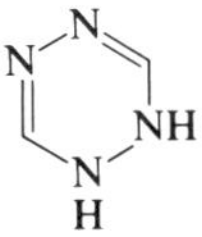

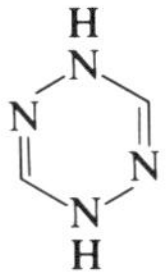

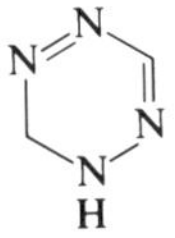

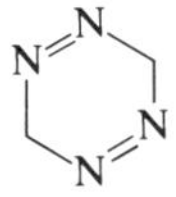

1,2-	1,4-	1,6-	3,6-
(16a)	(16b)	(16c)	(16d)

B. Structure

1. *Constitution and Stability*

One can formally divide isomeric dihydroazines (see Section III,A) into two groups of compounds: those in which a lone pair of *p* electrons from a nitrogen can conjugate with the four π electrons of the diene portion of the molecule, e.g., cyclic enamines[9] (structures A, B, C, and D), and those in which it cannot, e.g., cyclic imines (structures E, F, and G).

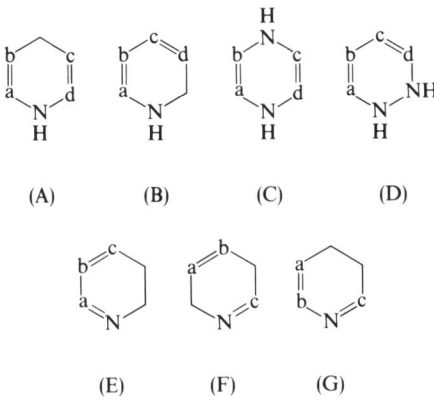

where a, b, c, d = C or N

The isosteric and isoelectronic nature of the various molecules formed by permuting the carbon and nitrogen atoms of each individual type leads to "families" of compounds whose members show certain similarities in physical and chemical properties. Common experience in the field, for example, indicates that cyclic enamines of types A and B are usually more stable than

---

[9] O. Cervinka, *in* "Enamines: Synthesis, Structure and Reactions" (A.G. Cook, ed.), Chapter 7, p. 253. Dekker, New York, 1969.

cyclic imines E, F, and G (in contrast to acyclic compounds[10]), whereas C and D are less stable. These differences may be explained in terms of aromaticity,[11] that is, stabilization due to cyclic electron delocalization within the planar ring depends on the number of, and the orbital interaction between, the participating electrons. The cyclic p conjugation present in structures A and B produces a six-electron delocalized system, which obeys the Hückel $4n + 2$ rule, and is potentially homoaromatic.[12] The degree of aromaticity exhibited by these compounds will depend on the energies of the orbitals containing the p lone pair and the π system as well as, obviously, the degree of planarity of the molecule. On the other hand, the analogues of C and D exhibit potential antiaromaticity[13] as a consequence of cyclic conjugation of a total of eight ($4n$) electrons, giving rise to antiaromatic destabilization. Thus the calculated[14] antiaromatic destabilization for the unsubstituted, heteroconjugated 1,4-dihydropyrazine (**10b**) is about 2.9 kcal mol^{-1}, much smaller than that observed for the analogous isoconjugated system. The only stable parent compound isolated to date is 1,2-dihydro-1,2,4,5-tetrazine (**16a**);[15] 1,4-dihydropyridine (**7b**)[16] was isolated and found to be extremely unstable, and while tautomeric 1,4-(**9b**) and 1,6-dihydropyrimidine (**9c**) were prepared,[17] they could be identified only by spectroscopic means. The majority of other parent compounds (possible isomers) were investigated theoretically. Thus energies of all possible unsubstituted dihydropyridines (**7a–e**) were calculated via MINDO/3,[18] as well as *ab initio* 4-31G techniques.[19] Indeed, the unsubstituted 1,4-dihydropyridine (**7b**) was predicted to be the most stable isomer. Experimental confirmation of these calculations was provided by Fowler,[20] who prepared the stable 1-methyl derivatives of **7a** and **7b** and found the 1,4 derivative to be more stable than the 1,2 derivative by 2.29 ± 0.01 kcal mol^{-1}. He suggested the increased stabilities of the former compound to be due to "favorable electronic interactions" (hyperconjugation or homoaromaticity).

Gaussian 70 *ab initio* calculations of the energy of unsubstituted dihydropyrimidines (**9a–e**)[21] yielded the following order of stability: **9c** > **9b** > **9a** > **9d** > **9e**. These results agree with the experimentally observed behavior of

---

[10] P. W. Hickmott, *Tetrahedron* **38**, 1975, 3363 (1982).
[11] P. J. Garratt, "Aromaticity," McGraw-Hill, London, 1971.
[12] L. A. Paquette, *Angew. Chem., Int. Ed. Engl.* **17**, 106 (1978).
[13] R. Breslow, *Angew. Chem., Int. Ed. Engl.* **7**, 565 (1968); *Acc. Chem. Res.* **6**, 393 (1973).
[14] N. Trinajstic, *J. Mol. Struct.* **8**, 236 (1971).
[15] C. Grundmann and A. Kreutzberger, *J. Am. Chem. Soc.* **79**, 2839 (1957).
[16] T. Koenig and H. Longmaid, *J. Org. Chem.* **39**, 560 (1974).
[17] A. Weis and D. Zamir, unpublished results (1983).
[18] N. Bodor and R. Pearlman, *J. Am. Chem. Soc.* **100**, 4946 (1978).
[19] S. Bohm and J. Kuthan, *Collect. Czech. Chem. Commun.* **46**, 2068 (1981).
[20] F. W. Fowler, *J. Am. Chem. Soc.* **94**, 5926 (1972).
[21] A. L. Weis and U. Degan, unpublished results (1982).

these compounds. The total energies of the parent molecules 4,5-dihydropyridazine (**8f**) and 1,4-dihydropyridazine (**8b**) were also calculated by the Gaussian 70 and MINDO/3 techniques.[22] The importance of carrying out more sophisticated Gaussian 70 calculations (STO-3G, STO-6G, STO-4-31G) was confirmed by this pair of compounds: **8f** was found to be slightly more stable at lower levels of calculation, while **8b** was significantly more stable at higher levels.

Despite a paucity of theoretical calculations, those available agree with experimental data. Thus cyclic enamines of types A and B are the best known compounds among the DHA. In addition, when the energy difference between the imine and enamine is large, corresponding cyclic imines E, F, and G usually isomerize spontaneously via proton transfer to the theoretically more stable A or B. When it is small, the two isomers exist in tautomeric equilibrium. Tautomerism can also be achieved by proper selection of solvent and substituents. (For a summary of isomerism and tautomerism in DHA, see Section III,D.) On the other hand, antiaromtic DHA (C and D) are usually less stable than the corresponding cyclic imines. With regard to cyclic imines E, F, and G, only those containing disubstitution have so far been isolated because geminal disubstitution prevents the above-mentioned isomerization from taking place. For types C and D, stabilization can also be achieved by gem disubstitution, but for these compounds substituents on the nitrogen (**1**) stabilize as well.

Moreover, by comparison of all available data on DHA, one can conclude that, in general, β substitution of cyclic enamines A, B, C, and D by electron-withdrawing substituents, such as COR, CO_2R, CN, NO_2, and SO_2R, enhances their chemical stability as has been observed, for example, in "Hantzsch dihydropyridines" (**17a**), 5-functionalized pyrimidines (**17b**), and pyridazine (**17c**).

(17a)   (17b)   (17c)

Similarly, introduction of electron-donating substituents such as NR_2, SR, or OR in the α or γ position of usually unstable cyclic imines E, F, and G enhances chemical stability and enables these compounds to be easily isolated in each DHA family, for instance, 3,4-dihydropyridines (**18a**), 2,5-dihydropyrimidines (**18b**), 4,5-dihydropyrimidines (**18c**), and 4,5-dihydropyridazines (**18d**).

---

[22] W. L. Hedges, Ph. D. Thesis, University of New Orleans (1981). J. Baker, W. Hedges, J. W. Timberlake, and L. M. Tzefomas, *J. Heterocycl. Chem.* **20**, 855 (1983).

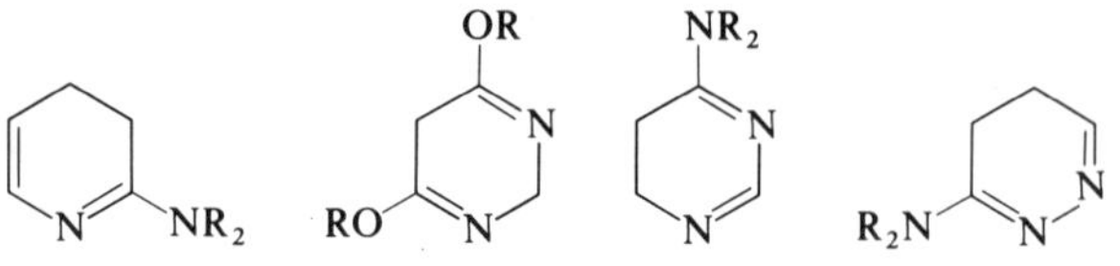

(18a) (18b) (18c) (18d)

2. Molecular Geometry and Conformation

The molecular geometry of a number of solid DHA have been determined by X-ray diffraction crystallography. Naturally, the majority of structures are available for the most stable dihydropyridines (**7a,b**).[2] These data indicate that, generally, the ring systems of 1,2-dihydropyridines show slight deviations from strict planarity as well as significant conjugation between the double bonds and nitrogen lone pair of electrons. The conformations of the extensively studied Hantzch's 1,4-dihydropyridines were found to be planar, boatlike, or envelopelike, depending on ring substitution.[2]

Recent X-ray examinations of the molecular structures of different dihydropyrimidinic isomers were carried out primarily to obtain unambiguous structural assignments. Of course, important information was also obtained regarding bond lengths and angles, i.e., conformation of these molecules in the solid state. Thus an X-ray diffraction study of 4,6,6-trimethyl-2-phenyldihydropyrimidine (**19**)[23] clearly showed the 1,6-dihydro structure (**19b**), where, similar to the 1,2-dihydropyridines, slight deviations from strict planarity were observed (N-1 and C-6 atom out of ring), while the lone pair of N-1 undoubtedly takes part in conjugation with the double bonds.

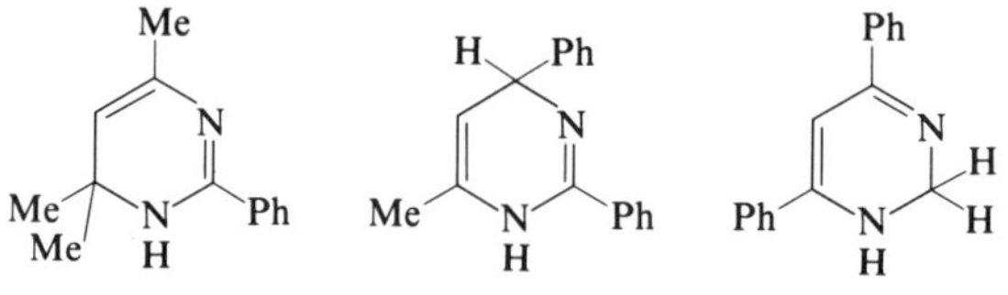

(19b) (20a) (21a)

The structure, geometry, and conformation of solid 2,4-diphenyl-6-methyldihydropyrimidine (**20**) were determined by X ray diffraction.[24] The analysis has shown clearly that the crystalline form of this compound has a 1,4-dihydro structure (**20a**) with a nonplanar ring. Atoms N-1 and C-4 are out

---

[23] A. L. Weis and F. Frolow, *Heterocycles* **19**, 493 (1982).
[24] A. L. Weis and F. Frolow, *Chem. Commun.*, 89 (1982).

of the plane formed by the two double bonds C-2=N-3 and C-5=C-6 by about 0.10 and 0.25 Å, respectively, thus forming a ring with a flat-boat conformation. Particularly interesting data were obtained from the X-ray diffraction analysis of 4,6-diphenyl-1,2-dihydropyrimidine (**21a**) in which only C-2 and C-5 were found to be out of plane and N-1 was nearly coplanar. The N-1 showed a high degree of conjugation with the double bonds, but does not possess the structure of a homoaromatic species.[25] This structure is very similar to that formed for dihydrotetrazine **22**, in which only C-6 is out of plane. The distance between the two nitrogens in **21a** precludes the existence of N–N interaction, and therefore also of "homoaromaticity."

Van der Plas and co-workers[26] elucidated the crystal structure of homoaromatic[27] 6-ethyl-3-phenyl-1,6-dihydro-1,2,4,5-tetrazine (**22**) by X-ray structure analysis. The molecule is in a boat conformation with C-6 and C-3 pointing upward, with dihedral angles of 49.3 (0.51 Å) and 26.7° (0.26 Å), respectively. N-1 was found to be sp^2 hybridized, and the N-1—N-2, N-2—C-3, C-3—N-4, and N-4—N-5 bond distances were intermediate between those of single- and double-bond lengths. These intermediate bond lengths are in agreement with delocalization of the π electrons over this ring. It is striking that the character of the N-1—N-2 (1.32 Å) bond is so aromatic.

(**22**)   (**23**)

Olofson and co-workers[28] obtained evidence for homoaromaticity of the 1,1,4-trimethyl-1,4-dihydro-1,2,4,5-tetrazinium cation (**23**) from a crystal-structure determination. The quaternary nitrogen N-1 is tetrahedral, causing the ring to assume the boat shape. There is evidence for substantial delocalization of the N-4 electron pair (hybridization between sp^2 and sp^3), whereas N-1 is 0.51 Å, while N-4 is only 0.26 Å above the N-2—C-3—N-5—C-6 plane. It should be noted that the crystal structures of **22** and **23** have remarkable similarities. Both molecules are in the boat conformation with their prows twice as high as their sterns. Also, carbons occupy these apices in

---

[25] A. L. Weis, F. Frolow, and R. Vishkaucan, unpublished results (1983).
[26] C. H. Stam, A. D. Counotte-Pottman, and H. C. van der Plas, *J. Org. Chem.* **47**, 2856 (1982).
[27] A. Counotte-Pottman, H. C. van der Plas, and A. van Veldhuizen, *J. Org. Chem.* **46**, 2138 (1981).
[28] D. H. Hoskin, G. P. Wooden, and R. A. Olofson, *J. Org. Chem.* **47**, 2858 (1982).

22, while nitrogens occupy these sites in 23. Molecule 23 is slightly flatter than 22, and subsequently the increased "boat character" of the latter molecule is reflected in a decreased rate of ring inversion from the one boat form to the other. It should be noted that all attempts to decrease ring inversion in 1,2-dihydropyrimidine (21) under the conditions reported[24] failed. The reason for this failure probably lies in the close proximity of C-2 to the plane, which acts to increase the rate of inversion.

Streitwieser,[29] employing molecular orbital calculations, has predicted the thermodynamic destabilization brought about by placing the last two electrons of 1,4-dihydropyrazine in antibonding orbitals. However, the degree of antiaromaticity must depend critically upon the conformation of the dihydropyrazine ring, which controls the ability of the nitrogen lone pairs to interact with the π system. The degree of antiaromaticity may be controlled by the substituents on the dihydropyrazine ring, and the reason for unexpected stability of some potentially antiaromatic 1,4-dihydropyrazines can result only from a special conformation opposing cyclic delocalization of the eight electrons. This was confirmed by X-ray structural analysis of 1,4-bis(p-chlorophenyl)-2,6-diphenyl-1,4-dihydropyrazine (24).[30]

The four carbon atoms of the dihydropyrazine skeleton form a plane, while the two nitrogen atoms N-1 and N-4 are arranged at different distances above this plane, forming a quasi-boat conformation. The position of the aryl group on N-1 apparently leads to separation of the electronic interactions, yielding a six-π-electronic delocalized homoaromatic system (C-2—C-3—N-4—C-5—C-6 plane) and an almost orthogonal lone pair on N-1. According to Schmidt,[4] this structural observation supports the notion that six-membered, formally antiaromatic ring systems can counteract electronic destabilization by assuming a conformation of substituents, which substantially reduces or eliminates delocalization of the eight π electrons. Moreover, this substituent control over stability of 1,4-dihydropyrazines should be more important than the deviations of the cyclic skeleton from planarity.[31,32]

To confirm the structures of NaBH$_4$ reduction products of 1,2,4-triazines, Japanese workers recently published an X-ray study, showing that the product has the 2,5-dihydro structure (25). The dihydrotriazine ring was found to take the boat conformation, which, because of the delocalization of the π electrons and lone-pair electrons of N-2, was significantly squashed.[33]

---

[29] A. Streitwieser, Jr., "Molecular Orbital Theory for Organic Chemistry," p. 275. Wiley, New York, 1961.
[30] J. J. Sterowski, *Cryst. Struct. Commun.* **4**, 21 (1975).
[31] J. W. Lown, M. H. Akhtar, and R. S. McDaniel, *J. Org. Chem.* **39**, 1998 (1974).
[32] H. Kohn and R. A. Olofson, *J. Org. Chem.* **37**, 3504 (1972).
[33] H. Ayato, I. Tanaka, T. Yamane, T. Ashida, T. Sasaki, K. Minamoto, and K. Harada, *Bull. Chem. Soc. Jpn.* **54**, 41 (1981).

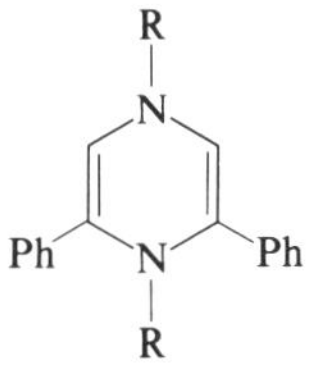

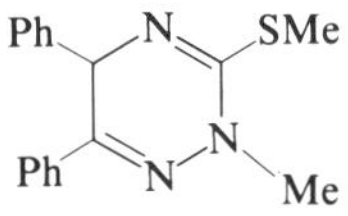

(24) R = p-ClC$_6$H$_4$ **(25)**

C. Preparation of Dihydroazines

The various methods of synthesizing dihydroazines fall broadly into two main divisions: cyclizations and modification of a preexisting azine nucleus. Besides these two main types, some more specific approaches have also been designed. As will be shown below, the cyclization methods involve mainly the following combinations of fragments:

⟨5 + 1⟩   ⟨4 + 2⟩   ⟨3 + 3⟩   ⟨2 + 1 + 2 + (1)⟩   ⟨3 + 1 + (1) + 1⟩

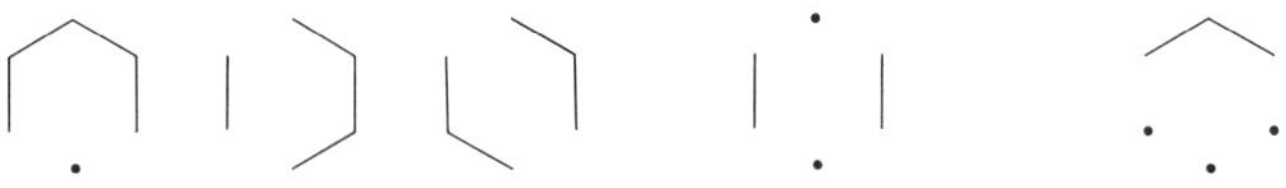

The most common approach to obtain dihydroazines usually involves Schiff-base chemistry. For example, ⟨4 + 2⟩ combinations have found wide application in the preparation of 4,5- **(8f)** and 1,4- **(8b)** dihydropyridazines by condensation of 1,4-diones with hydrazine (Scheme 1).

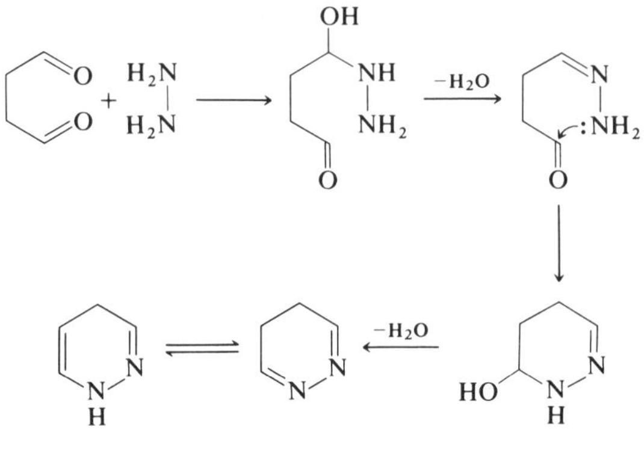

(8b)   **(8f)**   **(26a)**

SCHEME 1

On the other hand, 2,3-dihydropyrazines (**10c**) are formed most readily by a condensation of 1,2-diones with bisamines (Scheme 2).

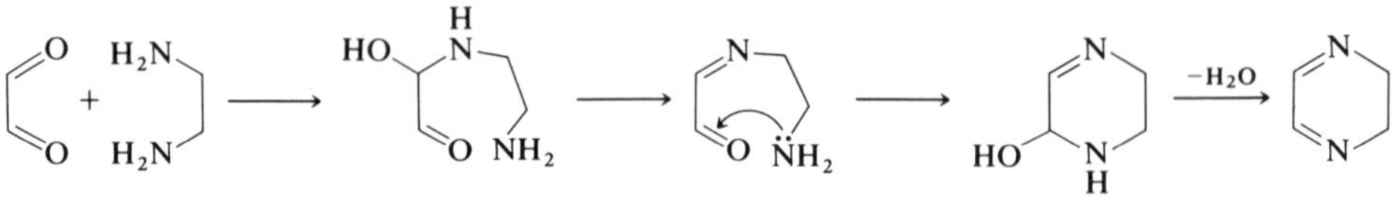

(**26b**) (**10c**)

SCHEME 2

The most direct synthesis of 1,4-dihydropyridines (**7b**) utilizes a procedure with either ⟨5 + 1⟩, ⟨3 + 2 + 1⟩, or ⟨2 + 1 + 2 + (1)⟩ fragment combination. The latter represents the widely used classical Hantzsch's synthesis, which combines two molecules of an active α-methylenic carbonyl compound, an aldehyde, and an ammonia source (Scheme 3).

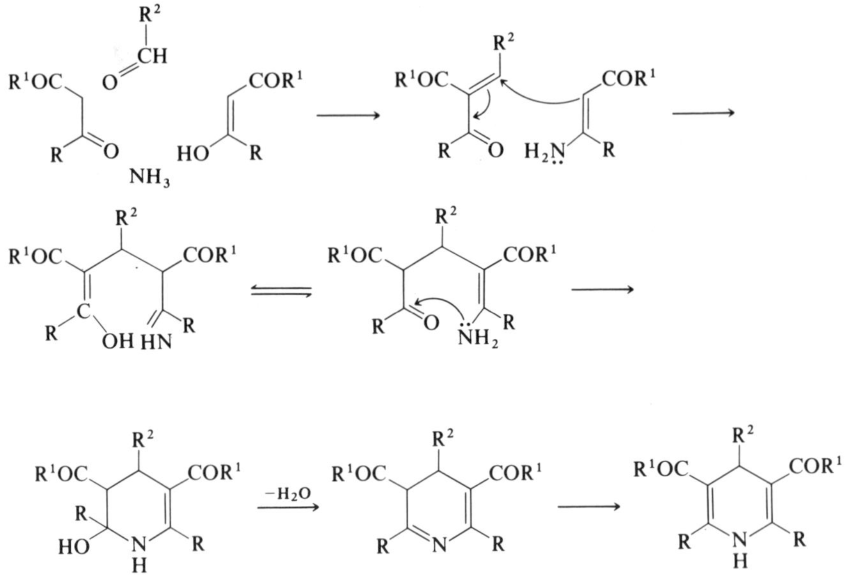

(**26c**) SCHEME 3

The reaction of easily available α,β-unsaturated carbonyl compounds and amidines is an attractive ⟨3 + 3⟩ combination for the preparation of tautomeric 1,4- (**9b**) and 1,6- (**9c**) dihydropyrimidines (Scheme 4).

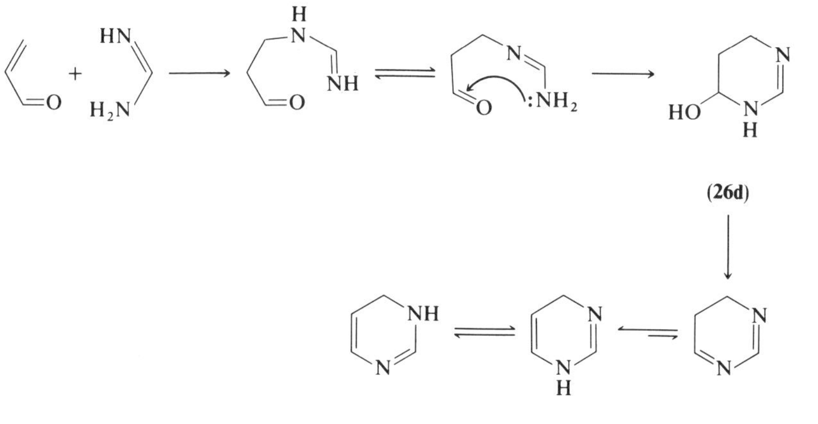

Sec. III.C]   DIHYDROAZINES   17

SCHEME 4

We chose the four above-mentioned reaction schemes specifically to indicate that, despite the wide use of these synthetic approaches and the relatively well-known chemistry of Schiff bases, they actually remain widely unstudied. For over a century, the mechanisms by which these reactions were suggested had no experimental justification. It is easy to note the similarities in formation of the intermediate cyclic N,O hemiacetals (carbinolamines) (**26a–d**). These compounds are usually found to be relatively stable and isolable[34] (see also Section V,B,1,a), in contrast to the unstable acyclic carbinolamines. The dehydration step also requires further investigation because the initial dehydration product may be stable, it may exist in tautomeric equilibrium, or it may be converted spontaneously to the more stable isomer ring, depending on the relative stabilities of the various isomers. If a detailed mechanism for these reactions is known, control of experimental conditions can be used to direct reaction products and improve yields.

As the name suggests, dihydroazines may be, and some actually are, formed by the addition of two hydrogens to the azine nucleus. According to Evans,[35] reductions of organic species can be classified according to the active entities involved:

1. Hydrogen atoms are implicated or thought to be implicated in (a) catalytic reduction, (b) electrochemical reduction at a low-overvoltage electrode, (c) photochemical reduction, (d) reduction involving complex metal

---

[34] K. L. Marsi and K. Thorre, *J. Org. Chem.* **29**, 3102 (1964).
[35] R. F. Evans, *in* "Modern Methods in Organic Chemistry" (C. J. Timmons, ed.), Chapter 1. Van Nostrand-Reinhold, Princeton, New Jersey, 1969.

hydrides derived from Group IVB metals, (e) diimide reductions, and (f) homogeneous reductions by dihydro derivatives of Group VIII metals.

2. Hydrogen is transferred as hydride ion in (g) reduction involving complex metal hydrides containing boron or aluminum, (h) homogeneous hydrogenation involving certain monohydride complexes of Group VIII metals, and (i) hydrogen transfer from one substrate to another.

3. Finally, hydrogen is transferred as protons in (j) reduction brought about by dissolving metals, (k) electrochemical reduction at a high-overvoltage electrode, and (l) reduction, the first stage of which is attack by an anion.

However, it is symptomatic of the present state of DHA chemistry that only a few of the possible reductions have been used systematically to produce azine compounds. Reduction chemistry is best known for the pyridine series.

The addition of organometallic reagents to six-membered nitrogen heterocyclic aromatic compounds (azines) is fairly general, leading to the corresponding dihydro derivatives. It should be noted, however, that organomagnesium compounds are less reactive toward these substrates, as shown by the requirement of more vigorous reaction conditions. Therefore, their use is often subject to complications, and they generally give results inferior to those of organolithium compounds. Both organolithium and organomagnesium compounds add to α,β-ethylenimine fragment of azines. In general, both 1,2- (or 1,6-) and 1,4-addition occur, but under kinetic control organolithium compounds show a greater tendency to 1,2-addition than do organomagnesium compounds. However, the reversibility of 1,2-addition can lead to the eventual formation of 1,4-adducts under thermodynamic control.[36] Certainly, many less general methods are found in the literature.

D. Tautomerism and Rearrangements

Existence of a variety of isomeric DHA (Section III,A) with different (ground-state energy) stabilities gives rise to various isomerization processes, which are usually reversible or irreversible migrations of a group from one site to another. The study of these isomerizations is still in its infancy. What is known will be presented below. The thermal suprafacial [1,3]-alkyl shift of 1,4-dialkyl-1,4-dihydropyrazines reported by Lown and co-workers[31,37–39] is the only known migration of a group different from hydrogen in the DHA

---

[36] J. B. Wakefield, in "Comprehensive Organometallic Chemistry" (G. Wilkinson, ed.), Vol. 7, p. 15, Pergamon, Oxford, 1982.
[37] J. W. Lown and M. H. Akhtar, *Chem. Commun.*, 829 (1972).
[38] J. W. Lown and M. H. Akhtar, *Chem. Commun.*, 511 (1973).
[39] J. W. Lown and M. H. Akhtar, *Tetrahedron Lett.*, 179 (1974).

series. Thermal isomerizations, including hydrogen transfer in DHA, may be divided into those involving, formally, either [1,3]- (amidinic and imine–enamine tautomerism) or [1,5]-hydrogen shifts.

In the chemical literature it is common practice to classify hydrogen migrations as either rearrangements or tautomerisms, depending on their kinetic and thermodynamic parameters, the former being reserved for irreversible or slow processes, while the latter are used to describe fast reversible exchanges. Minkin and co-workers have provided criteria for distinguishing between tautomerism and rearrangement, which are summarized in Eqs. (1) and (2).[40]

$$\Delta G°_{25} < 6 \text{ kcal mol}^{-1} \qquad (1)$$

$$\Delta G^{\ddagger}_{25} < 25 \text{ kcal mol}^{-1} \qquad (2)$$

Equation (1) is dictated by the sensitivity of current techniques to detect a minor tautomer in an equilibrium mixture. Equation (2) serves to define tautomeric rearrangements as those transformations in which the lifetimes of the isomers are too short to enable the preparative separation of the two forms. Obviously, the above criteria should be regarded as no more than guidelines. Nevertheless, they do form a useful basis for distinguishing tautomeric reactions from rearrangements.

1. *Amidinic Tautomerism*

The prototropic tautomerism of compounds containing an amidine moiety has been studied extensively.[41-43] However, the data obtained on this type of equilibrium are rather qualitative and have not been explored systematically. The difficulties encountered in this work stem primarily from the common experience that proton transfer between electronegative atoms, such as nitrogen, is very fast.[43,44]

The structure of an amidinic compound has usually been given in the literature as a presumed, predominant tautomer, without supporting evidence for a definitive structure. According to the recently published "systematization" of tautomeric processes,[45] the amidinic tautomerism is related to isodesmic [1,3]-sigmatropic tautomerism and, in the case of prototropism,

---

[40] V. I. Minkin, L. P. Olekhnovich, and Y. A. Zhdanov, *Acc. Chem. Res.* **14**, 210 (1981).
[41] G. Schwenker and K. Bosl, *Pharmazie* **24**, 653 (1969).
[42] A. R. Katritzky and J. M. Lagowski, *Adv. Heterocycl. Chem.* **1**, 311; **2**, 1 (1969).
[43] J. Elguero, C. Marzin, A. R. Katritzky, and P. Linda, *Adv. Heterocycl. Chem., Suppl.* **1** (1976).
[44] A. J. Kresge, *Acc. Chem. Res.* **3**, 354 (1975).
[45] N. S. Zefirov and S. S. Tratch, *Chem. Scr.* **15**, 4 (1980).

may be expressed by the following general scheme:

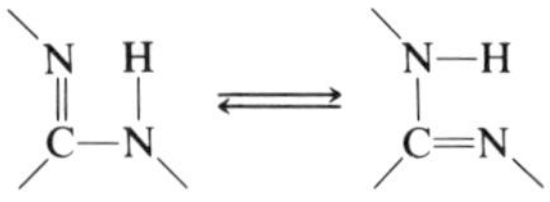

$$\begin{matrix} \diagdown \\ N \\ \parallel \\ C-N \\ \diagup \quad \diagdown \end{matrix} \begin{matrix} H \\ | \\ \\ \end{matrix} \rightleftharpoons \begin{matrix} \diagdown \\ N-H \\ | \\ C=N \\ \diagup \quad \diagdown \end{matrix}$$

Three classes of tautomeric equilibria can be distinguished according to the molecular structure of the amidinic compounds:

1. Acyclic:

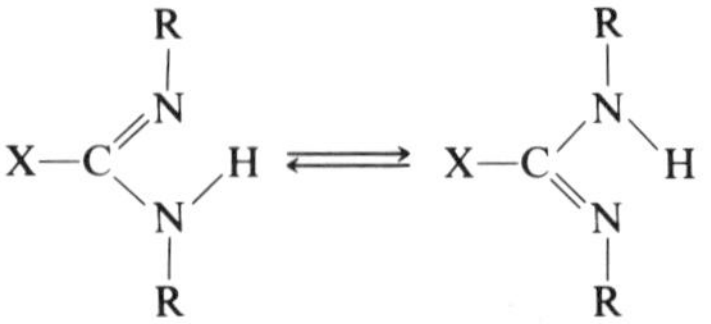

$$X-C\begin{matrix} R \\ | \\ \diagup^N \\ \diagdown_N \\ | \\ R \end{matrix} H \rightleftharpoons X-C\begin{matrix} R \\ | \\ N \\ \diagdown \\ \diagup^N \\ | \\ R \end{matrix} H$$

2. Semicyclic:

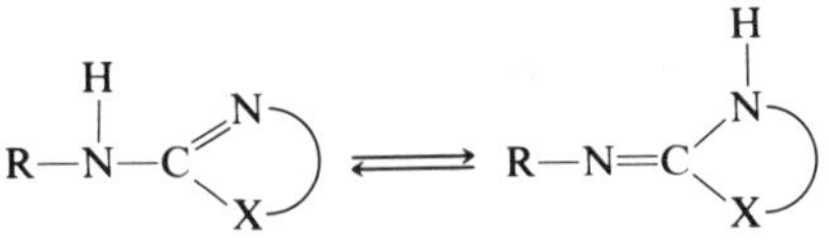

$$R-N-C\begin{matrix} H \\ | \\ \diagup^N \diagdown \\ \diagdown_X \diagup \end{matrix} \rightleftharpoons R-N=C\begin{matrix} H \\ | \\ N \diagdown \\ \diagdown_X \diagup \end{matrix}$$

3. Cyclic or annular:

$$X-C\begin{matrix} H \\ \diagdown \\ N \diagdown \\ \diagup^N \diagup \end{matrix} \rightleftharpoons X-C\begin{matrix} N \diagdown \\ \diagdown_N \diagup \\ | \\ H \end{matrix}$$

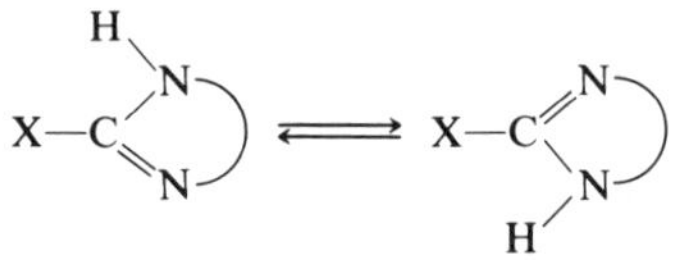

Among the many publications on tautomerism of heterocyclic compounds, annular tautomerism is one of the most widely studied, despite the strong hydrogen bonding and usually low activation energies, which complicate precise quantitative measurements. Most efforts were devoted to investigation of the amidinic tautomeric equilibrium in imidazoles or cyclic systems containing the imidazole fragment moiety (histidine, purine, etc). Experimental evidence concerning annular tautomerism in six-membered cyclic amidines is very limited. Moreover, in the dihydroazine series no data were available until recently.

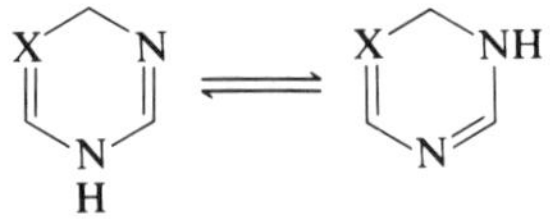

$$X = C \text{ or } N$$

Discovery of the amidinic tautomerism series of 1,4- (**9b**) and 1,6- (**9c**) dihydropyrimidines (Section V,E) and 1,2- (**13a**) and 1,4- (**13b**) dihydro-1,3,5-

triazines (Section VIII,C), together with the ability to observe individual tautomers and measure their concentrations and kinetic parameters under different experimental conditions by NMR, IR, and other instrumental methods, expands the horizons of proton-transfer chemistry. Moreover, it is to be expected that amidinic tautomerism could serve as a convenient model for investigating, in general, the factors governing tautomerism in cyclic molecules and could also be used for comparative studies on the factors controlling corresponding tautomeric effects in noncyclic analogues.

2. *Imine–Enamine Tautomerism*

Imine–enamine tautomeric equilibrium was observed in different substituted dihydroazines by NMR spectroscopy. Pinson and co-workers[46] reported, for example, that a solution of 3,6-diphenyl-1,2-dihydropyrazine (**27b**) in $CDCl_3$ contains a tautomeric mixture of 2,5- (**27a**) and 1,2- (**27b**) dihydro derivatives in the ratio 30:70.

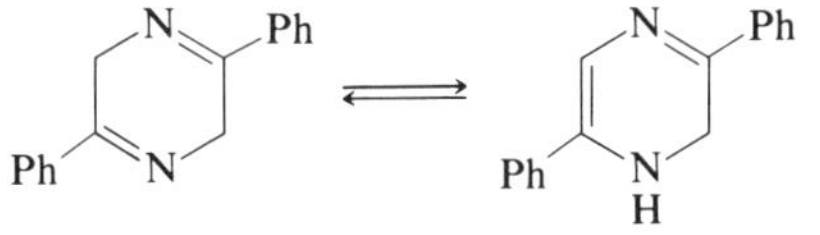

(27a)   ⇌   (27b)

This result shows that the two isomers are in thermodynamic equilibrium. Moreover, when deuterated acetone is added to the chloroform solution, equilibrium is displaced toward the 1,2-dihydro derivative (**27b**).

A similar type of tautomerism was found in a series of dihydropyridazines (see Section IV,C) and dihydropyrimidines (see Section V,C,2). For example, on the basis of NMR measurements in $CDCl_3$, it has been shown that 3,6-diphenyl-4,5-dihydropyridazine (**28a**) exists in tautomeric equilibrium with the corresponding 1,4-dihydro compound (**28b**) in the ratio 1:8.

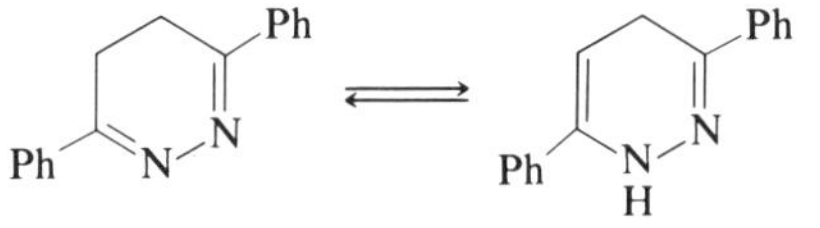

(28a)   ⇌   (28b)

Furthermore, ^1H- and ^{13}C-NMR investigations of 1,2-dihydropyrimidine **21a** in $CDCl_3$ have demonstrated the existence of tautomeric equilibrium with the corresponding 2,5-dihydro isomer (**21b**) in the ratio 2:1 (See Section V,C,2).

[46] J. Armand, K. Chekir, and J. Pinson, *Can. J. Chem.* **52**, 3971 (1974).

(21a)   (21b)

Imine–enamine equilibrium depends on the difference between the two isomers, solvent polarity, and substituent effects on the dihydroazine ring; and according to the definition of tautomerism, where the energy difference is not overwhelmingly large [Eq. (1)], tautomeric equilibria are observed experimentally for various dihydroazines. Whereas in the case of large energy differences, fast rearrangement to the thermodynamically more stable isomer occurs, for instance, the rapid transformation of 4,5-dihydropyrimidines to tautomeric 1,4- and 1,6-dihydropyrimidines (see Section V,C,1) or the formation of 1,4-dihydropyridazines by cycloaddition to 1,2,4,5-tetrazines (see Section IV,B,2).

Some other examples of rearrangements have been published. 2,3-Dihydropyridazines are rearranged thermally or by base to the 1,4-dihydro isomers.[47] Upon heating, the 5,6-dihydropyridazine derivative undergoes a thermally allowed suprafacial [1,5]-sigmatropic hydrogen rearrangement to give the 1,6-dihydro derivative.[48] Moreover, a [1,5]-sigmatropic hydrogen rearrangement of unsymmetrically substituted 1,2-dihydropyrimidines was reported.[49]

Undoubtedly, the above-mentioned examples are just a start toward investigating migrations in dihydroazines. Detailed research of these processes will certainly provide a deeper understanding of the problems of tautomerism and rearrangements as a whole as well as of the energetics, reactivity, and mechanisms of the formation of a variety of dihydroazines.

E. Physicochemical Properties and Reactivity

Most of the available data are found in the experimental sections of the original preparative work, giving the standard IR, UV, and NMR spectral data. To date no systematic study of the physical properties have been undertaken, nor have any measurements of theoretical interest been carried out, such as photoelectron emission, mass spectrometry, and electrochemical determinations.

[47] E. E. Schweizer and C. M. Kopay, *J. Org. Chem.* **37**, 1561 (1972).
[48] P. De Mayo and M. C. Usselman, *Can. J. Chem.* **51**, 1724 (1973).
[49] S. Hoffmann and E. Muehle, *Z. Chem.* **9**. 66 (1969).

IV. Dihydropyridazines

A. General Comments

Although no review dedicated entirely to dihydropyridazine chemistry has yet appeared, short surveys of synthetic approaches can be found in all reviews of pyridazine chemistry.[50-52] Naturally, this information is presented from the point of view of chemists desiring to synthesize pyridazine rather than as an account of dihydropyridazine chemistry per se, and the pertinent material is, therefore, scattered throughout the reviews and not presented systematically.

Since those dihydropyridazines with unsubstituted ring nitrogens exist almost exclusively as isomeric pairs **8f** and **8b**—often in tautomeric fast equilibrium with one another[22]—it has sometimes been difficult to differentiate clearly the chemistry of the 4,5-dihydropyridazines (**8f**) from that of the 1,4-dihydropyridazines (**8b**). Before the advent of modern spectroscopic techniques, workers found it difficult to assign a given tautomeric form to a particular dihydropyridazine. Thus many compounds originally reported as 4,5-dihydropyridazines may in fact exist predominantly as the corresponding 1,4-dihydro isomer. The reader should, therefore, note that in the following discussion, we have assumed that tautomeric forms of the various dihydropyridazines, as assigned by the original authors, are correct, unless subsequent data have contradicted their validity.

B. Synthesis

1. Cyclization Methods

One of the most common and versatile methods for synthesizing dihydropyridazines is the condensation of saturated 1,4-dicarbonyl compounds with hydrazines (Scheme 1, Section III,C).

The first example of this type, provided by Curtius[53] in 1894, involves the reaction between diethyl 1,2-diacetylsuccinate and hydrazine hydrate. Following this work, several dihydropyridazines were synthesized by other early

---

[50] T. L. Jacobs, in "Heterocyclic Compounds" (R. C. Erdelfield, ed.), Vol. 6, p. 101. Wiley, New York, 1957.
[51] M. Tisler and B. Stanovnik, *Adv. Heterocycl. Chem.* **9**, 211 (1968).
[52] M. Tisler and B. Stanovnik, *Adv. Heterocycl. Chem.* **24**, 363 (1979).
[53] T. Curtius, *J. Prakt. Chem.* [2] **50**, 508 (1894).

workers[54-58] at the beginning of this century. The results obtained by different authors gave rise to much controversy, since not all condensations of hydrazine with saturated diketones yield single products (mono- and bis-hydrazones, aminopyrroles, polymers, etc. can all be formed). Therefore, the reaction has been thoroughly investigated, especially with regard to reaction conditions, since different by-products can be produced under different reaction environments. Moreover, subsequent studies[59,60] showed that dihydropyrazine is formed only when equimolar amounts of diacetylsuccinic ester and hydrazine are used, and the originally proposed 4,5-dihydro structure (29a) for the reaction product has been corrected to the 1,4-dihydro structure 29b.[60]

        COOEt   COOEt
EtOOC   Me   EtOOC   Me

Me   N—N   Me   N—N
                                             H

(29a)   (29b)

Over the following half century, investigators have reported the formation of two types of dihydropyridazines, depending on the structures of the starting materials. Thus 4,5-dihydropyridazines (**8f**) are produced in the reaction of hydrazine with di- and polysubstituted saturated 1,4-diketones,[61-74] while 1,4-dihydropyridazines (**8b**) are formed from the re-

[54] C. Paal and E. Dencks, *Chem. Ber.* **36**, 491 (1903).
[55] A. Smith, *Justus Liebigs Ann. Chem.* **289**, 316 (1896).
[56] C. Paal and G. Kuhn, *Chem. Ber.* **40**, 4598 (1907).
[57] C. Bulow and H. Filchner, *Chem. Ber.* **41**, 1886 (1908).
[58] G. Korshun, *Chem. Ber.* **37**, 2183 (1904).
[59] R. G. Jones, *J. Am. Chem. Soc.* **78**, 159 (1958).
[60] W. L. Mosby, *J. Chem. Soc.*, 3997 (1957).
[61] G. Korshun and C. Roll, *Gazz. Chim. Ital.* **41**, 186 (1911).
[62] E. E. Blaise, *C. R. Hebd. Seances Acad. Sci.* **170**, 1324 (1920).
[63] B. G. Zimmerman and H. L. Lochte, *J. Am. Chem. Soc.* **60**, 2456 (1938).
[64] S. Gapuano, *Gazz. Chim. Ital.* **68**, 521 (1931).
[65] W. Borsche and A. Klein, *Justus Liebigs Ann. Chem.* **548**, 74 (1941).
[66] K. Alder and C. H. Schmidt, *Ber. Dtsch. Chem. Ger. B* **76**, 183 (1943).
[67] H. Heller, R. Pasternak, and H. von Halban, *Helv. Chim. Acta* **29**, 512 (1946).
[68] C. G. Overberger, N. R. Byrd, and R. R. Mesrobian, *J. Am. Chem. Soc.* **78**, 1961 (1956).
[69] C. L. Arcus and P. A. Hallgarten, *J. Chem. Soc.*, 3407 (1957).
[70] S. G. Cohen, S. H. Hsiao, E. Saklad, and C. H. Wang, *J. Am. Chem. Soc.* **79**, 4400 (1957).
[71] R. G. Jones, *J. Org. Chem.* **25**, 956 (1960).
[72] S. Fatutta, *Ann. Chim.* (*Rome*) **51**, 252 (1961).
[73] S. Fatutta, *Ann. Chim.* (*Rome*) **52**, 365 (1962).
[74] W. M. Williams and W. R. Dolbier, Jr., *J. Am. Chem. Soc.* **94**, 3955 (1972).

action of a monosubstituted hydrazine with 1,4-diketones, 1,4-ketoaldehydes, or 1,4-dialdehydes.[75-80] However, not all reactions with saturated 1,4-diketones proceed as simply as assumed. For example, acetonylacetone[61,68] or 1,2-dibenzoylethane[54,70] disproportionation leads to a mixture of the corresponding pyridazine and tetrahydropyridazine. Different reaction conditions may also influence the reaction course.

More recently, Russian workers have shown that 1-alkyl-3,6-dimethyl-1,4-dihydropyridazines (**30**) are produced in the condensation of acetonylacetone with alkyl hydrazines.[81]

(30a) R = Me   (31a) R = Pr, $R^1 = R^2 = H$
(30b) R = Et   (30b) R = t-Bu, $R^1 = R^2 = H$
(30c) R = Pr   (30c) R = i-Pr, $R^1 = H, R^2 = Me$
(30d) R = Bu   (30d) R = t-Bu, $R^1 = H, R^2 = Me$
                    (30e) R = Me, $R^1 = Me, R^2 = H$
                    (30f) R = Pr, $R^1 = Me, R^2 = H$

The same authors[82] suggested a mechanism for the reaction of 1,4-dicarbonyl compounds with monosubstituted hydrazines, which affords, depending on reaction conditions and structural factors, the mono- and bishydrazones, N-aminopyrroles, and 1,4-dihydropyridazines. Some relatively simple 1,4-dihydropyridazines (**31**) were prepared by condensation of monosubstituted hydrazines with succinic or levulinic aldehyde.[83]

At least two cases reporting the condensation of hydrazine with 1,4-dialdehydes are known. Timberlake et al.[84,85] found that unsubstituted 4,5-dihydropyridazine (**8f**), produced from succinaldehyde and hydrazine, trimerizes to give **32** [Eq. (3)].

---

[75] W. Schlenk, H. Hillemann, and J. Rodloff, *Justus Liebigs Ann. Chem.* **487**, 135 (1931).
[76] D. Desaty, O. Hadzija, and D. Keglevic, *Croat. Chem. Acta* **37**, 227 (1965).
[77] B. Helferich and O. Lecher, *Ber. Dtsch. Chem. Ger. B* **54**, 930 (1921).
[78] G. O. Schlenk, *Justus Liebigs Ann. Chem.* **584**, 156 (1953).
[79] M. Verzele and F. Govaert, *Bull. Soc. Chim. Belg.* **58**, 432 (1949).
[80] G. Westphal, *Z. Chem.* **9**, 339 (1965).
[81] K. N. Zelenin and J. Dumpis, *Khim. Geterotsikl. Soedin.* **7**, 1566 (1971).
[82] J. N. Zelenin and J. Dumpis, *Zh. Org. Khim.* **9**, 1295 (1973).
[83] K. N. Zelenin and J. Dumpis, *Khim. Geterotsikl. Soedin.* **7**, 400 (1971).
[84] B. K. Bandlish, J. N. Brown, J. W. Timberlake, and L. M. Trefonas, *J. Org. Chem.* **38**, 1102 (1973).
[85] J. Dodge, W. Hedges, J. W. Timberlake, and L. M. Trefonas, *J. Org. Chem.* **43**, 3615 (1978).

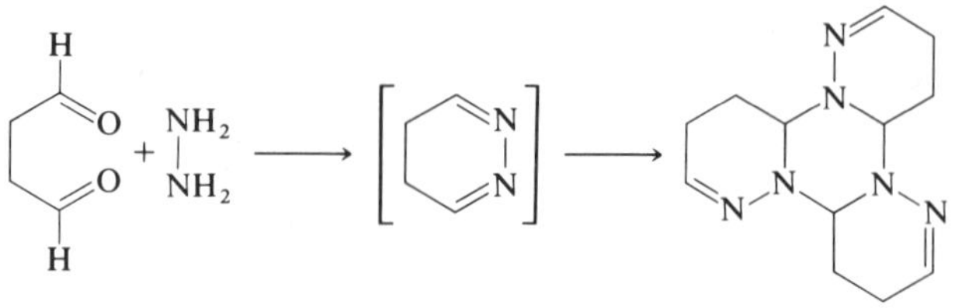

[Structural scheme for Eq. (3): succinaldehyde + hydrazine → dihydropyridazine → trimer (32)]

Schlenck and co-workers[75] showed that tetrapheylsuccinaldehyde reacted with hydrazine to give **33**. These authors did not report any trimerization of the dihydropyridazine. Apparently, the steric requirements of the phenyl groups preclude this eventuality.

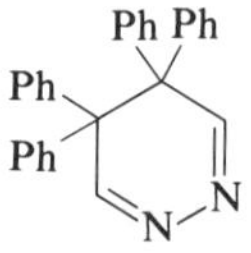

[Structure (33): tetraphenyl-dihydropyridazine]

The single example of the preparation of 1,2-dihydropyridazines (**34**) was reported by Zelenin and Dumpis,[86] namely, the condensation of 1,2-dialkylhydrazines with acetonylacetone [Eq. (4)].

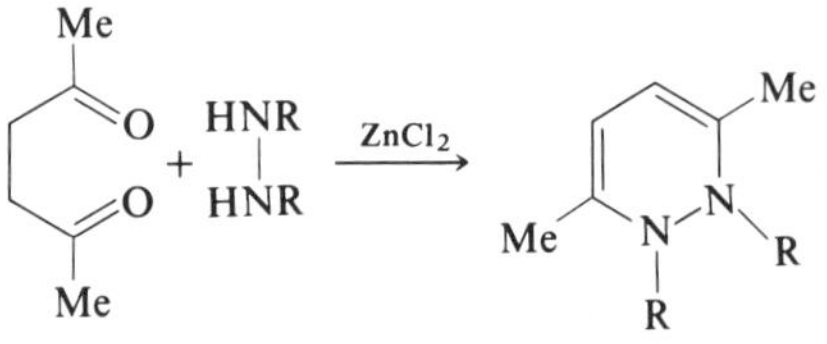

[Eq. (4): acetonylacetone + HNR–HNR, ZnCl$_2$ → 1,2-dihydropyridazine **34**]

(**34a**) R = Me
(**34b**) R = Et

Other methods for synthesizing dihydropyridazines have been recorded but have not been generally applied. Hauser and co-workers,[87] for example, have reported the synthesis of dihydropyridazines **28a** and **35** by an intramolecular coupling reaction [Eq. (5)]. However, it has been shown that **35** exists in the

---

[86] K. N. Zelenin, V. A. Nikitin, N. M. Anodina, and Z. M. Matvejeva, *Zh. Org. Khim.* **8**, 1438 (1972).
[87] F. E. Henoch, K. G. Hampton, and C. R. Hauser, *J. Am. Chem. Soc.* **91**, 676 (1969).

1,4-dihydro form, probably because of more effective conjugation stabilization.[22]

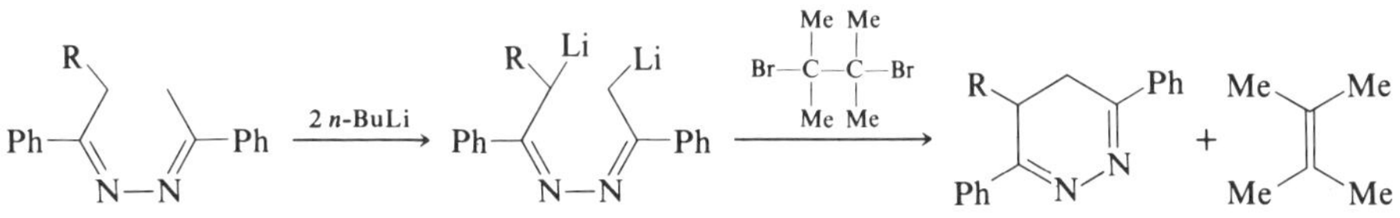

$$\text{(5)}$$

(35) R = Ph
(28a) R = H

1,6-Dihydropyridazine (36a) was prepared in moderate yield by treatment of substituted α-imino ketones with vinyl triphenylphosphonium bromide [Eq. (6)].[88]

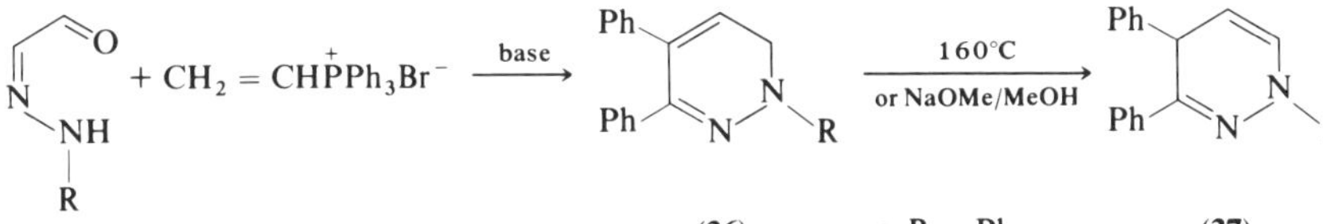

$$\text{(6)}$$

(36)   a. R = Ph   (37)
       b. R = Me
       c. R = CO_2Et

The latter undergoes thermal (44%) or base-catalyzed (100%) rearrangement, affording 1,4-dihydropyridazine (37a). Severin et al.[89] found that 1-phenyl-4-phenylhydrazono-2-buten-1-ol undergoes cyclization to 1,2-dihydropyridazine (38) in the presence of p-toluenesulfonic acid.

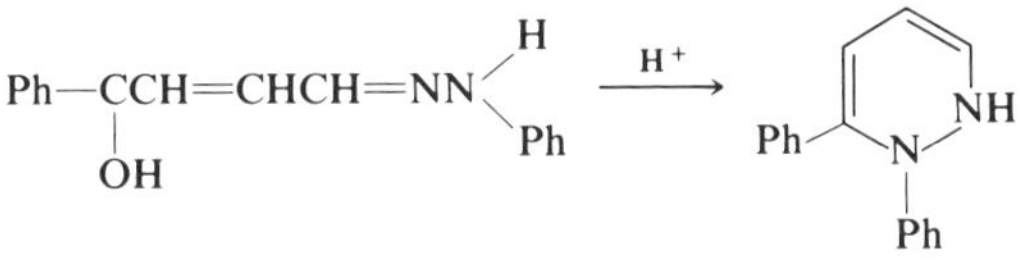

(38)

2. Cycloaddition Reactions

The cycloaddition reaction of 1,2,4,5-tetrazines with olefins developed by Carboni and Lindsey[90] represents a second major route to dihydropyridazines (Scheme 5).

---

[88] E. E. Schweizer, C. S. Kim, C. S. Labaw, and W. P. Murray, *Chem. Commun.*, 7 (1973).
[89] T. Severin, R. Adam, and H. Lerche, *Chem. Ber.* **108**, 1756 (1975).
[90] R. A. Carboni and R. V. Lindsey, *J. Am. Chem. Soc.* **81**, 4342 (1959).

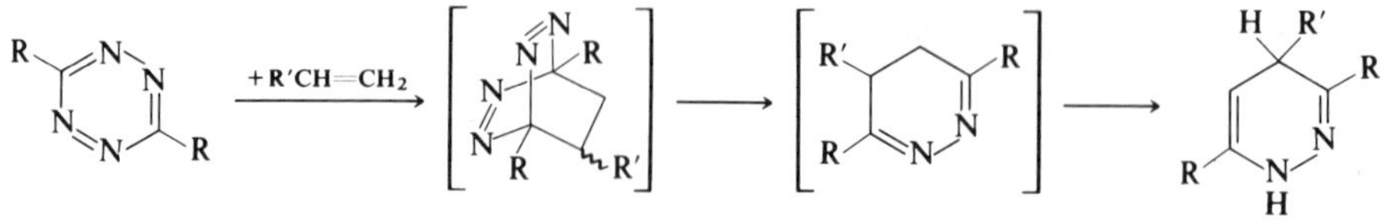

SCHEME 5

According to the mechanism proposed, the 1,4-cycloaddition (Diels–Alder reaction) of the diene part of the tetrazine to the dienophile (olefin) gives an unstable bicyclic intermediate, which spontaneously eliminates a molecule of nitrogen, forming a 4,5-dihydropyridazine, which in the absence of appropriate stabilization by gem disubstitution in positions 4 and 5, easily isomerizes to the thermodynamically more stable 1,4-dihydropyridazine (38). The structure of the latter intermediates were unambiguously proved using NMR.[91,92]

Noteworthy is the necessity to activate the tetrazines with electron-withdrawing groups and the olefins with electron-donating substituents for this reaction to occur. Thus tetrazines bearing electron-withdrawing substituents such as methoxycarbonyl, polyfluoroalkyl, and 2-pyridyl react with an olefinic component more rapidly than do those substituted with methyl or phenyl groups. Conversely, electron-rich olefins such as vinyl ethers show much greater reactivity than do electron-deficient olefins such as acrylonitrile and acrolein. Sauer and Heinrichs[93] found, for example, that N-morpholinostyrene (a very electron-rich olefin), ethylene, and acrolein showed relative reaction rates of 470,000:36,000:1, respectively, with 3,6-dicarbomethoxytetrazine.

Because of the ease, speed, and quantitative course (in most cases) of the reaction of olefins with 3,6-dicarbomethoxy-1,2,4,5-tetrazine, Nenitzescu et al.[94] proposed this reaction as a titrimetric method for the determination of these hydrocarbons.

Carboni and Lindsey[90] have also pointed out the significant role played by steric restrictions, which may be important for the reactivity of dienophiles in the synthesis of dihydropyridazines. For instance, 2,3-dimethyl-2-butene (an electron-rich but sterically hindered olefin) reacts only with the polyfluoroalkyl substituted tetrazines 39a and 39b, giving correspondingly 4,5-dihydropyridazine 40a[95] and 40b.[90] Even with these highly reactive tetrazines, the rate

[91] J. Sauer, A. Mielert, D. Lang, and D. Peter, *Chem. Ber.* **98**, 1435 (1965).
[92] M. Avram, G. R. Bedford, and A. R. Katritzky, *Recl. Trav. Chim. Pays-Bas* **82**, 1053 (1963).
[93] J. Sauer and G. Heinrichs, *Tetrahedron Lett.*, 4979 (1966).
[94] M. Avram, J. G. Dinulescu, E. Marica, and C. D. Nenitzescu, *Chem. Ber.* **95**, 2248 (1962).
[95] M. G. Barlow, R. N. Haszeldine, and J. A. Pickett, *J. C. S. Perkin I*, 378 (1978).

of cycloaddition is extremely slow, requiring about 2 weeks for the formation of **40a**. At the same time, however, tetrazine **39a** reacts rapidly with styrene to give 1,4-dihydropyridazine **41**.[95]

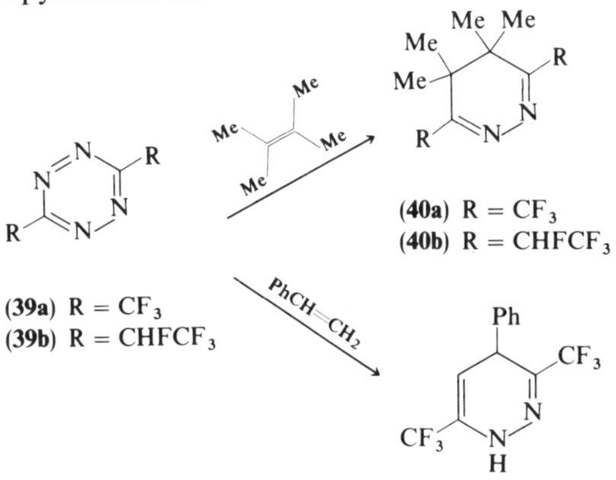

(39a) R = CF$_3$
(39b) R = CHFCF$_3$

(40a) R = CF$_3$
(40b) R = CHFCF$_3$

(41)

The number of different tetrazines which have been used in this type of reaction (Scheme 5) is small and is essentially limited to the species already mentioned. However, a large number of olefins have been treated with these few tetrazines, yielding a large variety of dihydropyridazines.

Cycloadditions are also known with diazoalkanes and related compounds. Diazomethane reacts with a cyclopropene, and the pyrazoline cycloadduct **42** is rearranged by acid to the dihydropyridazine **43**.[96,97] Similarly, the

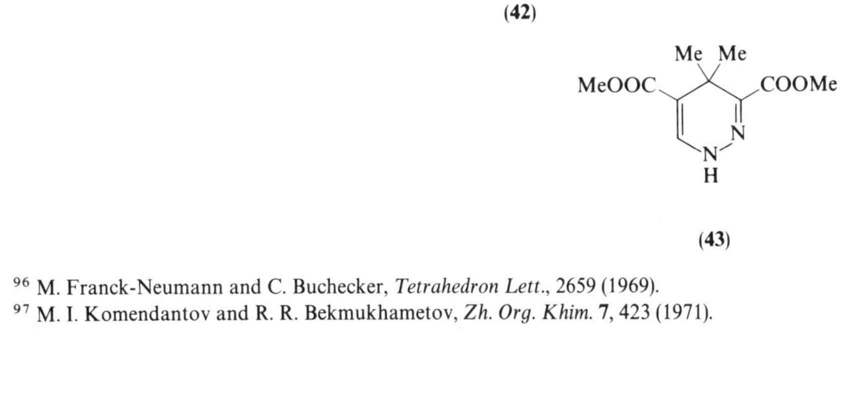

(42)

(43)

---

[96] M. Franck-Neumann and C. Buchecker, *Tetrahedron Lett.*, 2659 (1969).
[97] M. I. Komendantov and R. R. Bekmukhametov, *Zh. Org. Khim.* **7**, 423 (1971).

cycloadduct **44** is converted by alkali or acid to 3,5-diphenylpyridazine via the corresponding dihydropyridazine intermediates **45**.[98–100]

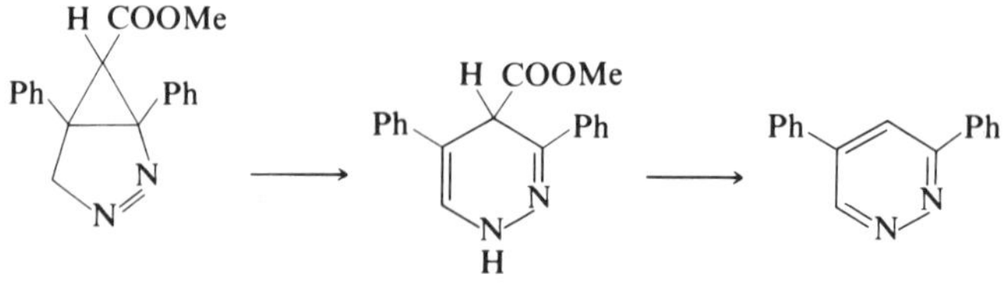

(44)   (45)

Using thermolysis in benzene at 80°C for 12 hr, Padwa[100] converted 3,6-diphenyl-1,2-diazabicyclo[3.1.0]hexene to 1,6-dihydropyridazine **46**; further heating of **46** resulted in the formation of 1,4-dihydropyridazine **28b**.

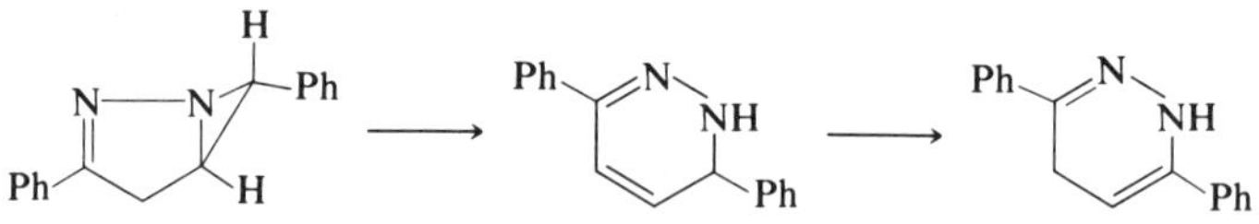

(46)   (28b)

One of the products isolated by treatment of cyclopropene with methyl diazoacetate is dihydropyridazine **48**,[101] apparently formed via an unstable adduct (**47**) [Eq. (7)].

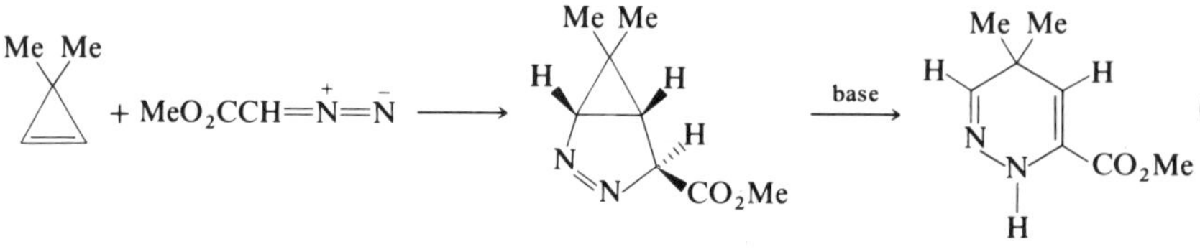

$$\text{Me Me} + \text{MeO}_2\text{CCH}=\overset{+}{\text{N}}=\overset{-}{\text{N}} \longrightarrow (47) \xrightarrow{\text{base}} (48) \quad (7)$$

(47)   (48)

Diazomethane reacts in ether with the cyclopropenes **49** to yield the corresponding cycloadducts, which may be transformed by base catalysis or by chromatography on silica gel to 1,4-dihydropyridazine **50**.[102]

---

[98] M. I. Komendantov, R. R. Bekmukhametov, and V. G. Novinskii, *Zh. Org. Khim.* **12**, 801 (1976).
[99] L. G. Zaitseva, I. B. Avezov, O. A. Subbotin, and I. G. Bolesov, *Zh. Org. Khim.* **11**, 1415 (1975).
[100] A. Padwa and H. Ku, *Tetrahedron Lett.*, 1009 (1980).
[101] D. H. Aue, R. B. Lorens, and G. S. Helwig, *J. Org. Chem.* **44**, 1202 (1979).
[102] M. Regitz, W. Welter, and A. Hartmann, *Chem. Ber.* **112**, 2509 (1979).

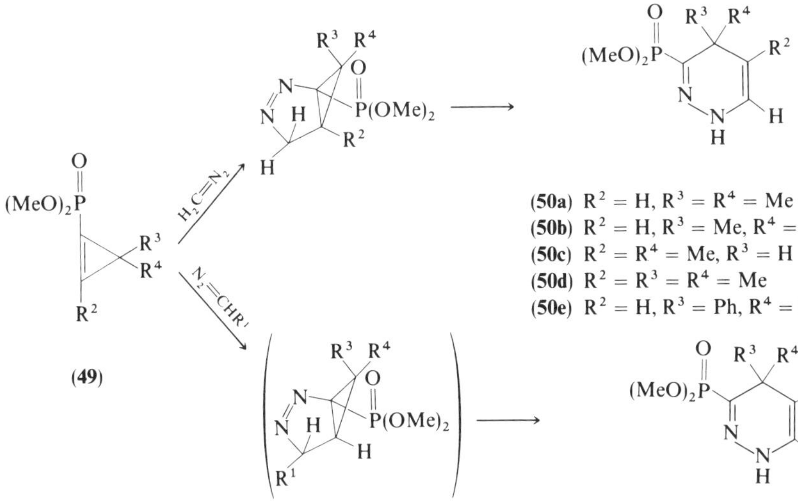

(50a) $R^2 = H, R^3 = R^4 = Me$
(50b) $R^2 = H, R^3 = Me, R^4 = t\text{-Bu}$
(50c) $R^2 = R^4 = Me, R^3 = H$
(50d) $R^2 = R^3 = R^4 = Me$
(50e) $R^2 = H, R^3 = Ph, R^4 = Me$

(51a) $R^1 = PO(OMe)_2, R^3 = R^4 = Me$
(51b) $R^1 = PO(OMe)_2, R^3 = Me, R^4 = t\text{-Bu}$
(51c) $R^1 = CO_2Et, R^3 = R^4 = Me$
(51d) $R^1 = CO_2Et, R^3 = Me, R^4 = t\text{-Bu}$

Analogously, use of the dimethyl ester of diazomethylphosphonic acid or the ethyl ester of diazoacetic acid instead of diazomethane gives the corresponding dihydropyridazine **51**.[102] 1,2-Diaza-1,3-butadienes undergo thermal dimerization according to a [4 + 2] cycloaddition mechanism, affording 1,4-dihydropyridazine **52**.[86,103]

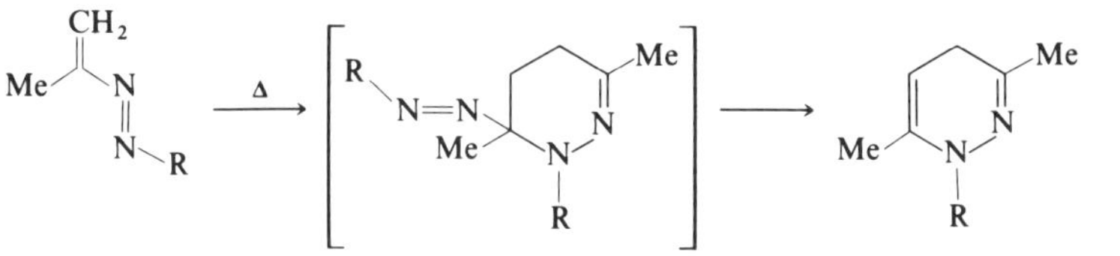

Aromatic diazonium salts undergo cycloaddition to dienes to produce N-substituted 1,6-dihydropyridazines **53** [Eq. (8)].

Carpenter and co-workers[104] have shown that 2,3-diazabicyclo[2.2.0]hex-2-ene opens thermally to give unsubstituted 4,5-dihydropyridazine (**8f**)

---
[103] S. Stickler and W. C. Hoffmann, *Angew. Chem.* **82**, 254 (1970).
[104] E. A. Wildi, D. van Engen, and B. K. Carpenter, *J. Am. Chem. Soc.* **102**, 7994 (1980).

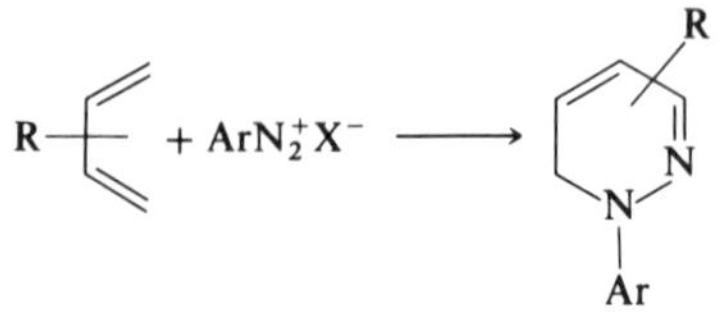

$$R\text{—}\diagup\diagdown + ArN_2^+X^- \longrightarrow \underset{Ar}{\underset{|}{\text{[pyridazine with R]}}}$$ (8)

(53)

[Eq. (9)]. Under the conditions of their experiment, they found that a dimeric species (which could be characterized only partially because of its instability) was formed along with trimer **32**.

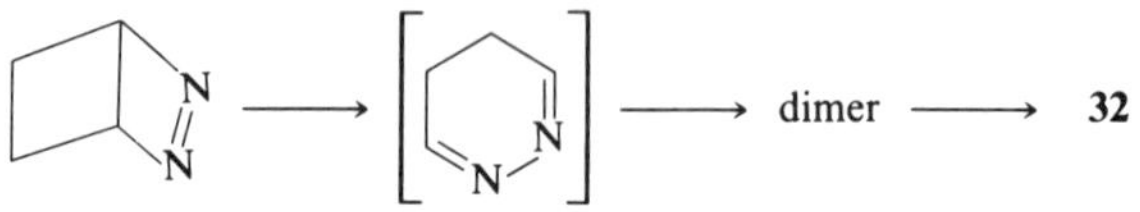

$$\text{[bicyclic N=N]} \longrightarrow [\text{pyridazine}] \longrightarrow \text{dimer} \longrightarrow \textbf{32} \qquad (9)$$

3. From Pyridazine and its Derivatives

a. *Addition of Organometallic Reagents.* A third major route to dihydropyridazines involves the addition of Grignard reagents or organolithium compounds to pyridazine or its derivatives. These reactions proceed via nucleophilic attack at the electron-deficient positions of the pyridazine ring, followed in many cases by 1,2-, 1,4-, or 1,6-addition, and further hydrolysis to give the desired dihydropyridazine. Letsinger and Lasco reported[105] that dihydropyridazines formed by the action of Grignard reagents or organolithium compounds on pyridazines are too unstable to permit characterization and thus underwent spontaneous aromatization. This reaction was reinvestigated by van der Stoel and van der Plas,[106] who managed to isolate the dihydropyridazines resulting from phenyllithium addition. However, these also proved too unstable for complete analysis.

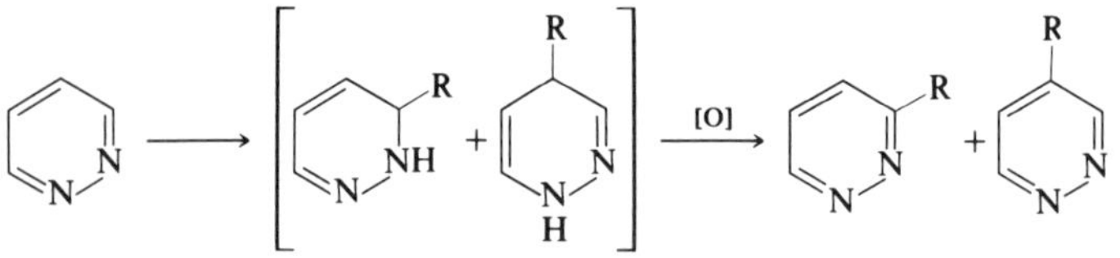

$$\text{pyridazine} \longrightarrow \left[\text{dihydro-R} + \text{dihydro-R} \right] \xrightarrow{[O]} \text{(54)} + \text{(55)}$$

(54)   (55)

---

[105] R. L. Letsinger and R. Lasco, *J. Org. Chem.* **21**, 812 (1956).
[106] R. E. van der Stoel and H. C. van der Plas, *Recl. Trav. Chim. Pays. Bas* **97**, 116 (1978).

On the basis of a detailed analysis of the resultant akylated and arylated pyridazines **54** and **55** produced by this reaction, Letsinger and Lasco found that the ratio of addition at position 3 to that at position 4 was sensitive to the metal in the organometallic reagent[105]: Grignard species attacked preferentially at position 4, while organolithiums produced the 3-substituted pyridazines. Both workers[106,107] also indicated that the solvent has a marked effect upon orientation, noting that ether promoted formation of **54**, while tetrahydrofuran increased the yield of **55**.

In addition, an interesting example of nucleophilic 1,2- and 1,4-addition of thienyllithium to pyridazine, with the formation of corresponding 1,6- (**56**) and 1,4-dihydropyridazine (**57**), has been published [Eq. (10)].[107]

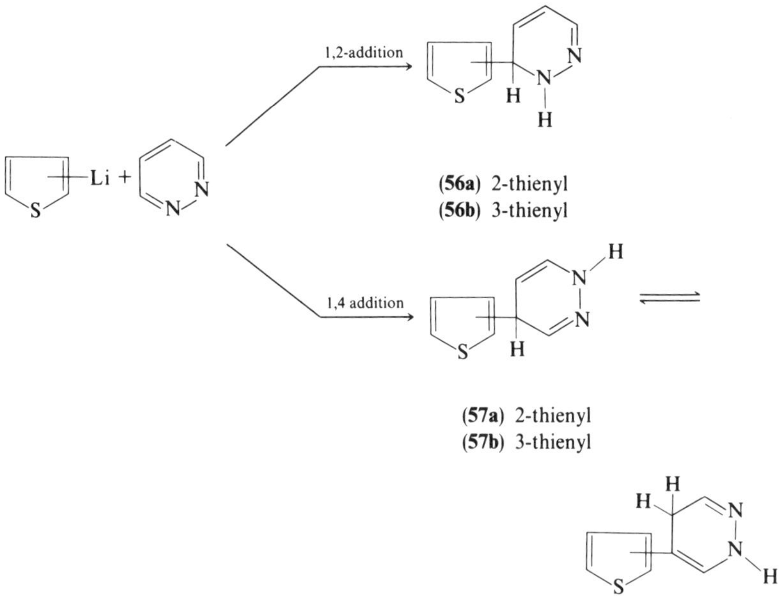

(**56a**) 2-thienyl
(**56b**) 3-thienyl

(**57a**) 2-thienyl
(**57b**) 3-thienyl

(10)

(**58a**) 2-thienyl

The product 1,4-dihydro-4-(2-thienyl)pyridazine (**57**) undergoes thermal rearrangement to 1,4-dihydro-3-(2-thienyl)pyridazine (**58**), which is induced by stronger stabilization conjugation between the two rings.

---

[107] J. Bourguignon, C. Becue, and G. Queguiner, *J. Chem. Res. Miniprint*, 1401 (1981); *J. Chem. Res., Synop.*, 104 (1981).

The addition of organometallics to substituted pyridazines was studied extensively.[108–121] Symmetrically 3,6-disubstituted pyridazines yield homogeneous 4-substituted products, but if the substituents differ, the formation of distinct 4- and 5-substituted products is expected.

Thus Crossland and co-workers found that organolithium and Grignard reagents add to symmetrical 3,6-disubstituted pyridazines **59** at position 4 to give dihydropyridazines **60** and **61** [Eq. (11)].[108,110,113,117] For example, 3,6-dimethoxypyridazine (**59a**) reacts with *tert*-butyl, phenyl, isopropyl, and ethyl Grignards to give the corresponding compound **60**,[108] although the ethyl adduct is isolated only with difficulty.[113] When methyl Grignard is used in the reaction, **59a** is recovered unchanged.

$$
\text{(59)} \longrightarrow \text{(60)} + \text{(61)} \qquad
\begin{array}{l}
\textbf{a.}\ R = OMe \\
\textbf{b.}\ R = Cl \\
\textbf{c.}\ R = \text{alkyl} \\
\textbf{d.}\ R = \text{aryl}
\end{array}
\qquad (11)
$$

On the other hand, pyridazines **59b, c**, and **d** were found to react with *tert*-butyl Grignard to give primarily the 1,4-dihydro compounds **61b, c**, and **d**.[108,110,117]

In unsymmetrical, 3,6-disubstituted pyridazines, the course of addition depends on both electronic and steric effects. Electronic considerations favor addition of organometallics such that the resultant negative charge is localized on the side of the more electron-withdrawing substituent.

The formation of dihydropyridazines by the reaction of *tert*-butyl Grignard with trisubstituted pyridazines has attracted particular attention. 3,4,6-Trichloropyridazine reacts with *tert*-butylmagnesium chloride, giving a

---

[108] I. Crossland, *Acta Chem. Scand.* **16**, 1877 (1962).
[109] A. Christensen and I. Crossland, *Acta Chem. Scand.* **17**, 1276 (1963).
[110] L. Avellen and I. Crossland, *Acta Chem. Scand.* **23**, 1887 (1969).
[111] I. Crossland and H. Kofod, *Acta Chem. Scand.* **24**, 751 (1970).
[112] I. Crossland, *Acta Chem. Scand.* **26**, 3257 (1972).
[113] I. Crossland, *Acta Chem. Scand.* **26**, 4183 (1972).
[114] I. Crossland, *Acta Chem. Scand.* **22**, 2700 (1968).
[115] L. Avellen, I. Crossland, and K. Lund, *Acta Chem. Scand.* **21**, 2104 (1967).
[116] I. Crossland and L. K. Rasmussen, *Acta Chem. Scand.* **19**, 1652 (1965).
[117] I. Crossland, *Acta Chem. Scand.* **18**, 1653 (1964).
[118] I. Crossland and E. Kelstrup, *Acta Chem. Scand.* **22**, 1669 (1968).
[119] A. K. Fateen and N. A. K. Shams, *J. Chem. U.A.R.* **11**, 301 (1968).
[120] E. Kelstrup, *Acta Chem. Scand.* **23**, 1797 (1969).
[121] F. G. Baddar, M. H. Nosseir, N. L. Doss, and N. N. Messiha, *J.C.S. Perkin I*, 1091 (1972).

mixture of 4-*tert*-butyl- (**62**) and 4,5-di-*tert*-butyl-1,4-dihydro derivatives (**63**), together with a small amount of the *trans*-4,5-di-*tert*-butyl-4,5-dihydropyridazine (**64**) and other products [Eq. (12)].[114] The formation of **64** probably occurs by nucleophilic displacement of chlorine at position 4, followed by Grignard addition at position 5.

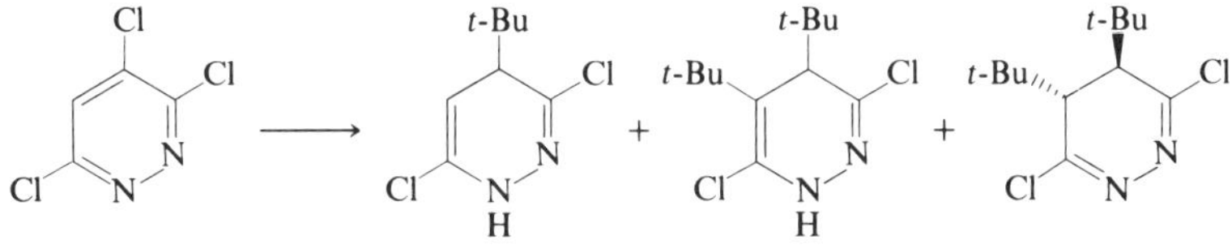

The addition of *tert*-butyl Grignard to 4-*tert*-butyl-3,6-dimethoxypyridazine is even more interesting [Eq. (13)].[108]

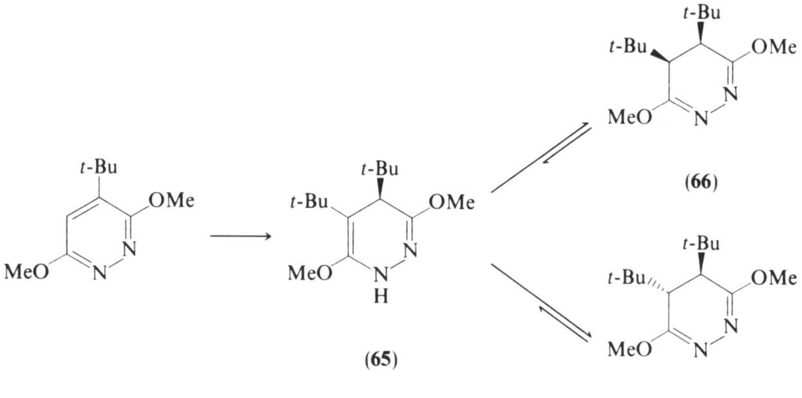

The product initially formed on workup with water is 4,5-di-*tert*-butyl-3,6-dimethoxy-1,4-dihydropyridazine (**65**). Upon treatment with acid, **65** converts rapidly to *cis*-4,5-di-*tert*-butyl-3,6-dimethoxy-4,5-dihydropyridazine (**66**), owing to the proton entering at the least hindered face of **65**. The cis compound is then slowly transformed to the more stable trans isomer (**67**), presumably through equilibrium with **65**.

b. *From 6-Oxo-1,4,5,6-tetrahydropyridazines.* The reaction of 6-oxo-1,4,5,6-tetrahydropyridazines with Grignard reagents or LiAlH$_4$ leads to dihydropyridazines.

For example, mixing 6-oxo-3-styryl-1,4,5,6-tetrahydropyridazine with a three- to fourfold molar excess of certain aryl and α-naphthylmagnesium bromides has been reported to yield the 4,5-dihydropyridazine **68**

[Eq. (14)].[122] Similarly, the addition of phenyl Grignard to 6-oxo-3-phenyl-1,4,5,6-tetrahydropyridazine gives the aromatic 3,6-diphenylpyridazine, presumably through the intermediacy of the dihydropyridazine.[123,124]

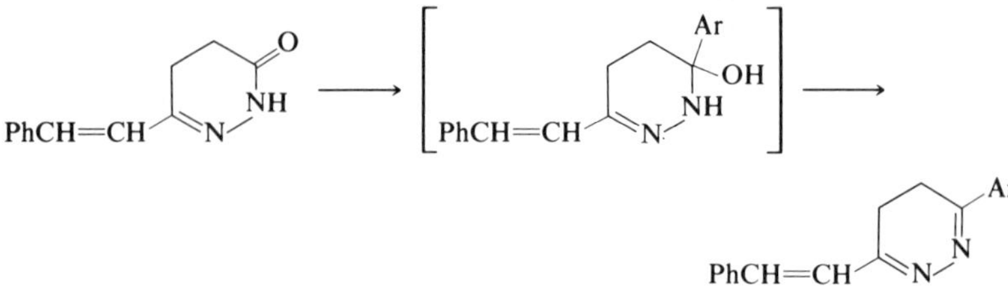

(68)

It was claimed that reaction of phenylmagnesium bromide with 1-phenyl-6-oxodihydropyridazines affords 1,6-dihydropyridazines (69) [Eq. (15)].[125]

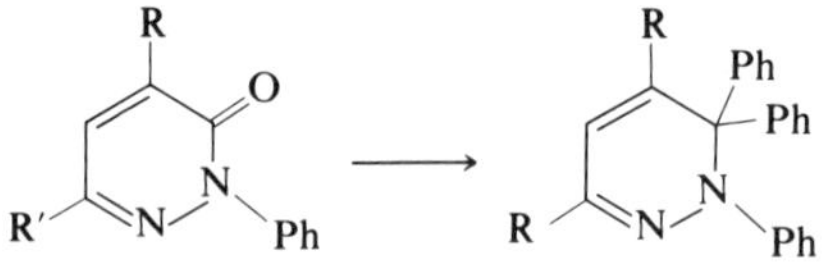

(69)

Aubagnac et al. reported the isolation of 1,4-dihydropyridazines 70 among the products of the reduction of 6-oxo-1,4,5,6-tetrahydropyridazines with LiAlH$_4$ [Eq. (16)].[126]

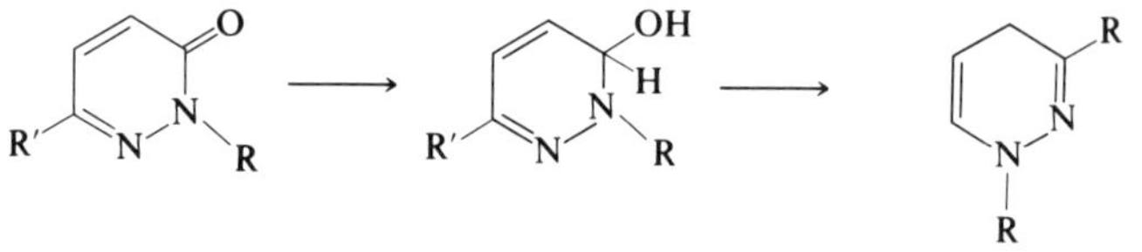

(70a) R = R' = Me
(70b) R = Me, R' = Ph
(70c) R = Ph, R' = Me
(70d) R = R' = Ph

---

[122] A. Mustafa, W. Asker, A. H. Harhash, K. M. Hoda, H. H. Jahine, and N. A. Kassab, *Tetrahedron* **20**, 531 (1964).
[123] F. G. Baddar, A. El-Habashi, and A. K. Fateen, *J. Chem. Soc.*, 3342 (1965).
[124] F. G. Baddar, N. Latif, and A. A. Nada, *J. Chem. Soc.*, 7005 (1965).
[125] M. A. F. El-Kaschef, F. M. E. Abdel-Megeid, and M. A. Michael, *Egypt. J. Chem.* **20**, 117 (1977).
[126] J. L. Aubagnac, J. Elguero, R. Jacquier, and R. Robert, *Bull. Soc. Chim. Fr.*, 2859 (1972).

c. *Reduction with Complex Hydrides.* The majority of known 1,4- and 1,6-dihydropyridazines are highly substituted or have strong electron-withdrawing groups on the pyridazine ring. Simple 1,6-dihydropyridazines (**71**) have now been prepared from 1-methylpyridazinium salts and sodium borohydride.[127] These dihydropyridazines gradually decompose in air at room temperature. If, however, the reduction is performed in the presence of methyl chloroformate, stable 1-methoxycarbonyl-1,6-dihydro compounds (**72**) were obtained, accompanied by a small amount of the 1,4-dihydro isomers (**73**) [Eq. (17)].[127]

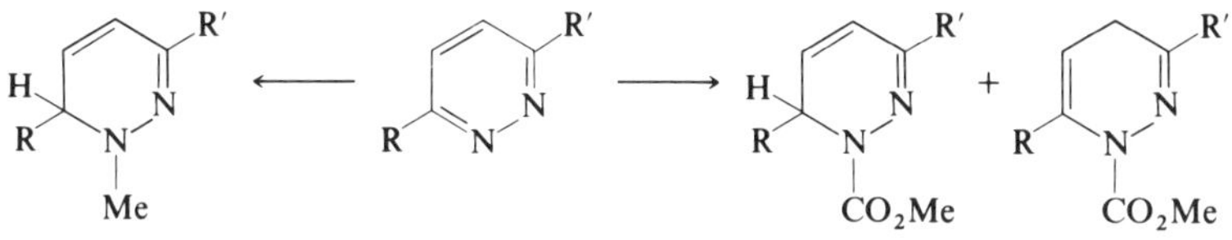

Dihydropyridazines have also been produced by the reduction of aromatic pyridazines. For instance, the action of lithium aluminum hydride on the diethyl ester of 3,6-dimethylpyridazine-4,5-dicarboxylic acid produces **74** in addition to other compounds [Eq. (18)].[128]

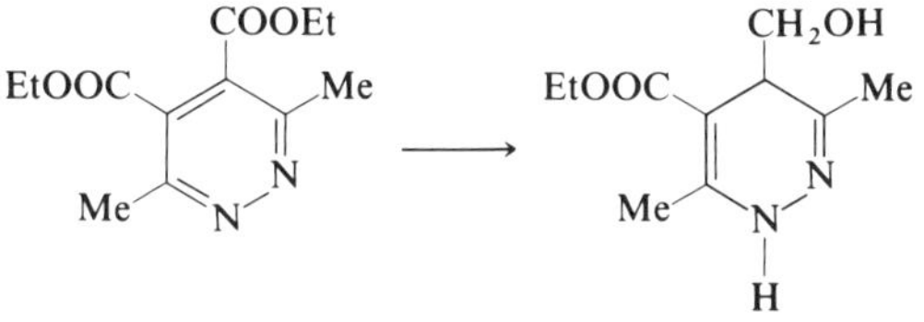

d. *Electrochemical Reduction.* Attention has been given to the electrochemical reduction of pyridazines. The following mechanism in Eq. (19) has been proposed by Klatt and Rouseff for the reduction of pyridazine in aqueous media.[129] The first reduction wave appears to be due to a two-electron reduction with subsequent protonation of the reduced ring occurring on the nitrogens; a 1,2-dihydropyridazine is formed, which decomposes to nitrogen and an unsaturated hydrazino aldehyde that polymerizes. A second reduction wave is also observed, but it is reported to be due to the reduction of a ring-opened species.

[127] C. Kaneko, T. Tsuchiya, and H. Igeta, *Chem. Pharm. Bull.* **22**, 2894 (1974).
[128] G. Adembri, F. De Sio, R. Nesi, and M. Scotton, *J. Heterocycl. Chem.* **12**, 95 (1975).
[129] L. N. Klatt and R. L. Rouseff, *J. Electroanal. Chem. Interfacial Electrochem.* **41**, 411 (1973).

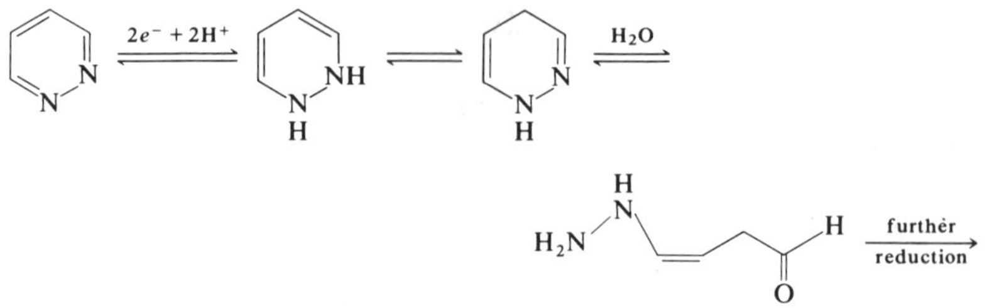

$$\text{(19)}$$

Lund studied the electrochemical reduction of substituted pyridazines and reported that reduction of 3,6-diphenylpyridazine gave the 1,4,5,6-tetrahydro derivative **75**.[130] This is probably produced through the intermediacy of 4,5-dihydropyridazine **28a** since, according to Lund, neither the 1,2- nor the 1,4-dihydro tautomers are expected to be reduced further under the reaction conditions used [Eq. (20)].

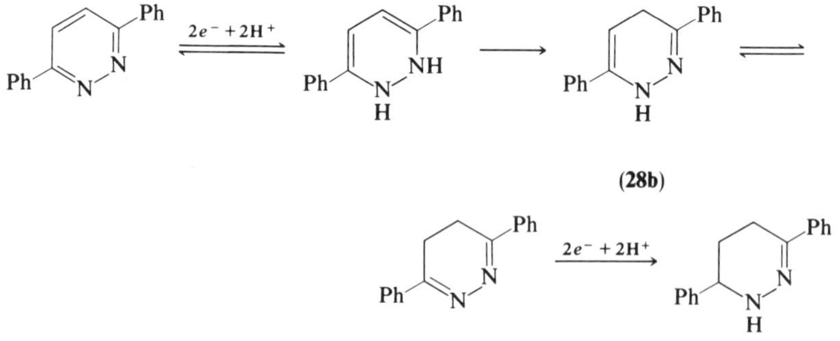

$$\text{(20)}$$

On the other hand, 3-phenyl-6-dimethylaminopyridazine **76a** is reduced smoothly to a 4,5-dihydropyridazine (**77a**). 3-Phenyl-6-methoxypyridazine (**76b**) and 3-methyl-6-chloropyridazine (**76c**) are also reduced to dihydro compounds but these are hydrolyzed to dihydropyridazinones (**77b,c**) at the low pH used in the reduction [Eq. (21)].[131]

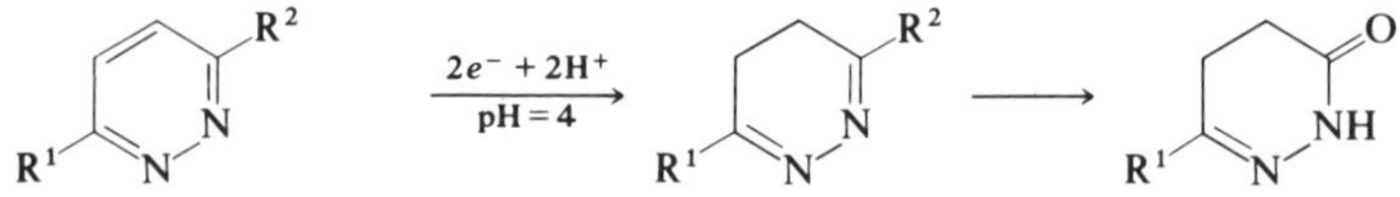

$$\text{(21)}$$

(**76a**) $R^1 = Ph, R^2 = N(Me)_2$
(**76b**) $R^1 = Ph, R^2 = OMe$
(**76c**) $R^1 = Me, R^2 = Cl$

---

[130] H. Lund, *Faraday Discuss. Chem. Soc.* **45**, 193 (1968).
[131] H. Lund., *Adv. Heterocycl Chem.* **12**, 272 (1970).

Electroreduction of pyridazines in the presence of acetic anhydride gives the acylated open-chain diamines.[132]

4. *Miscellaneous*

The synthesis of 3,6-dimethyl-6-*n*-butyl-5,6-dihydropyridazine (**79**), the first representative of 5,6-dihydropyridazines, using halogenation and dehydrohalogenation of the corresponding tetrahydro derivative (**78**), was described by de Mayo and Usselmann.[48] Upon heating to 125°C, the degassed benzene or *n*-hexane solution of **79** undergoes a suprafacial [1,5]-sigmatropic rearrangement to yield quantitatively the 1,6-dihydro derivative (**80**) [Eq. (22)].

$$\begin{array}{ccc} \text{(78)} & \text{(79)} & \text{(80)} \end{array} \quad (22)$$

3,6-Diphenylpyridazine is reduced with sodium and ethanol, affording the 1,2-dihydro derivative.[133] Since 3,6-diphenyl-4,5-dihydropyridazine is more stable than the isomeric 1,2-dihydro compound, the latter is isomerized in the presence of alkali to the 4,5-dihydro compound.[134] The 1,2-diethoxycarbonyl-1,2-dihydropyridazine is formed by selenium dioxide oxidation[135] or via allylic bromination–dehydrobromination[136,137] of the corresponding 1,2,3,6-tetrahydro compound. 1-Ethoxycarbonyl- or 1,2-diethoxycarbonyl-1,2-dihydropyridazines were obtained similarly from alkali treatment of 1,2-diethoxycarbonylhexahydropyridazines.[138]

C. TAUTOMERISM OF DIHYDROPYRIDAZINES

Dihydropyridazines without substituents on the nitrogens exist as 1,4- and/or 4,5-dihydro tautomers, often in equilibrium with one another.

As mentioned in Section III,D,2, it was found, for instance, that the ratio of **28a** to **28b** in deuteriochloroform is 1:8. In bromobenzene, the ratio is approximately 1:9.5.

---

[132] H. Lund and J. Simonet, *C.R. Hebd. Seances Acad. Sci., Ser. C* **277**, 1387 (1973).
[133] G. Rosseels, *Ing. Chim. (Brussels)* **46**, 7 (1964).
[134] G. Rosseels, *Ing. Chim. (Brussels)* **42**, 285 (1960).
[135] K. Alder, H. Niklas, R. Aumuller, and B. Olsen, *Justus Liebigs Ann. Chem.* **585**, 81 (1954).
[136] L. J. Altman, M. F. Semmelhack, R. B. Hornby, and J. C. Vederas, *Chem. Commun.*, 686 (1968).
[137] L. J. Altman, M. F. Semmelhack, R. B. Hornby, and J. C. Vederas, *Org. Prep. Proced. Int.* **7**, 35 (1975).
[138] M. Rink, S. Mehta, and K. Grabowski, *Arch. Pharm. (Weinheim, Ger.)* **292**, 225 (1959).

Crossland and Avellen[110] observed that addition of trifluoroacetic acid to a chloroform solution of **28a** and **28b** shifts the equilibrium entirely toward the 4,5-dihydro tautomer (**28a**). Neutralizing the solution, however, restores the original equilibrium. Equilibrium positions were also determined for two other 3,6-diaryl compounds, and in both cases the equilibrium favored the 1,4-dihydro tautomer. The ratio of 4,4-dimethyl-3,6-diphenyl-1,4-dihydropyridazine (**81**) to the 4,5-dihydro tautomer is 3.4:1, the 3,5,6-triphenyl compound **35** is found solely in the 1,4-dihydro form, probably because of conjugation stabilization.

(**81**)   (**82**)

As expected, the position of equilibrium varies significantly with the substituents at the diene termini. 3,6-Di-*tert*-butyldihydropyridazine (**82**) shows 3.4:1 for the ratio between the 4,5-dihydro tautomer to the 1,4-dihydro species and, as was noted earlier, the 3,6-dimethoxy- and 3,6-bis(dimethylamino)dihydropyridazines exist completely in the 4,5-dihydro form.

The difference in free energy, $\Delta G°$, between the two tautomers can be calculated from the ratio of the two forms in solution using Eq. (23), in which $K_T = [4,5\text{-dihydropyridazine}]/[1,4\text{-dihydropyridazine}]$.

$$\Delta G° = RT \ln K_T \quad (23)$$

In the case of the 3,6-diphenyldihydropyridazines **28a** and **28b**, a value for $\Delta G°$ of 1.24 kcal mol^{-1} in deuteriochloroform is indicated. For the 3,6-di-*tert*-butyl tautomers, a value of only 0.73 kcal mol^{-1} is obtained.

D. Physicochemical Properties

1. *Infrared Spectra*

The correlation of dihydropyridazine structure with absorption maxima in the IR region has not been systematically studied. Information presented here has been gleaned from data contained in original works. Dihydropyridazines usually give characteristic bands in the following regions: 3200–3500 cm^{-1} (ν_{NH}), 3000–3100 cm^{-1} ($\nu_{H-C=N}, \nu_{H-C=C}$), 1400–1700 cm^{-1} ($\nu_{C=C}, \nu_{C=N}$), and 700–950 cm^{-1} ($\nu_{H-C=C}, \nu_{H-C=N}$).[48,74,81,82,88,89,95,101,102,107,126] All dihydropyridazines show absorption in the 1400–1700 cm^{-1} region which is assigned to the C=C or C=N stretching modes. Two intense absorption bands have been reported for 1,4-dihydropyridazines at 1590–1630 and 1650–1670 cm^{-1}. However, some discrepancy exists in the assignment of these bands. Some workers[80,81,126] designate the former as the C=N stretching mode and the latter as the C=C stretching mode. Other authors[95] reverse the assignment. According to Zelenin et al.,[80,81] the first band falls in the region representing unconjugated hydrazones of ketones, while the second is common to enamines and enehydrazines.

Interestingly, the reported stretching mode of the conjugated C=N—N=C for 4,5-dihydropyridazine (**40a**)[95] and C=C—C=N—N for 2,3-dihydropyridazine (**36**)[88] fragments were below 1600 cm^{-1}. The measured C=C stretching mode in the C=C—N=N fragment of 5,6-dihydropyrazine (**79**) is 1675 cm^{-1}.[48] All N-unsubstituted dihydropyridazines absorb in the 3200–3450 cm^{-1} region and show the characteristic stretching frequencies for bonded and nonbonded NH groups.[48,89,95,101,102,107] The number and position of these depend on structural factors and on the conditions of measurements. Absorption in this region has been used for structural determinations.[48,89]

2. Ultraviolet Spectra

Only isolated UV spectra appear in the original literature. For the 5,6-dihydro compound (**79**) in hexane, two absorption maxima have been reported, presumably the $\pi \to \pi^*$ transition at 268 nm (ε 2100) and the $n \to \pi^*$ transition at 447 nm (ε 150).[48] 4,5-Dihydropyridazines (**69**) have one maximum at 285–287 nm,[121] whereas **40a** absorbs at 263 (ε 700) and 232 nm (ε 1250).[95] 6-Methoxycarbonyl-4,4-dimethyl-1,4-dihydropyridazine (**48**) has two maxima,[101] at 278 (ε 1600) and 247 nm (ε 4600), the long wavelength absorption normally being attributed to the $n \to \pi^*$ transition. 2,4-Diphenyl-1,2-dihydropyridazine (**37**)[89] showed three absorption bands in methanol: at 232 (ε 12300), 274 (ε 9500), and 370 nm (ε 220).

3. Nuclear Magnetic Resonance Spectra

Nuclear magnetic resonance is an invaluable tool for investigating dihydroazines and, in particular, the dihydropyridazines. Unfortunately, until recently, most investigators have reported only proton resonance data. No doubt, in the future, more attention will be paid to ^{13}C and ^{15}N, which can contribute much information.

The most useful applications of NMR have been in structural determinations and in identifying the presence of tautomeric equilibria.

Table I summarizes some typical dihydropyridazine spectra. The chemical shifts of the ring protons at unsaturated centers range from δ 4.5–7.0. As might

TABLE I
SELECTIVE ^{1}H-NMR DATA OF DIHYDROPYRIDAZINES

Structure	δ_1 (N)H^1	δ_3 H^3	δ_4 H^4	δ_5 H^5	δ_6 H^6	$J_{3,4}$	$J_{4,5}$	$J_{5,6}$	$J_{4,6}$	Solvent	Reference
1,2-Dihydro-											
34b	—	—	5.53	5.53	—	—	—	—	—	CCl$_4$	86
38	1.34	—	5.70–7.67	5.70–7.67	5.70–7.67	—	—	—	—	CDCl$_3$	89
1,4-Dihydro-											
30b	—	—	2.63	4.08	—	—	—	—	—	—	81
31a	—	5.98	2.66	4.36	6.32	—	3.0	6.7	—	CCl$_4$	83
31b	—	6.18	2.52	4.30	6.35	—	3.3	7.0	—	CCl$_4$	83
31c	—	6.06	2.50	4.04	—	—	3.3	—	—	CCl$_4$	83
31d	—	5.98	2.50	3.97	—	—	3.1	—	—	CCl$_4$	83
31e	—	—	2.43	4.12	6.00	—	3.2	6.9	—	CCl$_4$	83
31f	8.15	—	2.53	4.20	5.95	—	3.1	6.7	—	CCl$_4$	83
28b	—	—	3.35	4.95	—	—	4.0	—	—	CDCl$_3$	22,87,100
35	—	—	3.56	—	6.93	—	—	—	—	CDCl$_3$	22,87
37a	—	—	4.85	5.31	6.93	—	6.0	7.5	—	CDCl$_3$	88
37b	—	—	2.60	4.91	6.13	—	5.3	7.5	—	CDCl$_3$	88
48	8.24	6.13	—	5.40	—	—	—	—	2.5	CCl$_4$	101
50a	8.92	—	—	4.45	6.25	—	—	7.5	—	CDCl$_3$	102
50b	8.58	—	—	4.40	6.38	—	—	7.5	—	CDCl$_3$	102
50c	7.85	—	3.13	—	6.10	—	—	—	—	CDCl$_3$	102
50d	7.85	—	—	6.08	—	—	—	—	—	CDCl$_3$	102
50e	8.60	—	—	4.61	6.31	—	—	8.0	—	CDCl$_3$	102
51a	8.20	—	—	5.35	—	—	—	—	—	CDCl$_3$	102
51b	8.27	—	—	5.52	—	—	—	—	—	CDCl$_3$	102
51c	8.55	—	—	5.55	—	—	—	—	—	CDCl$_3$	102
51d	8.40	—	—	5.50	—	—	—	—	—	CDCl$_3$	102
57a	—	6.56	3.80	4.70	6.34	—	—	—	—	CDCl$_3$	107

Compound									Solvent	Ref
58a	—	—	—	3.10	—	—	—	—	CDCl$_3$	107
70a	—	6.72	—	—	6.72	—	—	—	CDCl$_3$	126
70b	—	—	—	3.05	4.50	—	—	—	CDCl$_3$	126
70c	—	—	—	2.78	4.62	—	—	—	CDCl$_3$	126
70d	—	—	—	3.18	4.73	—	3.8	7.7	1.1	—
73b	—	—	—	2.78	4.83	—	—	—	CDCl$_3$	127
73c	—	—	—	2.65	4.91	—	4.0	10.0	1.2	CDCl$_3$	127
73d	—	—	—	2.81	4.75	—	3.6	—	—	CDCl$_3$	127
73e	—	—	—	3.02	4.91	—	4.0	8.0	1.5	CDCl$_3$	127
					4.90		4.0	8.5	1.2	CDCl$_3$	127
1,6-Dihydro-
36a	—	—	—	—	5.99	—	—	5.0	—	CDCl$_3$	88
36b	—	—	—	—	6.10	—	—	5.3	—	CDCl$_3$	88
36c	—	—	—	—	6.22	—	—	5.0	—	CDCl$_3$	88
46	6.34	—	6.75	—	6.02	—	10.0	4.0	2.0	CDCl$_3$	100
56a	6.20	6.95	5.94	—	5.94	—	—	—	—	CDCl$_3$	107
71a	—	6.66	5.72	—	5.90	—	9.5	3.6	—	CDCl$_3$	127
71b	—	—	5.64	—	5.94	—	9.5	4.0	—	CDCl$_3$	127
71c	—	—	5.48	—	5.48	—	—	—	—	CDCl$_3$	127
71d	—	—	5.77	—	6.14	—	9.5	4.0	—	CDCl$_3$	127
71e	—	—	6.35	—	5.98	—	10.0	4.0	0.9	CDCl$_3$	127
71f	—	—	6.33	—	5.88	—	9.5	4.5	—	CDCl$_3$	127
72a	—	6.70–6.95	5.60–5.90	—	5.92–6.41	3.5	10.0	4.0	—	CDCl$_3$	127
72b	—	—	5.74	—	6.18	—	10.0	4.0	1.5	CDCl$_3$	127
72d	—	—	5.82	—	6.38	—	10.0	4.0	1.5	CDCl$_3$	127
72e	—	—	5.85–6.35	—	5.85–6.35	—	—	—	—	CDCl$_3$	127
80	5.70	—	5.80	—	5.60	—	10.0	—	—	CDCl$_3$	48
4,5-Dihydro-
28a	—	—	2.71	2.71	—	—	—	—	CDCl$_3$	22
82	—	—	2.02	2.02	—	—	—	—	CDCl$_3$	22
5,6-Dihydro-
79	—	—	5.2	1.60–2.00	—	—	—	—	CDCl$_3$	48

be expected, proximity of an electron-withdrawing substituent on a ring nitrogen results in shifts to lower fields. Ring protons at saturated centers produce signals at δ 2.4–3.2. Vicinal coupling constants across a double bond (CH=CH, J = 6–10 Hz) are generally larger than those across a single bond (=CHCH, J = 2–4 Hz). The NH protons have been found for 1,2-dihydropyridazine at δ 1.34, whereas for 1,6-dihydro compounds at δ 5.7–6.3 and for the 1,4-dihydro compounds at δ 8.1–8.8. The ring methylene protons are usually equivalent, indicating, apparently, a rapidly occurring interconversion. Long-range coupling constants across the ring are frequently observed (see Table I).

E. Reactivity

The chemical properties of dihydropyridazines were not studied systematically until now, and information about these compounds is almost completely absent. The only known property of most of the N-unsubstituted dihydropyridazines is that they can be easily oxidized (very often spontaneously in air) to the corresponding aromatic pyridazines.

Padwa and Gehrlein reported[139] that excited-state behavior of 1,6-dihydropyridazines differs from that of other diazacyclohexadienes. Thus photooxidation of 2,5,6-triphenyl-1,6-dihydropyridazine (**36a**) in 95% ethanol gave pyridazinone (**83**). However, when irradiation of dihydropyridazines (**36**) was carried out in cyclohexene in the presence of fumaronitrile, a [2 + 2] cycloadduct (**84**) was obtained. Furthermore, the authors noted that the dihydropyridazine system underwent a deep-seated skeletal rearrangement when treated with aqueous acid. Thus reaction of **36a** with aqueous hydrochloric acid results in the formation of 1,3,4-triphenylpyrazole (**85**) [Eq. (24)].

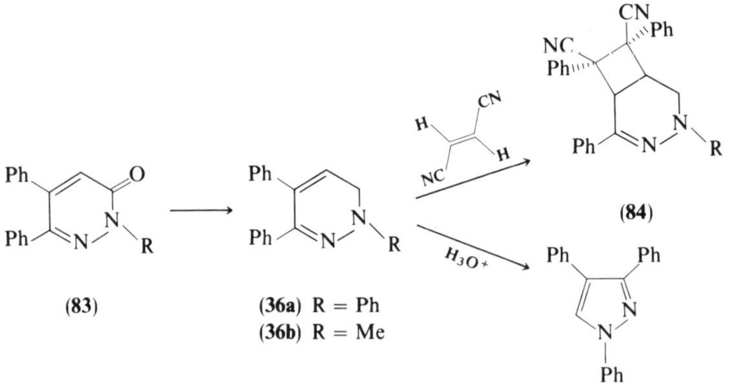

(24)

(83)　　(36a) R = Ph
　　　　(36b) R = Me

(84)

(85)

[139] A. Padwa and L. Gehrlein, *J. Heterocycl. Chem.* **12**, 589 (1975).

As mentioned above, unsubstituted 4,5-dihydropyridazine and its simple derivatives readily dimerize or trimerize.[84,104,140]

V. Dihydropyrimidines

A. General Comments

Formally, five structures can be drawn in two dimensions for the isomeric dihydropyrimidines. However, most of the known dihydropyrimidines have either the 1,2- (**9a**) or the tautomeric 1,4- (**9b**) and 1,6- (**9c**) dihydro structures (Section III,B).

The chemistry of dihydropyrimidines is virtually a closed book owing to an absence of efficient synthetic methods. Moreover, there is also a dearth of quantitative data on the stabilization–destabilization effects of substituents on these ring systems. In addition, structures and relative stabilization of the various isomers and the tautomeric 1,6- and 1,4-dihydropyrimidines, in particular, were unknown until recently. In most published reports, suggested structures for products have been proposed with insufficient information to justify the given assignments. Early work on dihydropyrimidines is scattered throughout the exhaustive monographs of Brown on pyrimidines.[141,142] This survey, therefore, will aim to highlight the latest synthetic achievements in the preparation of dihydropyrimidines, along with unambiguous spectral assignments and confirmed tautomeric exchanges characteristic of these compounds.

B. Synthesis

1. *Cyclization Methods*

a. *From α,β-Unsaturated Carbonyls.* The reaction of amidines with readily available α,β-unsaturated carbonyl compounds is an attractive ⟨3 + 3⟩ approach to dihydropyrimidines. However, until recently, only a few reactions of this type were known, the result of a fortuitous combination of reaction conditions and starting materials. Traube and Schwarz prepared

---

[140] P. De Mayo, J. B. Stothers, and M. C. Usselman, *Can. J. Chem.* **50**, 612 (1972).
[141] D. J. Brown, *in* "The Chemistry of Heterocyclic Compounds" (A. Weissberger, ed.). Wiley (Interscience), New York, 1962.
[142] D. J. Brown, *in* "The Chemistry of Heterocyclic Compounds" (A. Weissberger, ed.), Suppl. 1. Wiley, New York, 1970.

dihydropyrimidine (**19**) by heating benzamidine with mesityl oxide, to which they assigned the 4,5-dihydro structure **19c**[143] [Eq. (25)]. Silversmith has shown that the spectral data (IR and NMR) contradict the proposed structure, but are consistent with either of the tautomeric 1,4- and 1,6-dihydropyrimidines.[144] It has been demonstrated that the product obtained by Traube and Schwarz exists in the solid state as a 2-phenyl-4,6,6-trimethyl-1,6-dihydropyrimidine (**19b**), whereas in solution a tautomeric equilibrium mixture of the 1,4 and 1,6 forms is observed.

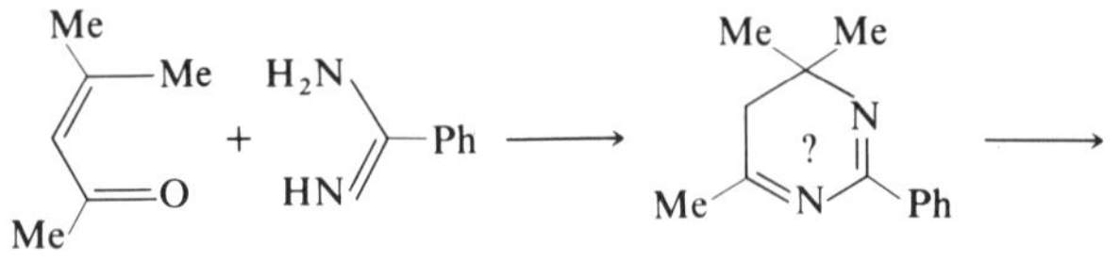

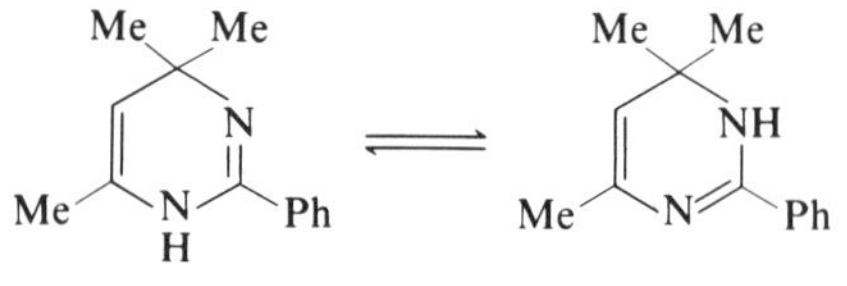

In a second example, Ruhemann[145] found that the reaction of benzamidine with benzilideneacetylacetone at 100°C results in the loss of an acetyl group with formation of 6-methyl-2,4-diphenyldihydropyrimidine (MDHP) (**20**) [see Eq. (31)].

The reason for the lack of follow-up on this synthetic route may be attributed to literature reports of failures to isolate or even observe the expected dihydropyrimidines, particularly in the case of the simple α,β-ethylenic aldehydes.[143,146] Dihydropyrimidine formation from α,β-unsaturated carbonyl compounds and amidines occurs via nucleophilic attack by amidine at the activated double bond (Michael-type addition), followed by ring closure and dehydration (see Scheme 4). In the course of confirming this reaction scheme, the intermediacy of tetrahydropyrimidines and dihydropyrimidines was demonstrated.

Further improvements that facilitated each step of this reaction (using molecular sieves, DMSO, and air) led to a new one-pot synthesis of

---

[143] W. Traube and R. Schwarz, *Ber. Dtsch. Chem. Ger.* **32**, 3163 (1899).
[144] E. F. Silversmith, *J. Org. Chem.* **27**, 4090 (1962).
[145] S. Ruhemann, *J. Chem. Soc.* **83**, 1371 (1903).
[146] R. M. Dodson and J. K. Seyler, *J. Org. Chem.* **16**, 461 (1951).

Sec. V.B] DIHYDROAZINES 47

$$R^3\text{-CH=CR}^2\text{-C(R}^4\text{)=O} + HN=C(R^1)NH_2 \xrightarrow[\text{molecular sieves, air}]{\text{DMSO, 130–140°C}} \text{pyrimidine} \quad (26)$$

R^1 = H, Me, Ph, NH_2, $NHCOCH_3$, OCH_3, SCH_3
R^2, R^3, R^4 = H, Me, Ph

pyrimidines[147–150] [Eq. (26)]. Moreover, this understanding has led to selective preparation of cyclic adducts of benzamidine with chalcone and cinnamic aldehyde.[151] MDHP (**20**), identical to the samples obtained by Ruhemann[145] and Heyes and Roberts,[152] was easily prepared from benzalacetone and benzamidine in benzene with azeotropic removal of water.[153] However, attempts to synthesize other dihydropyrimidines, particularly those with fewer and smaller substituents, failed. Therefore, in the author's laboratory, a new strategy to prepare these compounds was proposed, which executes the two key steps in the reaction mechanism in separate stages: a preparation of hydroxytetrahydropyrimidine (**86**) [Eq. (27)], and dehydration of **86** to dihydropyrimidines (**88**). Usually a nearly quantitative yield of 6-hydroxyl-1,4,5,6-tetrahydropyrimidines was achieved under mild conditions, and optimal methods of dehydration were developed. This gives a new and versatile synthetic route, enabling the preparation of a large variety of dihydropyrimidines (see Table II).

First step:

$$\text{(}\alpha,\beta\text{-unsaturated ketone)} + HN=C(R^4)NH_2 \longrightarrow \text{(86)} \quad (27)$$

(**86**)

[147] V. P. Mamaev and A. L. Weis, *Khim. Geterotsikl. Soedin.*, 1555 (1975).
[148] A. L. Weis and V. P. Mamaev, *Izv. Sib. Otd. Akad. Nauk SSSR, Ser. Khim. Nauk*, 91 (1975) [*CA* **84**, 121755t (1976)].
[149] A. L. Weis, V. M. Shirina, and V. P. Mamaev, *Izv. Sib. Otd. Akad. Nauk SSSR, Ser. Khim. Nauk*, 144 (1975) [*CA* **84**, 105528r (1976)].
[150] A. L. Weis and V. P. Mamaev, *Izv. Sib. Otd. Akad. Nauk SSSR, Ser. Khim. Nauk*, 147 (1975) [*CA* **84**, 121763u (1976)].
[151] A. L. Weis and V. P. Mamaev, *Khim. Geterotsikl. Soedin.*, 674 (1977).
[152] T. D. Heyes and J. C. Roberts, *J. Chem. Soc.*, 328 (1951).
[153] A. L. Weis and V. P. Mamaev, *Izv. Sib. Otd. Akad. Nauk SSSR, Ser. Khim. Nauk*, 148 (1975) [*CA* **84**, 121764v (1976)].

TABLE II

PREPARATION OF 6-HYDROXYTETRAHYDROPYRIMIDINES (**86**) AND 1,4(1,6)-DIHYDROPYRIMIDINES (**88**)

Substances	R^1	R^2	R^3	R^4	86 Yield (%)[a]	86 Melting point (°C)	88 Yield (%)[a]	88 Melting point (°C)
a	H	H	Ph	Ph	98	140–141	96	144–145
b	H	H	Ph	Me	96	162–163	94	137–138
c	H	H	Ph	H	88	119–120	96	—[c]
d	H	H	H	Ph	96	156–158	87	131–132
e	H	H	H	Me	87	—[c]	30(80)[b]	—[c]
f	H	H	H	H	84	—[c]	0.5(27)[b]	—[c]
g	Me	COMe	Ph	Ph	98[d]	137–138	82	83–84
h	Me	COMe	Ph	Me	70[d]	188–189	76	104–105
i	Me	CO$_2$Me	Ph	Me	83[d]	182–184	67	135–137
j	Me	CO$_2$Et	Ph	Ph	89[d]	217–219	78	106–108
k	Me	CO$_2$Et	Ph	Me	79[d]	168–170	73	147–149
l	Me	CO$_2$Et	p-NO$_2$C$_6$H$_4$	Me	72[d]	214–215	—	—
m	Me	CO$_2$Et	m-NO$_2$C$_6$H$_4$	Me	74[d]	207–208	—	—
n	Me	CO$_2$Et	o-NO$_2$C$_6$H$_4$	Me	71[d]	148–149	68	182–184
o	Me	CO$_2$Me	o-NO$_2$C$_6$H$_4$	Me	68[d]	152–153	69	200–202

[a] The isolated yield of analytical sample is indicated.
[b] Yield in parentheses according to NMR.
[c] Unstable compounds.
[d] Isomeric mixture.

Second step:

$$86 \xrightarrow{-H_2O} \underset{(88a)}{\underset{H}{\overset{R^2}{\underset{R^1}{\bigvee}}\overset{R^3}{\underset{N}{\bigvee}}R^4}} \rightleftharpoons \underset{(88b)}{\underset{R^1}{\overset{R^2}{\bigvee}}\overset{R^3}{\underset{N}{\bigvee}}{\underset{R^4}{\overset{NH}{\bigvee}}}} \qquad (28)$$

Armed with these encouraging results, we attempted the preparation of the parent compounds in this series, a goal long sought by synthetic chemists. In the past, attempts to synthesize unsaturated dihydropyrimidines from the parent α,β-unsaturated carbonyl compound acrolein failed,[143,146,154,155] probably because of fast polymerization of acrolein in the presence of strong bases. Therefore, if one could interfere with this polymerization, the Michael-type addition, leading to the desired product, could then proceed unhindered. Hydroquinone was found to provide the best such protection against polymerization of acrolein.[156]

By this method, 6-hydroxytetrahydropyrimidines (86) were prepared in high yield by dropwise addition of a solution of acrolein to a stirred solution of corresponding amidines in acetone at −5°C under an inert atmosphere. Dehydration of these products was carried out by boiling in dimethoxyethane (or other solvents such as acetonitrile and DMF), giving 2-phenyl- and 2-methyldihydropyrimidine in reasonable yields (Table II). Unsubstituted dihydropyrimidine was also prepared by this method; however, high instability did not allow isolation in a pure state, and therefore only spectral verification could be provided. Good yields were also obtained by dehydration, using acid catalysis or acetic acid as a solvent. All dihydropyrimidines synthesized by this method exist in solution in tautomeric equilibrium of 1,4- and 1,6-dihydro compounds (Section V,C,1).

Besides the high yields of dihydropyrimidines, another valuable advantage of this method is that one can usually follow the transformation to dihydropyrimidine visually by choosing a suitable solvent, namely, one with a low solubility for hydroxytetrahydropyrimidine (86). Moreover, the reaction is very clean, and evaporation of the solvent alone usually gives an analytically pure dihydropyrimidinic product.

---

[154] A. Le Berre and C. Renault, *Bull. Soc. Chim. Fr.*, 3146 (1969).
[155] A. L. Weis, Candidate's Thesis, Institute of Organic Chemistry, Siberian Division of the USSR Academy of Sciences, Novosibirsk, 1975.
[156] A. L. Weis, F. Frolow, D. Zamir, and M. Bernstein, *Heterocycles* **22**, 657 (1984).

The initial product of dehydration of hydroxytetrahydropyrimidine **86** is the unstable 4,5-dihydropyrimidine **87**, which presumably undergoes either thermal suprafacial [1,5]-hydrogen migration to the corresponding 1,6-

(**87**) (**88b**)

(**87**) (**88a**)

(**89**) (**90**)

(**91**) (**92a**) (**92b**)

R = Me R = Ph

(29)

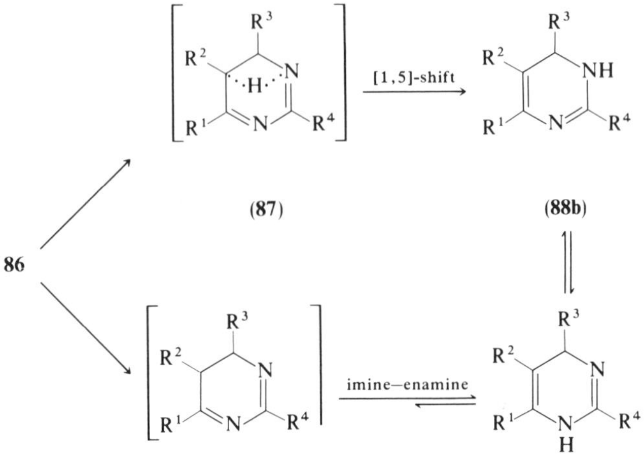

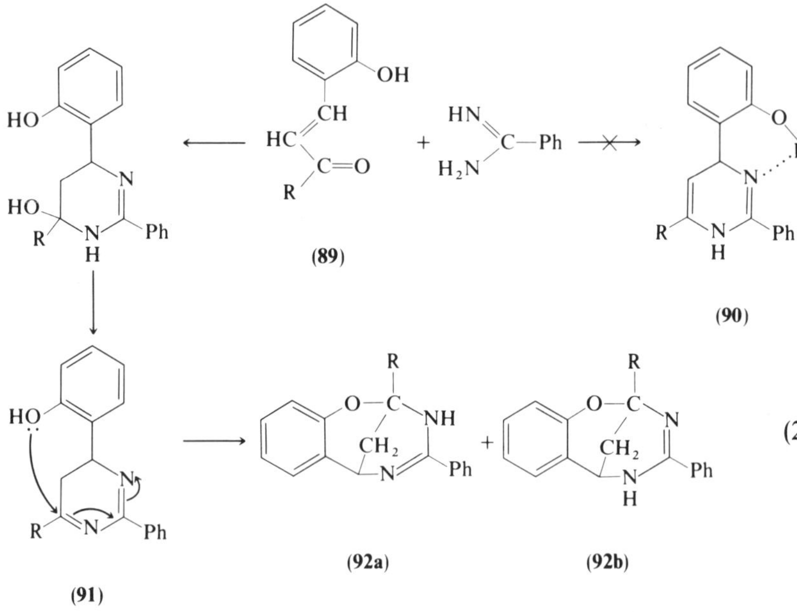

dihydropyrimidine **88b** or imine–enamine tautomerism to the 1,4-dihydropyrimidine **88a**.

Apparently, the rate of hydrogen migration is fast, since 4,5-dihydropyrimidines were neither isolated nor detected spectroscopically. However, some indirect proofs for intermediacy of 4,5-dihydropyrimidines were obtained. For example, in the reaction of o-hydroxyphenyl-α,β-unsaturated carbonyl compounds **89** with benzamidine, instead of the desired dihydropyrimidine (**90**) (which is stabilized by intramolecular hydrogen bonding), the corresponding bridgehead compounds **92** were isolated. The key step in this reaction is likely the formation of highly reactive 4,5-dihydropyrimidine **91**, which is rapidly attacked by the phenolic hydroxy group before it undergoes proton migration to the stable dihydropyrimidine [Eq. (29)].[157]

Moreover, in the course of structural investigations, the deuterated dihydropyrimidine was prepared for specific assignments of protons at C-4 and C-5 by the sequence given in Eq. (30).

$$PhCHO + CD_3CCD_3 \longrightarrow$$

(30)

66%   34%

The ratio of deuteration at position 5 clearly indicates a kinetic isotope effect during the transformation of 4,5-dihydropyrimidine to the more stable tautomeric 1,4- and 1,6-dihydro compounds.

The mechanism of this hydrogen migration needs further study. It would be interesting to compare the rates of and substituent influences on this migration with the similar processes in the last step of Hantzsch's synthesis of dihydropyridines (Scheme 3) and in hydrogen migration in the dihydropyridazine series (Section IV,C).

A reinvestigation of Ruhemann's reaction has clearly shown that—in conformity with the proposed mechanism—the initial product formed is hydroxytetrahydropyrimidine (**86g**), which, depending on reaction conditions, may be transformed to 5-acetyldihydropyrimidine (**88g**) [Eq. (31)].

[157] A. L. Weis, F. Frolow, M. Bernstein, and J. Fahima, *Isr. Chem. Soc., 49th Annu. Meet., 1982*, p. 55 (1982); *J. Org. Chem.* **49**, 3635 (1984).

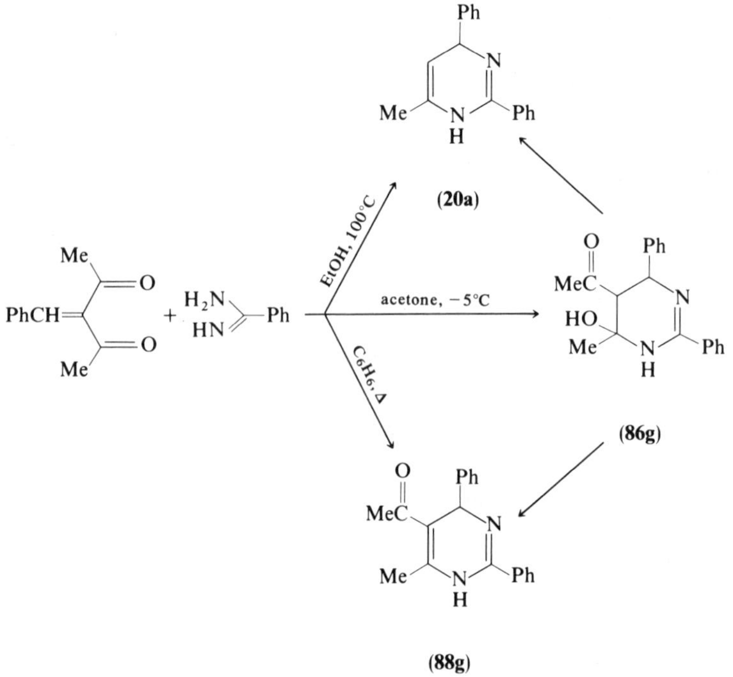

(88g)

The loss of the acetyl group in ethanol may be explained in terms of solvent participation under base catalysis.

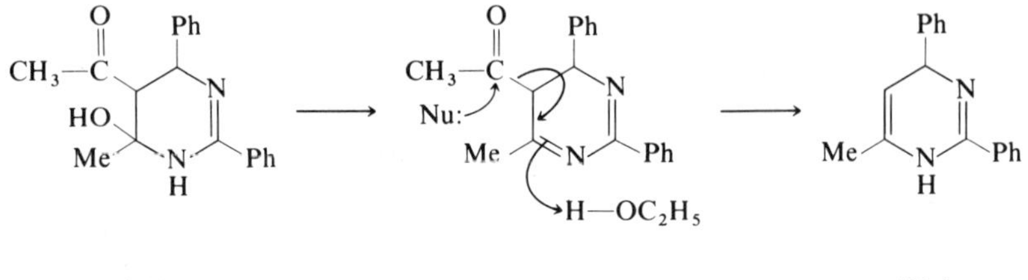

However, in the case of benzylideneacetoacetic esters, which have two distinct carbonyl groups and, therefore, two different possibilities for ring closure, the more reactive ester group usually participates in the pyrimidine ring formation. By using the reaction pathway developed in our laboratory, reaction of the ester function can be avoided, and high yields of dihydropyrimidines are obtained from hydroxytetrahydropyrimidines. This is an attractive route for preparation of 5-functionalized dihydropyrimidines,[158]

[158] A. L. Weis, Israel Patent Application pending. A. L. Weis and R. Vishkautsau, *Isr. J. Chem.* (1985) (in press).

which are physiologically active compounds with interesting antioxidant and membranotropic properties, as well as low toxicity, and which are also analogues of the calcium antagonist nifedipine (see Section I).

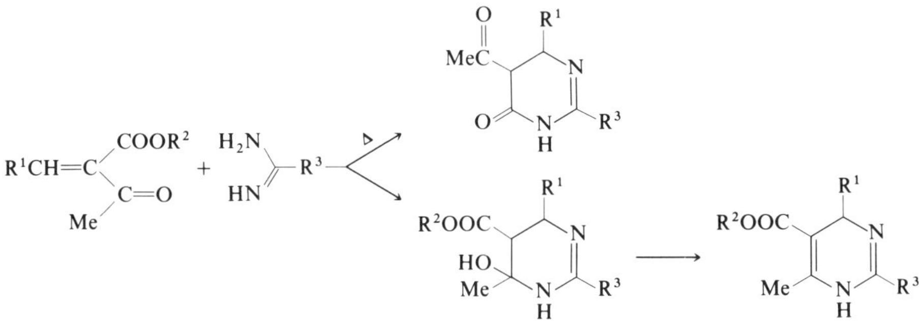

b. *From β-Dicarbonyl Compounds.* The four-component condensation reaction of β-dicarbonyl compounds with ammonia (or ammonium salts) and carbonyl-containing substances is a convenient and rather promising ⟨3 + 1 + (1) + 1⟩-fragment approach for the preparation of 1,2-dihydropyrimidines (Scheme 6).

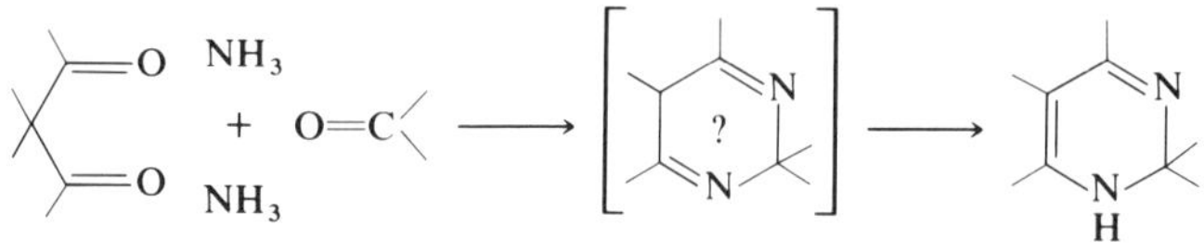

SCHEME 6

Formally, one can also propose the formation of the 2,5-dihydro isomer, but even if this isomer does form during the reaction, it is easily transformed to the thermodynamically more stable 1,2-dihydro compound. This reaction is nearly identical to the classical Hantzsch's synthesis of dihydropyridines, with only the molar ratio of starting materials being changed. However, in contrast to the widely investigated Hantzsch reaction, there are few examples in the literature, and it needs detailed study.

This original approach was first used in 1964 by Kröhnke and coworkers[159] for preparing dihydropyrimidines, but these spontaneously decomposed to the corresponding pyrimidines and could not be isolated.

---

[159] F. Kröhnke, E. Schmidt, and W. Zacher, *Chem. Ber.* **97**, 1163 (1964).

Hoffmann and co-workers[49,160] first successfully isolated 1,2-dihydropyrimidines **93** from a condensation of the dimethylacetal of acetoacetic aldehyde with different carbonyl-containing compounds in the presence of ammonia and ammonium nitrate [Eq. (32)].

$$\begin{array}{c}\text{Me}\\ \diagdown\text{C=O}\\ \text{H}_2\text{C}\\ \diagup\text{CH}\\ \text{MeO}\text{OMe}\end{array} + \text{O=C}\diagup^{R^1}_{R^2} \longrightarrow \text{(93)} \rightleftharpoons \quad (32)$$

Usually, these compounds are rather unstable at room temperature, although they could be stored in a freezer. Several authors have used this method for the preparation of 1,2-dihydropyrimidines.[161-163] However, the majority of the compounds isolated were disubstituted in position 2, preventing further oxidation to corresponding pyrimidines.

Salts of 1,2-dihydropyrimidines **95** were also prepared by a clever modification of this approach, which instead of a mixture of ammonia and carbonyl derivative, used a gem diamine (**94**) in condensation with acetylacetone or β-methoxyacrolein under acidic conditions[164] [Eq. (33)].

$$(CF_3)_2C\diagup^{NH_2}_{NH_2} \xrightarrow[\text{MeOCH=CHCH(OMe)}_2]{\text{MeCOCH}_2\text{COMe}} \quad (95) \quad X^- \quad (33)$$

(**94**)

(**95a**: R = Me, X = ClO$_4$)
(**95b**: R = H, X = Cl)

c. *Other Cyclizations.* Barluenga *et al.* have shown that N-substituted 1,2-dihydropyrimidines **96** are obtained with high yields in an aluminum chloride-catalyzed reaction of diimines, carbonyl compounds, or their acetals[165,166] [Eq. (34)].

[160] S. Hoffmann and W. Shultze, Patent, DDR 48617 (1966) [*CA* **66**, 28793 (1967)].
[161] D. Lloyd and H. McNab, *J.C.S. Perkin I*, 1784 (1976).
[162] H.-J. Teuber, G. Schutz, and W. Kern, *Chem. Ber.* **108**, 383 (1975).
[163] A. L. Weis and V. Rosenbach, *Tetrahedron Lett.*, 1453 (1981).
[164] G. A. Reynolds, G. H. Hawks, and K. H. Drexhage, *J. Org. Chem.* **41**, 2783 (1976).
[165] V. G. Aranda, J. Barluenga, V. Gotor, and S. Fustero, *Synthesis*, 720 (1974).
[166] J. Barluenga, M. Tomas, S. Fustero, and V. Gotor, *Synthesis*, 346 (1979).

Sec. V.B] DIHYDROAZINES 55

$$\begin{array}{c} \text{Ph} \\ R^2 \diagdown\!\!\!\!\diagup \text{NHR}^1 \\ R^3 \diagup\!\!\!\!\diagdown \text{NH} \end{array} + X{=}C\diagup\!\!\!\!\diagdown\begin{array}{c} R^4 \\ H \end{array} \longrightarrow \begin{array}{c} R^2 \diagdown\!\!\!\!\diagup R^3 \diagdown N \\ \text{Ph} \diagup N \diagdown R^4 \\ R^1 \end{array} \quad (34)$$

X = NH, O

(96)

Recently, it was reported that imidoylketene imines (**97**) rearrange at room temperature to form corresponding N-substituted 1,2-dihydropyrimidines **98** via a [1,5]-hydrogen shift[167] [Eq. (35)].

$$\begin{array}{c} R^1 \\ R^2 \end{array}\!\!\!\!\diagup\text{CHN}{=}\overset{\text{Me}}{\underset{|}{C}}{-}\overset{\text{Ph}}{\underset{|}{C}}{=}C{=}\text{NR}^3 \longrightarrow \begin{array}{c} \text{Ph}\diagdown\!\!\!\!\diagup\overset{\text{Me}}{\diagdown} N\diagdown R^1 \\ \diagup N\diagdown R^2 \\ R^3 \end{array} \quad (35)$$

(97) (98)

Takamisawa and co-workers described an unusual cyclization with formation of the ethyl ester of 2-methyl-1,6-dihydropyrimidine-5-carboxylic acid (**99**), observed by interaction of 2-ethoxymethoxymethyl-3-ethoxypropionitrile with acetamidine[168] [Eq. (36)].

$$\begin{array}{c} \text{EtOCH}_2\text{CHCN} \\ | \\ \text{CH} \\ \text{MeO} \diagup \diagdown \text{OEt} \end{array} + \begin{array}{c} \text{H}_2\text{N}\diagdown \\ \diagup \!\!\!\!-\text{Me} \\ \text{HN} \end{array} \xrightarrow{\text{HCl, EtOH}} \begin{array}{c} \text{EtOOC}\diagdown\!\!\!\!\diagup \text{NH} \\ \diagup N\diagdown \text{Me} \end{array} \cdot \text{HCl} \quad (36)$$

(99)

In Poland, Kuran and co-workers demonstrated the formation of dihydropyrimidines by reaction of methyl aluminates with acrylonitriles.[169,170]

2. From Pyrimidine Derivatives

a. *Addition of Organometallic Reagents.* Several reactions of organolithium compounds or Grignard reagents with pyrimidine and derivatives have been reported to give a mixture of 2- and 4-substituted

[167] J. Goerdeler, C. Lindner, and F. Zander, *Chem. Ber.* **114**, 536 (1981).
[168] A. Takamisawa, S. Hayashi, and K. Tori, *Yakugaku Zasshi* **78**, 1166 (1958).
[169] W. Kuran and S. Posynkiewicz, *J. Organomet. Chem.* **23**, 343 (1970).
[170] W. Kuran, *Pr. Nauk, Politech. Warsz., Chem.* **7**, 72 (1972).

pyrimidines[152,171-173] (via spontaneous or KMnO$_4$ oxidation of the corresponding unstable dihydropyrimidines). Most of these early reactions were aimed at preparing the substituted pyrimidines, and the dihydropyrimidine intermediates were never isolated and seldom identified. Heyes and Roberts[152] described the first preparation of a compound formulated as 1,6-dihydro-4-methyl-2,6-diphenylpyrimidine (**20b**) by reaction of 4-methyl-2-phenylpyrimidine with phenyllithium.

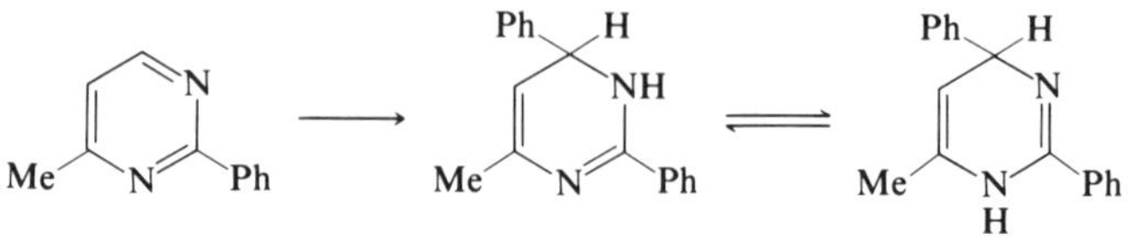

(20b)   (20a)

It has been shown that this compound in the solid state has the 1,4-dihydropyrimidine structure **20a**,[24] whereas in solution it exists in a tautomeric equilibrium of **20a** and **20b**[23,24,153] (see Section V,C,1). Unpublished data on the preparation of 1,6-dihydropyrimidine by treatment of pyrimidine with *tert*-butyllithium was mentioned earlier.[174] However, the action of

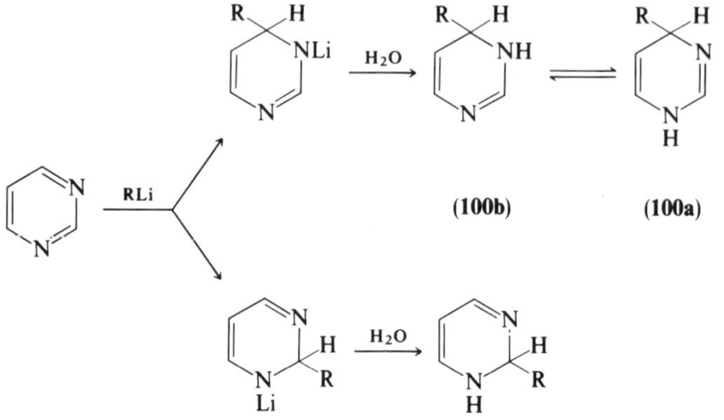

(100b)   (100a)

(37)

(101)

R = Ph, 2-thienyl, 3-thienyl, 2-furyl, 1-methyl-2-pyrrolyl, 3-pyridyl

[171] H. Bredereck, R. Gompper, and H. Herlinger, *Angew. Chem.* **70**, 571 (1958).
[172] H. Bredereck, R. Gompper, and H. Herlinger, *Chem. Ber.* **91**, 2832 (1958).
[173] S. Gronowitz and J. Roe, *Acta Chem. Scand.* **19**, 1471 (1965).
[174] R. F. Evans, unpublished results, from Brown.[142]

organolithium reagents (particularly, aryl- and hetaryllithiums) on pyrimidine and derivatives has been developed more recently by van der Plas et al. as a useful synthetic method for the preparation of dihydropyrimidines.[106,175,176]

This reaction usually occurs at unsubstituted positions. Thus in the case of pyrimidines, a mixture of 1,6- (or 1,4-) (**100**) and 1,2-dihydropyrimidines (**101**) is formed [Eq. (37)]. However, the formation of 1,6- or 1,4-dihydropyrimidines predominates. On the other hand, when only one unsubstituted or sterically less hindered position is available, a single product results, as in the case of 5-cyano-4,6-dimethoxy-2-ethyl-2-methyl-1,2-dihydropyrimidines (**102**).[177]

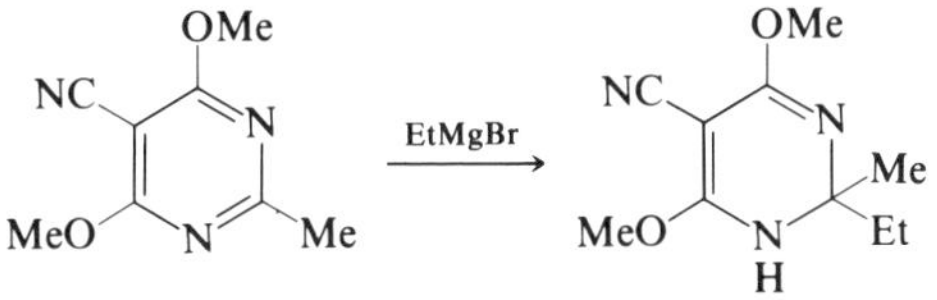

(**102**)

Another interesting example of this type of reaction is the nucleophilic addition of butyllithium to 4,6-dialkoxypyrimidines to give the 2-butyl-4,6-dialkoxy-2,5-dihydropyrimidine (**104**), probably via the intermediate 1,2-dihydropyrimidine **103**[178] [Eq. (38)]. A series of dihydropyrimidines of this type was prepared analogously in the author's laboratory.[179]

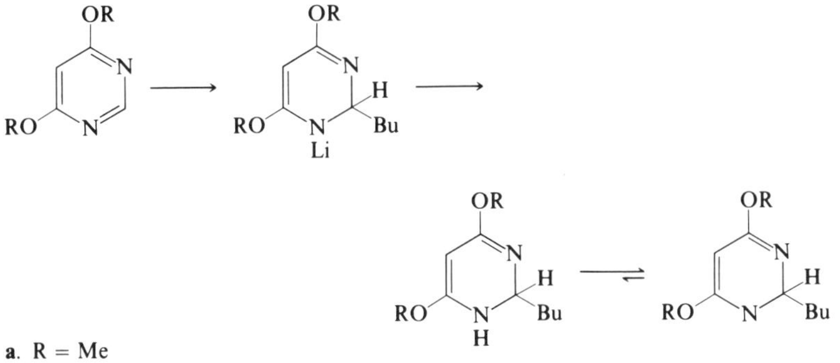

a. R = Me
b. R = Et
c. R = Bu

(**103**)   (**104**)   (38)

---

[175] R. E. van der Stoel and H. C. van der Plas, *J.C.S. Perkin I*, 1288 (1979).
[176] R. E. van der Stoel and H. C. van der Plas, *J.C.S. Perkin I*, 2393 (1979).
[177] H. Yamanaka, K. Edo, and S. Konno, *Yakugaku Zasshi* **97**, 726 (1977).
[178] M. D. Mehta, D. Miller, and E. F. Mooney, *J. Chem. Soc.*, 6695 (1965).
[179] A. L. Weis and G. Kaminas, unpublished results (1983).

The dihydropyrimidines **104** are the only known isolated compounds with 2,5-dihydro structures that can undergo oxidation to the corresponding pyrimidines. The existence of this structure may probably be explained by stabilization by the two alkoxy groups in positions 4 and 6.

b. *Reduction with Complex Metal Hydrides.* Despite the fact that a large number of dihydropyridine derivatives have been prepared by reduction of the corresponding pyridines or pyridinium salts with complex metal hydrides, no systematic investigation of this synthetic approach, with regard to the dihydropyrimidines, has been carried out. May and Sykes[180] reduced a number of ethylpyrimidine-5-carboxylates and 5-cyano- and 5-carbamoylpyrimidines with lithium aluminum hydride and obtained dihydropyrimidines of unknown configurations. The influence of groups occupying the 2-, 4-, 5-, and 6-positions upon reduction with complex hydrides is unclear, although an electron-withdrawing moiety (CO_2Et, CN, etc.) in the 5-position appears to facilitate ring reduction.

When ethyl 4-methyl-2-methylthiopyrimidine-5-carboxylate or ethyl 2,4-dimethylpyrimidine-5-carboxylate was treated with lithium aluminum hydride at $-70°C$, preferential reduction of the ester group occurred, yielding the 5-hydroxymethyl derivative rather than ring reduction.[181]

Shadbolt and Ulbricht reported[182] reduction of a number of 2-methylthiopyrimidines by $LiAlH_4$ in tetrahydrofuran and by $NaBH_4$ in ethanol to form the corresponding 1,6-dihydro derivatives (**105**) [Eq. (39)].

$$\text{EtOOC} \underset{N}{\overset{R}{\diagdown}} \text{N} \diagup \text{SMe} \longrightarrow \text{EtOOC} \underset{\underset{H}{N}}{\overset{R}{\diagdown}} \text{N} \diagup \text{SMe} \tag{39}$$

R = CH=NOH, CN, Cl   (**105**)

Mamaev and Gracheva suggested[183] that a yellow by-product, which was isolated from the reduction of 4,6-diphenylpyrimidin-2-one, is 1,2-dihydropyrimidine **21a**. Although attempts to prepare an analytically pure sample of this material failed. Reinvestigation and consequent optimization of this reaction enabled the preparation of **21a** in 78% yield.[184] The material was stable, could

---

[180] C. D. May and P. Sykes, *J. Chem. Soc. C*, 649 (1966).
[181] R. S. Shadbolt and T. L. V. Ulbricht, *J. Chem. Soc. C*, 733 (1968).
[182] R. S. Shadbolt and T. L. V. Ulbricht, *J. Chem. Soc. C*, 1203 (1968).
[183] V. P. Mamaev and E. A. Gracheva, *Khim. Geterotsikl. Soedin.*, 596 (1968).
[184] A. L. Weis and R. Vishkaucan, *Chem. Lett.*, 1773 (1984).

be purified, and provided single crystals for X-ray diffraction.[25] In addition, NMR investigation has shown that enamine–imine tautomerism is present in $CDCl_3$ (Section V,C,2).

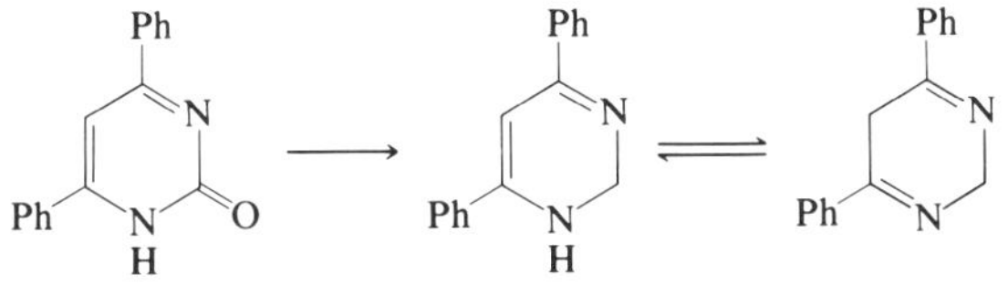

(40)

(21a)   (21b)

The mechanism of this $LiAlH_4$ reduction of pyrimidine-2-ones should be similar to that of amides (Scheme 7). Therefore, we supposed that one should also be able to obtain 1,2-dihydropyrimidine (**21a**) by reduction of the corresponding 4,6-diphenylpyrimidine (**106**), and, indeed, $LiAlH_4$ reduction of **106** in tetrahydrofuran gives 30–70% (depending on conditions) transformation of **106** to **21a**. The reaction is very clean and only unchanged **106** was isolated. The reason for incomplete transformation of **106** is still unclear.[184]

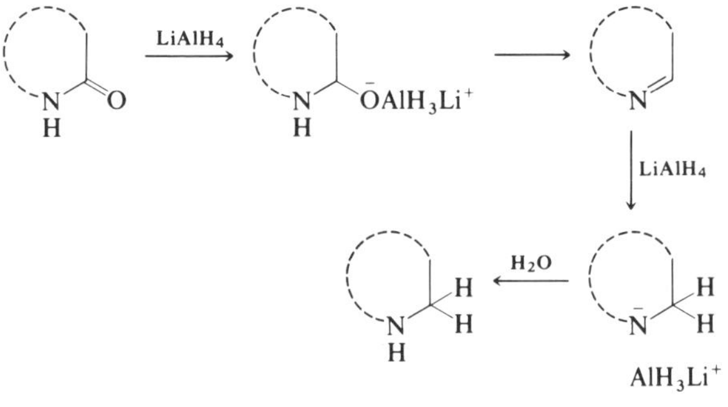

SCHEME 7

This reaction was extended to other derivatives. Thus 2-phenyl-1,6-dihydropyrimidine was prepared by $LiAlH_4$ reduction of 2-phenylpyrimidin-4-one or 2-phenylpyrimidine. Undoubtedly, there is great potential in the reduction of pyrimidines with complex hydrides, and this approach should attract wide attention in the future.

c. *From 4-Halopyrimidines.* 1,6-Dihydropyrimidines **107** were isolated as a side-product in the zinc–acetic acid-catalyzed ring contraction of pyrimidines to pyrroles. Thompson and co-workers[185,186] demonstrated that dechlorination is the initial step in reductive dehalogenation of 4-dichloropyrimidines with zinc in 50% aqueous acetic acid. According to the mechanism they proposed, further reductive attack of zinc occurs at the 1,6 site of the pyrimidine ring, as shown in Eq. (41).

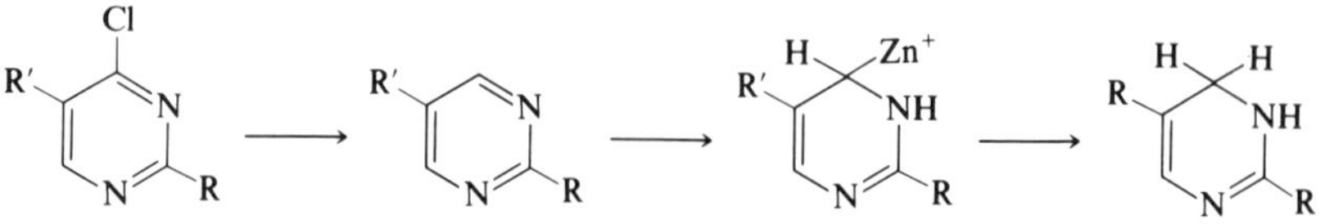

$$\text{(41)}$$

a: R = Ph, R' = Me   (107)
b: R = Me, R' = p-ClC$_6$H$_4$
c: R = Me, R' = CH$_2$COOEt

d. *Electrophilic Aromatic Substitution with Protonated Pyrimidines.* Girke[187] showed that, in the presence of trifluoroacetic acid, pyrimidine and 5-methylpyrimidine react with a number of active aromatic compounds to form stable 4-aryl-substituted dihydropyrimidium salts (108). Further, from these salts the dihydropyrimide bases 109a and/or 109b can be obtained. The mechanism for this attractive and novel reaction is apparently best described as an aromatic, electrophilic substitution by the protonated pyrimidine derivative at the electron-rich position of the active aromatic compound[187] [Eq. (42)].

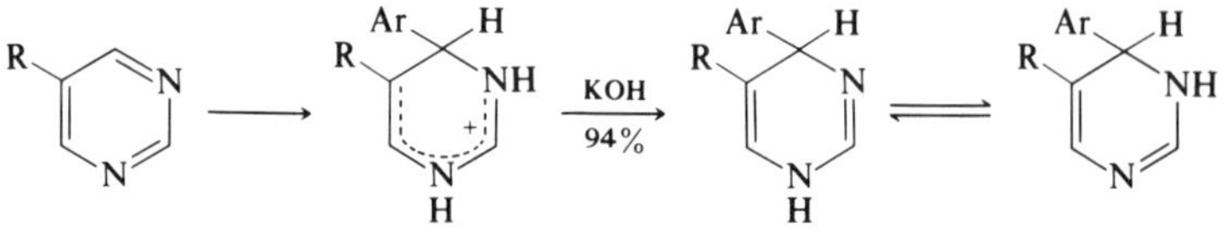

$$\text{(42)}$$

(108)   (109a)   (109b)

R = H, Me

e. *Desulfurization of Pyrimidine-2(1H)-thiones.* Kashima et al.[188,189] reported an independent preparation of three types of N-substituted 1,2-

[185] T. W. Thompson, *Chem. Commun.*, 532 (1968).
[186] J. L. Longridge and T. W. Thompson, *J. Chem. Soc. C*, 1658 (1970).
[187] W. Girke, *Chem. Ber.* **112**, 1 (1979).
[188] C. Kashima, M. Shimizu, A. Katoh, and Y. Omote, *Tetrahedron Lett.*, 209 (1983); *J.C.S. Perkin I*, 1799 (1983).
[189] C. Kashima, unpublished results (1983).

(**110**), 1,4- (**111**), and 1,6- (**112**) dihydropyrimidines by the desulfurization of pyrimidine-2(1H)-thiones and their dihydro derivatives with Raney nickel catalyst [Eq. (43)].

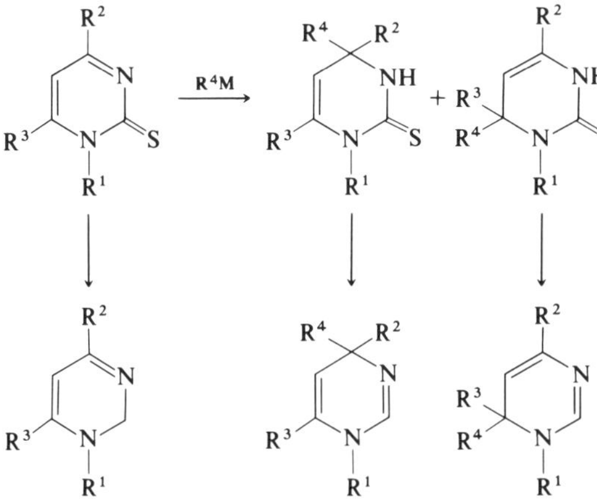

$$R^4M = R^4MgI, R^4Li, NaBH_4, \text{ or } LiAlH_4$$

Preparation of three different types of N-substituted stable dihydropyrimidines allows for an excellent comparative study of the spectral properties of these compounds, enabling unambiguous structural determinations of less stable congeners.

f. *Electrochemical Reduction.* Pyrimidine and some of its derivatives have been investigated by classical and *ac* polarography, as well as by cyclic voltammetry.[190-195] Usually, pyrimidine gives rise to two one-electron waves in the acid region. These waves merge, in slightly alkaline solution, into a single two-electron wave, followed by a second two-electron wave, which results in the formation of a tetrahydropyrimidine. The structures of most of the electrochemical reduction products have been deduced from their UV spectra.

---

[190] D. L. Smith and P. J. Elving, *J. Am. Chem. Soc.* **84**, 2741 (1962).
[191] D. L. Smith and P. J. Elving, *Anal. Chem.* **34**, 930 (1962).
[192] J. E. O'Reilly and P. J. Elving, *Electroanal. Chem.* **21**, 169 (1969).
[193] P. J. Elving, S. J. Pase, and J. E. O'Reilly, *J. Am. Chem. Soc.* **95**, 647 (1973).
[194] D. J. Trenevot, *Electroanal. Chem.* **46**, 89 (1973).
[195] P. Martigny and H. Lund, *Acta Chem. Scand. Ser. B* **B33**, 575 (1979).

Interesting results by Martigny and Lund[196] have shown that 2-phenylpyrimidines (113) are reduced in a four-electron reaction with ring contraction to 2-phenylpyrroles [Eq. (44)]. The site of reduction depends significantly on the ring substituents. Thus, for instance, the presence of a phenyl group at position 2 of the pyrimidine ring changes the usually observed 1,6- to a 1,2-reduction. The latter was proved by the observation that if the reduction of 113 in DMF is halted after the consumption of two electrons, hydrolysis of the product gives benzaldehyde.

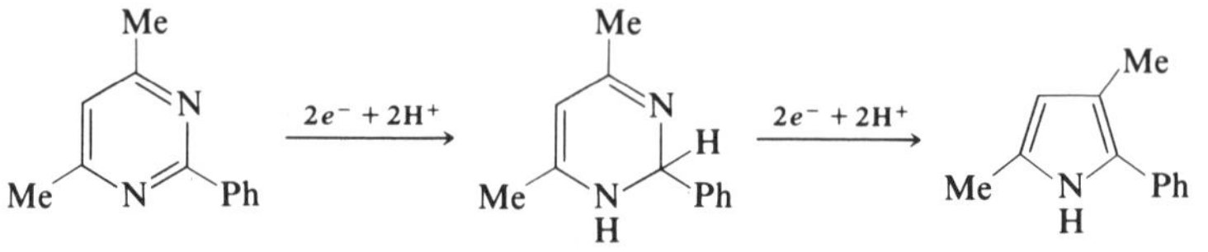

$$\text{(113)} \xrightarrow{2e^- + 2H^+} \xrightarrow{2e^- + 2H^+} \tag{44}$$

(113)

3. Miscellaneous

Polychlorinated 2,5-dihydropyrimidine 114 was obtained by reaction of phosphorus pentachloride with 5,5-diethylbarbituric acid. Further treatment with zinc in neutral aqueous solution gave 2,5-dihydropyrimidine 115 dehalogenated at position 2[197,198] [Eq. (45)].

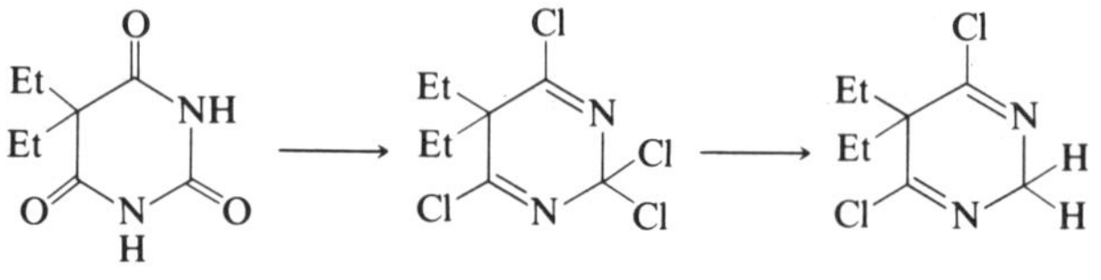

$$\tag{45}$$

(114)   (115)

When a methanolic solution of 2,6,8-triphenyl-1,5-diazabicyclo[5.1.0]octa-3,5-diene (116) was treated with sodium methoxide in methanol, a near-quantitative yield of *trans*-2-styryl-4,6-diphenyl-1,2-dihydropyrimidine (117) was obtained.[199]

Among the products formed by the 1,3-dipolar cycloaddition reaction of benzonitrile-*p*-nitrobenzylide to 3-phenyl-2*H*-azirines (in benzene under basic

---
[196] A. W. Dox, *J. Am. Chem. Soc.* **53**, 1559 (1931).
[197] A. W. Dox, *J. Am. Chem. Soc.* **53**, 2741 (1931).
[198] K. Folkers and T. B. Johnson, *J. Am. Chem. Soc.* **55**, 2886 (1933).
[199] A. Padwa and L. Gehrlein, *J. Am. Chem. Soc.* **94**, 4933 (1972).

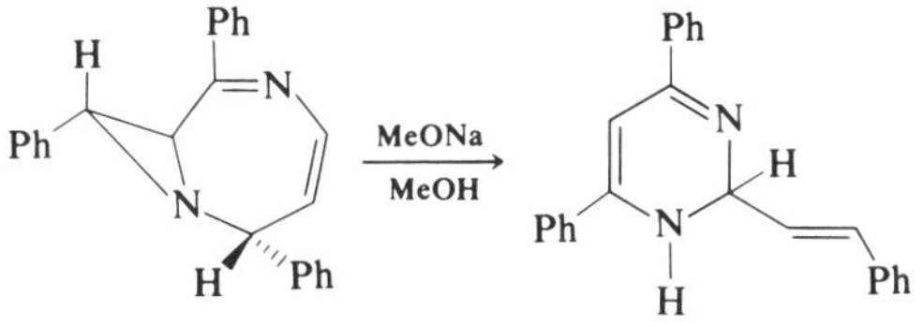

Sec. V.C] DIHYDROAZINES 63

$$(116) \xrightarrow{\text{MeONa}}_{\text{MeOH}} (117) \quad (46)$$

conditions), 2-(p-nitrophenyl)-4,5,6-triphenyl-1,6-dihydropyrimidine (118) was isolated[200] [Eq. (47)].

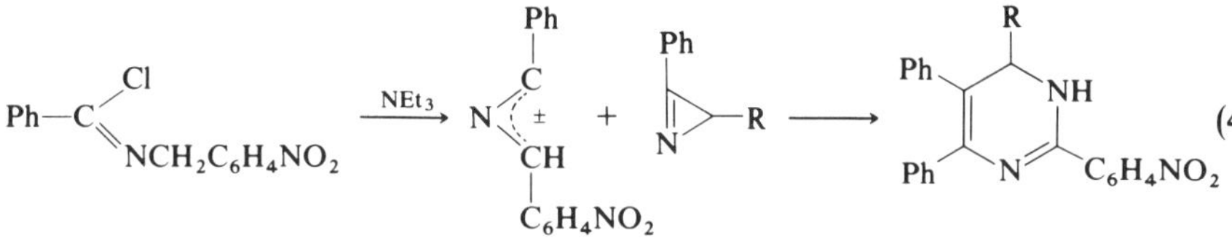

$$\text{(47)}$$

(118: R = Ph)

An attractive method for the preparation of 1,2-dihydropyrimidines 119 by rearrangement of 1-benzyl-3,5-dimethylpyrazole in the presence of sodium amide has been reported[201] [Eq. (48)].

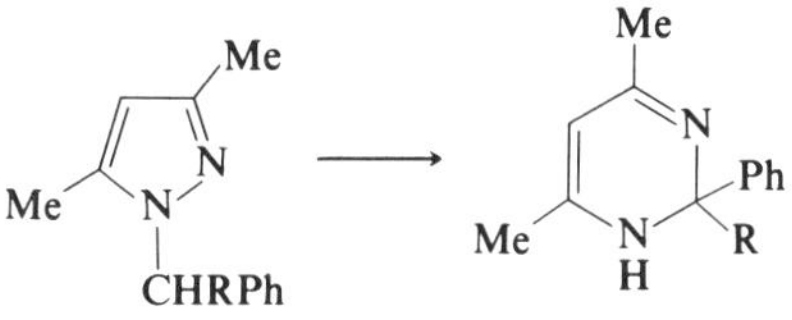

$$\text{(48)}$$

(119)

R = H, Ph

C. Tautomerism of Dihydropyrimidines

1. *Tautomerism of 1,4- and 1,6-Dihydropyrimidines*

a. *Structures of the Tautomers.* Any fruitful discussion of tautomeric equilibria must be predicated upon a firm knowledge of the structures of the

---

[200] N. S. Narashimhan, H. Heimgartner, H.-J. Hansen, and H. Schmidt, *Helv. Chim. Acta* 56, 1351 (1973).
[201] B. A. Tertov and Yu. G. Bogachev, *Khim. Geterotsikl. Soedin.*, 119 (1981).

isomers involved. However, the structures and relative stabilities of the tautomeric 1,4- and 1,6-dihydropyrimidines have been unresolved until recently. The available literature on the structure of 1,6- and 1,4-dihydropyrimidines has been quite speculative. Particular assignments for both forms have been proposed with insufficient information to justify the suggested preference. The majority of publications describe products as being in one of their possible tautomeric forms, usually the 1,6 form (B), without relating at all to the possibility of a tautomeric mixture.[145,152,173,185,186] Others suggest either a completely delocalized structure X^{200} or a time average of rapidly interchanging tautomers A and B,[144,174,175,187] without assignments for the individual isomers.

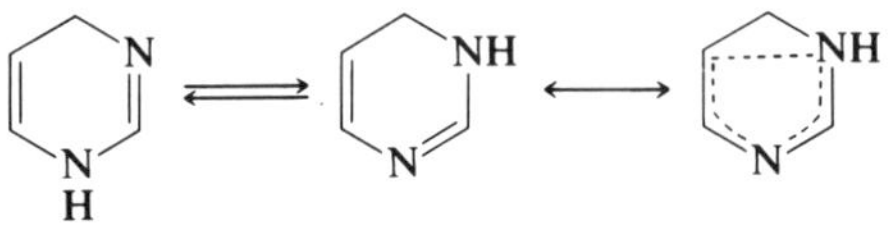

A   B   X
1,4   1,6

In 1975, Weis and Mamaev[153] showed that the 6-methyl-2,4-diphenyl dihydropyrimidine (MDHP) (**20**), which was obtained by condensation of benzylidene, acetone, and benzamidine, exists in solution in a tautomeric equilibrium of 6-methyl-2,4-diphenyl-1,4-dihydropyrimidine (**20a**) and 6-methyl-2,4-diphenyl-3,4-dihydropyrimidine (**20b**) [although the systematic nomenclature for structure **20b** would be 4-methyl-2,6-diphenyl-1,6-dihydropyrimidine (see Section III,B), we have retained the original nomenclature to provide a convenient comparison of the two tautomers]. This tautomerism was detected by spectral studies (NMR, IR, and UV) of solutions of **20**. Thus the NMR spectra of the two individual tautomeric structures can be observed in dipolar aprotic solvents such as dimethyl sulfoxide (DMSO) and hexamethylphosphoramide (HMPA).[153]

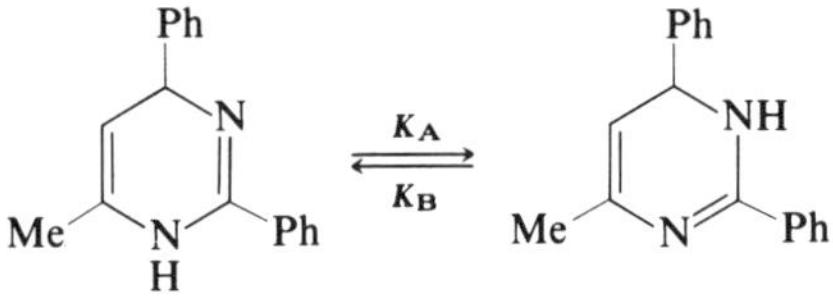

20a   **20b**

(1,4-MDHP)   (1,6-MDHP)

(49)

However, when the IR and UV spectra of **20** were recorded, using solid samples (KBr), only one tautomer was present, but the final assignment of its structure could not be made using these data alone. X-Ray diffraction, however, showed that in the crystalline state compound **20** exists in the 1,4-dihydro form **20a**.[24]

The combined X-ray, NMR, IR, and UV data enabled definite spectroscopic assignments for each of the tautomers of compound **20**. Thus the IR v_{max} band at 1700 cm^{-1} was attributed to the stretching mode of the C=C—NH—C=N fragment of **20a**, whereas the new band at 1645 cm^{-1}, which appears in solution, is due to the C=C—N=C—NH fragment of **20b**. The relative intensities of these absorption maxima, which depend on solvent polarity, give a qualitative estimate of the relative concentrations of each tautomer in a given solvent.

Usefulness of IR and NMR data in assigning tautomer structure was further validated by structural elucidation of dihydropyrimidine **19** based on Silversmith's data.[144] Its structure was assigned as the 1,6-tautomer **19b**, based on a reinvestigation of its IR band at 1652 cm^{-1}, similar to the IR spectra of **20b**. This was later confirmed by X-ray analysis.[23]

The infrared spectra of a series of these compounds were of particular interest. They showed a minimum of three important absorption bands at 1550–1620 ($v_{C=C}$), 1600–1700 ($v_{C=N}$), and 3200–3500 (v_{N-H}) cm^{-1}. However, characteristic absorptions in the 1600–1700 cm^{-1} region provided an excellent tool for differentiation between the tautomers. It has been consistently observed in a large number of newly prepared compounds that the band assigned to the C=C—NH—C=N fragment of the 1,4 tautomers appears 30–60 cm^{-1} higher in frequency than the corresponding band for the C=C—N=C—NH fragment of the 1,6 tautomer. Intensity of these peaks correlates well with the ratio of tautomers in solution.

b. *External Parameters Affecting the Rate of Tautomerism.* van der Plas et al.[174,175] and Girke[187] have also studied the behavior of a variety of dihydropyrimidines, but attempts to obtain the NMR spectra in deuteriochloroform of each of the tautomers failed, even at $-88°C$. Experiments in our laboratory confirmed this, demonstrating that in CDCl$_3$ solutions only the averaged spectra could be seen.

This observation stimulated further investigations directed toward identifying the factors, enabling observation of annular tautomerism in dihydropyrimidine systems. The easily available and reasonably stable MDHP (**20a**) proved to be a convenient model for carrying out these measurements.

In the course of NMR studies on **20** in DMSO-d_6 solution, tautomerism was affected by the following factors:

1. Dryness of solvent. This was confirmed by the addition of small amounts of water to a sample in which slow exchange was observed. This causes an increase in the rate of proton exchange as indicated by the broadening of the

H-4 and H-5 signals of both tautomers and a decrease in the separation between them.

2. Microconcentrations of H^+ ions in solution. The rate of exchange is greatly accelerated by the addition of catalytic amounts of H^+; in each case, an "average" spectrum was obtained.

3. Paramagnetic impurities in solution also accelerate the rate of exchange.

4. Concentration dependence. It has been shown that the observation of the tautomers by ^{1}H NMR is only possible at 0.001–0.003 M concentrations. Thus this factor plays an extremely crucial role in our ability to detect tautomerism.

Knowing that proton transfer between two heteroatoms is usually rapid (on the NMR time scale),[43,44] it may be expected that by using an aprotic dipolar solvent (DMSO, HMPA), a decrease in the rate of tautomerism should occur because of strong intermolecular hydrogen bonding with the solvent, enabling observation of the individual tautomers.

The fact that in $CDCl_3$ only average spectra were observed may be attributed both to a lack of hydrogen-bonding stabilization and to minute amounts of acid which are usually present in this solvent. However, when the spectra were determined at low temperature ($-50°C$) on a dilute sample (0.001 M) in $CDCl_3$ (initially filtered through alumina in order to remove traces of acid and degassed with argon), tautomerism is observed.[202]

Further, based on the above-mentioned encouraging results, a quantitative analysis of the kinetics of the tautomeric equilibrium is possible, and the external parameters that influence the rate of reaction can be studied.[203,204]

c. *Description of NMR Spectra.* In Fig. 1, two ^{1}H-NMR spectra of MDHP in $CDCl_3$ are shown. The lower spectrum, at $-60°C$, corresponds to the slow-exchange limit where signals due to the tautomeric forms **20a** and **20b** are completely separated. The upper spectrum in this figure was taken at room temperature and in a commercial, deuterated solvent. Here, the rate of tautomerism is fast and the observed peaks are at the weighted average positions of the corresponding tautomers. The NH peak in this spectrum, however, was not observed, owing to exchange with an impurity whose signal is strongly shifted from that of the NH. At $-60°C$ (and in specially purified $CDCl_3$), the positions of the NH signal were somewhat concentration dependent, and when water was present, so was the H_2O peak (see Fig. 1,

---

[202] A. L. Weis, *Tetrahedron Lett.*, 499 (1982).
[203] A. L. Weis and Z. Porat, *Isr. Chem. Soc., 49th Annu. Meet., 1982*, p. 46 (1982).
[204] A. L. Weis, Z. Porat, and Z. Luz, *J. Am. Chem. Soc.* **106**, 8021 (1984).

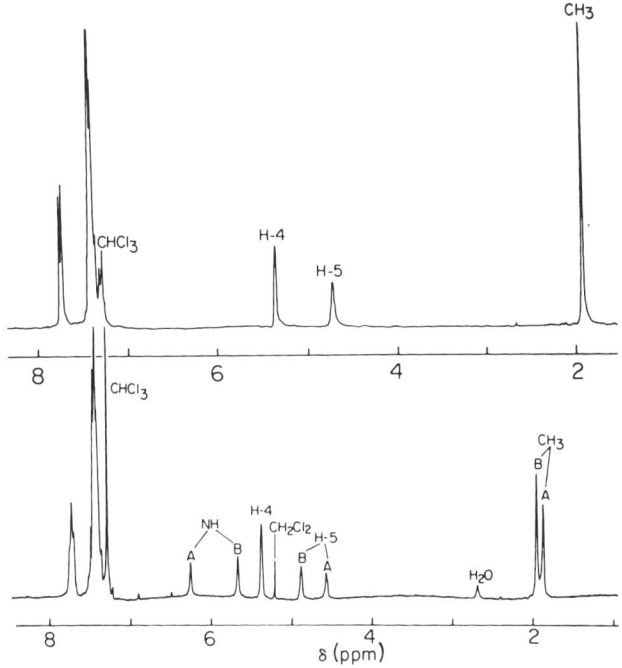

FIG. 1. ^{1}H-NMR spectra at 270 MHz of a 0.026 M solution of MDHP in CDCl$_3$. The lower and upper spectra were taken at $-60°$C and room temperature, respectively. The solvent used for the low-temperature measurement was purified by method d while for the room-temperature spectrum untreated commercial CDCl$_3$ solvent was utilized.

lower). All these peaks shifted to low field with increasing MDHP concentration, apparently owing to the formation of hydrogen-bonded complexes involving these molecules.

MDHP spectra in DMSO at room temperature are shown in Fig. 2. The lower spectrum corresponds to a purified solvent, while the upper one corresponds to a regular commercial grade. As before, the two spectra correspond, respectively, to the slow and fast limits of proton exchange. Between this spectrum and the one in CDCl$_3$, there is a large low-field shift of the NH peaks, apparently because of strong hydrogen bonding with the solvent. Also, the equilibrium constant $K = [20b]/[20a]$ has decreased to 0.67, as compared with 1.27 in CDCl$_3$. Two other minor differences are (a) that splitting of the H-4 signals is observed in DMSO but not in CDCl$_3$, and (b) that the two methyl signals are degenerate in DMSO but not in CDCl$_3$. Thus solvent polarity seems to affect chemical shifts of the protons, in particular the NH, as well as the equilibrium distribution of the two isomeric species.

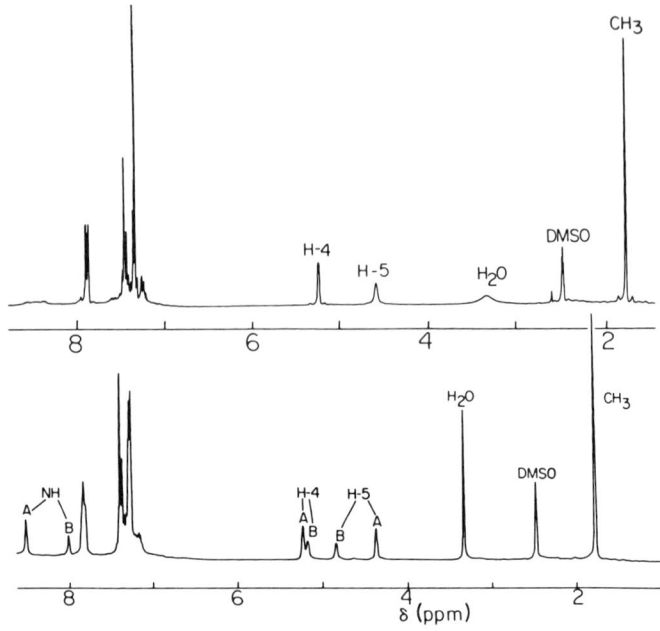

FIG. 2. ^{1}H-NMR spectra (270 MHz) of a 0.04 M solution of MDHP in DMSO-d_6. The spectra were taken at room temperature using a purified (lower) and untreated (upper) solvent.

TABLE III
CHEMICAL SHIFTS OF THE NH PROTONS AND EQUILIBRIUM CONSTANTS FOR MDHP TAUTOMERISM IN SEVERAL ORGANIC SOLVENTS

Solvent	Temperature (°C)	Concentration (M)	NH(A) (ppm)	NH(B) (ppm)	K = [B]/[A]
CDCl$_3$	−60	0.013	6.18	5.56	1.27
CDCl$_3$	−60	0.073	6.30	5.74	1.27
CDCl$_3$	−60	0.121	6.54	6.08	1.27
Dioxane-d_8	15	0.01	7.19	6.75	0.82
DMSO-d_6	25	0.01	8.56	8.05	0.67
HMPA-d_{18}	25	0.01	9.35	8.90	0.50

In Table III, chemical shift data for the NH peak and equilibrium constants in four solvents are summarized, and the results are arranged according to increasing solvent polarity: CDCl$_3$ < dioxane < DMSO < HMPA. There is a significant low-field shift of the NH resonance of both tautomers with increasing polarity, apparently owing to formation of hydrogen-bonding complexes with the solvent. In $CDCl_3$, there is also a low-field shift of the NH signal with increasing MDHP concentration, indicating the formation of dimers or higher clusters. This concentration effect on the NH shift exists only in $CDCl_3$ and is essentially absent in the other solvents studied. The equilibrium distribution [**20b**]/[**20a**] also decreases monotonically with increasing polarity, suggesting that in MDHP form A is more polar than form B.

In general, the tautomeric transformation between the two forms A and B is slower in the more polar solvents. Thus separate signals for the two tautomers can readily be observed even in unpurified HMPA, whereas in $CDCl_3$ it can be seen only in purified solvent and very dilute solutions. The proton-transfer process is hindered by the formation of the hydrogen-bonded complexes with the polar solvents.

Finally, ^{13}C-NMR spectra in $CDCl_3$ (at $-60°C$) and DMSO, both corresponding to the slow-exchange limit, are shown in Fig. 3. Rationalization of these spectra is similar to that discussed above for the ^{1}H-NMR results.

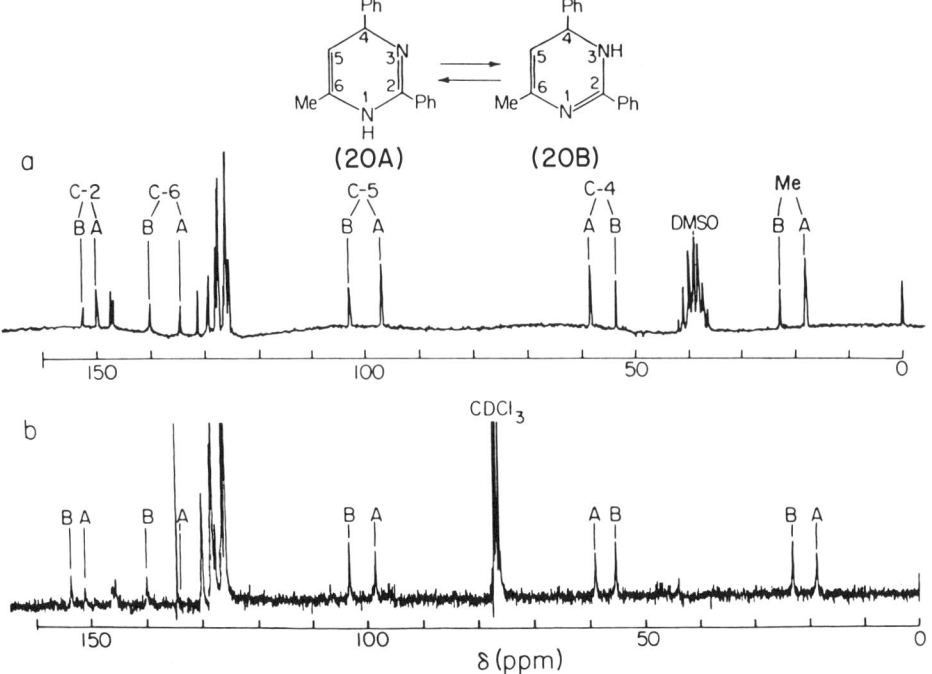

FIG. 3. ^{13}C-NMR spectra (270 MHz) of 0.06 M 1,4-MDHP (**20a**) and 1,6-MDHP (**20b**) in $CDCl_3$ (lower) and DMSO (upper). Purified solvents were used and the specta recorded at $-60°C$ for $CDCl_3$ and at room temperature for DMSO.

d. *Kinetic Measurements.* Analysis in detail of the kinetics of this tautomeric equilibrium suggests a mechanism for the reaction and allows comparison of the behavior of analogous cyclic and acyclic tautomeric systems.

For these measurements, 6-methyl-2,4-diphenyl-1,4-dihydropyrimidine (MDHP) was used because of its relative stability and the near unity of its tautomeric equilibrium constant in aprotic solvents.

Preliminary kinetic NMR measurements performed in dried, ultrapure commercial DMSO-d_6 were not reproducible, owing to oxidation to 6-methyl-2,4-diphenylpyrimidine by the DMSO.[147–150] Purified $CDCl_3$ is more stable, giving reproducible kinetic measurements.

Kinetic results obtained by proton-NMR studies of MDHP in $CDCl_3$ are based on measurements made by means of standard lineshape analysis applied to H-5.[205]

Discussed below are the effects of solute and solvent purity on the rate of the tautomerism, as well as a systematic study of the temperature and concentration dependence of the rate constant.

i. *Effect of solvent purity.* The presence of even minute amounts of impurities in the solvent strongly catalyzes the tautomeric equilibration described in Eq. (49) as seen in the so-called "coalescence temperature" (t_c) of the H-5 peaks of the A and B species. Various combinations of the solute and solvent were examined, and the coalescence temperatures were determined as a function of MDHP concentration (see Fig. 4). Clearly, purification of both solvent and MDHP has a marked effect on t_c. Further recrystallization of the MDHP and distillation of $CDCl_3$ produced no further increase of t_c. Therefore, such solutions were used for the final kinetic studies.[204]

The addition of trace amounts of ammonium salts, HCl, or even a drop of commercial $CDCl_3$ to the purified solutions of DMSO or $CDCl_3$ resulted in complete coalescence of the spectra of tautomers A and B. On the other hand, in $CDCl_3$ the addition of equal amounts of water with respect to solute had almost no affect on the rate of tautomerism.

ii. *Concentration and temperature dependence of* k_A. The temperature dependence of the ^{1}H-NMR spectra of MDHP in $CDCl_3$ containing different concentrations of solute in the range 10–60 mM and between $-60°$ and $+60°C$ is shown in Fig. 5. To determine the effect of concentration on the rate, plots were made of k_A versus concentration of MDHP at each

---

[205] H. Gunther, "NMR Spectroscopy—An Introduction," Chapter VIII. Wiley, Chichester, 1980.

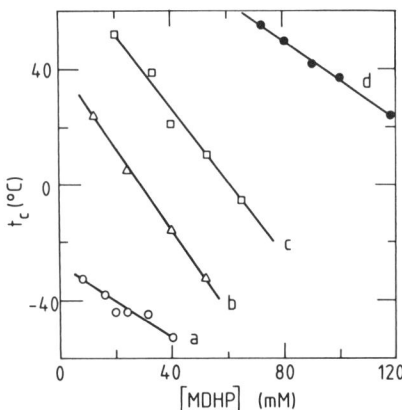

FIG. 4. The coalescence temperature t_c for the proton peaks H-5 of MDHP in CDCl$_3$ solution as a function of solute concentration. Curves a–d represent successively more highly purified solutions.

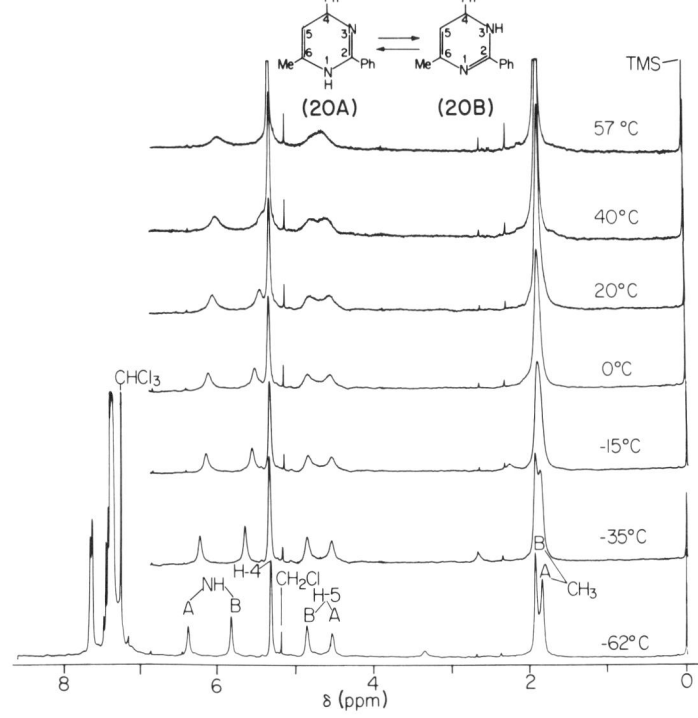

FIG. 5. ^{1}H-NMR spectra (270 MHz) of 0.073 M 1,4-MDHP (**20a**) and 1,6-MDHP (**20b**) in CDCl$_3$ (highly purified) at different temperatures.

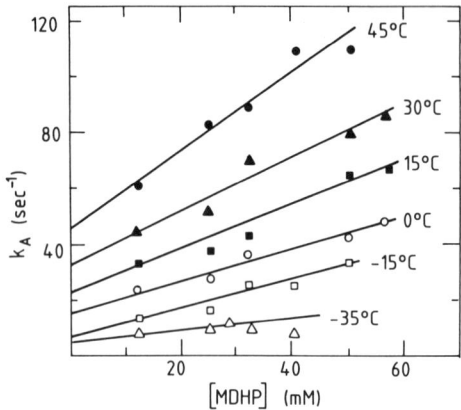

FIG. 6. Plots of k_A as a function of MDHP concentration for different temperatures.

temperature (see Fig. 6). At a fixed temperature, k_A can be written as the sum of two terms:

$$k_A = k_1 + k_2[\text{MDHP}] \quad (50)$$

where k_1 corresponds to a monomolecular reaction yielding A from MDHP, while k_2 corresponds to a bimolecular path. At concentrations greater than 60 mM MDHP, k_A plots deviated from linearity; this may be due to solubility and aggregation effects. Values for k_1 and k_2 of Eq. (50) were obtained for six temperatures in the range -35 to $+45°$C, and Arrhenius plots of these rate constants are given in Fig. 7. The kinetic parameters are:

1. For k_1: $k_1(300 \text{ K}) = 31 \text{ sec}^{-1}$, $\Delta H^{\ddagger} = 3.9 \text{ kcal mol}^{-1}$, and $\Delta S^{\ddagger} = -43$ e.u.
2. For k_2: $k_2(300 \text{ K}) = 1.0 \times 10^3 \, M^{-1} \text{ sec}^{-1}$, $\Delta H^{\ddagger} = 2.7 \text{ kcal mol}^{-1}$, and $\Delta S^{\ddagger} = -41$ e.u.

Two mechanisms are involved in the transformation between tautomers: one is first order with respect to MDHP concentration, while the other is second order. The first-order reaction involves the solvent in a proteolytic reaction:

$$\text{MDHP} + \text{Sol} \xrightleftharpoons{k_1} \text{Sol H}^+ + \text{MDHP}^- \quad (51)$$

where in the reverse reaction, the protonation may take place at the second nitrogen. This process may also proceed with remnant impurities still present in the solution; these may serve as conjugate bases to the MDHP molecules. Therefore, the observed k_1 must be considered as an upper limit for the reaction with $CDCl_3$.

The second-order reaction is a proton exchange between two MDHP molecules. Alternatively, it may involve proton transfer between MDHP

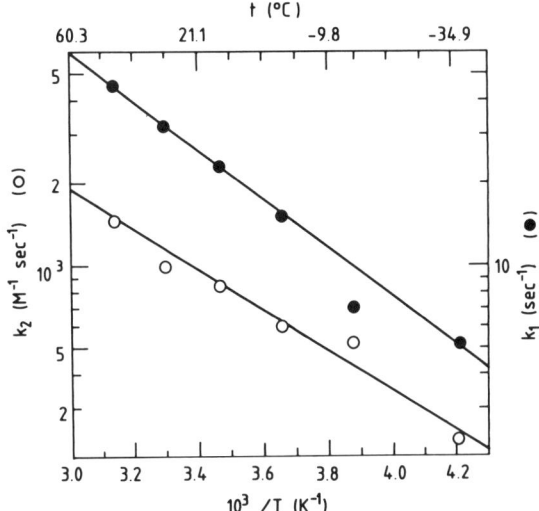

FIG. 7. Arrhenius plots for the monomolecular (k_1) and bimolecular (k_2) rate constants derived from the analysis of the results in Fig. 6.

molecules and either one of the disproportionation products of the self-ionization reaction.

$$2\text{MDHP} \rightleftharpoons \text{MDHP}^+ + \text{MDHP}^- \qquad (52)$$

$$\langle\text{MDHP}^\pm\rangle + \text{MDHP} \rightleftharpoons \langle\text{MDHP}\rangle + \text{MDHP}^\pm \qquad (53)$$

In concentrated solutions (above 20 mM MDHP), the bimolecular reaction dominates, whereas in dilute solutions, monomolecular protolysis prevails.

e. *Intra- and Intermolecular Mechanisms of Proton Transfer in Amidines.* The proton is easily transferred between atoms of similar electronegativities. In a case where the steric structure of a molecule prevents intramolecular transfer, a proton may take part in intermolecular migration. Migration from one center to another apparently depends on the ability of the molecules to form hydrogen-bonded bridges. Two situations may arise for amidines.

Free rotation about the C—N bond in acyclic amidines can place the hydrogen attached to the N(sp^3) in a very accessible position to the lone pair of the N(sp^2), producing intramolecular hydrogen bonding that enables proton migration.

In contrast, in cyclic amidines free rotation does not occur, and conformational changes of the ring system cannot lead to intramolecular hydrogen bonding. Only intermolecular proton transfer is possible. Confirming this interpretation is the observation that in cyclic amidines a concentration dependence is observed on the rates of tautomerism, whereas in acyclic systems the proton transfer is independent of concentration.

Despite the wide belief that hydrogen transfer between electronegative atoms, such as N or O, is a very fast process and that the presence of individual tautomers cannot, therefore, be detected on the NMR time scale, it is possible to slow down this equilibration rate by carefully controlling structural factors in a series of DHP.

1. The delocalization of the lone pair of nitrogen electrons will slow down proton transfer to this atom.[44] Compounds in which phenyl groups at position 2 of dihydropyrimidines were replaced by methyls or those formed by reducing the conjugated C=C bond to give the corresponding tetrahydropyrimidines (**120**) show drastically enhanced hydrogen exchange. On the other hand, introduction of a second phenyl at position 6 in compound **121** notably decreased the rate of tautomerization, and separate signals for both isomers were detectable by ^{1}H and ^{13}C NMR in CDCl$_3$ at rather high concentrations (0.1 M) and at ambient temperature. This is probably the reason that exchange occurs slowly, if at all, in dihydroquinazolines.

(**120**)   (**120**)   (**121a**)   (**121b**)

2. The steric environment of substituents adjacent to the amidine fragment of dihydropyrimidines (particularly the bulkiness of substituents at the neighboring sp^3-hybridized carbon atom) also affects tautomeric exchange. [An analogous situation was found in the dihydro-1,3,5-triazine series (Section VIII,C).] Further investigations are necessary to clarify the mechanisms, particularly ^{15}N-NMR studies of the annular tautomerism of dihydropyrimidines, the chemical shift data of which enable calculation of the electron

density on the nitrogens. A correlation of the latter with the rates of tautomerism for a number of variously substituted dihydropyrimidines would be a critical test of the "delocalization hypothesis."

2. *Tautomerism of 1,2-Dihydropyrimidines*

Successful preparation of 4,6-diphenyl-1,2-dihydropyrimidine (**21a**) (Section V,B) stimulated comprehensive spectral studies of this compound.[184] Of particular interest were NMR measurements to verify the presence of homoaromaticity as previously found for dihydrotetrazine.[24,26] To facilitate observation of two signals for the protons on carbon 2, measurements were made at $-60°$C. However, all attempts to detect homoaromaticity under

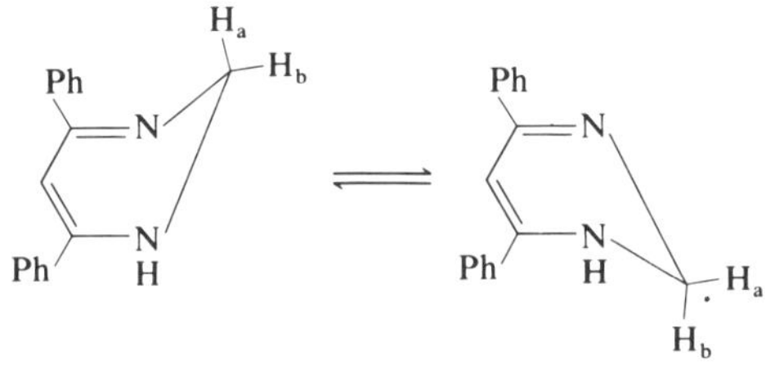

(21a)

conditions similar to those used for dihydrotetrazines failed. One likely explanation is a significantly faster flipping in the 1,2-dihydropyrimidine ring, which, to freeze, requires even lower experimental temperatures than those investigated.

Instead of homoaromaticity in the $CDCl_3$ spectra of 4,6-diphenyl-1,2-dihydropyrimidine, two new triplets were observed at δ 3.56 ($J = 6.6$ Hz) and δ 5.79 ($J = 6.6$ Hz). These were assigned to 4,6-diphenyl-2,5-dihydropyrimidine (**21b**) (Fig. 8), the second tautomeric form of the imine–enamine tautomeric equilibrium. The ratio of **21a** to **21b** is 2:1. In DMSO-d_6 solution, the equilibrium shifts completely toward 1,2-dihydropyrimidine **21a** because of strong intermolecular hydrogen bonding with the solvent. An analogous effect was observed in 1,6-dihydropyrazine.[46]

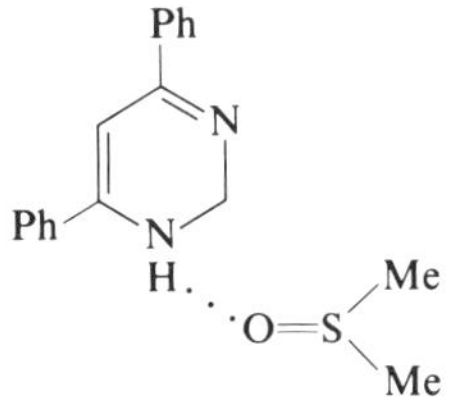

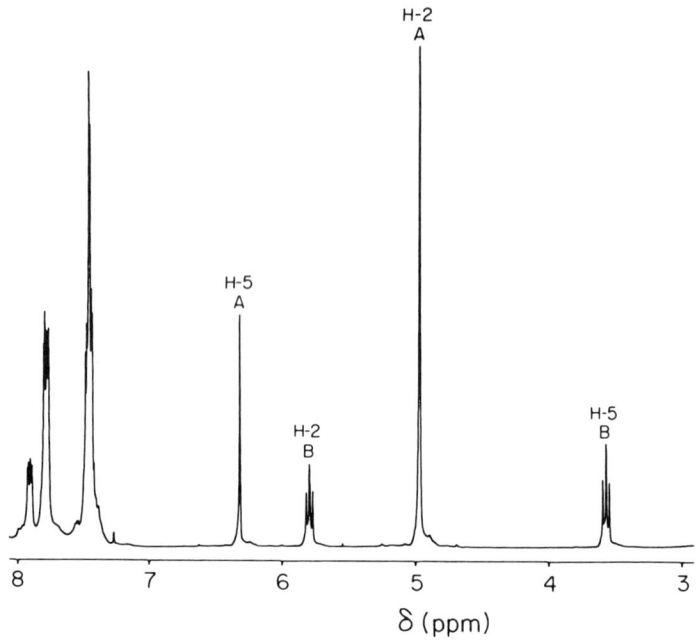

FIG. 8. 270-MHz ^{1}H-NMR spectrum of 4,6-diphenyl-1,2-dihydropyrimidine (**21a**) in CDCl$_3$.

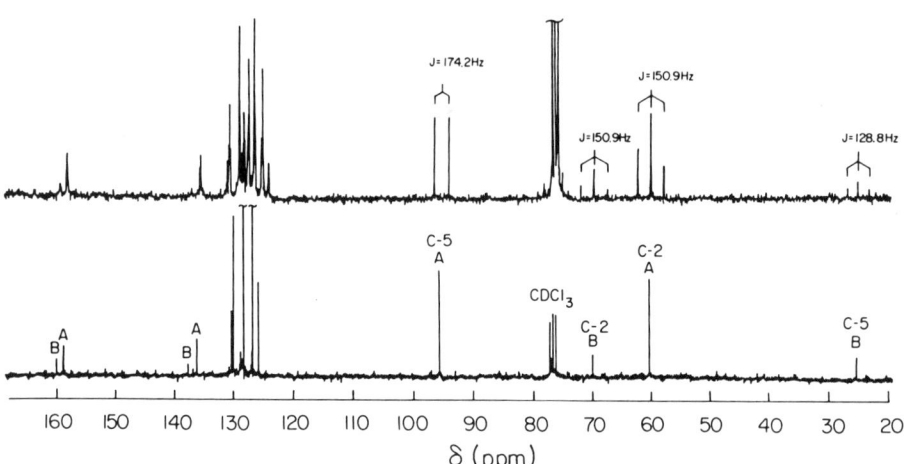

FIG. 9. 67.9-MHz ^{13}C-NMR decoupled spectrum (lower trace) and coupled spectrum (upper trace) of 4,6-diphenyl-1,2-dihydropyrimidine (**21a**) in CDCl$_3$. Peaks A correspond to the 1,2-dihydro and peaks B to the 2,5-dihydro tautomers.

Existence of this imine–enamine tautomerism is clearly indicated in the ^{13}C-NMR spectrum of **21a** in CDCl$_3$ (Fig. 9). A simple calculation using the concentrations of the two tautomers in CDCl$_3$ gives the value of $\Delta G°$ in this solvent as 0.41 kcal mol^{-1}.[184] Unsymmetrically substituted 1,2-dihydropyrimidines undergo tautomeric equilibration via [1,5]-hydrogen shift.[49]

D. PHYSICOCHEMICAL PROPERTIES

1. Infrared Spectra

All N-substituted dihydropyrimidines absorb in the 3100–3500 cm^{-1} region and show the characteristic stretching modes for bonded and nonbonded N—H groups. The number and position of these depend on structural factors and on the conditions of measurements.

The N—H stretching frequencies have been reported for a wide variety of N-substituted tautomeric 1,4- and 1,6-dihydropyrimidines; the absorptions, which are of medium intensity, appear for solutions in the 3360–3490 cm^{-1} region.[23,24,144,152,156,175,176,187] IR spectroscopy has been used to study the association of such compounds.[206] Thus in dilute chloroform solution (0.005 M), the dihydropyrimidine **20** shows a sharp absorption at 3450 cm^{-1} attributable to nonassociated N—H; in a more concentrated solution (0.5M) and in the solid state, the compound displays in the former a broad signal at 3320 cm^{-1} and in the latter at 3100–3240 cm^{-1}, owing to intermolecularly hydrogen-bonded N—H.[206] The N—H stretching frequencies of N-unsubstituted 1,2-dihydropyrimidines generally appear in the region 3380–3450 cm^{-1}.[184]

All dihydropyrimidines show absorptions in the region 1500–1700 cm^{-1}, which are assigned to the C═C or C═N stretching modes. As was mentioned in Section V,C,1,a, the IR absorption bands in the region 1600–1700 cm^{-1} provide an excellent tool for differentiating between tautomeric 1,4- and 1,6-dihydropyrimidines[23,24,202] since the C═N stretching band of the 1,4 tautomers has been consistently observed 30–60 cm^{-1} higher than the corresponding band of the 1,6 tautomer, and the intensity of these peaks correlates with the ratio of tautomers in solution. By introducing a phenyl group at position 2 of these dihydropyrimidines, the C═N absorptions of both tautomers are shifted to higher frequencies by $\sim$20–30 cm^{-1}. On the basis of these data, the existence of two tautomers was found for all synthesized N-unsubstituted 1,4- and 1,6-dihydropyrimidines by analysis of published IR data.[175,176,187] The C═N absorption of 1,2-dihydropyrimidines usually appears in the 1595–1630 cm^{-1} region.[184] A similar IR

---

[206] A. L. Weis, unpublished results.

tendency was observed by Kashima and co-workers[188,189] for isomeric compounds **110**, **111**, and **112** in the region 1600–1700 cm^{-1}, with the frequencies decreasing in the order **111** > **112** > **110**.

Undoubtedly, IR spectroscopy provides important structural information and, therefore, needs more systematic study.

2. Ultraviolet Spectra

Surprisingly few publications contain UV spectral data on dihydropyrimidines.[153,188,189,200] However, despite the paucity of examples, one can easily distinguish the following characteristic absorption bands for different types of dihydropyrimidines. Thus 1,4-dihydropyrimidines display $n \to \pi^*$ transitions for the C=C—N—C=N chromophore, which appears as a medium intensity band ($\varepsilon \sim 6500$) in the 250–260 nm region. The introduction of an aryl group at positions 2 and/or 4, as well as an alkoxycarbonyl group at position 5, shifts the absorption to a longer wavelength and increases its intensity.

The UV spectra of 1,6-dihydropyrimidines usually appear at a higher wavelength than their corresponding 1,4 isomers and display an $n \to \pi^*$ transition for the C=C—N=C—N chromophore, which appears as a medium intensity absorption ($\varepsilon \sim 5000$) in the 300–320 nm region.[188,189] This absorption also depends on substituents and the solvents. This difference in UV absorption of tautomeric 1,4- and 1,6-dihydropyrimidines may be used for quantitative study of amidinic tautomerism of dihydropyrimidines, as well as for assigning structures.

The longest wavelength absorption appears in 1,2-dihydropyrimidines,[164,184,188,189] which is clearly the result of an $n \to \pi^*$ transition involving the N—C=C—C=N chromophore. In general, such compounds show absorption maxima at 340–350 nm, and are usually of medium intensity ($\varepsilon \sim 6000$). 1,4-Didhydropyrimidines are usually colorless substances, whereas most of the 1,6- and 1,2-dihydropyrimidines are yellow.

3. Nuclear Magnetic Resonance Spectra

The use of ^{13}C NMR has as yet received only sporadic attention,[106,189,203,204] although this area, as well as ^{15}N NMR, is expected to develop rapidly over the next few years. The olefinic proton at position 5 of 1,4- and 1,6-dihydropyrimidines resonate in the region δ 4.20–5.00.[23,24,144,152,156,175,176,185–189,200] When separate ^{1}H-NMR signals of both tautomers or N-substituted isomers can be observed, the signal for the 1,4-dihydro compound appears 0.4–0.6 ppm upfield from that for the 1,6-dihydro isomer. As can be seen in Figs. 1 and 2, signals for protons at position 4 are not usually affected by hybridization of the adjacent N atom. The protons at position 5 in 1,2-dihydropyrimidines usually appear at δ 5.2–5.4 lower than corresponding 1,4- and 1,6-dihydro isomers.[49,161,162,164–167,184,188,189,199] NH proton shifts depend on the polarity of the solvent used and usually appear in CDCl$_3$ at δ 5.6–6.8, in dioxane at δ 6.5–7.5, in acetone and DMSO at δ 7.8–8.6, and in HMPA at δ 9.0–9.6.

Interesting regularities can be deducted from ^{13}C-NMR spectra. Thus the sp^3 carbons in dihydropyrimidines usually appear in the 50–70 ppm region, whereas the carbons at position 5 appear at δ 100–108.[188,189,203,204] Comparison of the available data for ^{13}C-NMR shifts in tautomeric dihydropyrimidines (see Section V,C) and in isomeric N-phenyl-1,4-, -1,6-, and -1,2-dihydropyrimidines indicate[203,204] that C-5 in the 1,4 isomer appears at a higher field than that in the 1,6 isomer, but at a lower field than that in 1,2-dihydro compounds. Signals of saturated (sp^3) carbons, on the other hand, give shifts to higher field in the order 1,6- > 1,4- > 1,2-dihydropyrimidines.

A detailed analysis of ^{1}H- and ^{13}C-NMR spectra of 1,2-dihydropyrimidinium salts was reported.[161,184] Both authors assumed that the 2-position in these salts lies out of the plane of the conjugated portion of the molecule. However, attempts to observe possible inversion in this system produced no change in the ^{1}H-NMR spectrum of **122** down to $-60°$C [Eq. (54)].

(54)

(**122**)

4. *Crystal Structure*

X-Ray diffraction examination of the molecular structure of 1,6-,[23] 1,4-,[24] and 1,2-dihydropyrimidines[25] not only provides the bond lengths, angles, and conformations of these isomers, but also serves as an instrument for unambiguous structural assignment (Section III,E,1).

The crystal structure of each of these compounds demonstrates the presence of an intermolecular NH⋯N hydrogen bond. Molecular association through the NH⋯N hydrogen-bond bridge is undoubtedly one of the major factors responsible for the fact that, as a rule, N-1-unsubstituted

1,6-, 1,4-, and 1,2-dihydropyrimidines are solids, whereas N-substituted derivatives,[188,189] as well as 2,5-dihydropyrimidines,[178,179] are liquids.

Dihydropyrimidines in which the NH hydrogen is available for intermolecular hydrogen bonding are soluble in polar and rather insoluble in nonpolar solvents (as CCl_4, hexane, or cyclohexane). When dihydropyrimidines are substituted in position 1, the solubility characteristics are reversed.

E. REACTIVITY

The most important reaction of dihydropyrimidines is their oxidation to the corresponding pyrimidines (via dehydrogenation, hydrogen transfer, or disproportionation). However, although many such oxidations have been carried out, they were aimed at enabling identification of dihydropyrimidines from the pyrimidine formed, rather than a study of the kinetics and mechanism of the oxidation reactions themselves. Thus besides spontaneous oxidation in air, 1,4(1,6)-dihydropyrimidines were oxidized by $KMnO_4$,[152,155,156,171-173] DMSO,[147-151,153] or potassium hexacyanoferrate(III)[187]; 1,2-dihydropyrimidines were oxidized by $KMnO_4$,[175,176,178,183,184] DMSO,[163] or 2,3-dichloro-5,6-dicyanobenzoquinone (DDQ).[199]

1,4-Dihydropyrimidine (**20**) easily reduces some α,β-unsaturated ketones to saturated ketones (for example, chalcone, benzylacetone, benzilideneacetylacetone, and maleic anhydride), as well as some activated carbonyls, such as trifluoroacetone, pyruvic acid, and phenyl glyoxylate. This important property of dihydropyrimidines needs further thorough quantitative investigation.

Tautomeric 1,4- and 1,6-dihydropyrimidines easily undergo nucleophilic addition. For example, upon prolonged contact with moisture, the addition of water across the C=C bond, with quantitative formation of 6-hydroxy-1,4,5,6-tetrahydropyrimidines, was observed [Eq. (55)]. The products are identical to those obtained by interaction of α,β-unsaturated carbonyl compounds with amidines and indicate the reversibility of the dehydration step in the course of dihydropyrimidine formation for 6-hydroxytetrahydropyrimidines (Section V,B,1).

$$\text{(55)}$$

Very little work has been published on the thermolytic reactions of dihydropyrimidines, a field which should produce interesting results. When

1,6-dihydropyrimidine **122** was heated at 210°C, ammonia was evolved, and pyrolytic rearrangement gave 2,4,6-triphenylpyridine.[207] A plausible mechanism for this rearrangement has been proposed (see Section VIII,E).

A new synthesis of pyrimidines substituted at position 5 was reported by van der Stoel and van der Plas[175,176,208] via photochemical rearrangement of 4-substituted 1,4(1,6)-dihydropyrimidines (**100**) to the corresponding 5-substituted 1,2-dihydropyrimidines (**125**), which on oxidation yielded 5-substituted pyrimidines [Eq. (56)].

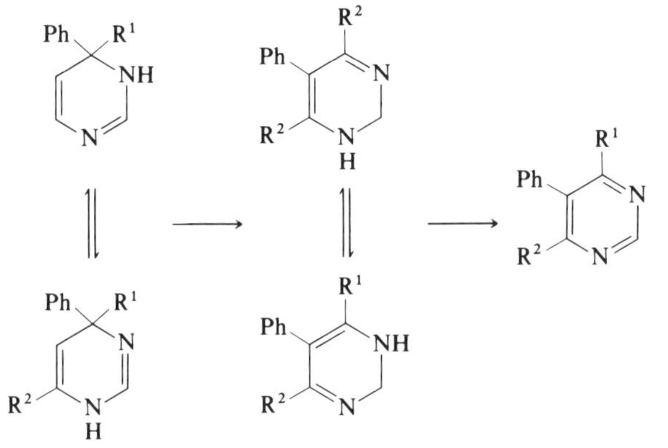

The authors postulated a di-π-methane mechanism for the photochemical isomerization, with the formation of 2,4-diazabicyclo[3.1.0]hex-2(3)-ene (**123**) as an intermediate. Subsequent opening of the three-membered ring with a concomitant [1,5]-hydrogen shift from nitrogen to C-2 gave 2,5-dihydropyrimidine **124** [Eq. (57)].

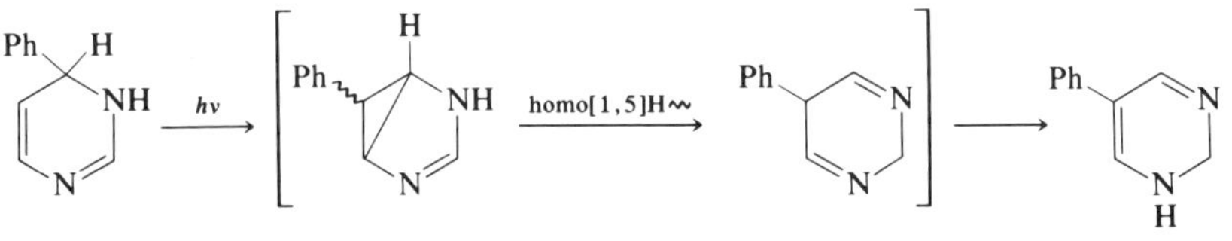

Although the 2,5-dihydropyrimidines (**124**) could be neither isolated nor detected by spectroscopic means, the migration of hydrogen attached to nitrogen in **123** to the adjacent C-2 (**124**) was demonstrated by irradiating

---

[207] L. S. Cook and B. J. Wakefield, *Tetrahedron Lett.*, 1241 (1979).
[208] R. E. van der Stoel, H. C. van der Plas, and G. Geurtsen, *J. Heterocycl. Chem.* **17**, 1617 (1980).

an N-1-deuterated analog of 123. Oxidation of the photolysis mixture gave 4,5-diphenylpyrimidine containing 56.5% deuterium at C-2, the amount expected for a homo[1,5]-deuterium shift from nitrogen to C-2. Moreover, spectroscopic evidence for the intermediacy of a 6-substituted 2,4-diazabicyclo[3.1.0]hex-2(3)-ene in this photoisomerization was presented.[208] The isolation of 1,2-dihydropyrimidines (125) failed because of their instability, and, therefore, evidence for structure 125 was based on spectral and chemical data. Thus heating of the crude irradiation mixture in air or its oxidation with potassium permaganate gave 5-substituted pyrimidines.[175,176] Finally, by irradiation of 1,4(1,6)-dihydropyrimidines (100) containing electron-withdrawing substituents at position 4, a photochemical ring contraction to imidazoles 126 was observed.[208]

$$\text{(100)} \xrightarrow{h\nu} \text{(126)} + \text{(125)} \quad (58)$$

(100)　　　(126)　　　(125)

a: R^1 = 2-thiazolyl, R^2 = H
b: R^1 = 2-pyridyl, R^2 = H
c: R^1 = 4-pyridyl, R^2 = H
d: R^1 = 2-pyridyl, R^2 = Ph

A 2-thiazolyl or 4-pyridyl group in position 4 causes an exclusive photoinduced ring contraction to an imidazole (126). 2-Pyridyl substitution leads to both an imidazole (126) and a 5-substituted 1,2-dihydropyrimidine (125), whereas a phenyl group in position 4 yields only 125.

Lloyd and McNab investigated the behavior of 1,2-dihydropyrimidinium salts in reactions with different electrophilic and nucelophilic reagents and concluded that these dihydropyrimidinium compounds present an intriguing example of dihydro "aromatic" compounds, which show much greater reactivity toward electrophilic substitution than do the parent aromatic compounds.[161]

VI. Dihydro-1,2,3-triazines

At present, available data on dihydro-1,2,3-triazines is limited. Although reduction of 4,5,6-triphenyl-1,2,3-triazine (127a) with zinc in acetic acid or with hydrogen over palladium affords 3,4,5-triphenylpyrazole (130),[209] the

[209] E. A. Chandross and G. Smolinsky, *Tetrahedron Lett.*, 19 (1960).

reduction of triaryl-1,2,3-triazines (**127a**) with lithium aluminum hydride,[210] as well as catalytic hydrogenation (Pd–C)[211] of the triazines (**127b–d**), yields, in an almost quantitative yield, dihydro-1,2,3-triazines, which are best identified as the 2,5-dihydro isomers (**129a–d**) [Eq. (59)].

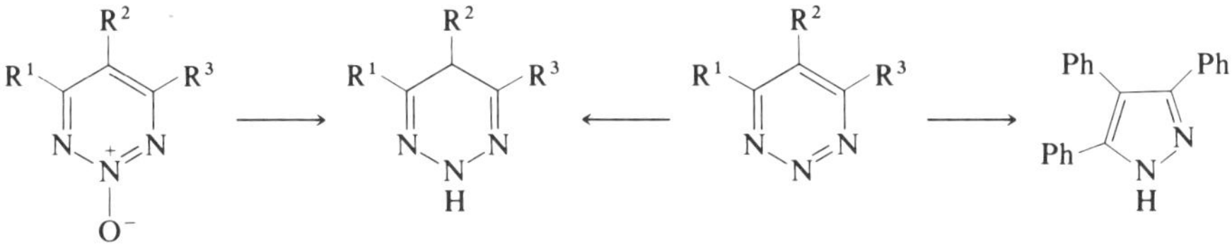

(128)   (129)   (127)   (130)

a $R^1 = R^2 = R^3 = Ph$
b $R^1 = R^3 = Me, R^2 = H$
c $R^1 = Ph, R^2 = H, R^3 = Me$
d $R^1 = R^2 = R^3 = Me$

(59)

2,5-Dihydro-1,2,3,-triazines (**129b,c**) were also prepared in reasonable yield by sodium borohydride[4] (in MeOH) or catalytic (H_2, Pd–C) reduction[211,212] of 4,6-disubstituted 1,2,3-triazine 2-oxides (**128b,c**).

The structures of 2,5-dihydro-1,2,3-triazines were supported by ^{1}H-NMR spectroscopy. Thus the proton attached to N-2 appears in the range δ 8.3–9.02 and methylene protons at position 5 appear at δ 2.45–2.80 ppm. The singlet form of the signal of these protons indicates fast ring inversion of the compound.

VII. Dihydro-1,2,4-triazines

A. General Comments

Of the eight isomeric structures theoretically possible for dihydro-1,2,4-triazines (Section III,A), the best known are the 4,5- (**12h**) and 2,5-dihydro derivatives (**12e**). Surprisingly, no thorough study of the structures of

---

[210] A. Neunhoefer and P. F. Wiley, *in* "The Chemistry of Heterocyclic Compounds" (A. Weissberger and E. C. Taylor, eds.), Vol. 33. Wiley (Interscience), New York, 1978.
[211] A. Ohsawa, H. Arai, H. Ohnishi, and H. Igeta, *Chem. Commun.*, 1182 (1980).
[212] H. Arai, A. Ohsawa, H. Ohnishi, and H. Igeta, *Heterocycles* **17**, 317 (1982).

dihydro-1,2,4-triazines is available.[210] Therefore, the structures of dihydro compounds in the following discussion are written as proposed by the original authors (usually, the 4,5-dihydro isomer), but it should not be assumed that these structures have been unambiguously established. In addition, data on possible amidinic tautomerism of 4,5- and 2,5-dihydro isomers have not yet been reported in the literature. The parent compounds of dihydro-1,2,4-triazine are also unknown.

B. Synthesis

1. Cyclization Methods

a. *From N-Acylamino Ketones.* N-Acylamino ketones (**131**) can be treated with hydrazine to form dihydro-1,2,4-triazines (**132**) directly[213–216] or to produce N-acylaminohydrazones (**133**),[217,218] The latter are easily transformed to **132** by treatment with hydrochloric acid [Eq. (60)].

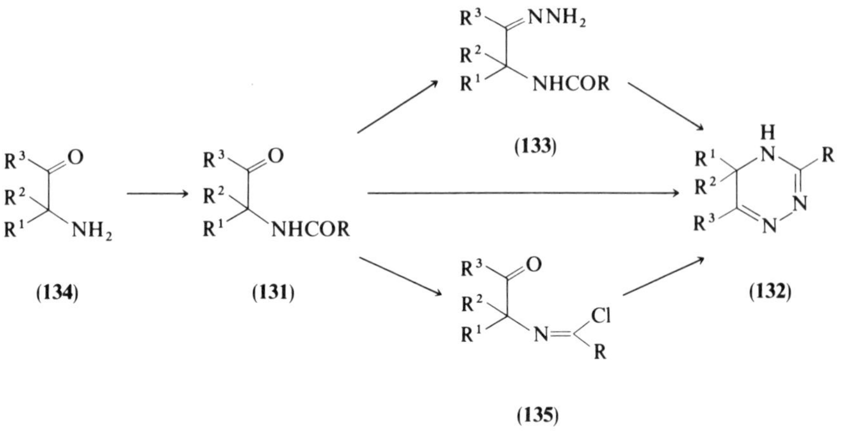

---

[213] C. M. Atkinson and H. D. Cossey, *J. Chem. Soc.*, 1805 (1962).
[214] R. Metze, *Chem. Ber.* **91**, 1863 (1958).
[215] V. Sprio and P. Madonia, *Gazz. Chim. Ital.* **87**, 992 (1957).
[216] V. Sprio and P. Madonia, *Ann. Chim. (Rome)* **49**, 731 (1959).
[217] H. Hagiwara, Y. Oka, Y. Hara, S. Yuguri, J. Susuki, K. Furuno, and I. Iida, *Takeda Kenkyusho Nempo* **22**, 1 (1963) [*CA* **60**, 12011d (1964)].
[218] K. Masuda, *Yakugaku Zasshi* **81**, 533 (1961).

The N-acylamino ketones (131) can be converted by reaction with phosphorous pentachloride to the chlorimine derivative 135, which undergoes further cyclization to 132 by reaction with hydrazine.[219]

b. *Other Cyclizations.* 2,5-Dihydro-3,5,6-triphenyl-1,3,5-triazine (137a) was isolated from the reaction of benzaldehyde azine (136) with potassium *tert*-butoxide in toluene.[220] A self-condensation mechanism has been proposed involving 1,4-addition of an azine molecule to the C=N bond of a second azine.

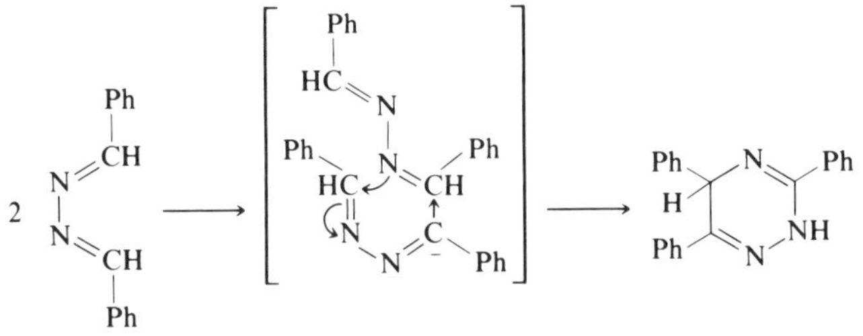

(61)

(136)   (137a)

Tsuge *et al.* reported the thermal cyclization of 1,6-diazido-2,5-diphenyl-3,4-diazahexadi-2,4-ene (138) to 2-azidomethyl-3,6-diphenyl-2,3-dihydro-1,2,4-triazine (139) in boiling toluene.[221] Irradiation of the latter gives 3,6-diphenyl-4,5-dihydro-1,2,4-triazine (140) [Eq. (62)].

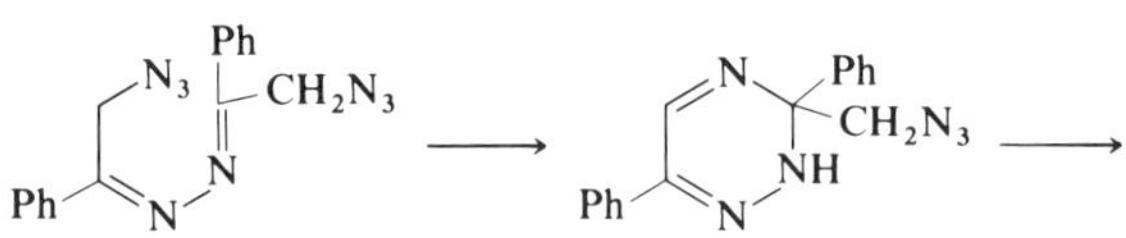

(62)

(138)   (139)   (140)

2. *From 1,2,4-Triazine Derivatives*

a. *Reduction of 1,2,4-Triazines.* Early publications report the reduction of 1,2,4-triazines to dihydro-1,2,4-triazines, using zinc and acetic

[219] Japanese Patent 65/4,638 (1965) [*CA* **63**, 5661h (1965)].
[220] J.T.A. Boyle, M. F. Grundon, and M. Scott, *J.C.S. Perkin I*, 207 (1976).
[221] O. Tsuge, H. Samura, and M. Tashiro, *Chem. Lett.*, 1185 (1972).

acid[222-224] or catalytic hydrogenation on Raney nickel. More recently, quantitative reduction with sodium borohydride of 3-methoxy- or 3-methylthio-5,6-diaryl-1,2,4-triazines to the corresponding dihydro-1,2,4-triazines was described.[225-228] Although in the initial publication,[225] the authors described the products as 4,5-dihydro-1,2,4-triazine derivatives, later, based on X-ray analysis,[33] they revised this assignment to the 2,5-dihydro-1,2,4-triazine structure (141) [Eq. (63)].

$$\begin{array}{c} R^2 \diagdown N \diagdown XME \\ \| \\ R^1 \diagup N \diagup N \end{array} \longrightarrow \begin{array}{c} R^2 \diagdown N \diagdown XMe \\ H \\ R^1 \diagup N \diagup NH \end{array} \quad (63)$$

(141)

b. *From 1,2,4-Triazines or 1,2,4-Triazinones and Grignard Reagents.* 4,5-Dihydro-1,2,4-triazines were also obtained from the reaction of 1,2,4-triazines with Grignard reagents by nucleophilic addition across the C=N bond.[229-231] On the other hand, the dihydro-1,2,4-triazines isolated from a reaction of 1,2,4-triazin-3-one or 1,2,4-triazine-3,5-diones[232] with a Grignard reagent were ascribed as 2,3-dihydro isomers (142) [Eq. (64)].

$$\begin{array}{c} O \diagdown N \diagdown O \\ H \\ Ph \diagup N \diagup NH \end{array} \longrightarrow \begin{array}{c} Ph \diagdown N \diagdown Ph \\ Ph \\ Ph \diagup N \diagup NH \end{array} \longleftarrow \begin{array}{c} Ph \diagdown N \diagdown O \\ Ph \diagup N \diagup NH \end{array} \quad (64)$$

(142)

c. *Electrochemical Reduction.* Pinson and co-workers reported that electrochemical reduction of 1,2,4-triazines leads to unstable 1,4-dihydro derivatives (143) which rearrange either to a 1,2- (144) or a 4,5-dihydro

---

[222] P. V. Laakso, R. Robinson, and H. P. Van drewala, *Tetrahedron* **1**, 103 (1957).
[223] R. Metze and G. Scherowsky, *Chem. Ber.* **92**, 2481 (1959).
[224] C. M. Atkinson and H. D. Cossey, *J. Chem. Soc.*, 1628 (1963).
[225] T. Sasaki, K. Minamoto, and K. Harada, *Heterocycles* **10**, 93 (1978).
[226] T. Sasaki, K. Minamoto, and K. Harada, *Tetrahedron Lett.*, 1529 (1980).
[227] T. Sasaki, K. Minamoto, and K. Harada, *J. Org. Chem.* **45**, 4587 (1980).
[228] T. Sasaki, K. Minamoto, and K. Harada, *J. Org. Chem.* **45**, 4594 (1980).
[229] J. Daunis and C. Pigiere, *Bull. Soc. Chim. Fr.*, 2493 (1973).
[230] J. Daunis and R. Jacquier, *J. Heterocycl. Chem.* **10**, 559 (1973).
[231] J. Daunis, L. Djouai-Hifdi, and H. Lopez, *J. Heterocycl. Chem.* **16**, 427 (1979).
[232] A. Mustafa, A. K. Mansour, and H. A. A. Zaher, *J. Prakt. Chem.* **313**, 699 (1971).

compound (**137b**).²³³ These derivatives can be further reduced to an imidazole (**145**) or a tetrahydro-1,2,4-triazine (**146**) [Eq. (65)].

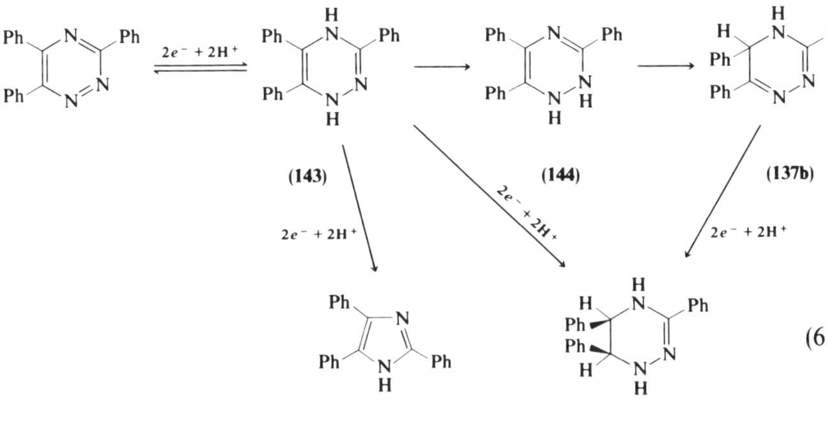

(65)

d. *Other Methods.* Busch and Küspert have shown that hydrolysis of 4-amino-2,3,4,5-tetrahydro-1,2,4-triazines by treatment with hydrochloric acid produces dihydro compounds, which were formulated as derivatives of 2,3-dihydro-1,2,4-triazine (**147**).²³⁴

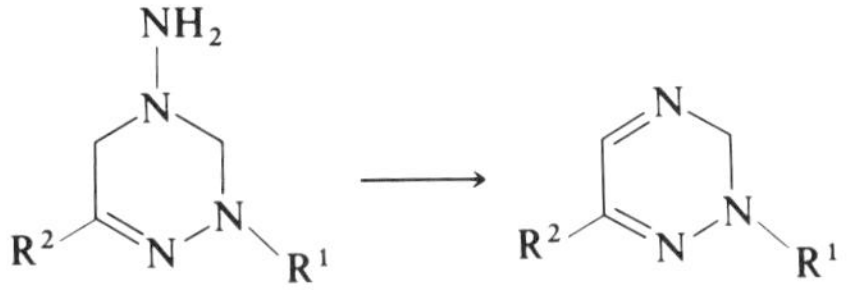

(66)

(**147**)

A number of dihydro-1,2,4-triazines were prepared by alkylation of the corresponding 1,2,4-triazinones.²³⁰,²³⁵

3. *From Other Heterocycles*

Shvaika and Fomenko demonstrated the formation of 4,5-dihydro-1,2,4-triazines (**149**) from recyclization reactions of 1,3-oxazolium salts (**148a**),²³⁶

---

²³³ J. Pinson, J.-P. M'Packo, N. Vinot, J. Armand, and P. Bassinet, *Can. J. Chem.* **50**, 1581 (1972).
²³⁴ M. Busch and K. Küspert, *J. Prakt. Chem.* **144**, 273 (1936).
²³⁵ A. Comparini, A. M. Celli, F. Ponticceli, and P. Tedeschi, *J. Heterocycl. Chem.* **15**, 1271 (1978).
²³⁶ O. P. Shvaika and B. F. Lipnickii, *Zh. Obshch. Khim.* **51**, 1842 (1981).

1,3-thiazolium (**148b**),[237,238] or 1,3-selenazolium salts (**148c**)[239] with hydrazine [Eq. (67)].

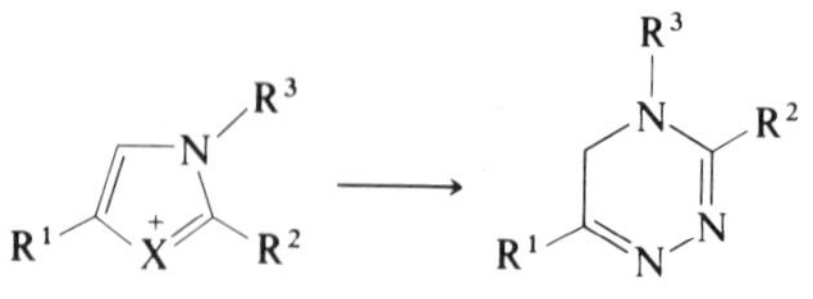

$$\underset{(\mathbf{148})}{\overset{R^3}{\underset{R^1\overset{+}{X}R^2}{\bigwedge}}} \longrightarrow \underset{(\mathbf{149})}{\overset{R^3}{\underset{R^1}{\bigwedge}}\overset{R^2}{\underset{N}{\bigvee}}} \tag{67}$$

a: X = O
b: X = S
c: X = Se

An interesting ring contraction to form dihydro-1,2,4-triazine (**151**) was reported by Halladin and Hassner[240] by the reaction of 1,4-oxazepin-7-ones (**150**) with hydrazine [Eq. (68)].

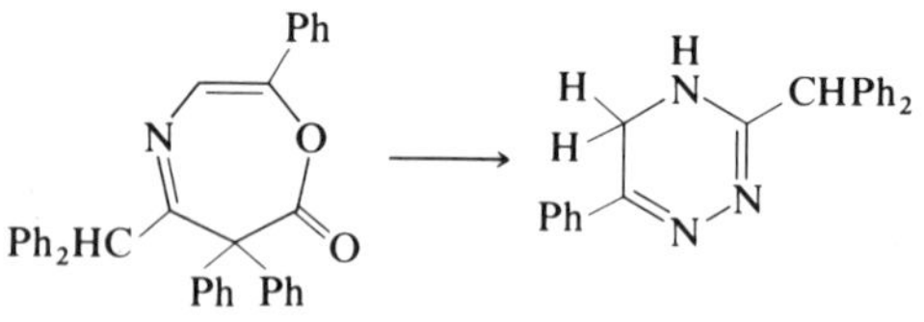

$$(\mathbf{150}) \longrightarrow (\mathbf{151}) \tag{68}$$

C. Physicochemical Properties

1. Infrared Spectra

The N—H stretching vibrations of 4,5-dihydro-1,2,4-triazines usually appear at 3250–3260 cm^{-1},[233,239] with the C=N absorptions being found in the 1620–1680 cm^{-1} region.[239] The N—H stretching band of the 1,2-dihydro compound (**144**) displays two peaks at 3180 and 3380 cm^{-1}.[233] The C=N absorption bands in the 2,5-dihydro compounds were found at 1645–1650 cm^{-1}.

---

[237] O. P. Shvaika and V. I. Fomenko, *Dokl. Akad. Nauk SSSR* **200**, 134 (1971).
[238] O. P. Shvaika and V. I. Fomenko, *Khim. Geterotsikl. Soedin.*, 635 (1976).
[239] O. P. Shvaika and V. I. Fomenko, *Zh. Org. Khim.* **10**, 2429 (1974).
[240] M. J. Halladin and A. Hassner, *J. Org. Chem.* **38**, 3466 (1973).

Sec. VII.C] DIHYDROAZINES 89

2. *Ultraviolet Spectra*

The UV spectra of a large number of 3-methoxy- and 3-methylthio-4,5-dihydro-1,2,4-triazines (**152**) have been determined.[227,229] In general, such compounds show two or three absorption maxima at $\sim$ 214–235, 252–296, and 326–360 nm. The short-wavelength absorption is usually of medium intensity ($\varepsilon \sim 10,000$), although it is sometimes the strongest band. The second band, at ~ 270 nm, is less intense ($\varepsilon \sim 6000$) and obviously is the result of $n \to \pi^*$ transitions in the MeX—C=N—N=C structure. The long-wavelength absorptions that sometimes appear at ~ 340 nm are usually the weakest band ($\varepsilon \sim 2000$), the result of $n \to \pi^*$ transitions in the N—C(OMe)=N—N=C chromophore. These latter absorptions strongly depend on delocalization of the ring π electrons and the lone-pair electrons of N—H, i.e., on the planarity of the N—C=N—N=C fragment of the potentially homoaromatic molecule.

Recently, the UV spectra of a series of 3-methoxy- and 3-methylthio-2,5-dihydro-1,2,4-triazines have been reported.[227,228,231,235] In general, these compounds show only one absorption maximum, in the region 240–270 nm ($\varepsilon \sim 2000$–5000). According to Sasaki *et al.*,[227,228] the hypsochromic shift in the UV absorptions of the similarly substituted 2,5-dihydro-1,2,4-triazines, relative to that of the 4,5-dihydro-1,2,4-triazines, indicates less extensive conjugation of the former, and confirms the insulation of both double bonds. Interestingly, introduction of a phenyl group in position 6 of 2,5-dihydro-1,2,4-triazines produces a new maximum in the UV spectra, at ~ 300 nm and with medium intensity ($\varepsilon \sim 1200$). In addition, a hypsochromic shift of the first absorption band was observed (230–250 nm), with a significant increase in intensity ($\varepsilon \sim 20,000$).

3. *Nuclear Magnetic Resonance Spectra*

The NMR spectra of a number of 4,5-dihydro-1,2,4-triazines have been reported, the proton at position 6 absorbing in the δ 6.57–7.12 region.[227–229,239] The chemical shift of the proton at position 5 depends on the substituent at this position; in the case of methyl, it resonates at δ 3.48–3.66, whereas phenyl induces a significant downfield shift to δ 4.50–4.70.

The proton at position 6 of 2,5-dihydrotriazines appears at δ 6.66–6.80,[227,228] whereas that at position 5 in the case of methyl appears at δ 3.70–3.90, and in the case of phenyl at δ 4.70–5.20.[222,235] The downfield shifts of the resonance of position-5 protons in 2,5-dihydro compounds, relative to 4,5-dihydro, results from the presence of adjacent double bonds in the former molecules.[227] The NH signals of 2,5-dihydro-1,2,4-triazines appear as a broad singlet at δ 7.70–8.20. The vicinal coupling constants of the 5- and 6-protons of both 2,5- and 4,5-dihydro derivatives are between 1.9 and 2.5 Hz.[227]

D. REACTIVITY

When 2,5-dihydro-3,5,6-triphenyl-1,2,4-triazine (**137a**) was refluxed in toluene with potassium *tert*-butoxide, rearrangement occurred to give 2,3,4-triphenyl-1,3,5-triazine (**153**) and its 1,2-dihydro derivative (**154b**). However, if less than 1 equivalent of base is used, the 1,2,4-triazine is obtained along with the 1,3,5 isomer (**153**).[220] A novel photochemical ring contraction by protonation and photolysis of **173a** to yield 3,4,5-triphenylpyrazoles (**155**) was reported [Eq. (69)].[241]

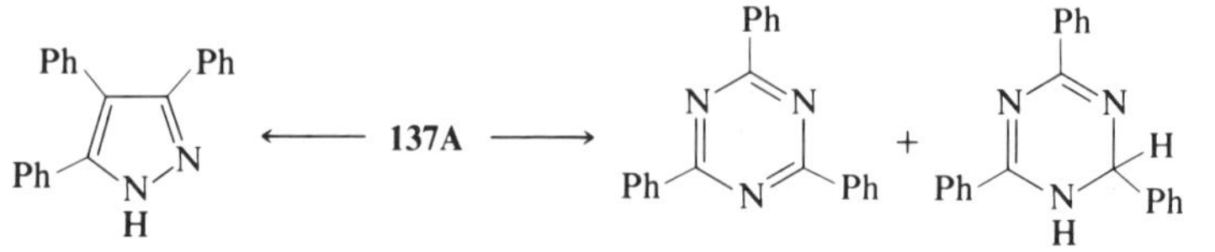

(69)

(**155**)   (**153**)   (**154b**)

Sasaki and co-workers have reported the thermal reaction of 5,6-diaryl-2,5-dihyro-1,2,4-triazines (**141**) with dimethyl acetylenedicarboxylate.[242] Following an initial incorrect assignment of the products,[225] detailed X-ray diffraction has shown that the substances produced are derivatives of 3,4,7-triazabicyclo[3.3.0]octa-2,7-dienes (**156**).[33]

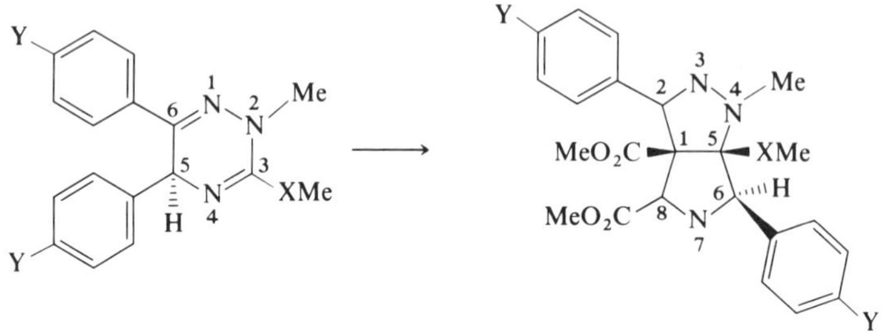

(70)

(**141**)   (**156**)

a: X = O; Y = H   e: X = O; Y = OMe
b: X = S; Y = H   f: X = S; Y = OMe
c: X = O; Y = Me   g: X = O; Y = Cl
d: X = S; Y = Me

---

[241] J. Nagy and J. Nyitrai, *Acta Chim. Acad. Sci. Hung.* **109**, 1 (1982).
[242] T. Sasaki, S. Eguchi, K. Minamoto, and M. Ohno, *Kenkyu Hokoku–Asahi Garasu Kogyo Gijutsu Shoreikai* **39**, 97 (1981) [*CA* **97**, 216069e (1982)].

VIII. Dihydro-1,3,5-triazines

A. GENERAL COMMENTS

Because of the high degree of symmetry of 1,3,5-triazines, only two dihydro isomers are possible: 1,2- (**13a**) and 1,4- (**13b**). It should be noted that the majority of (if not all) known dihydro-1,3,5-triazines bear at least two aryl or polyfluoroalkyl groups, substituents which probably stabilize the compounds. The simple dihydro-1,3,5-triazines containing only alkyl substituents, or just hydrogen, are at present unknown. Despite the recent discovery that N-unsubstituted dihydro-1,3,5-triazines usually exist in a tautomeric equilibrium of 1,2- and 1,4-dihydro derivatives, for purposes of covenience, the dihydrotriazines discussed here will be presented as the structures given in the original literature, still require proper structural study, using modern instrumental methods.

B. SYNTHESIS

1. From Nitriles

One of the first-known synthetic preparations of 1,2-dihydro-1,3,5-triazines involves trimerization of benzonitrile in the presence of alkali metals or organometallic reagents.

In 1896, Lottermoser reported[243] that in the presence of sodium, benzonitrile in refluxing benzene produces 2,2,4,6-tetraphenyl-1,2-dihydro-1,3,5-triazine (**157**), sodium cyanide, and 2,4,6-triphenyl-1,3,5-triazine (**153**) [Eq (71)].

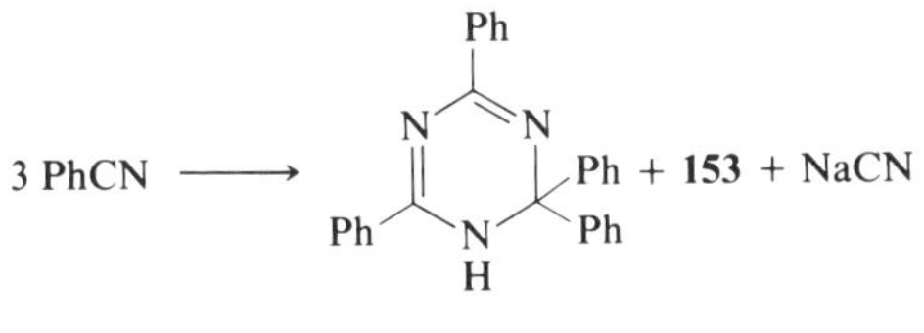

$$3 \text{ PhCN} \longrightarrow \underset{(157)}{\text{[structure]}} + 153 + \text{NaCN} \qquad (71)$$

(**157**)

In support of this structure for **157**, it was found that alkaline hydrolysis of the compound gives two molecules of benzoic acid and one molecule of

---

[243] A. Lottermoser, *J. Prakt. Chem.* [2] **54**, 132 (1896).

benzophenone. Cook and Jones obtained **157** as the sole product and proposed the following mechanism: initial formation of a phenylsodium intermediate, followed by reaction with three molecules of benzonitrile in a complex addition and cyclization reaction.[244] On the basis of this mechanism, Anker and Cook studied the action of organolithium reagents on benzonitrile and succeeded in preparing a series of dihydro-1,3,5-triazines (**158**) [Eq. (72)].[245]

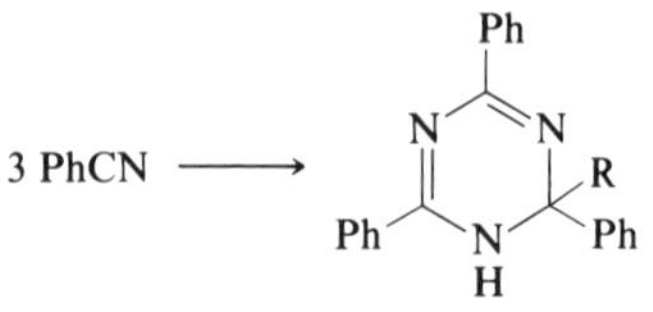

$$3 \text{ PhCN} \longrightarrow \underset{(\textbf{158})}{\text{dihydro-1,3,5-triazine}} \quad (72)$$

Swamer et al.[246] described the formation of 1,2-dihydro-2,4,6-triphenyl-1,3,5-triazine (**154b**) (yield 14%) from benzonitrile in the presence of sodium hydride and proposed a detailed mechanism (Scheme 8), which is, presumably, valid for all trimerizations of nitriles in the presence of organometallic reagents.[243–246]

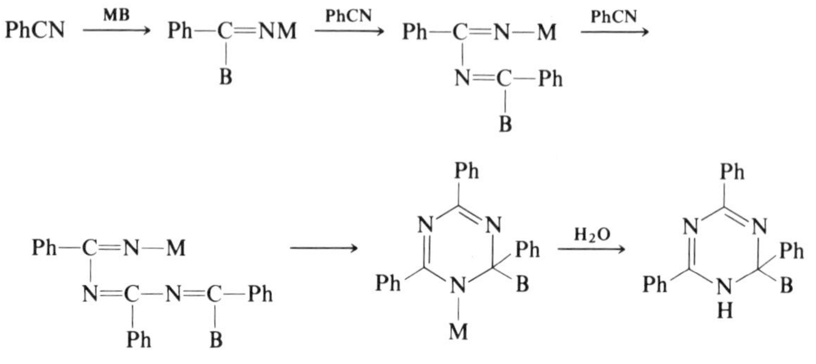

$$\text{PhCN} \xrightarrow{\text{MB}} \underset{\text{B}}{\text{Ph—C=NM}} \xrightarrow{\text{PhCN}} \underset{\underset{\text{B}}{\text{N=C—Ph}}}{\text{Ph—C=N—M}} \xrightarrow{\text{PhCN}}$$

$$\underset{\underset{\text{Ph B}}{\text{N=C—N=C—Ph}}}{\text{Ph—C=N—M}} \longrightarrow \text{(triazine-M)} \xrightarrow{\text{H}_2\text{O}} \text{(triazine-H)} \quad (73)$$

M = Metal; B = H, alkyl, aryl

SCHEME 8

The optimization of reaction conditions used by Swamer et al.[246] enabled us to prepare **154** in ~80% yield.[247] A spectroscopic study has shown that in

---

[244] A. H. Cook and D. G. Jones, *J. Chem. Soc.*, 278 (1941).
[245] R. M. Anker and A. H. Cook, *J. Chem. Soc.*, 323 (1941).
[246] F. W. Swamer, G. A. Reynolds, and C. R. Hauser, *J. Org. Chem.* **16**, 43 (1951).
[247] A. L. Weis, *Is. Chem. Soc., 49th Ann. Meet.*, 1982, p. 41 (1982).

solution compound **154** exists in a tautomeric equilibrium between **154a** and **154b** (see Section VIII,C). The preparation of **154b** from aluminum benzaldimine and benzonitrile has been reported[248] and rationalized according to Scheme 8.

Researchers in Japan have claimed to produce 1,2-dihydro-1,3,5-triazines by high-pressure reactions between aromatic or heteroaromatic nitriles and formaldehyde,[249] acetaldehyde,[250] or ketones[251] in the presence of ammonia.

2. From Amidines

In 1969, Cherkasov and co-workers[252] showed that the condensation of benzaldehyde with benzamidine gives 2,4,6-triphenyl-1,2-dihydro-1,3,5-triazine (**154b**) rather than benzylidenebenzamidine (**159**) (as was reported incorrectly in the early work of Pinner[253]) [Eq. (74)].

$$PhCHO + \underset{HN}{\overset{H_2N}{>}}\!\!=\!\!Ph \longrightarrow PhCH=N-\underset{Ph}{C}=NH \longrightarrow \quad (74)$$

(159)   (154b)

These Soviet workers also prepared a series of 1,2-dihydro-1,3,5-triazines by reaction of aryl aldehydes,[252,254] Shiff bases,[255-257] or ketimines[258] with amidines of different acids (usually benzoic, trichloro-, or trifluoroacetic). They proposed that the dihydrotriazines form via intermediate azadienes, which undergo cyclization via [4 + 2] cycloaddition, accompanied by elimination of an akylidenimine [Eq. (75)].

---

[248] H. Hoberg and J. Barluenga-Mur, *Justus Liebigs Ann. Chem.* **751**, 86 (1971).
[249] Japanese Patent 8053,275 (1980) [*CA* **94**, P15775w (1981)].
[250] Japanese Patent 8053,276 (1980) [*CA* **93**, P239469s (1980)].
[251] Japanese Patent 8226,668 (1982) [*CA* **97**, P23825q (1982)].
[252] V. M. Cherkasov, N. A. Kapran, and V. N. Zavatzkii, *Khim. Geterotsikl. Soedin.*, 350 (1969).
[253] A. Pinner, *Ber. Dtsch. Chem. Ger.* **22**, 1610 (1889).
[254] V. M. Cherkasov, N. A. Kapran, V. N. Zavatzkii, and V. T. Tsyba, *Khim. Geterotsikl. Soedin.*, 704 (1971).
[255] V. M. Cherkasov and N. A. Kapran, *Khim. Geterotsikl. Soedin.*, 281 (1973).
[256] N. A. Kapran and V. M. Cherkasov, *Khim. Geterotsikl. Soedin*, 995 (1976).
[257] N. A. Kapran and V. M. Cherkasov, *Ukr. Khim. Zh.* **45**, 862 (1979) [*CA* **92**, 6505a (1980)].
[258] V. M. Cherkasov, N. A. Kapran, and Y. L. Yagupolskii, *Khim. Geterotsikl. Soedin.*, 842 (1982).

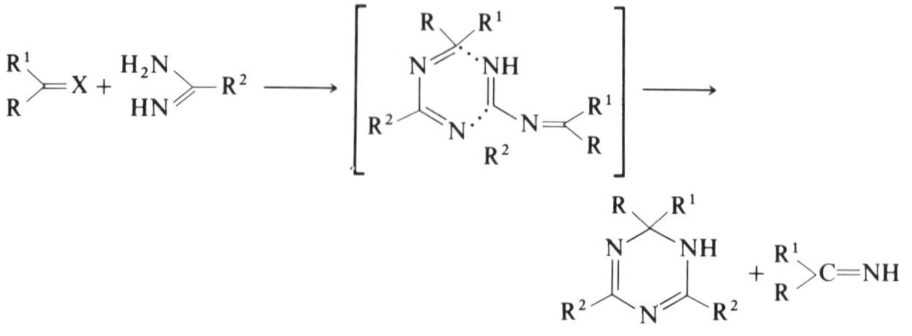

[Scheme for Eq. (75)]

$$X = O, NH, NR$$

N-Substituted 1,2-dihydro-1,3,5-triazine (160) was prepared by Hunter and Sim via condensation of benzaldehyde with N-benzylbenzamidine [Eq. (76)].[259]

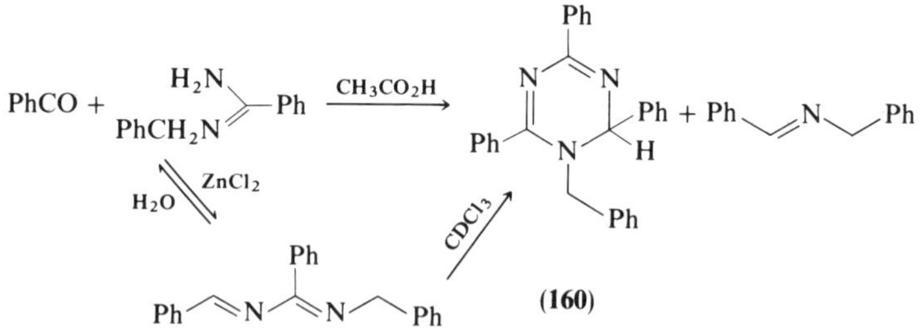

[Scheme for Eq. (76) showing formation of (160)]

Smith et al. demonstrated the formation of 2-aryl-4,6-diphenyl-1,2-dihydro-1,3,5-triazines (154b, 162) by thermolysis of ylides 161 [Eq. (77)].[260,261]

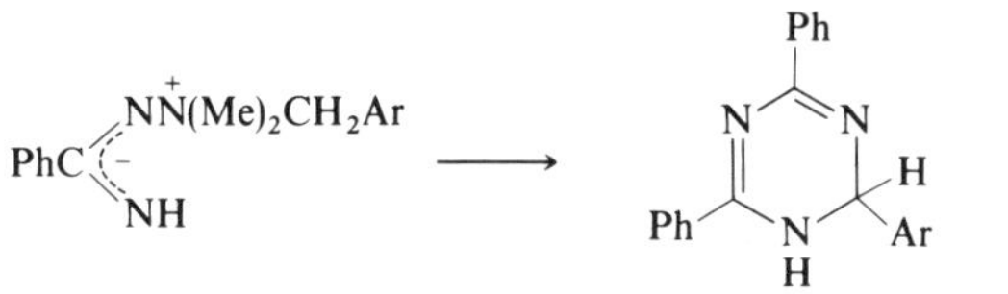

[Scheme for Eq. (77)]

(161)   (154b: Ar = Ph)
        (162: Ar = p-CH$_3$C$_6$H$_4$)

Moreover, considering the mechanistic similarities between dihydro-1,2-triazine formation via ylide thermolysis and the condensation of aryl aldehydes with benzamidine or N-alkylbenzamidines, the [4 + 2] cycloaddition

---

[259] D. H. Hunter and S. K. Sim, Can. J. Chem. **50**, 669 (1972).
[260] R. F. Smith, A. S. Craig, L. A. Buckley, and R. R. Solech, Tetrahedron Lett., 4193 (1979).
[261] R. F. Smith, R. R. Soelch, T. P. Feltz, M. J. Martinelli, and S. M. Geer, J. Heterocycl. Chem. **18**, 319 (1982).

proposed by Cherkasov et al. for dihydrotriazine formation is unlikely.[262] Thus the mechanistic interpreation first proposed by Hunter and Sim[259] is preferable, although the formation of triazine through cyclization of intermediate labile *gem*-diamidine 163 cannot be excluded [Eq. (78)].

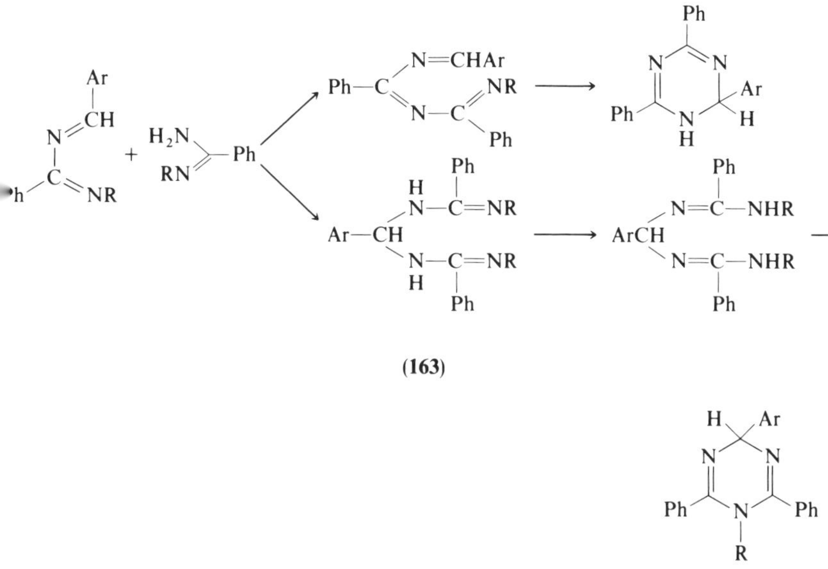

(78)

The unusual reaction of N-haloamidines with enamines to form 1,4-dihydro-1,3,5-triazines (164) was investigated and a possible mechanism suggested [Eq. (79)].[262]

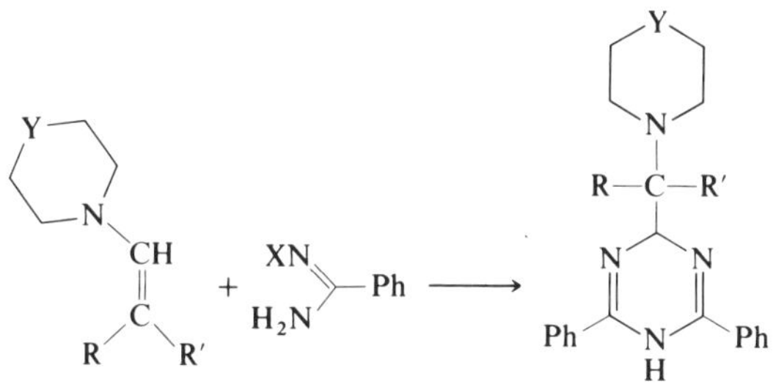

(79)

X = Cl, Br, I; Y = O, CH$_2$ (164)

1,2-Dihydro-1,3,5-triazines (166) were obtained by interaction of N-alkenylimidoyl chlorides (165) with benzamidines followed by chlorination [Eq. (80)].[263]

---

[262] L. Citerio, D. Pocar, R. Stardi, and B. Gioia, *Tetrahedron* 35, 69 (1979).
[263] T. P. Popovich, A. I. Sedlov, and B. S. Drach, *Zh. Org. Khim.* 17, 2626 (1981).

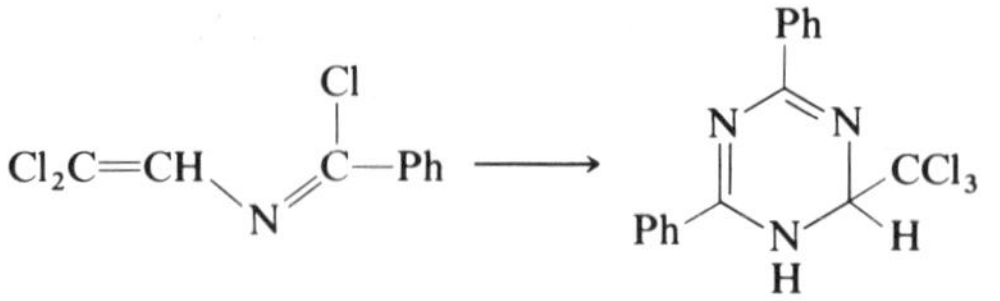

$$Cl_2C=CH\diagdown_N\diagup\overset{Cl}{\underset{}{C}}-Ph \longrightarrow \underset{\underset{H}{\overset{Ph}{\diagup}}}{\overset{Ph}{\diagdown}}\text{-triazine-CCl}_3 \quad (80)$$

(165)   (166)

3. From 1,3,5-Triazines

Starting from the 1,3,5-triazines, few syntheses of dihydro-1,3,5-triazines have been reported. Although many versatile synthetic approaches are available, most of them are yet to be tried. The known reactions involve preparation of dihydro-1,3,5-triazine derivatives from 2,4,6-triphenyl-1,3,5-triazine (153) via (a) lithium aluminum hydride reduction[264]; (b) treatment with potassium in tetrahydrofuran, followed by hydrolysis[265]; and (c) reaction with organolithium compounds and hydrolysis[266] [Eq. (81)].

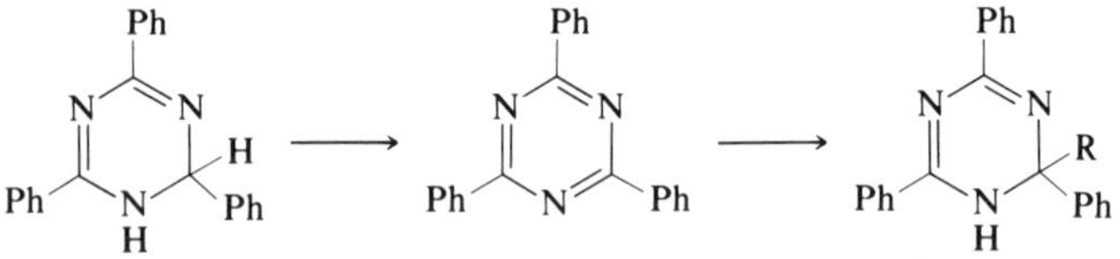

(154b)   (153)   (158)   (81)

Wakefield and Cook have indicated that the high yield and relative absence of side-products by using organolithium reagents make this reaction advantageous in comparison to trimerization of benzonitrile in the presence of organolithium reagents (Section VIII,A,1).

An interesting photochemical formation of different 1,4-dihydro-1,3,5-triazines (168-171) was demonstrated by Kobayashi et al.,[267] who irradiated substituted 1,3,5-triazines (167) of low aromaticity in pentane or cyclohexane, utilizing compounds with two different fluoroalkyl groups on the triazine ring. Two or three distinct products were isolated, the ratio of these dihydro compounds depending on the nature of both substituents and the solvent (which also serves as a reagent) [Eq. (82)].[267]

[264] H. L. Nyquist, *J. Org. Chem.* **31**, 784 (1966).
[265] P. Weil and N. Collignon, *Bull. Soc. Chim. Fr.*, 258 (1974).
[266] L. S. Cook, G. Prudhoe, N. D. Vernayak, and B. J. Wakefield, *J. Chem. Res., Synop.*, 113 (1982); *J. Chem. Res., Miniprint*, 1357 (1982).
[267] Y. Kobayashi, A. Ohsawa, and M. Honda, *Chem. Pharm. Bull.* **21**, 1575 (1973).

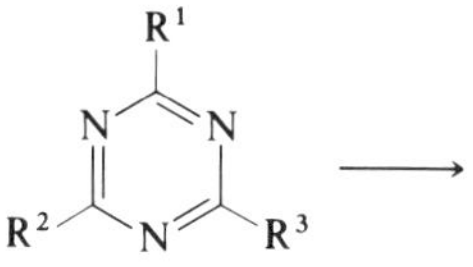

Sec. VIII.B] DIHYDROAZINES 97

(167)

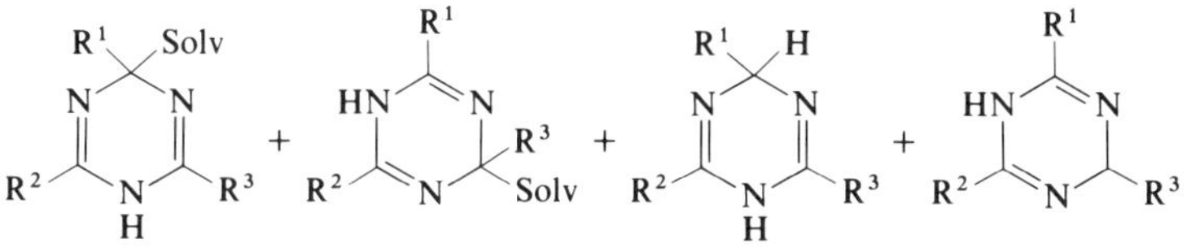

(82)

	R^1	R^2	R^3
a	CF_3	CF_3	CF_3
b	CF_3	CF_3	CF_2Cl
c	CF_3	CF_2Cl	CF_2Cl
d	CF_2Cl	CF_2Cl	CF_2Cl

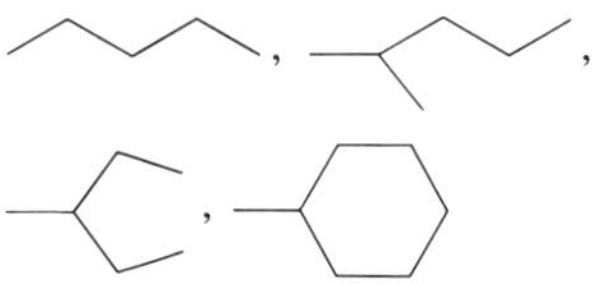

4. *From 1,3,5-Oxadiazines*

Gambaryan and Zeifman synthesized 1,4-dihydro-1,3,5-triazine (**173**) by reaction of 4,4-bis(trifluoromethyl)-2,6-diphenyl-1,3,5-oxadiazine (**172**) with ammonia.[268]

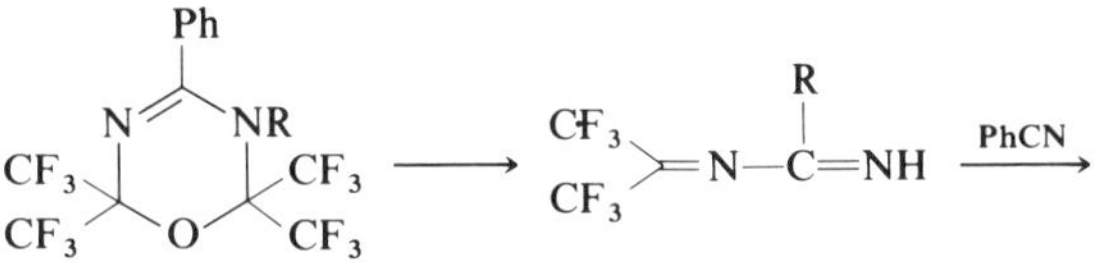

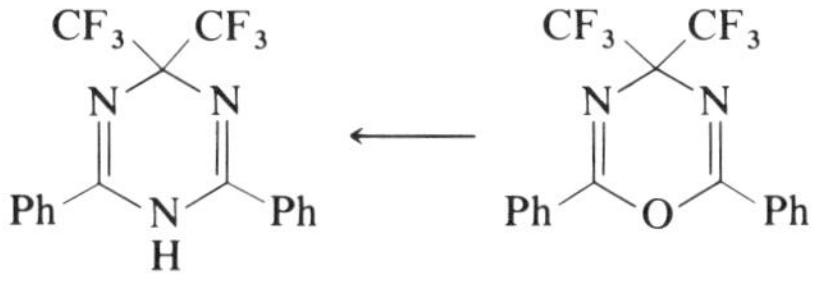

(83)

[268] N. P. Gambaryan and Y. V. Zeifman, *Izv. Akad. Nauk SSSR*, 2059 (1969).

Burger et al. have shown that upon thermolysis dihydro-1,3,5-oxadiazines (**174**) undergo a retro Diels–Alder reaction with elimination of hexafluoroacetone.[269,270] The intermediate product, 4,4-bis(trifluoromethyl)-1,3-diaza-1,3-diene, was transformed to the corresponding 1,4-dihydro-1,3,5-triazine (**173**) [Eq. (83)].

C. Tautomerism in Dihydro-1,3,5-triazines

Dihydro-1,3,5-triazines are of fundamental interest because of their ability to undergo amidinic tautomerism. Furthermore, as these are nitrogen-containing analogues of dihydropyrimidines (methylene at position 5 replaced by N) it would certainly be interesting to compare the effects of nitrogen substitution on the structual stability and tautomeric behavior of these compounds.

Kobayashi et al.[271] examined chemical shifts in the ^{19}F-NMR spectra of [polyfluoro(chloro)methyl]-dihydro-1,3,5-triazines (**168-171**) in deuteriochloroform and in methanol, concluding that, in these solvents, these compounds exist as an equilibrium between 1,4-dihydro (**168a**) and 1,2-dihydro tautomers (**168b,b'**).

(**168a**)   (**168b**)   (**168b'**)

During our investigation on tautomerism in dihydropyrimidines, we measured the IR and ^{1}H- and ^{13}C-NMR spectra of several known dihydro-1,3,5-triazines,[147] which exist in the solid state as the 1,2-dihydro isomers. The NMR measurements were carried out in two solvents, CDCl$_3$ and DMSO, and signals of the two tautomers A and B were always observed. The rate of tautomeric exchange was rather slow, very similar to that in **121**, obviously owing to the presence of two phenyls at positions 2 and 4 (see Section V,C,1,d).

All the factors affecting amidinic tautomerism in dihydropyrimidines (see Section V,C,1,b) apply equally to dihydro-1,3,5-triazine derivatives. In CDCl$_3$ the rate of tautomerism was clearly affected by the bulkiness of the

---

[269] K. Burger and S. Penninger, *Synthesis*, 524 (1978).
[270] K. Burger, S. Penninger, M. Greisel, and E. Daltrozzo, *J. Fluorine Chem.* **15**, 1 (1980).
[271] Y. Kobayashi, A. Ohsawa, and M. Honda, *Chem. Pharm. Bull.* **21**, 1586 (1973).

A ⇌ B

(154) R = H
(158a) R = Me
(158b) R = Bu

substituents on the saturated (sp^3) carbon. Thus a change of substituents in the order Bu, Me, H (other reaction conditions being held constant) clearly increases the rate of tautomerism as measured by NMR.

IR spectra of dihydro-1,3,5-triazines showed the absorptions similar to those of dihydropyrimidines. Thus two C=N absorption peaks were found for the 1,2- and 1,4-dihydro-1,3,5-triazines at 1610 and 1640 cm^{-1}, respectively. In addition, a reanalysis of the IR data given by Kobayashi[271] clearly reveals two signals for C=N absorption, at 1680 and 1740 cm^{-1}, which indicates the presence of two tautomeric forms.

D. Physicochemical Properties

1. Infrared Spectra

Stretching frequencies for the N—H bond have been reported for a wide variety of N-unsubstituted 1,2- and 1,4-dihydro-1,3,5-triazines, the absorptions of which are of medium intensity and appear in the 3380–4430 cm^{-1} region.[247,271] Of particular interest, however, are the C=N stretching vibrations, the wave numbers of which are influenced by the substitutents on the doubly bonded carbons. Thus when substitutents are polylfluoroalkyl groups, a strong band is observed in the region 1680–1730 cm^{-1},[267,271] whereas in aryl derivatives the C=N absorption (which is also strong) shifts to a lower frequency (1605–1680 cm^{-1}).[247,258,269,270] Moreover, reminiscent of the dihydropyrimidines, IR spectra for solutions of potentially tautomeric 1,2- and 1,4-dihydro-1,3,4-triazines display two absorptions due to C=N stretching frequencies differing by 30–60 cm^{-1}.[247,267,271] These, too, can serve as an excellent tool for differentiating between the two tautomers. Thus the absorption at higher frequency relates to the 1,4-dihydro isomer, while the absorption at lower frequency is attributed to the 1,2-dihydrotriazine. The IR spectra of N-methyl derivatives of both tautomers of **173** support this observation.[270] Since in the solid state, dihydro-1,3,5-triazines usually exist in only one of the two possible tautomeric forms, a comparison of the IR spectrum in the solid with those in solution is usually a facile method for unambiguous structural assignments. Finally, intensity of the absorptions for the two tautomers provides qualitative information about the relative concentrations of the tautomers in solution.

2. Ultraviolet Spectra and Photochromism

The UV spectra of 2,4,6-triphenyl-1,2-dihydro-1,3,5-triazines are very similar to those of the corresponding 1,2-dihydropyrimidines (Section V,D,2) and show a strong band at ~ 260 nm ($\varepsilon \sim 46{,}000$).[265] It was briefly reported by Lottermoser,[243] and then confirmed by von Walthers,[272] that TPDT (157) exhibits photochromism in the solid state. Hayashi,[273] who reinvestigated this phenomenon, has shown that TPDT in the solid state has two absorption bands, $\lambda_{max} = 257$ and 325 nm (broad, $n \to \pi^*$ transition). On irradiation at 320–380 nm (sunlight or Xenon lamp), TPDT turns brownish pink and shows, in the solid state, $\lambda_{max} = 430, 447, 487, 502, 533, 565,$ and 584 nm. Moreover, photochromism of TPDT is temperature dependent. According to the author,[273] these results suggest that the mechanism of photochromism of TPDT is of a new type which he attributes to intermolecular interaction in a crystalline structure, presumably between NH and excited C=N, rather than intramolecular change of the molecular structure.

3. Nuclear Magnetic Resonance Spectra

The NMR spectra of several dihydro-1,3,5-triazines have been reported.[247,258–263,266–270] Since most of the known compounds of this type bear either aromatic, polychloroalkyl, or polyfluoroalkyl substituents on the triazine ring, ^{1}H NMR is used for the former and ^{19}F NMR for the latter derivatives. The main information obtained from ^{1}H NMR concerns the NH and H or alkyl substituents on the sp^3 carbon. Thus the NH signals of tautomeric 1,2- and 1,4-dihydro-1,3,5-triazines were found in the region δ 5.1–8.2 in CDCl$_3$ and δ 8.0–11.0 in DMSO-d_6. The downfield shift of the NH signal in DMSO is caused by intermolecular hydrogen bonds formed between the polar solvent and the solute. Similar to the situation in the dihydropyrimidines, the NH signal of the 1,4-dihydro isomer appears at a lower field than that of the 1,2-dihydro isomer. The proton attached to the saturated

---

[272] R. von Walthers, *J. Prakt. Chem.* [2] **67**, 446 (1905).
[273] T. Hayashi, *Bull. Chem. Soc. Jpn.* **50**, 2489 (1977).

carbon appears in the region δ 5.2–6.7, depending on the second substituent, which, in the case of an aryl as the second substituent, resonates at 6.2 ppm.

E. REACTIVITY

In common with other dihydroazines, dihydro-1,3,5-triazines are easily oxidized to the corresponding aromatic triazines when heated or oxidized by various agents such as chloranil.[260,261] Alkylation of dihydro-1,3,5-triazines with diazomethane[270] and with benzyl chloride[259] have been reported.

Interesting results were obtained by reaction of substituted 1,2-dihydro-1,3,5-triazines with esters of acetylenedicarboxylic acid (DMAD) and of propiolic acid.[254] In the case of the reaction with DMAD, formation of the different products **175** and **176** has been reported,[254] presumably due to differences in the nucleophilicity of the secondary nitrogen atom of the triazines.

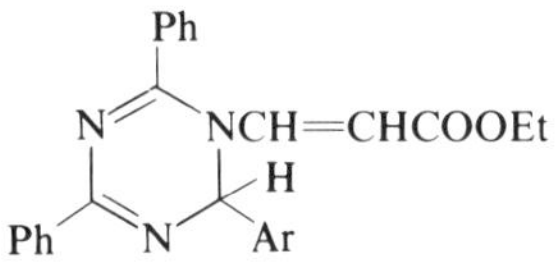

(175) 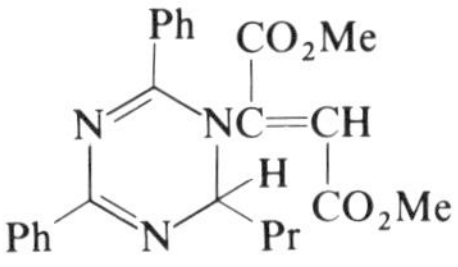　　　　　(176)

Furthermore, it was demonstrated that 2-phenyl-4,6-diaryl-1,3,5-triazine reacts with hexafluoro-2-butyne as a dienophile to give 4,5-bis(trifluoromethyl)-2,6-diarylpyrimidines (**177**)[274] [Eq. (84)].

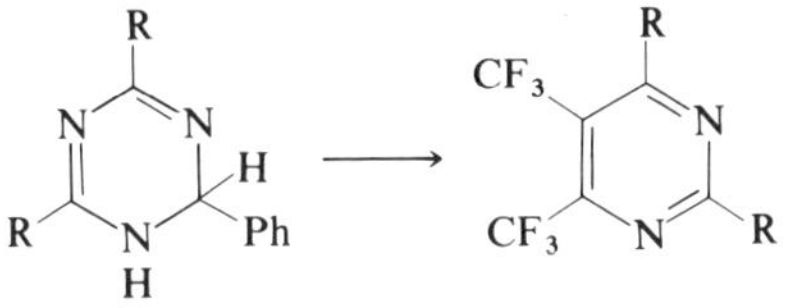

(84)

(177)

R = Ph, p-CH$_3$C$_6$H$_4$, p-ClC$_6$H$_4$, p-CH$_3$OC$_6$H$_4$, CCl$_3$

In 1941, Anker and Cook reported that thermolysis of certain alkyldihydrotriazines (**158**) at 200–250°C gave pyrimidines, with loss of ammonia.[245] Wakefield and co-workers[207,266] repeated this reaction, confirming the results obtained earlier. They also proposed a mechanism

---

[274] N. A. Kapran, V. G. Lukamanov, L. M. Yagupol'skii, and V. M. Cherkasov, *Khim. Geterotsikl. Soedin.*, 122 (1977).

[Eq. (85)] for this transformation [Eq. (85)], involving ring opening accompanied by a 1,3-hydrogen shift, followed by electrocyclic ring closure and elimination of ammonia.[207,266]

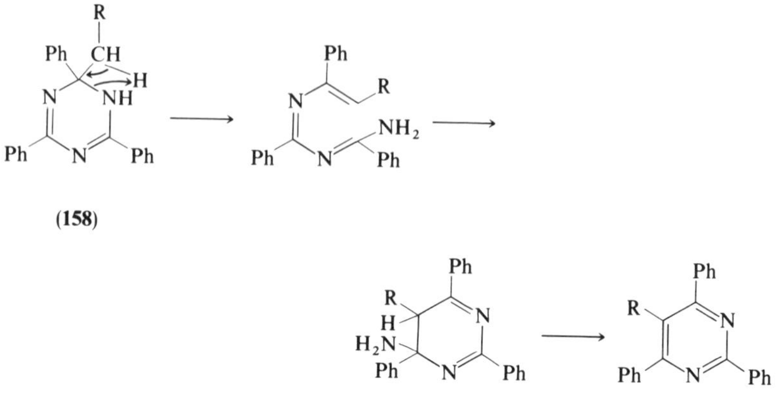

(158)

(85)

R = H, Pr

IX. Conclusions

This survey demonstrates clearly that, despite the full century that has passed since the first synthesis of dihydroazines, the chemistry of these compounds remains largely unexplored. Historically, this neglect resulted from insufficiently sophisticated physical and synthetic techniques by which to study these compounds and equally from a widely accepted, though inaccurate, view that these materials are unstable. We now know that, depending on ring substituents, many of these compounds can be highly stable and exhibit interesting variations in chemical reactivity. Because of the multiplicity of possible isomeric forms, depending on the energy state of the system, they may undergo molecular rearrangement, valence isomerizations, or tautomeric equilibria, all of which are of fundamental interest to synthetic and theoretical chemists alike. Moreover, basic properties of these materials, such as stability (redox potentials) or basicity, await detailed examination. Dihydroazines also participate in intra- and intermolecular cyloaddition reactions, which have great potential utility in synthesis and can provide versatile precursors of both known and rare heterocyclic molecules.

Some dihydroazines are found in living organisms, and various synthesized analogues can serve as useful models for studying redox reactions involving NADH and flavines. Dihydroazine derivatives also serve as calcium channel antagonists and thus have a considerable future in the area of pharmaceutical chemistry. Thus dihydroazines offer a rich source of compounds possessing biological activity as well as interesting chemical and physical properties. We shall probably hear a great deal about them in the years to come.

ACKNOWLEDGMENT

The author thanks Dr. Choji Kashima for communicating experimental results in advance of publication.

Recent Advances in the Chemistry of Benzisothiazoles and Other Polycyclic Isothiazoles

MICHAEL DAVIS

Department of Organic Chemistry, La Trobe University, Bundoora, Victoria, Australia

I. Introduction	106
II. 1,2-Benzisothiazoles	107
A. Formation	107
1. From Mercaptophenyl Aldehydes, Ketones, or Acids	107
2. From 2,2'-Dithiobis(benzonitrile)	108
3. From o-Chlorophenyl Derivatives with Sulfur and Ammonia	108
4. From Benzo[b]thiophen-3(2H)-ones and from Thiochromanones	109
5. Addition of Benzyne to 1,2,5-Thiadiazole and 1,2,5-Selenadiazole	109
6. Reported Formation from an N-Benzylsulfonamide and Trichloromethanesulfenyl Chloride	110
7. From 1H,3H-1,2-Benzisothiazoline 1-Oxides	110
B. Physical Properties	111
C. Chemical Properties	111
1. Electrophilic Substitution	111
2. Reactions of 3-Chloro-1,2-benzisothiazole and Related Compounds	112
3. Photochemistry and Photocycloaddition Reactions	113
4. Miscellaneous Reactions Involving Ring Opening	114
D. 1,2-Benzisothiazolin-3(2H)-ones	114
E. Applications	116
III. 1,2-Benzisothiazolin-3(2H)-one 1,1-Dioxide (Saccharin) and Derivatives	116
A. Formation	116
B. Physical Properties	117
C. Chemical Properties	117
D. Biological Properties	120
E. Related Compounds	121
IV. 2,1-Benzisothiazoles	122
A. Formation	122
1. From the Corresponding Benzisoxazole	122
2. By Hydrogen Peroxide Oxidation of o-Aminothiobenzamides	122
3. From the Reaction of Thionyl Chloride or Related Compounds with o-Toluidines	123
4. From Other Nitrogen–Sulfur Transfer Reagents	123

5. Cycloaddition Reactions of Triphenylthieno[3,4-c]isothiazole 124
6. Preparation of 2,1-Benzisothiazole 2-Oxides and 2,2-Dioxides 125
B. Physical Properties 125
C. Chemical Properties. 126
  1. Diels–Alder Addition Reactions 126
  2. Photochemistry and Thermolysis 126
  3. Decomposition of 2,1-Benzisothiazolium Quaternary Salts 127
  4. Reactions of 3-Chloro-2,1-benzisothiazole 128
D. 2,1-Benzisothiazolin-3(1H)-one 128
  1. Preparation of 2,1-Benzisothiazolin-3(1H)-ones 128
  2. Reactions of 2,1-Benzisothiazolin-3(1H)-ones 128
E. Applications 129
V. Other Isothiazole Systems 130
A. Naphthoisothiazoles 130
B. Benzobisisothiazoles and Benzotris[c]isothiazole 131
VI. Selenium and Tellurium Analogues 132

I. Introduction

This review, intended to continue the earlier review of benzisothiazoles[1] and of benzisosulfonazoles (Saccharin and related compounds),[2] covers the period between about 1970 and early 1984. In general, attention has been focused on new results; however, an attempt has been made to be comprehensive. Some parts of this review have received attention elsewhere. For example, a review has appeared on the biological activity of 1,2-benzisothiazole derivatives,[3] thus that topic has received less attention.

Until 1980, The Royal Society of Chemistry (previously the Chemical Society) published a section on benzisothiazoles in *Specialist Periodical Reports* in the volumes on the *Organic Chemistry of Sulfur, Selenium, and Tellurium*;[4] it was then transferred to *Heterocyclic Chemistry*.[5]

Saccharin's use as a sweetening agent has just achieved its centenary. During the 1970s the alleged carcinogenicity in humans of saccharin was the subject of a prolonged controversy. The interested reader should refer to one of the excellent summaries written by Lepkowski.[6]

---

[1] M. Davis, *Adv. Heterocycl. Chem.* **14**, 43 (1972).
[2] H. Hettler, *Adv. Heterocycl. Chem.* **15**, 233 (1973).
[3] A. De, *Prog. Med. Chem.* **18**, 117 (1981).
[4] D. H. Reid, "Organic Compounds of Sulphur, Selenium and Tellurium," Reporter, Vol. 1. Chemical Society, London, 1970; Vol. 2, 1973; Vol. 3, 1975; Vol. 4, 1977; D. R. Hogg, Reporter, Vol. 5, 1979; Vol. 6. Royal Society of Chemistry, London, 1981.
[5] H. Suschitzky and O. Meth-Cohn, "Heterocyclic Chemistry", Reporters, Vol. 1. Royal Society of Chemistry, London, 1980; Vol. 2, 1981; Vol. 3, 1982.
[6] W. C. Lepkowski, *Chem. Eng. News*, April 11, p. 17 (1977).

II. 1,2-Benzisothiazoles

A. FORMATION

Since 1970, several new syntheses of 1,2-benzisothiazoles have been described, and older syntheses have been extended.

1. *From Mercaptophenyl Aldehydes, Ketones, or Acids*

Lawson[7] discussed the factors influencing the yields of 1,2-benzisothiazoles from 2-(alkylthio)phenyl-substituted oximes. This reaction, one of the standard synthetic methods, has been used extensively by a group at the University of Hull in their syntheses of substituted 1,2-benzisothiazoles.[8] A variation involves the formation of a seven-membered heterocyclic ring (**1**) which is then pyrolyzed to the 1,2-benzisothiazole (**2**).[9]

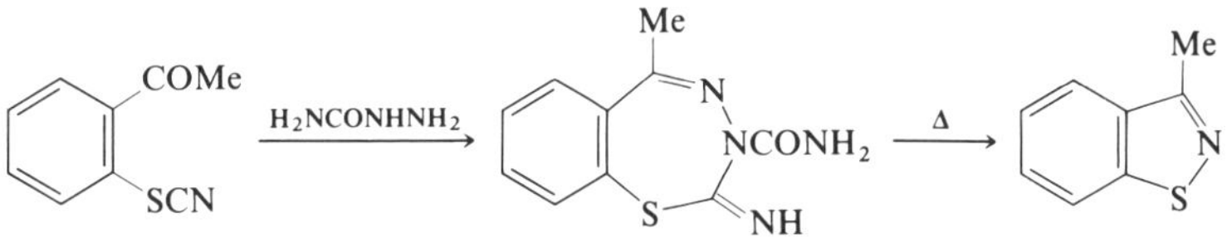

Meth-Cohn and colleagues[10] used the *tert*-butylthio derivatives **3** as useful synthons; the reaction of **3** (R = H) with hydroxylamine, followed by cyclization with polyphosphoric acid, gave good yields of 1,2-benzisothiazoles. However, in using the nitro-substituted derivative **3** (R = NO_2) the *tert*-butylthio group was cleaved too readily. The resulting thiol odor, being similar to the stenching agent used in town gas, "caused the evacuation of a school half a mile away." The group was "encouraged to discontinue" their experiments.[11]

Chloramine is a valuable alternative to hydroxylamine in these reactions and has been used to prepare 7-methoxy-1,2-benzisothiazole (**4**; R = MeO) in 70% yield from 2-mercapto-3-methoxybenzaldehyde.[12]

---

[7] A. J. Lawson, *Phosphorus Sulfur* **12**, 357 (1982).
[8] K. Clarke, C. G. Hughes, and R. M. Scrowston, *J.C.S. Perkin I*, 356 (1973).
[9] K. Clarke, B. Gleadhill, and R. M. Scrowston, *J. Chem. Res., Synop.*, 395 (1979).
[10] O. Meth-Cohn and B. Tarnowski, *Synthesis*, 58 (1978).
[11] L. K. A. Rahman and R. M. Scrowston, *J.C.S. Perkin I*, 385 (1984).
[12] L. K. A. Rahman and R. M. Scrowston, *J.C.S. Perkin I*, 2973 (1983).

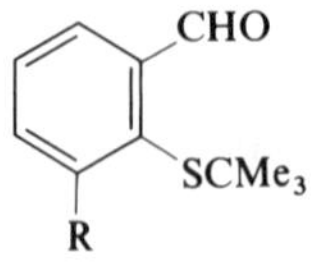

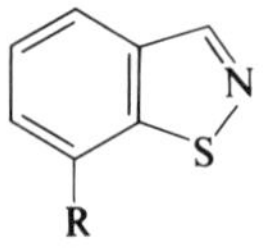

(3) (4)

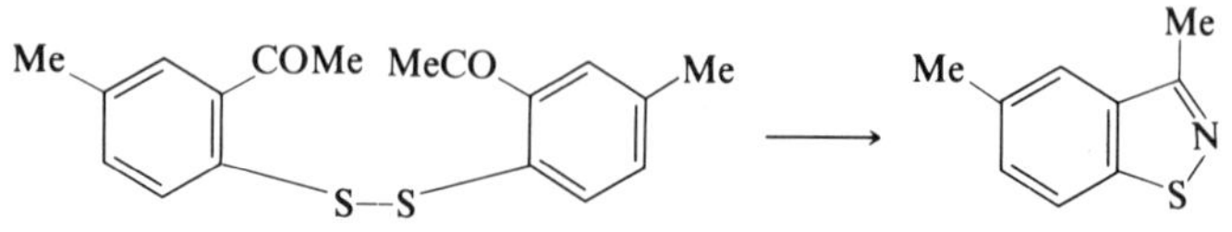

(5) (6)

The treatment of a disulfide with ammonia, silver nitrate, and a ketone is a general one-step synthesis of sulfenimines.[13] The disulfide **5**, in an intramolecular reaction, affords[13] the 1,2-benzisothiazole **6**.

2. From 2,2'-Dithiobis(benzonitrile)

Oxidation of 2,2'-dithiobis(benzonitrile) (**7**) with chlorine affords 3-chloro-1,2-benzisothiazole (**8**).[14] This product is a valuable source of other 1,2-benzisothiazoles.

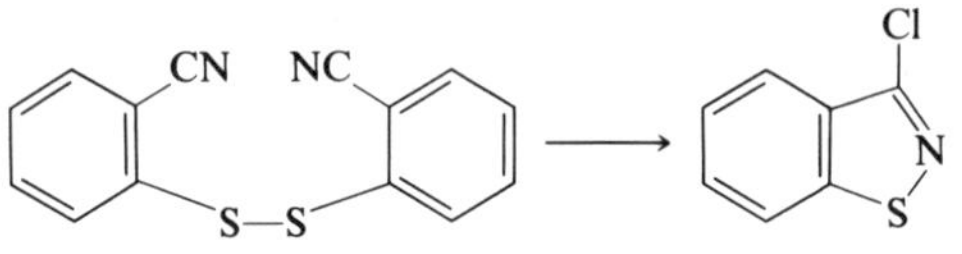

(7) (8)

3. From o-Chlorophenyl Derivatives with Sulfur and Ammonia

The previously described synthesis[1] from o-chlorobenzylidene chlorides has been extended to the use of o-chlorobenzonitriles,[15] o-chlorobenzaldehydes,[16] and o-chlorophenacyl compounds.[17] An example is the preparation

---

[13] F. A. Davis, W. A. R. Slegeir, S. Evans, A. Schwartz, D. L. Goff, and R. Palmer, *J. Org. Chem.* **38**, 2809 (1973).
[14] J. R. Beck and J. A. Yahner, *J. Org. Chem.* **43**, 1604 (1978).
[15] H. Fleig and H. Hagen, Ger. Offen. 2,609,864 (1977) [*CA* **88**, 6867 (1978)].
[16] H. Hagen and H. Fleig, Ger. Offen. 2,503,699 (1976) [*CA* **85**, 177401 (1976)].
[17] J. Markert and H. Hagen, *Liebigs Ann. Chem.* 768 (1980).

of 5-chloro-1,2-benzisothiazole (9) in 90% yield from 2,5-dichlorobenzaldehyde.[16]

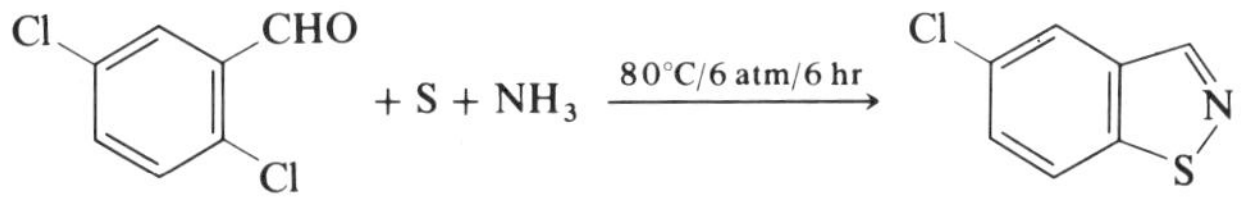

$$\text{2,5-dichlorobenzaldehyde} + S + NH_3 \xrightarrow{80°C/6\text{ atm}/6\text{ hr}} \text{(9)}$$

4. From Benzo[b]thiophen-3(2H)-ones and from Thiochromanones

A completely new synthesis of 1,2-benzisothiazoles, developed by a group in Japan, involves the use of o-mesitylenesulfonylhydroxylamine in the presence of base, as in the formation of the 1,2-benzisothiazole 10 from the thiochromanone 11. The benzo[b]thiophen-3(2H)one 12 is an equally good precursor.[18]

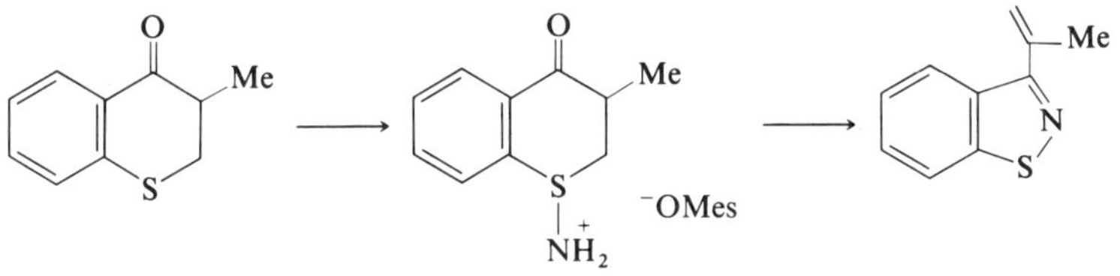

(11) → intermediate → (10)

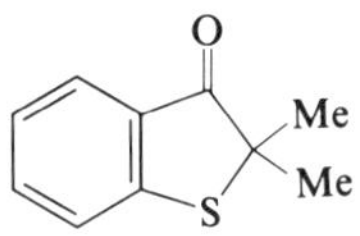

(12)

5. Addition of Benzyne to 1,2,5-Thiadiazole and 1,2,5-Selenadiazole

Benzyne adds to the 1,2,5-thiadiazole or 1,2,5-selenadiazole ring (13; Z = S or Se), yielding a 1,2-benzisothiazole, or benzisoselenazole, and other products.[19]

---

[18] Y. Tamura, S. M. Bayomi, C. Mukai, M. Ikeda, M. Murase, and M. Kise, *Tetrahedron Lett.* **21**, 533 (1980); Y. Tamura, S. M. M. Bayomi, C. Mukai, M. Ikeda, and M. Kise, *J.C.S. Perkin I*, 2830 (1980).

[19] M. R. Bryce, P. Hanson, and J. M. Vernon, *J.C.S. Chem. Commun.*, 299 (1982).

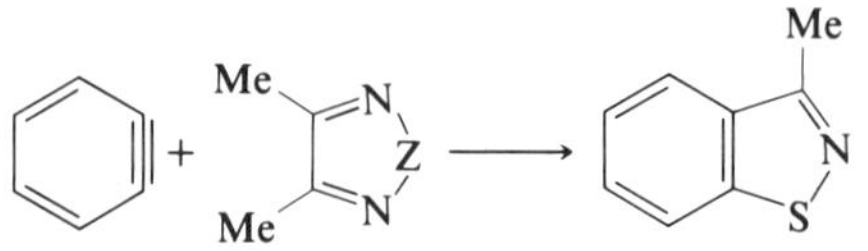

(13) (2)

6. Reported Formation from an N-Benzylsulfonamide and Trichloromethanesulfenyl Chloride

In the previous review[1] there was discussed a claim in the patent literature that a 1,2-benzisothiazoline could be produced from a reaction between an *N*-benzyl-substituted sulfonamide and trichloromethanesulfenyl chloride in the presence of base. Reexamination of this reaction has failed to confirm the production of benzisothiazolines.[20]

7. From 1H,3H-1,2-Benzisothiazoline 1-Oxides

1*H*,3*H*-1,2-Benzisothiazoline 1-oxides (14; X = O, NH) can be prepared from the ester (15; X = OMe) or the amide (15; X = NH$_2$).[21,22] If the benzyl chloride (16) is used, the product is 14, which as its hydrochloride yields 1,2-benzisothiazole (17) (94%) when heated to 155°C.[23]

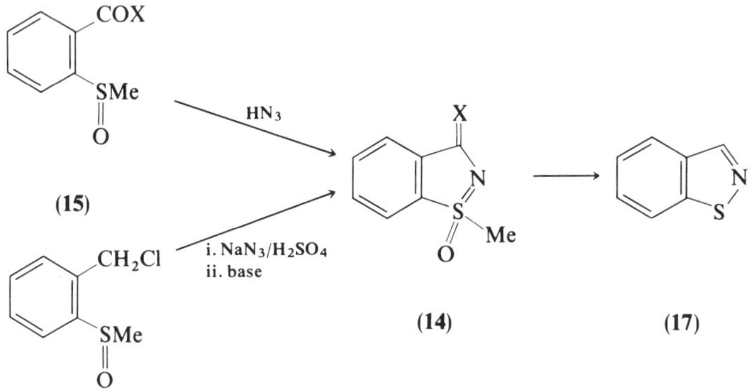

[20] M. Davis and E. Homfeld, *Aust. J. Chem.* **26**, 1365 (1973).
[21] P. Stoss and G. Satzinger, *Angew. Chem., Int. Ed. Eng.* **10**, 76 (1971); *Chem. Ber.* **108**, 3855 (1975).
[22] T. R. Williams and D. J. Cram, *J. Am. Chem. Soc.* **93**, 7333 (1971).
[23] R. H. Rynbrandt and D. P. Balgoyen, *J. Org. Chem.* **43**, 1824 (1978).

B. Physical Properties

Palmer[24] discussed the electronic structure of 1,2-benzisothiazole in the context of the other aza derivatives of indole, benzofuran, and benzothiophene. The infrared and Raman spectra of 1,2- and 2,1-benzisothiazoles have been recorded, and a complete assignment of the fundamental vibrations proposed.[25] The rates of N-methylation and the NMR chemical shifts of the resulting N-methyl salts of isothiazoles, pyrazoles, isoxazoles, and their benzofused derivatives have been examined.[26] The protonation equilibria of 1,2-benzisothiazoles have been measured.[27]

In the mass spectrum, breakdown of 1,2-benzisothiazoles usually results in loss of HCN and CS fragments[28]; there is some evidence that the molecular ions from 1,2-benzisothiazole and from benzothiazole possess a common structure.[29]

The crystal and molecular structures of 3-[(2-diethylammonio)ethoxy]-1,2-benzisothiazole tetrachlorocobaltate (**18**) have been determined. The N—S distance is 1.67 Å, indicating a small degree of multiple bonding; the N—S—C angle is a small 95.1°.[30] These results are typical of 1,2-benzisothiazoles.[31]

(**18**)

C. Chemical Properties

1. *Electrophilic Substitution*

1,2-Benzisothiazole undergoes nitration at its 5- and 7-positions.[1] Electrophilic substitution of 3-methyl-1,2-benzisothiazole (**2**) and its 5-amino, 5-

---

[24] M. H. Palmer and S. M. F. Kennedy, *J. Mol. Struct.* **43**, 203 (1978).
[25] T. El Jammal, M. Guiliano, J. D. Fourneron, and G. Mille, *Phosphorus Sulfur* **16**, 313 (1983).
[26] M. Davis, L. W. Deady, and E. Homfeld, *Aust. J. Chem.* **27**, 1221 (1974); *J. Heterocycl. Chem.* **11**, 1011 (1974).
[27] A. Braibanti and M. T. L. Mangia, *Gazz. Chim. Ital.* **111**, 71 (1981).
[28] A. Croisy, P. Jacquiqnon, R. Weber, and M. Renson, *Org. Mass Spectrom.* **6**, 1321 (1972).
[29] A. Selva and E. Gaetani, *Org. Mass Spectrom.* **7**, 327 (1973); A. Selva, U. Vettori, and E. Gaetani, *ibid.* **9**, 1161 (1974).
[30] A. C. Bonamartini, M. Nardelli, and C. Palmieri, *Acta Crystallogr. Sect. B* **B28**, 1207 (1972).
[31] E. Gaetani, T. Vitali, A. Mangia, M. Nardelli, ad G. Pelizzi, *J.C.S. Perkin II*, 2125 (1972).

acetamide, 5-hydroxy, 5-methoxy, 5-bromo, and 4-bromo derivatives have been examined in detail.[32] The 3-methyl group can be oxidized to an aldehyde by iodine in dimethyl sulfoxide, or to a carboxylic acid by thionyl chloride followed by alkali.[9] Electrophilic substitution of 1,2-benzisothiazole-3-acetic acid (19) has been studied.[33]

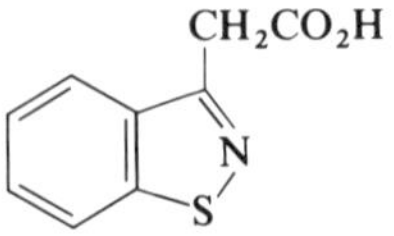

(19)

2. Reactions of 3-Chloro-1,2-benzisothiazole and Related Compounds

Nucleophilic displacement of the chlorine atom of 3-chloro-1,2-benzisothiazole has proved to be a popular procedure. Boeshagen and Geiger[34] have continued their earlier work on nitrogen nucleophiles, and now include carbon, oxygen, and sulfur nucleophiles.[35] In some cases, rearrangements occur, as in the formation of 3-amino-2-acylbenzo[b]thiophenes (20) from reaction of 21 with methyl ketones. Similar results are obtained from the reaction of other carbon nucleophiles, and it has been suggested that attack may be either at the 3-carbon or the sulfur atom.[36] The reaction of 3-chloro-1,2-benzisothiazole (8) with the anion of ethyl cyanoacetate, for example,

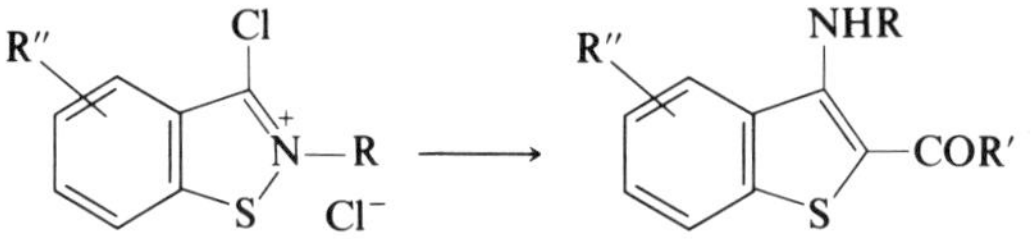

(21)   (20) 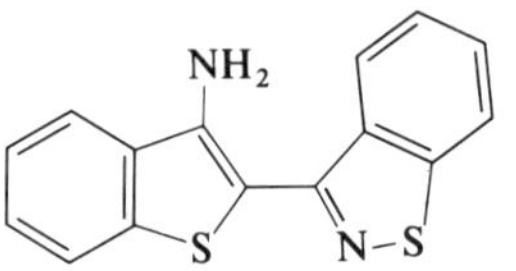   (22)

---

[32] K. Clarke, B. Gleadhill, and R. M. Scrowston, *J. Chem. Res. Synop.*, 197 (1980); *J. Chem. Res. Miniprint*, 2845 (1980).

[33] H. Uno and M. Kurokawa, *Chem. Phar. Bull.* **26**, 3888 (1978).

[34] H. Boeshagen and W. Geiger, *Chem. Ber.* **109**, 659 (1976); **112**, 3286 (1979); **113**, 2490 (1980).

[35] S. Watanabe, *Bull. Chem. Soc. Jpn* **42**, 1152 (1969); S. Huenig, G. Kiesslich, and H. Quast, *Justus Liebigs Ann. Chem.* **748**, 201 (1971); H. Boeshagen and W. Geiger, *ibid.* **764**, 58 (1972); *Chem. Ber.* **107**, 1667 (1974).

[36] D. E. L. Carrington, K. Clarke, and R. M. Scrowston, *Tetrahedron Lett.*, 1075 (1971); *J. Chem. Soc. C*, 3262 (1971); *ibid.*, 3903.

affords a 70% yield of the expected product and is a valuable route to other 3-substituted compounds.[37] A group at the University of Parma, Italy, has made considerable use of these and similar displacement reactions in the synthesis of biologically active compounds.[38] One interesting reaction is the formation of 3-(3-aminobenzo[b]thiophen-2-yl)-1,2-benzisothiazole (22) from the reaction of 3-chloro-1,2-benzisothiazole (8) with diethyl malonate in the presence of tetraalkylammonium salts. Structure 22 was confirmed by X-ray analysis.[39]

3. *Photochemistry and Photocycloaddition Reactions*

Photodecomposition of 1,2-benzisothiazole (17) affords the ring-opened disulfide 7.[40] If the 3-position is blocked by substitution, then more interesting products may result. Photocycloaddition reactions of 3-phenyl-1,2-benzisothiazole (23) with alkenes afford excellent yields of 2,3-dihydro-1,4-benzothiazepins such as 24.[41] With electron-rich alkynes, thiazabicycloheptadienes (25) are produced, but no reaction occurs with 2-butyne, 1-phenylpropyne, or dimethyl acetylenedicarboxylate.[42] An analogous S—N bond fission occurs with 2-aryl-1,2-benzisothiazolinones.[43]

(25a: R = OEt) (23) (24)
(25b: R = NEt$_2$)

[37] D. E. L. Carrington, K. Clarke, C. G. Hughes, and R. M. Scrowston, *J.C.S. Perkin I*, 3006 (1972).
[38] T. Vitali, P. Scrivani, G. Pellegrini, R. Ponci, F. Gialdi, and E. Arsura, Ger. Offen. 1,950,370 (1970) [*CA* **73**, 130991 (1971)]; M. Gianella, F. Gualtieri, and C. Melchiorre, *Phytochemistry* **10**, 539 (1971); T. Vitali, F. Mossini, R. M. Mingiardi, E. Gaetani, and V. Plazzi, *Ateneo Parmense, Acta Anat* **7**, 71 (1971) [*CA* **77**, 56391 (1972)], T. Vitali and L. Amoretti, *Boll. Soc. Ital. Biol. Sper.* **47**, 790 (1971) [*CA* **77**, 71265 (1972)]; C. Branca, D. Serafini-Fracossini, and N. Bagni, *Experientia* **30**, 105 (1974); P. V. Plazzi, M. Vitto, C. Branca, and L. Zappia, *Ateneo Parmense, Acta Nat.* **15**, 49 (1979) [*CA* **92**, 76378 (1980)]; P. V. Plazzi, M. Vitto, F. Bordi, and M. Chiavarini, *ibid.* **18**, 51 (1982) [*CA* **99**, 447 (1983)].
[39] F. Mossini, M. R. Mingiardi, E. Gaetani, M. Nardelli, and G. Pelizzi, *J.C.S. Perkin II*, 1665 (1979).
[40] M. Ohashi, A. Ezaki, and T. Yonezawa, *J.C.S. Chem. Commun.*, 617 (1974).
[41] M. Sindler-Kulyk, D. C. Neckers, and J. R. Blount, *Tetrahedron* **37**, 3377 (1981).
[42] M. Sindler-Kulyk and D. C. Neckers, *J. Org. Chem.* **48**, 1275 (1983).
[43] N. Kamigata, S. Hashimoto, S. Fujie, and M. Kobayashi, *J.C.S. Chem. Commun.*, 765 (1983).

114   MICHAEL DAVIS   [Sec. II.D

4. Miscellaneous Reactions Involving Ring Opening

Boeshagen and Geiger have shown that reduction of 3-dialkylamino-1,2-benzisothiazoles (26) with thiophenol affords a mercaptobenzamidine (27) which is in equilibrium with dipolar or 1,2-benzothioquinone methide (28).[44] Diazotization of 7-amino-1,2-benzisothiazole (29) yields 1,2,3-benzothiadiazole-7-carbaldehydes (30).[45]

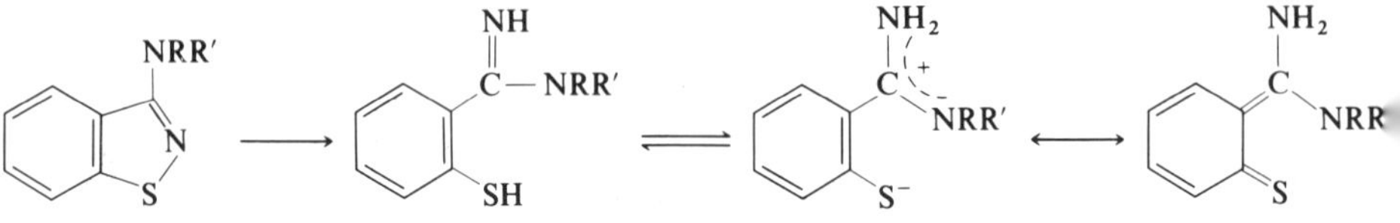

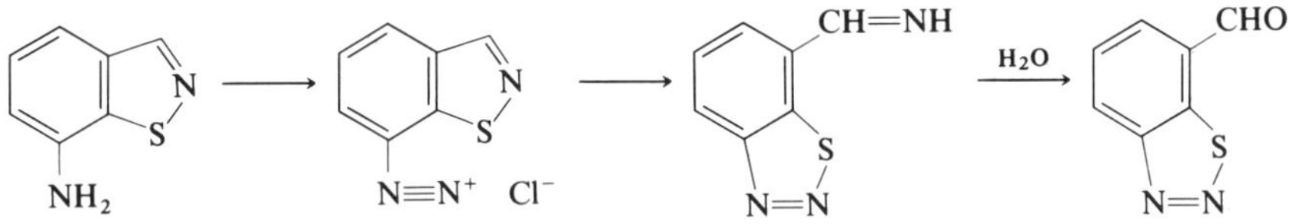

D. 1,2-BENZISOTHIAZOLIN-3(2H)-ONES

The crystal and molecular structure of 1,2-benzisothiazolin-3-one (31) have been determined. Both C—S and S—N bonds are single.[46]

Some novel methods of preparation have appeared. One of the most interesting is the Diels–Alder reaction of 4-isothiazolin-3-one 1-oxides (32;

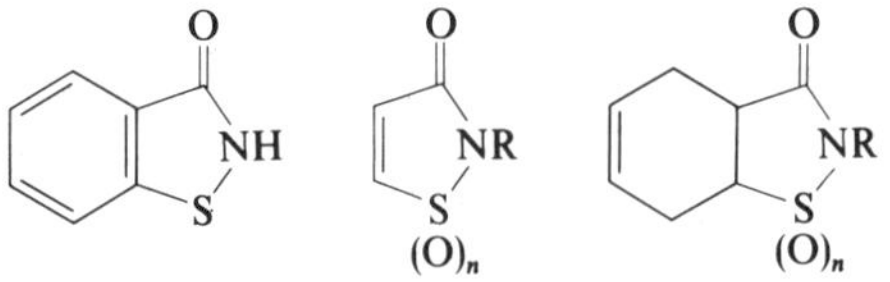

---

[44] H. Boeshagen and W. Geiger, *Liebigs Ann. Chem.*, 14 (1982).
[45] E. Haddock, P. Kirby, and A. W. Johnson, *J. Chem. Soc. C*, 3994 (1971).
[46] L. Cavalca, A. Gaetani, A. Mangia, and G. Pelizzi, *Gazz. Chim. Ital.* **100**, 629 (1970).

$n = 1$) and dioxides (**32**; $n = 2$), which afford reduced derivatives such as **33**.[47] Another is the use of thionyl chloride to cyclize 2-methylsulfenyl- or 2-methylsulfinylbenzamides (**34**; $n = 0, 1$) to 2-substituted 1,2-benzisothiazolinones (**35**) in quantitative yield.[48] If the diaryl sulfoxide (**36**) is used instead, then the interesting S(IV) derivative **37** is formed.[49] Workers in the Soviet Union have examined oxidative reactions of **31** and have also produced a number of sulfoxide derivatives.[50]

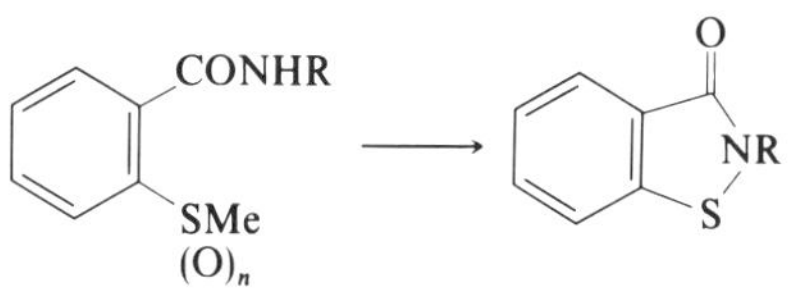

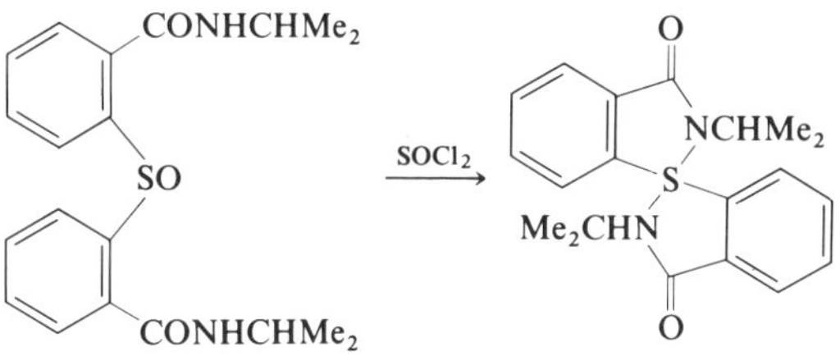

The nitrogen atom of **31** is readily alkylated. The 2-phenacyl derivative (**38**) rearranges to 2-benzoyl-2H-1,3-benzothiazin-4(3H)-one (**39**) in mild base,[51] illustrating once again the ease with which the S—N bond is broken.

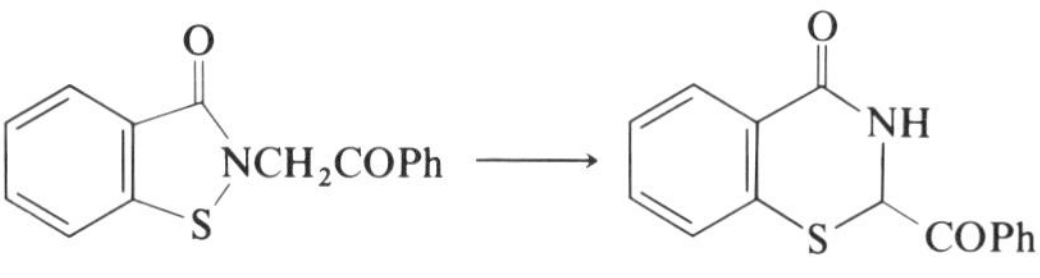

---

[47] E. D. Weiler and J. J. Brennan, *J. Heterocycl. Chem.* **15**, 1299 (1978).
[48] Y. Uchida and S. Kozuka, *J.C.S. Chem. Commun.*, 510 (1981); *Bull. Chem. Soc. Jpn.* **55**, 1183 (1982).
[49] L. J. Adzima, C. C. Chiang, I. C. Paul, and J. C. Martin, *J. Am. Chem. Soc.* **100**, 953 (1978).
[50] E. S. Levchenko and I. N. Berzina, *Zh. Org. Khim.* **6**, 2273 (1970); E. S. Levchenko and T. N. Dubinina, *ibid.* **14**, 862 (1978); V. N. Klyuev, A. B. Korzhenevskii, and B. D. Berezin, *Izv. Vyssh. Uchebn. Zaved., Khim. Khim. Tekhnol.* **21**, 31 (1978) [*CA* **88**, 170041 (1978)].
[51] J. C. Grivas, U.S. Patent 3,761,489 (1973) [*CA* **79**, 137126 (1973)]; *J. Org. Chem.* **40**, 2029 (1975); **41**, 1325 (1976).

E. APPLICATIONS

The biological activity of 1,2-benzisothiazole derivatives was reviewed in 1981 by De.[3] Since then, further claims have been made regarding the auxin-like activity of 3-acetic acid derivatives,[52] the analgesic activity of 3-benzyl derivatives,[53,54] the antithrombogenic activity of 5-chloro-2-alkyl compounds,[54,55] the fungicidal activity of 3-thio derivatives,[56,57] the antiinflammatory activity of 2-halophenyl derivatives,[58] and the diuretic effects[59] of 1,2-benzisothiazole-6-oxyacetic acid derivatives.

III. 1,2-Benzisothiazolin-3(2H)-one 1,1-Dioxide (Saccharin) and Derivatives

A. FORMATION

Few totally new syntheses have been reported. Autoclaving (165°C, 6 hr) of the sulfonyl chloride **40** with sulfur and ammonia gives[60] a moderate yield of saccharin (**41**). Photolysis of the oxime ether **42** affords N-alkoxysaccharins

(40)   (41)
              (30%)

---

[52] C. A. Maggiali, M. R. Mingiardi, M. T. Lugari Mangia, F. Mossini, and C. Branca, *Farmaco Ed. Sci.* **37**, 319 (1982) [*CA* **97**, 18952 (1982)]; C. Branca, C. A. Maggiali, M. R. Mingiardi, and D. Ricci, *J. Plant Growth Regul.* **1**, 243 (1982).
[53] F. Bordi, M. Vitto, P. V. Plazzi, and G. Morini, *Ateneo Parmense, Acta Nat.* **19**, 35 (1983) [*CA* **99**, 187589 (1983)]; G. Morini, M. Impicciatore, F. Bordi, P. V. Plazzi, and M. Vitto, *Farmaco, Ed. Sci.* **38**, 794 (1983) [*CA* **99**, 205623 (1983)].
[54] T. Vitali, M. Impicciatore, and P. V. Plazzi, *Farmaco, Ed. Sci.* **37**, 674 (1982) [*CA* **98**, 379 (1983)].
[55] Green Cross Corp., *Jpn. Kokai Tokkyo Koho* **58 177,915** (1983) [*CA* **100**, 56845 (1984)].
[56] H. Hagen, H. Ziegler, and E. H. Pommer, Ger. Offen. DE 3,202,298 (1983) [*CA* **99**, 175752 (1983)].
[57] P. Borgna, L. Vicarini, M. L. Carmellino, and G. Pagani, *Farmaco, Ed. Sci.* **38**, 801 (1983) [*CA* **100**, 19153 (1984)].
[58] A. Welter, H. H. Lauterschlagen, E. Etschenberg, and S. Leyck, European Patent Appl. EP 51,193 (1982) [*CA* **97**, 98360 (1982)].
[59] G. M. Shutske, R. C. Allen, M. F. Foersch, L. L. Setescak, and J. C. Wilker, *J. Med. Chem.* **26**, 1307 (1983).
[60] H. Feichtinger, *Chem. Ber.* **104**, 1697 (1971).

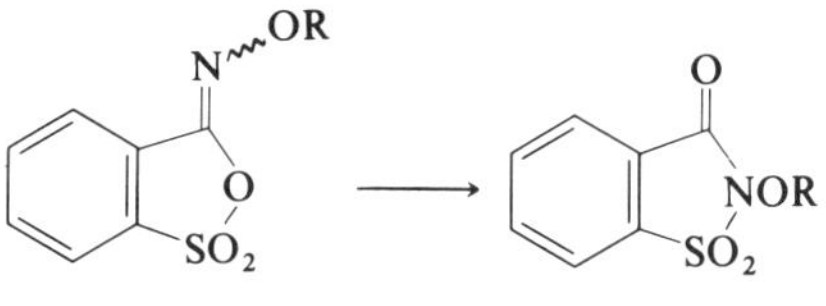

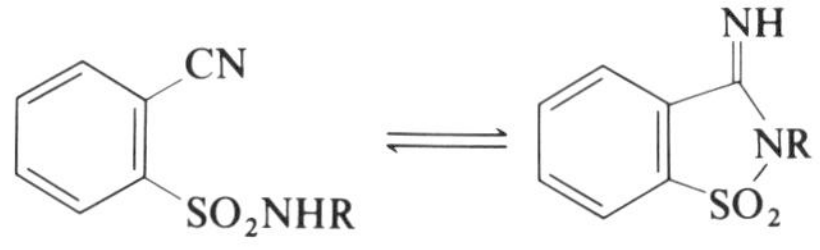

(43).[61] O-Cyanobenzenesulfonamides (44) are in equilibrium with 3-iminosaccharin (45), equilibrium being reached faster when R is an aryl group. Electron-donating groups in R push the equilibrium to the right.[62] Workers in Japan described ways of making saccharin in high purity, free of toluenesulfonamide impurities.[63]

B. Physical Properties

These were considered at length in the previous review.[2] A number of X-ray crystallographic structures of saccharin salts have been reported.[64]

C. Chemical Properties

3-Arylpseudosaccharin derivatives such as 46 are available from saccharin (41) by lithiation, followed by reaction with a halide[65] or by reaction between saccharin and a Grignard reagent or an organolithium compound[66]; a further

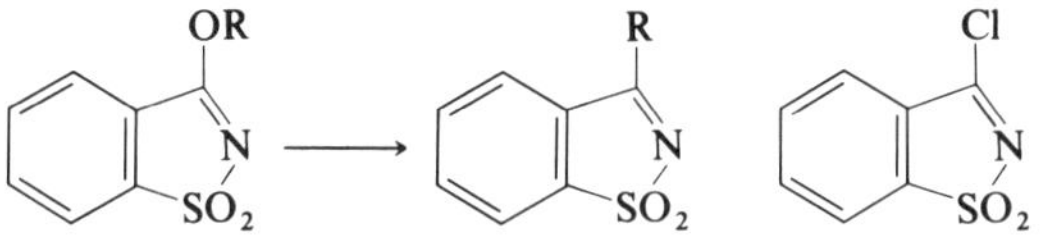

---

[61] L. A. Levy, *J. Heterocycl. Chem.* **8**, 873 (1971); *Tetrahedron Lett.*, 3289 (1972).
[62] D. Balode, R. Valtere, and S. Valtere, *Khim. Geterosikl. Soedin.*, 1632 (1978) [*CA* **90**, 120792 (1979)].
[63] W. Koike, T. Kimoto, and S. Matsui, Ger. Offen. 2,616,611 (1977) [*CA* **86**, 171437 (1977)]; Jpn. Kokai 77/71,464 (1977) [*CA* **87**, 168016 (1977)].
[64] N. Shimizu and S. Nishigaki, *Acta Crystallogr., Sect. B* **B38**, 1834 (1982); **C39**, 502 (1983).
[65] A. L. Borror, L. Circotta, E. W. Ellis, J. W. Foley, and M. M. Kampe, U.S. Patent 4,181,660 (1980) [*CA* **92**, 146747 (1980)].
[66] R. A. Abramovitch, E. M. Smith, M. Humber, B. Purtschert, P. C. Srinivasan, and G. M. Singer, *J.C.S. Perkin I*, 2589 (1974); P. C. Srinivasan, Ph. D. Thesis, University of Alabama (1975) [*CA* **85**, 192622 (1976)].

method is the heating of benzenesulfonamides that possess an aryl ketone moiety in the ortho position.[67] Pseudosaccharin ethers (**47**; R = alkyl) also react with phenyllithium to give 3-phenylpseudosaccharin (**46**, R = Ph). However, the nucleophilic displacement reaction of 3-chloropseudosaccharin (**48**) remains the method of choice for preparation of substituted saccharins, and recent work includes improved procedures for preparing **48** and related compounds.[68] Numerous examples have been given.[69]

Thermolysis[70] of saccharin itself yields the benzoxazinone **49**, which may be regarded as the dimer of the hypothetical imino ketene **50** produced by the elimination of sulfur dioxide from **41**. Flash vacuum pyrolysis of 3-arylpseudosaccharins (**46**; R = Ar) affords reasonable yields of 2-arylbenzoxazoles; 3-alkyl compounds (**46**; R = alkyl) yield aromatic nitriles only.[71]

(**49**)   (**50**)

Further examples of the formation of 3-aminopseudosaccharin derivatives (**46**; R = NHR′, NR′R″) from amines and saccharin have been reported.[72]

Ring expansion reactions of saccharin have proved fertile areas of investigation. Two distinct types of reaction are involved (Scheme 1). In the first, ring expansion can occur by base-catalyzed ring opening at the carbonyl

---

[67] O. B. T. Nielsen, H. Brunn, C. Bretting, and C. W. Feit, *J. Med. Chem.* **18**, 41 (1975).

[68] K. Enoki, T. Fukui, T. Yamamoto, T. Okada, and H. Iwajo, Jpn. Kokai 77/36,663 (1977) [*CA* **87**, 152177 (1977)]; M. Ito, T. Yamamoto, H. Iwashiro, T. Okada, and Y. Miyazaki, *Jpn. Kokai Tokkyo Koho* **78** 111,064 (1978) [*CA* **90**, 72172 (1979)]; P. C. Wade, B. R. Vogt, and T. P. Kissick, U.S. Patent 4,108,860 (1978) [*CA* **90**, 87473 (1979)]; P. C. Wade and T. P. Kissick, U.S. Patent 4,104,388 (1978) [*CA* **90**, 72174 (1979)]; U.S. Patent 4,104,387 (1978) [*CA* **90**, 87441 (1979)]; P. C. Wade and B. R. Vogt, U.S. Patent 4,178,451 (1979) [*CA* **92**, 110999 (1980)].

[69] A. Klemer, G. Uhlemann, S. Chahin, and M. N. Diab, *Liebigs Ann. Chem.*, 1943 (1973); N. Matsumura, Y. Otsuji, and E. Imoto, *Nippon Kagaku Kaishi*, 1532 (1974) [*CA* **82**, 4168 (1975)]; 1539 (1974) [*CA* **82**, 57595 (1975)]; G. L. Bachman, J. W. Baker, and D. P. Roman, *J. Pharm. Sci.* **67**, 1323 (1978).

[70] A. J. Barker and R. K. Smalley, *Tetrahedron Lett.*, 4629 (1971).

[71] R. A. Abramovitch and S. Wake, *J.C.S. Chem. Commun.*, 673 (1977).

[72] E. B. Pedersen and S. O. Lawesson, *Tetrahedron* **30**, 875 (1974); H. Kutlu, *Istanbul Univ. Eczacilik Fak. Mecm.* **11**, 254 (1975) [*CA* **87**, 68214 (1977)]; K. G. Jensen and E. B. Pedersen, *Z. Naturforsch., B: Anorg. Chem., Org. Chem.* **36B**, 1640 (1981); Y. Imai, H. Okunoyama, K. Hirata, and M. Ueda, *Nippon Kagaku Kaishi*, 111 (1982) [*CA* **96**, 104134 (1982)].

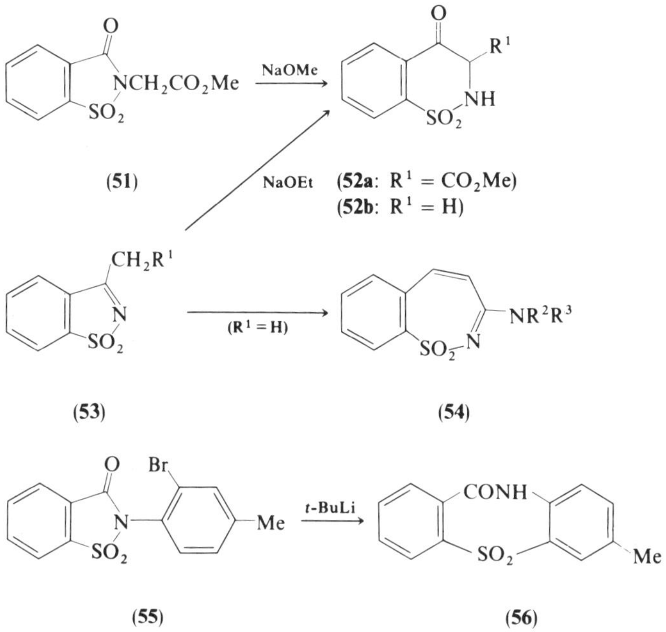

SCHEME 1

group, followed by a Dieckmann recyclization; this produces a 2H-1,2-benzothiazine-4(3H)-one 1,1-dioxide (**52**; R = CO_2Me).[73] A similar product (**52**, R = H) is produced[74] from 3-bromomethylpseudosaccharin (**53**, R' = Br). Reaction of 3-methylpseudosaccharin (**53**, R' = H) with a dialkylaminoacetylene produces a benzothiazepine dioxide (**54**) in which the seven-membered ring is not aromatic.[75] Thiadiazocines can be produced from 3(2-hydroxyethylamino)pseudosaccharin (**46**; R = $NHCH_2CH_2OH$) by reaction with thionyl chloride.[76] The second type of reaction, with rupture of

---

[73] J. D. Genzer and F. O. Fontsere, U.S. Patent 3,960,856 (1976) [*CA* **85**, 46728 (1976)]; A. C. Fabian, J. D. Genzer, C. F. Kasulanis, J. Shavel, and H. Zinnes, U.S. Patent 3,957,772 (1976) [*CA* **85**, 46725 (1976)]; U.S. Patent 3,978,073 (1976) [*CA* **86**, 16690 (1977)]; U.S. Patent 3,987,038 (1976) [*CA* **86**, 72679 (1977)]; H. Zinnes, N. A. Lindo, and J. Shavel, U.S. Patent 4,074,048 (1978) [*CA* **88**, 190868 (1978)]; C. B. Schapira, I. A. Perillo, and S. Lamdam, *J. Heterocycl. Chem.* **17**, 1281 (1980); I. A. Perillo, C. B. Schapira, and S. Lamdam, *ibid.* **20**, 155 (1983); W. Herrmann, W. Geibel, and G. Satzinger, Ger. Offen. DE 3,212,485 (1983) [*CA* **100**, 22665 (1984)].

[74] R. A. Abramovitch, K. M. More, I. Shinkai, and P. C. Srinivasan, *J.C.S. Chem. Commun.*, 771 (1976).

[75] R. A. Abramovitch, B. Mavunkel, and J. R. Stowers, *J.C.S. Chem. Commun.*, 520 (1983).

[76] J. Ashby, D. Griffiths, and D. Paton, *J. Heterocycl. Chem.* **15**, 1009 (1978).

the S—N bond, is exemplified by the base-catalyzed ring expansion of the N-arylsaccharin **55** to the dibenzothiazepine dioxide **56**.[77]

D. BIOLOGICAL PROPERTIES

The long-standing controversy over the alleged human carcinogenicity of saccharin (**41**) intensified during the 1970s. In 1977 the United States Food and Drug Administration announced that, as a result of Canadian tests showing that the compound (at massive dosage levels) induced bladder cancer in rats, it intended to prohibit sale for human usage of saccharin. The proposed ban was deferred by the United States Congress. Two excellent and authoritative articles on the controversy are available. The first[6] deals with the specific issue of saccharin, the second[78] is concerned with the practicality of the 1958 Delaney clause, which in United States legislation requires the banning from food and beverages of any additive found carcinogenic in man or in animals.

Neither saccharin nor any of the usual impurities present in commercial material is active in the Ames mutagenicity test.[79] The metabolism of saccharin,[80] of saccharin impurities,[81] and of other saccharin derivatives[82] has been studied.

Claims have been made for the fungicidal activity of saccharin itself[83] and of various halogeno derivatives,[84] for the diuretic action of 3-arylpseudosaccharin compounds[85,86] including 3-(o-fluorophenyl)pseudosaccharin derivatives,[59] for hypolipodemic activity of N-propionic and N-valeric acid derivatives,[87] and for gastric acid inhibition by 3-aminopseudosaccharin

---

[77] D. Hellwinkel, R. Lenz, and F. Laemmerzahl, *Tetrahedron* **39**, 2073 (1983).

[78] W. Lijinsky, F. Coulston, S. M. Wolfe, and J. G. Martin, *Chem. Eng. News*, 27 June, pp. 24–46 (1977).

[79] J. Ashby, J. A. Styles, D. Anderson, and D. Paton, *Food Cosmet. Toxicol.* **16**, 95 (1978) [*CA* **89**, 145167 (1978)].

[80] A. G. Renwick and R. T. Williams, *Xenobiotica* **8**, 475 (1978).

[81] A. G. Renwick, *Xenobiotica* **8**, 487 (1978).

[82] M. Uchiyama, H. Abe, R. Sato, M. Shimura, and T. Watanabe, *Agric. Biol. Chem.* **37**, 737 (1973); T. Koeda, M. Odaki, T. Hisamatsu, H. Sasaki, M. Yokota, T. Niizato, S. Uchida, and T. Watanuki, *Oyo Yakuri* **13**, 205 (1977) [*CA* **88**, 99161 (1978)]; R. Becker, E. Frankus, I. Graudums, W. A. Guenzler, F. C. Helm, and L. Flohe, *Arzneim.-Forsch.* 1101 (1982).

[83] Y. Miyashiro, K. Fujishima, F. Araki, T. Hattori, and T. Okada, Jpn. Kokai 77/105,216 (1977) [*CA* **88**, 17320 (1978)].

[84] I. Chiyomaru, E. Yoshinagu, and H. Ito, Jpn. Kokai 74/14,466 (1974) [*CA* **80**, 108506 (1974)]; F. G. Bollinger and J. J. D'Amico, U.S. Patent 4,006,007 (1977) [*CA* **87**, 5948 (1977)]; J. Drabek, European Patent Appl. EP 86,748 (1983) [*CA* **99**, 175753 (1983)]; R. J. Thiessen, U.S. Patent 4,410,353 (1983) [*CA* **100**, 34533 (1984)].

[85] O. B. T. Nielsen, C. K. Nielsen, and P. W. Feit, *J. Med. Chem.* **16**, 1170 (1973).

[86] P. W. Feit, O. B. T. Nielsen, and N. Rastrup-Andersen, *J. Med. Chem.* **16**, 127 (1973).

[87] J. M. Chapman, G. H. Cocolas, and I. H. Hall, *J. Med. Chem.* **26**, 243 (1983).

compounds.[88] N-Acylsaccharins are elastase inhibitors,[89] and the acetic acid derivative (57) is a potent sedative, hypnotic, and anticonvulsant compound.[90]

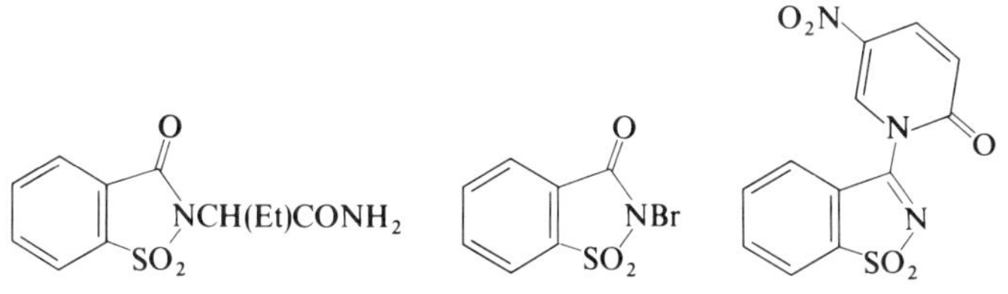

(57)   (58)   (59)

N-Bromosaccharin (NBSac, 58) has been proposed as an alternative to N-bromosuccinimide; in some instances it gives better yields in benzyl brominations.[91] 3-Substituted pseudosaccharin derivatives such as 59 may be used as condensing agents for peptide or ester synthesis.[92] The sodium salt of thiosaccharin (60) reacts with alkyl halides to give the 3-alkylthiopseudosaccharin (61), an odorless, crystalline thiolequivalent. The thiol may be liberated by reaction with piperidine.[93]

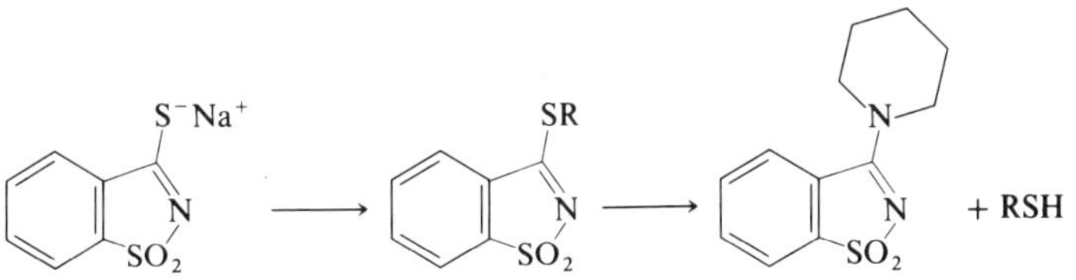

(60)   (61)

E. RELATED COMPOUNDS

Thiophenesaccharin (62) and its isomers (63 and 64) are about twice as sweet as saccharin and lack any bitter aftertaste.[94] Reduced compounds, such as 33, have already been mentioned.[47]

[88] G. A. Schiehser and D. P. Strike, European Patent Appl. EP 81,955 (1983) [CA 99, 158407 (1983)].
[89] D. Mulvey, H. Jones, and M. Zimmerman, Ger. Offen. 2,636,599 (1977) [CA 87, 102315 (1977)].
[90] C. M. Svahn and N. A. Jonsson, Acta Chem. Scand., Ser. B B32, 137 (1978).
[91] E. I. Sánchez and M. J. Fumarola, Synthesis, 736 (1976); J. Org. Chem. 47, 1588 (1982).
[92] A. Ahmed, H. Fukuda, K. Inomata, and H. Kotake, Chem. Lett., 1161 (1980); M. Ueda, N. Kawaharasaki, and Y. Imai, Synthesis, 933 (1982).
[93] K. Inomata, H. Yamada, and H. Kotake, Chem. Lett., 1457 (1981); H. Yamada, H. Kinoshita, K. Inomata, and H. Kotake, Bull. Chem. Soc. Jpn. 56, 949 (1983).
[94] O. Hromatka and D. Binder, Ger. Offen. 2,534,689 (1976) [CA 85, 5612 (1976)]; P. A. Rossy, W. Hoffmann, and N. Mueller, J. Org. Chem. 45, 617 (1980).

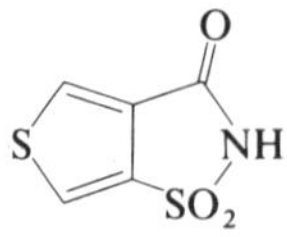

122   MICHAEL DAVIS   [Sec. IV.A

(62) 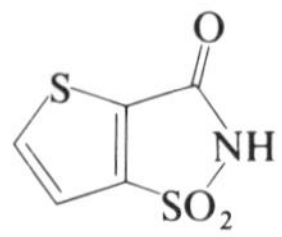   (63) 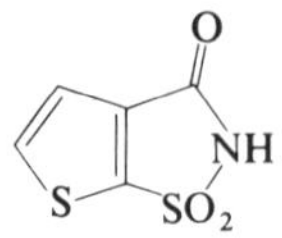   (64)

IV. 2,1-Benzisothiazoles

A. FORMATION

1. *From the Corresponding Benzisoxazole*

Reaction of 2,1-benzisoxazoles (65) with phosphorus pentasulfide affords the benzisothiazole 66.[95-97] The yields are increased by the addition of imidazole or of 2-methylimidazole.[98]

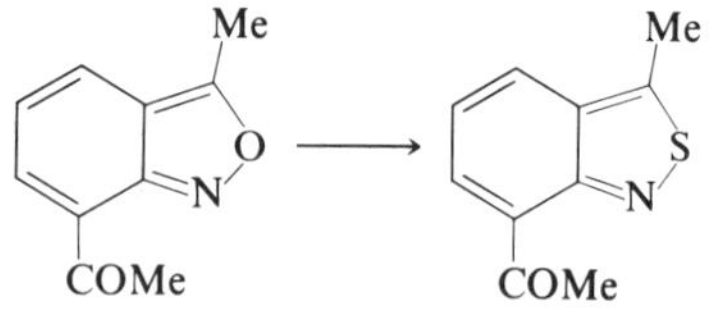

(65)   (66)

2. *By Hydrogen Peroxide Oxidation of o-Aminothiobenzamides*

Further examples of the hydrogen peroxide oxidation of *o*-aminothiobenzamides have been reported.[99] It is a reliable and reasonably versatile method, especially because the resulting 3-amino-2,1-benzisothiazoles can be diazotized and thus converted to a range of other 3-substituted compounds.[100]

---

[95] D. M. McKinnon and J. Y. Wong, *Can. J. Chem.* **49**, 2018 (1971).
[96] O. Aki, Y. Nakagawa, and K. Sirakawa, *Chem. Pharm. Bull.* **20**, 2372 (1972).
[97] M. S. Chauhan and D. M. McKinnon, *Can. J. Chem.* **53**, 1336 (1975).
[98] O. Aki, K. Nakagawa, M. Yamamoto, and K. Sirakawa, Japanese Patent 73/08,098 (1973) [*CA* **79**, 78779 (1973)].
[99] J. Norek, *Barwniki—Srodki Pomocnicze* **21**, 75 (1977) [*CA* **88**, 154310 (1978)]; J. Gray and D. R. Waring, *J. Heterocycl. Chem.* **17**, 65 (1980).
[100] M. Davis and K. S. L. Srivastava, *Curr. Sci.* **40**, 351 (1971); R. K. Buckley, M. Davis, and K. S. L. Srivastava, *Aust. J. Chem.* **24**, 2405 (1971).

(67)   (68a: R = R' = H)   (69)
         (68b: R = Cl, R' = H)
         (68c: R = R' = Cl)

$CH_3SO_2N{=}S{=}O$

3. From the Reaction of Thionyl Chloride or Related Compounds with o-Toluidines

Onaka and Oikawa[101] made an extensive study of the reaction between thionyl chloride and o-toluidine (67). They showed that, in addition to 1,2-benzisothiazole and 5-chlorobenzisothiazole,[102] 3-chloro- and 3,5-dichloro-2,1-benzisothiazole were formed in considerable quantity. The reaction of o-ethylaniline with thionyl chloride affords, after workup, a good yield of 2,1-benzisothiazole-3-carboxylic acid (68, R = CO_2H, R' = H) together with the 5-chloro acid (68, R = CO_2H, R' = Cl).[103,104] 3-Methyl-2,1-benzisothiazole, a presumed intermediate, is rapidly oxidized by thionyl chloride under the conditions of the reaction.[103]

Singerman[104,105] made a significant advance in the usefulness of this reaction by replacing thionyl chloride with N-sulfinylmethanesulfonamide (69). Singerman's reagent (69) is easily obtained from methanesulfonamide and thionyl chloride. By use of this reagent, any aromatic or heteroaromatic compound containing adjacent methyl and amino groups may have a fused isothiazole ring grafted onto it. Better yields are usually obtained than by the thionyl chloride procedure, and chlorinated by-products are not formed.[106]

4. From Other Nitrogen–Sulfur Transfer Reagents

Thermolysis of N-thiosulfinyl-2,4-di-*tert*-butyl-6-methylaniline (70) affords reasonable yields of the 2,1-benzisothiazole (71). The same product is formed

---

[101] T. Onaka and T. Oikawa, *Itsuu Kenkyusho Nempo*, 53 (1971) [*CA* 77, 48320 (1972)].
[102] M. Davis and A. W. White, *J.C.S. Chem. Commun.*, 1547 (1968); M. Davis, E. Homfeld, and T. G. Paproth, *Org. Prep. Proced. Int.* 5, 197 (1973).
[103] M. Davis, T. G. Paproth, and L. J. Stephens, *J.C.S. Perkin I*, 2057 (1973).
[104] G. M. Singerman, U.S. Patent 3,890,340 (1975).
[105] G. M. Singerman, *J. Heterocycl. Chem.* 12, 877 (1975).
[106] G. M. Singerman, U.S. Patent 3,929,815 (1975); U.S. Patent 3,997,548 (1976).

by irradiation of **70**, followed by heating to reflux of a benzene solution of the intermediate sulfur diimide compound **72** (R = Ar) then formed.[107]

A similar intermediate (**72**, R = Ph) can be used in an unusual synthesis of 3-cyano-2,1-benzisothiazole (**73**) in 67% yield from chlorocyanoketene (**74**).[108]

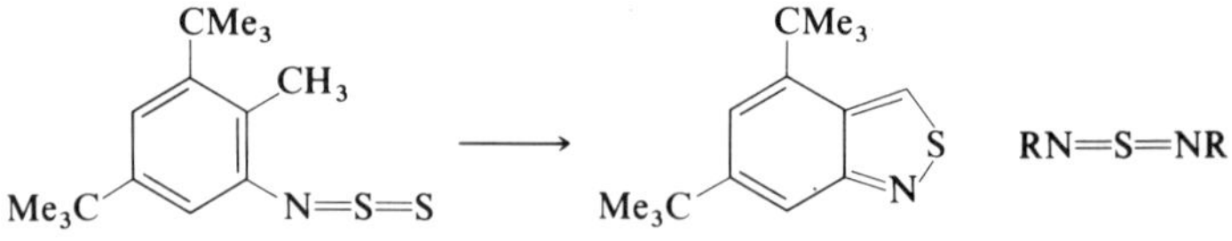

(**70**)   (**71**)   RN=S=NR (**72**)

(**74**) Cl(NC)C=C=O + **72** (R = Ph) ⟶ (**73**)

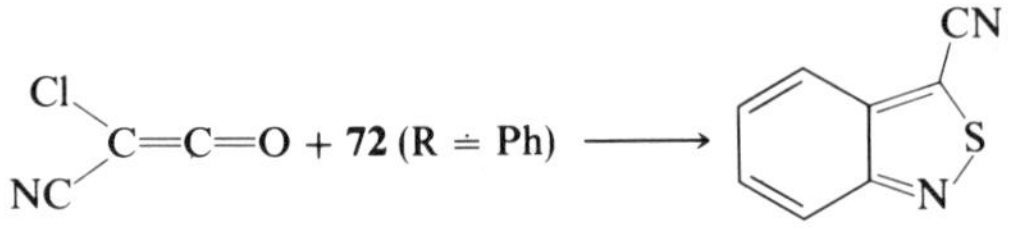

A related compound, the bis(trimethylsilyl)sulfur diimide (**75**) reacts with the lithiated silylamine **76** to afford a 70% yield of the benzisothiazole **77**.[109]

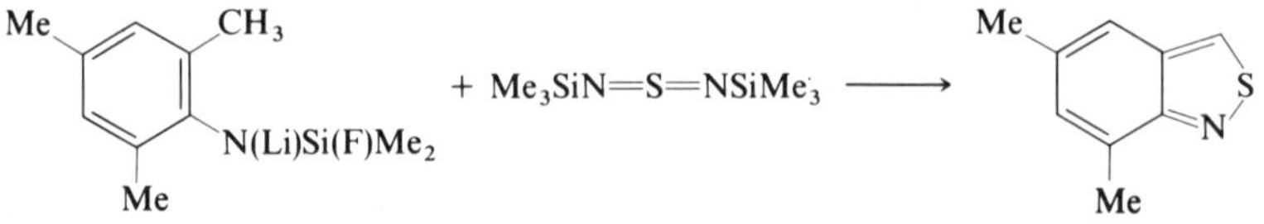

(**76**) + Me$_3$SiN=S=NSiMe$_3$ ⟶ (**77**)

(**75**)

5. Cycloaddition Reactions of Triphenylthieno[3,4-c]isothiazole

Acetylenes add to the 1,3-dipolar system of triphenylthieno[3,4-c]isothiazole (**78**), giving an unstable bridged intermediate, which eliminates sulfur and forms the 2,1-benzisothiazole **79**.[110]

---

[107] Y. Inagaki, R. Okazaki, and N. Inamoto, *Tetrahedron Lett.*, 293 (1977).
[108] D. M. Goldfish, B. W. Axon, and H. W. Moore, *Heterocycles* **20**, 187 (1983).
[109] U. Klingebiel and D. Bentmann, *Z. Naturforsch., B: Anorg. Chem., Org. Chem.* **34B**, 123 (1979); see also M. Davis, *ibid.* **35B**, 405 (1980).
[110] H. Gotthardt and F. Reiter, *Tetrahedron Lett.*, 2163 (1976); *Chem. Ber.* **112**, 266 (1979).

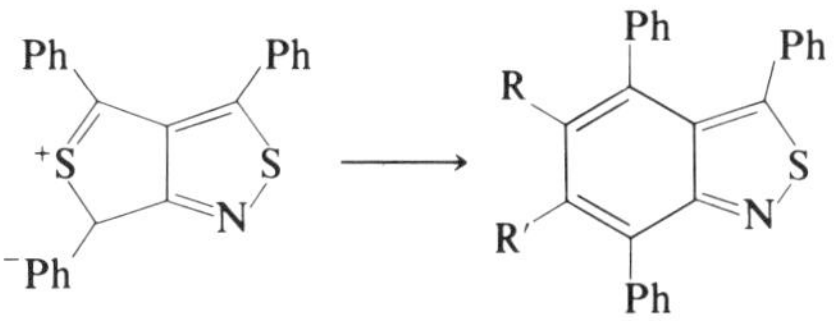

6. Preparation of 2,1-Benzisothiazole 2-Oxides and 2,2-Dioxides

Derivatives of 2,1-benzisothiazole 2-oxide (**80**) can be prepared by oxidation of *o*-aminophenylmethylthiolmethane (**81**) with *N*-chlorosuccinimide and alkali, followed by further oxidation with *m*-chloroperbenzoic acid.[111]

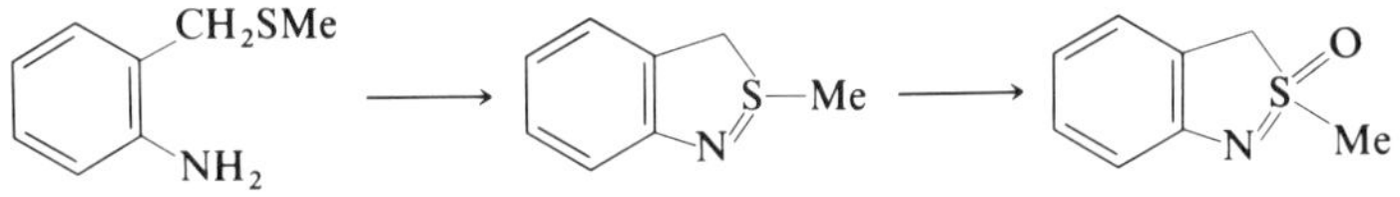

The sulfonamide **82** is readily cyclized by potassium amide in liquid ammonia, affording 1-methyl-2,1-benzisothiazoline 2,2-dioxide (**83**).[112–114]

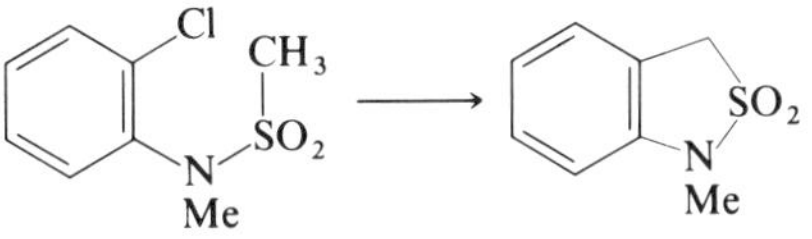

B. Physical Properties

The crystal and molecular structures of 5-chloro-2,1-benzisothiazole have been determined.[115] The C—S and S—N bond distances are 1.664 and

---

[111] T. E. Jackson, U.S. Patent 4,031,227 (1977) [*CA* **87**, 102318 (1977)].
[112] J. F. Bunnett, T. Kato, R. R. Flynn, and J. A. Skorcz, *J. Org. Chem.* **28**, 1 (1963).
[113] J. A. Skorcz, J. T. Suh, and C. I. Judd, U.S. Patent 3,528,989 (1970) [*CA* **73**, 130993 (1970)].
[114] J. A. Skorcz, J. T. Suh, and C. I. Judd, U.S. Patent 3,560,512 (1971) [*CA* **74**, 141766 (1971)];
J. A. Skorcz and J. T. Suh, *J. Heterocycl. Chem.* **9**, 219 (1972); J. A. Skorcz, J. T. Suh, and R. E. Germershausen, *ibid.* **10**, 249 (1973); **11**, 73 (1974).
[115] M. Davis, M. F. Mackay, and W. A. Denne, *J.C.S. Perkin II*, 565 (1972).

1.636 Å, respectively, indicating significant double-bond characteristics in each; the C—S—N angle is 97.7°, rather greater than in 1,2-benzisothiazoles. Proton[1,26,115] and ^{13}C-NMR[116] data have been reported for 2,1-benzisothiazole and substituted derivatives. Infrared and Raman assignments have been made for 2,1-benzisothiazole,[25] and the electronic structure of this compound has been discussed.[117]

C. CHEMICAL PROPERTIES

1. Diels–Alder Addition Reactions

2,1-Benzisothiazoles are not good Diels–Alder substrates; in early experiments only 3-alkylamino-2,1-benzisothiazoles (**84**) gave any evidence of reaction with maleic anhydride.[118] However, prolonged reaction (10 days, 90°C) of 2,1-benzisothiazole (**68**, R = R′ = H) with acetylenedicarboxylate and propiolate esters produces substituted quinolines (**85**).[119]

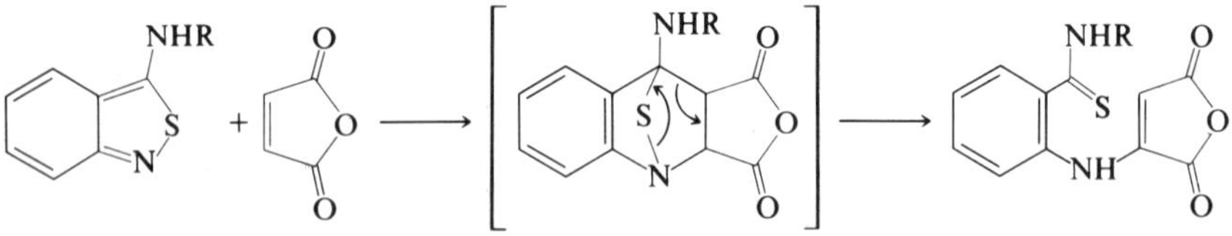

(**84**) R = Me, Et

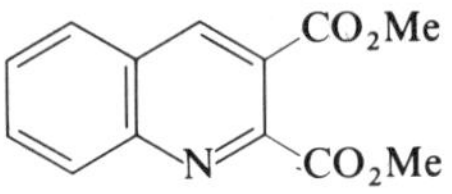

(**85**)

2. Photochemistry and Thermolysis

The photodecomposition of 2,1-benzisothiazoles removes the sulfur atoms. In methanol, o-aminoaryl aldehydes or o-aminoaryl ketones are formed,[40,120]

---

[116] N. Plavac, I. W. J. Still, M. S. Chauhan, and D. M. McKinnon, *Can. J. Chem.* **53**, 836 (1975).

[117] E. Chacko, J. Bornstein, and D. J. Sardella, *J. Am. Chem. Soc.* **99**, 8248 (1977); M. H. Palmer and S. M. F. Kennedy, *J. Mol. Struct.* **43**, 33 (1978).

[118] M. Davis and K. S. L. Srivastava, *J.C.S. Perkin I*, 935 (1972).

[119] M. R. Bryce, R. M. Acheson, and A. J. Rees, *Heterocycles* **20**, 489 (1983).

[120] B. Jackson, H. Schmidt, and H.-J. Hansen, *Helv. Chim. Acta* **62**, 391 (1979).

and in acetonitrile a dimer is produced.[40] Thionitroso (N=S) compounds are proposed intermediates.[121]

3. *Decomposition of 2,1-Benzisothiazolium Quaternary Salts*

2,1-Benzisothiazole and its derivatives readily form quaternary salts on treatment with alkyl or benzyl halides. These salts (**86**) are useful synthetic intermediates; the reactions they undergo are summarized in Scheme 2. The products include 2,1-benzisothiazolines,[96] quinolones,[122,123] benzodiazepinones (including a one-step synthesis of the tranquilizer Valium),[96,124] *o*-aminobenzaldehydes and *o*-aminoaryl ketones,[122,125] 2,1-benzisothiazole-3-thiones,[122] and cyanine dyes,[126] the latter resulting from the activation of the methyl group in the 3-position of the quaternary salt (**86**, R = R′ = Me).

R = H, alkyl, aryl
R^1 = alkyl, benzyl
R^2 = H, Cl, NO_2

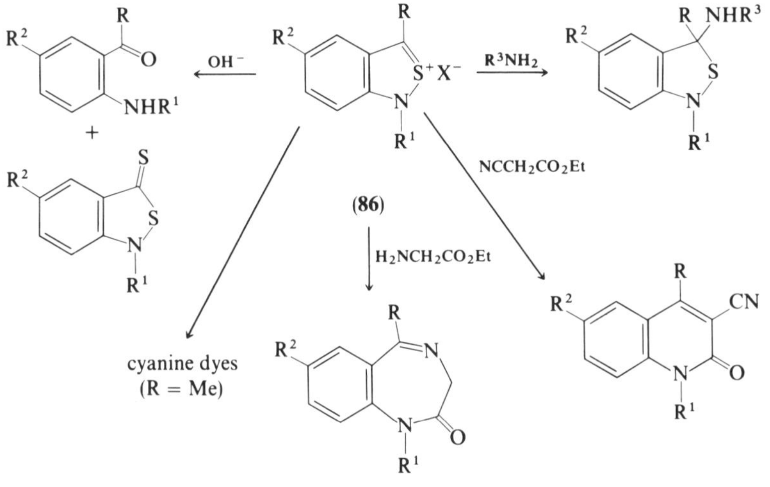

SCHEME 2

[121] M. F. Joucla and C. W. Rees, *J.C.S. Chem. Commun.*, 374 (1984).
[122] D. M. McKinnon, K. A. Duncan, and L. M. Millar, *Can. J. Chem.* **60**, 440 (1982).
[123] M. Davis and M. J. Hudson, *J. Heterocycl. Chem.* **20**, 1707 (1983).
[124] K. Shirakawa, O. Aki, K. Nagakawa, and M. Yamamoto, Japanese Patent 73/08, 097 (1973) [*CA* **79**, 92197 (1973)].
[125] M. Davis, E. Homfeld, and K. S. L. Srivastava, *J.C.S. Perkin I*, 1863 (1973).
[126] N. F. Haley, *J. Org. Chem.* **43**, 1233 (1978).

4. Reactions of 3-Chloro-2,1-benzisothiazole

3-Chloro-2,1-benzisothiazole (**68**, R = Cl, R' = H), which can be prepared from the 3-amino compound by a Sandmeyer reaction[100] or from 2,1-benzisothiazolin-3(1H)-one by treatment with phosphoryl chloride,[127] is similar to its 1,2 isomer (**8**) in that it is readily attacked by nitrogen, oxygen, sulfur, and carbon nucleophiles. This provides a good route to numerous 3-substituted 2,1-benzisothiazoles.[128–130]

D. 2,1-BENZISOTHIAZOLIN-3(1H)-ONE

1. Preparation of 2,1-Benzisothiazolin-3(1H)-ones

Treatment of isatoic anhydride (**87**) with sodium hydrosulfide solution yields[127] the unstable thioanthranilic acid **88**, which can be oxidized by hydrogen peroxide to 2,1-benzisothiazolin-3(1H)-one (**89**).[127,131,132] 2,1-Benzisothiazolin-3(1H)-thiones can be produced by the action of sodium sulfide on 2-aminobenzotrifluoride; the probable intermediate is the anion of dithioanthranilic acid.[133]

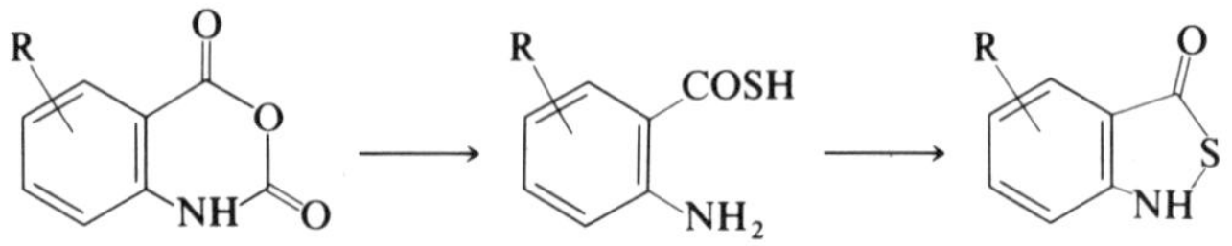

(**87**)   (**88**)   (**89**)

2. Reactions of 2,1-Benzisothiazolin-3(1H)-ones

2,1-Benzisothiazolin-3(1H)-one and substituted derivatives exist predominantly in the keto form **89**. Electrophilic substitution takes place readily, affording 5-substituted compounds. Like its 1,2 isomer, **89** is acidic and

---

[127] A. H. Albert, R. K. Robins, and D. E. O'Brien, *J. Heterocycl. Chem.* **10**, 413 (1973).
[128] M. Davis, L. W. Deady, E. Homfeld, and S. P. Pogany, *Aust. J. Chem.* **28**, 129 (1975).
[129] M. Davis, E. Homfeld, J. McVicars, and S. P. Pogany, *Aust. J. Chem.* **28**, 2051 (1975).
[130] A. H. Albert, D. E. O'Brien, and R. K. Robins, *J. Heterocycl. Chem.* **15**, 529 (1978).
[131] H. Boeshagen and W. Geiger, *Chem. Ber.* **106**, 376 (1973).
[132] J. Faust and R. Mayer, *J. Prakt. Chem.* **318**, 161 (1976).
[133] G. P. Jourdan and B. A. Dreikorn, *J. Org. Chem.* **47**, 5255 (1982).

dissolves readily in dilute base. Alkylation of the anion occurs exclusively at the nitrogen atom.[128]

Treatment of **89** (R = H) with acetic anhydride and pyridine brings about sulfur elimination and a rapid rearrangement to produce the benzoxazinone (**90**, R = H, R' = Me). A similar, slower reaction occurs with benzoyl chloride.[134] The latter reaction is autocatalytic, the "catalyst" being low-molecular-weight sulfur produced by the elimination process.[135] Similar elimination reactions take place when **89** (R = H) is treated with isocyanates.[136] Triethyl phosphite will also bring about sulfur elimination from **89** (R = H), affording the bright yellow benzoxazinone (**49**); in pyridine solution pyracridone (**91**) is also formed. Remarkably, **91** can be produced simply by boiling a solution of **89** (R = H) in a mixture of pyridine and water.[137] The iminoketene **50** is apparently not an intermediate in these reactions.[136,137]

E. APPLICATIONS

Numerous azo dyes derived from 3-amino-2,1-benzisothiazole have been claimed.[138] The University of Parma group has continued to develop auxin analogues from derivatives of 2,1-benzisothiazole-3-carboxylic acid,[139]

---

[134] M. Davis and S. P. Pogany, *J. Heterocycl. Chem.* **14**, 267 (1977).
[135] M. Davis and K. C. Tonkin, *Aust. J. Chem.* **34**, 755 (1981).
[136] J. Perronet and L. Taliani, *J. Heterocycl. Chem.* **17**, 673 (1980).
[137] M. Davis, R. J. Hook, and W. Y. Wu, *J. Heterocycl. Chem.* **21**, 369 (1984).
[138] R. Niess, Ger. Offen. 2,412,975 (1975) [*CA* **84**, 46061 (1976)]; R. J. Maner and M. A. Weaver, U.S. Patent 3,943,121 (1976) [*CA* **84**, 152215 (1976)]; G. Hansen, H. Kaack, and N. Grund, Ger. Offen. 2,507,460 (1976) [*CA* **86**, 157011 (1977)]; W. Deucker and R. Loewenfeld, Ger. Offen. 2,524,481 (1976) [*CA* **86**, 91660 (1977)]; R. J. Maner and M. A. Weaver, U.S. Patent 4,070,352 (1978) [*CA* **88**, 154319 (1978)]; S. Imahori, Y. Murata, K. Abe, and S. Suzuki, Jpn. Kokai 78/65,480 (1978) [*CA* **89**, 181173 (1978)]; H. Eilingsfeld, G. Hansen, G. Seybold, and G. Zeidler, Ger. Offen. 2,716,033 (1978) [*CA* **90**, 73289 (1979)]; J. Dehnert and G. Lamm, Ger. Offen. DE 3,201,268 (1983) [*CA* **99**, 177493 (1983)].
[139] E. Coghi, C. Branca, and G. Massimo, *Ateneo Parmense, Acta Nat.* **15**, 211 (1979) [*CA* **92**, 158857 (1980)]; E. Coghi and C. Branca, *ibid.*, 217 [*CA* **93**, 2100 (1980)].

of 2,1-benzisothiazole-3-acetic acid,[140] and from 3-acylamino-2,1-benzisothiazoles.[141] Local anesthetic or antiinflammatory activity is shown by amides and esters of 2,1-benzisothiazole-3-carboxylic acid[142] and by alkylaminoacetyl derivatives of 3-amino-2,1-benzisothiazole.[143] Azomethine derivatives of the latter compound have weak antifungal and antistaphylococcal activity.[144] An American company reports antiinflammatory activity in certain 3-(m-substituted p-hydroxyphenyl) derivatives of 2,1-benzisothiazole.[145] Some 1-methyl-2,1-benzisothiazolin-3(1H)-one derivatives are claimed to have anticoagulant properties.[146] The 3-aminoalkyl derivatives of the 2,1-benzisothiazole 1,1-dioxides, developed by Skorcz and colleagues, have hypotensive and CNS-depressant activity.[113]

V. Other Isothiazole Systems

Included among other isothiazoles are compounds closely related to benzisothiazoles: naphthoisothiazoles and systems with two or more isothiazole rings fused to a single benzene ring. Compounds which have a thiophene or pyridine ring, etc., fused to an isothiazole are omitted.

A review of steroidal thiazoles, isothiazoles, thiazolines, and thiazolidines has appeared.[147]

A. Naphthoisothiazoles

Three of the four possible "angular" isomers of the naphthoisothiazole system (**92, 93, 94**) have been prepared by methods similar to those used in

---

[140] E. Coghi, C. Branca, G. Massimo, and F. Zani, *Ateneo Parmense, Acta Nat.* **16**, 141 (1980) [*CA* **95**, 75268 (1981)]; A. Bellotti, C. Branca, E. Coghi, G. Massimo, and F. Zani, *ibid.* **18**, 25 (1982) [*CA* **97**, 67707 (1982)].

[141] A. Bellotti, C. Branca, and E. Coghi, *Ateneo Parmense, Acta Nat.* **13**, 601 (1977) [*CA* **89**, 101651 (1978)]; **14**, 249 (1978) [*CA* **90**, 1572 (1979)].

[142] L. Amoretti, A. Bellotti, E. Coghi, and L. Zappia, *Ateneo Parmense, Acta Nat.* **12**, 345 (1976) [*CA* **87**, 62380 (1977)]; E. Coghi, G. Massimo, and L. Zappia, *ibid.* **15**, 205 (1979) [*CA* **92**, 157721 (1980)].

[143] A. Bellotti, E. Coghi, and L. Zappia, *Ateneo Parmense, Acta Nat.* **14**, 411 (1978) [*CA* **91**, 49567 (1979)].

[144] A. Bellotti, E. Coghi, and O. Sgorbati, *Ateneo Parmense, Acta Nat.* **12**, 123 (1976) [*CA* **85**, 187599 (1976)].

[145] J. A. Carlson and M. R. Bell, U.S. Patent 4,122,105 (1978) [*CA* **90**, 137801 (1979)].

[146] Kyowa Hakko Kogyo Co. Ltd., *Jpn. Kokai Tokkyo Koho* JP **82 108,016** (1982) [*CA* **97**, 133584 (1982)].

[147] P. Catsoulacos and Ch. Camoutsis, *J. Heterocycl. Chem.* **18**, 1485 (1981).

Sec. V.B]   POLYCYCLIC ISOTHIAZOLES   131

syntheses of 1,2- and 2,1-benzisothiazoles.[17,148–150] The "linear" isomer **95** has not yet been made.

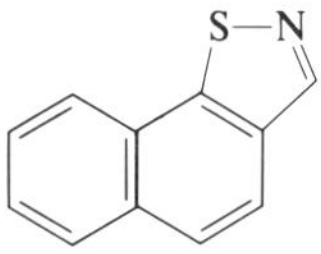

(92) 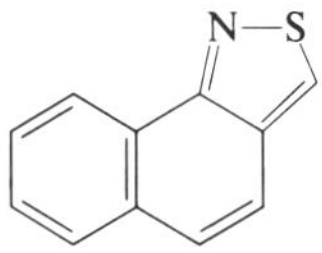   (93) 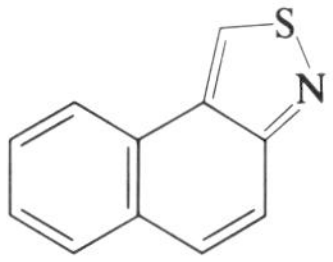   (94) 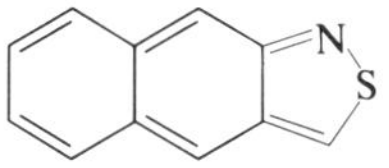   (95)

B. BENZOBISISOTHIAZOLES AND BENZOTRIS[c]ISOTHIAZOLE

Meth-Cohn and co-workers have described the preparation of the benzobis[d]isothiazole **96** from the dialdehyde **97**.[10]

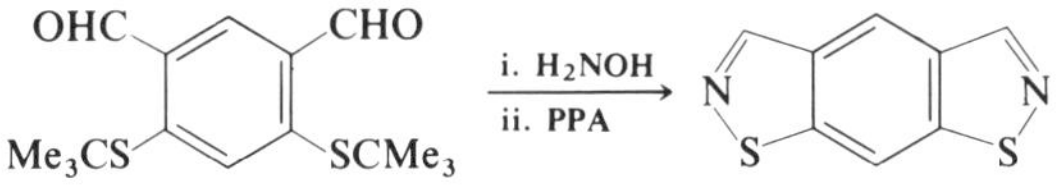

(97)   (96)

A particularly interesting reaction, capable of extension, is the addition of nitrile sulfides **98** to quinones. These nitrile sulfides are prepared by the thermolysis of 1,3,4-oxathiazol-2-ones (**99**), and the products (with quinone itself) are the bis(isothiazolo)benzoquinones **100**.[151] All of the possible angular benzo[c]isothiazoles, of which **101** is one, and the symmetrical

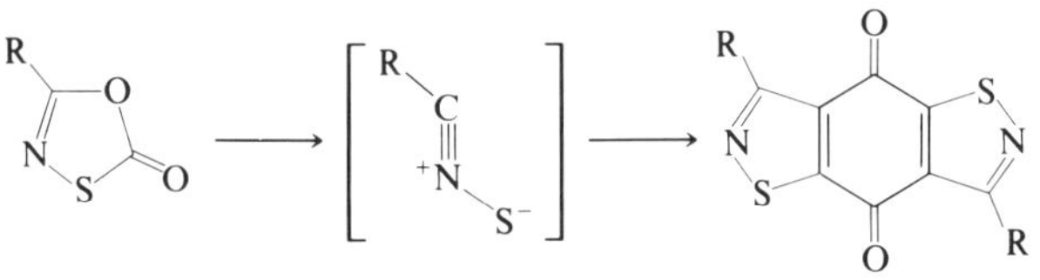

(99)   (98)   (100)

---

[148] M. Davis, G. C. Ramsay, and L. J. Stephens, *Aust. J. Chem.* **25**, 1355 (1972).
[149] H. Adolphi, H. Fleig, and H. Hagen, Ger. Offen. 2,626,967 (1977) [*CA* **88**, 136608 (1978)]; BASF A.-G., Belgian Patent 844,659 (1977) [*CA* **88**, 17341 (1978)].
[150] J. Faust, *J. Prakt. Chem.* **319**, 65 (1977).
[151] R. M. Paton, J. F. Ross, and J. Crosby, *J.C.S. Chem. Commun.*, 1194 (1980).

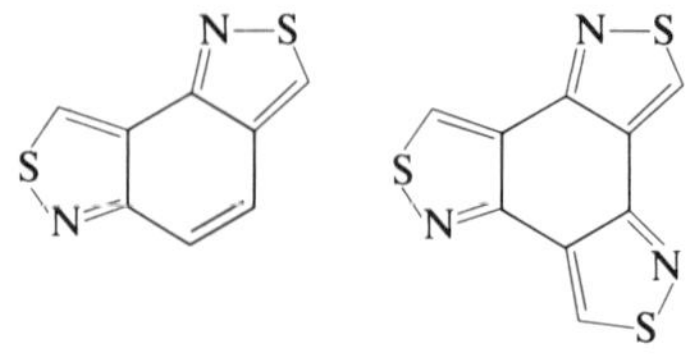

(101)   (102)

benzotris[c]isothiazole **102** have been prepared by repeated use of Singerman's reagent.[152]

VI. Selenium and Tellurium Analogues

A comprehensive review of selenium–nitrogen heterocycles has appeared.[153]

The selenium and tellurium analogues of 1,2-benzisothiazole can be prepared by procedures essentially similar to those used for 1,2-benzisothiazole. Thus 1,2-benzisoselenazolin-3(2H)-one (**103**) may be prepared from the amide

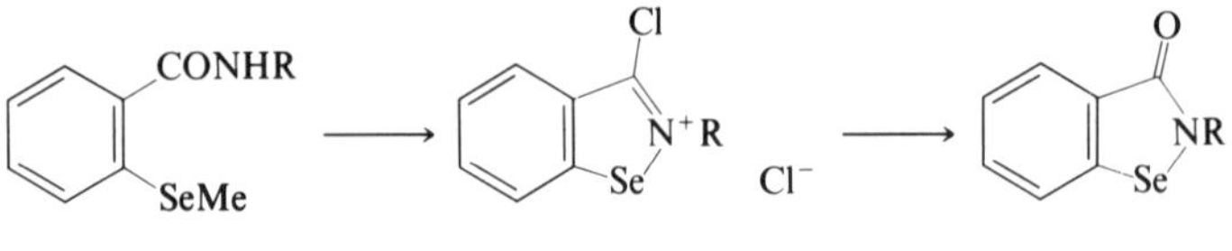

(104)   (103)

104 by treatment with phosphorus pentachloride, followed by hydrolysis.[154] 1,2-Benzisotellurazole (**105**) can be prepared in 74% yield from the telluryl bromide **106**[155]; the selenium analogue is prepared similarly.[156] An unusual

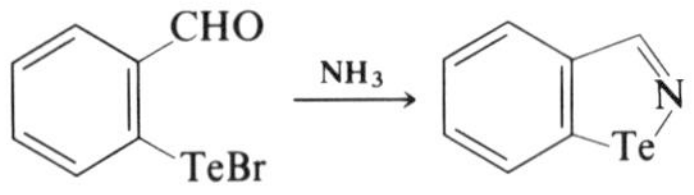

(106)   (105)

[152] B. Danylec and M. Davis, *J. Heterocycl. Chem.* **17**, 533 (1980).
[153] I. Lalezari and A. Shafiee, *Adv. Heterocycl. Chem.* **24**, 109 (1979).
[154] R. Weber and M. Renson, *J. Heterocycl. Chem.* **10**, 267 (1973); *Bull. Soc. Chim. Fr.*, 1124 (1976); *Bull. Soc. R. Sci. Liege* **48**, 146 (1979).
[155] R. Weber, J. L. Piette, and M. Renson, *J. Heterocycl. Chem.* **15**, 865 (1978).
[156] D. E. Ames, A. G. Singh, and W. F. Smyth, *Tetrahedron* **39**, 831 (1983).

preparation of a 1,2-benzisoselenazole (**107**) is that achieved by treatment of 2,1,3-benzoselenadiazole (**108**) with benzyne.[157] Naphtho[2,3-c]-[1,2,5]selenadiazole undergoes a similar reaction, affording 3-substituted benzisoselenazoles in low yields.[158]

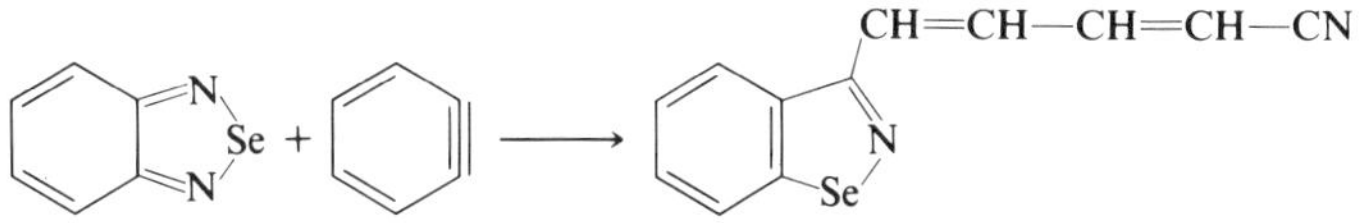

The crystal and molecular structures of **105** have been determined; it has an anomalous structure with very short intermolecular Te—N bonds of 2.4 Å.[159] Mass spectral,[28] ^{13}C- and ^{77}Se-NMR data[160] of 1,2-benzisoselenazoles have been reported. The intramolecular Se—N bond in these compounds is weaker than in 1,2-benzisothiazoles and is readily cleaved by organolithium reagents.[161,162]

---

[157] C. D. Campbell, C. W. Rees, M. R. Bryce, M. D. Cooke, P. Hanson, and J. M. Vernon, *J.C.S. Perkin I*, 1006 (1978); M. R. Bryce, C. D. Reynolds, P. Hanson, and J. M. Vernon, *ibid.*, 607 (1981).
[158] J. M. Vernon, M. R. Bryce, and T. A. Dransfield, *Tetrahedron* **39**, 835 (1983).
[159] H. Campsteyn, L. Dupont, J. Lamotte-Brasseur, and M. Vermeire, *J. Heterocycl. Chem.* **15**, 745 (1978).
[160] N. V. Onyamboko, M. Renson, S. Chapelle, and P. Granger, *Org. Magn. Reson.* **19**, 74 (1982).
[161] R. Weber and M. Renson, *J. Heterocycl. Chem.* **12**, 1091 (1975).
[162] N. V. Onyamboko, R. Weber, A. Fauconnier, and M. Renson, *Bull. Soc. Chim. Belg.* **92**, 53 (1983).

1,4-Benzothiazines, Dihydro-1,4-benzothiazines, and Related Compounds

C. BROWN AND R. M. DAVIDSON

The Chemical Laboratory, University of Kent at Canterbury, Canterbury, Kent, England

I. Introduction 135
II. Natural Occurrence 138
III. Pharmacology of 1,4-Benzothiazines and Their Derivatives 142
    A. Antiinflammatory and Antihistaminic Activity 143
    B. Ataractic Activity 144
    C. Miscellaneous Applications 145
IV. Agricultural Uses 145
V. Industrial Uses 146
VI. Synthesis 146
    A. Preparation from 2-Aminobenzenethiols 146
        1. Reaction with Unsaturated Carbon–Carbon Bonds 146
        2. Reactions with Epoxides 147
        3. Reaction with α-Halo Acids and α-Halo Esters 148
        4. Reaction with α-Halo Ketones and Derivatives 149
        5. Reactions with 1,3-Diketones and β-Keto Esters 150
        6. Reaction with Other Compounds 151
    B. Preparation by Ring Expansion 154
    C. Preparation from 2-Nitrobenzenethiols 159
        1. Reaction with Unsaturated Carbon–Carbon Bonds 159
        2. Reaction with α-Halo Acids and Esters 159
        3. Reaction with Halides 160
        4. Other Reactions with the SH Group 161
    D. Preparation from 2-Nitrobenzenesulfenyl Chlorides 161
    E. Preparation from Disulfides, Cysteine Derivatives, and Other Reagents . . . 163
    F. Other Routes 165
VII. Reactions of 1,4-Benzothiazines 167
    A. N-Alkylation and N-Acylation 167
    B. Electrophilic Substitution 169
    C. Nucleophilic Substitution 171
    D. Rearrangement Reactions 172
    E. Oxidative Coupling of 1,4-Benzothiazines 175

I. Introduction

The benzothiazines constitute a group of heterocyclic compounds which are especially interesting owing to their occurrence in nature as mammalian pigments and to their extensive pharmacological activity. In addition, they have found a variety of industrial uses and show promise as herbicides. Possible isomeric ring structures for this class of compounds are given in formulas **1–9**.

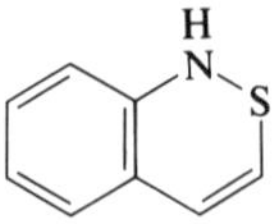

1H-2,1	2H-3,1	4H-3,1	2H-1,2
(1)	(2)	(3)	(4)

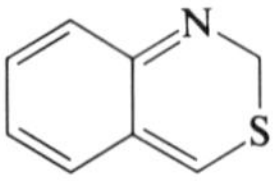

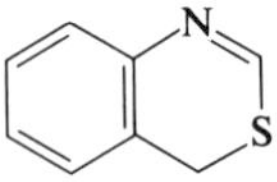

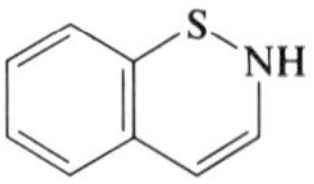

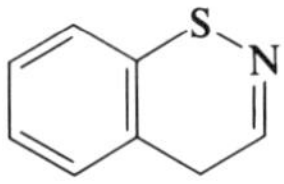

4H-1,2	4H-1,3	2H-1,3	2H-1,4	4H-1,4
(5)	(6)	(7)	(8)	(9)

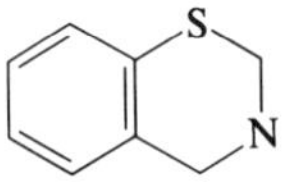

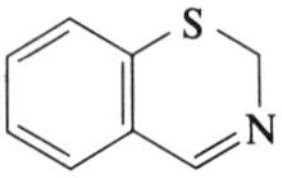

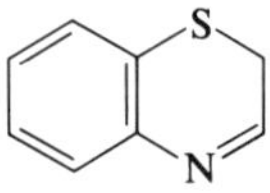

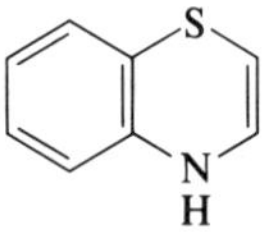

The chemistry of 1,3-benzothiazines has been reviewed as have the 1,2- and 2,1-benzothiazines and benzothiazinone dioxides.[1–3] The 1,4-benzothiazines have not been reviewed comprehensively, however, although a number of articles are available containing shorter resumés of their chemistry.[4–9] This review, which covers the period through mid-1981, including *Chemical*

---

[1] J. Szabo, *Khim. Geterosikl. Soedin.*, 291 (1979) [*CA* **91**, 39354 (1979)].
[2] P. Catsoulacos and C. Camoutsis, *J. Heterocycl. Chem.* **16**, 1503 (1979).
[3] J. G. Lombardino and D. E. Kuhla, *Adv. Heterocycl. Chem.* **28**, 73 (1981).
[4] M. Salisbury, *in* "Road's Chemistry of Carbon Compounds" (S. Coffey, ed.), 2nd ed., Vol. IVH, p. 509. Elsevier, Amsterdam, 1978.
[5] G. Prota, *In* "Organic Chemistry of Sulphur, Selenium and Tellurium" (Specialist Periodical Reports), Vol. 1, p. 454. Chemical Society, London, 1970.
[6] G. Prota, *In* "Organic Chemistry of Sulphur, Selenium and Tellurium" (Specialist Periodical Reports), Vol. 2, p. 758. Chemical Society, London, 1973.
[7] G. Prota, *In* "Organic Chemistry of Sulphur, Selenium and Tellurium" (Specialist Periodical Reports), Vol. 3, p. 708. Chemical Society, London, 1975.
[8] G. Prota, *In* "Organic Chemistry of Sulphur, Selenium and Tellurium" (Specialist Periodical Reports), Vol. 4, p. 453. Chemical Society, London, 1977.
[9] J. Piechaczek, *Wiad. Chem.* **20**, 233 (1966) [*CA* **65**, 3864 (1966)].

Abstracts, Vol. 94, is primarily concerned with 2*H*- and 4*H*-1,4-benzothiazines, (**8** and **9**), and their dihydro derivatives, though some references will be made to 1,4-benzothiazine 1,1-dioxides and 1,4-benzothiazin-3-ones, since these compounds have a common ring skeleton together with, in certain cases, comparatively similar chemical behavior.

To date, the unstable parent ring of the 1,4-isomers (**8** and **9**) has not been isolated, although it has been observed by ^{1}H NMR as a transient species existing only in solution, predominantly as the 2*H* tautomer.[10]

The first synthesis of a derivative was reported by Unger,[11] who claimed that reaction of phenacyl bromide with 2-aminobenzenethiol gave 3-phenyl-2*H*-1,4-benzothiazine (**10a**). Nearly 70 years later Friedrich *et al.* proposed the 4*H* structure (**10b**).[12]

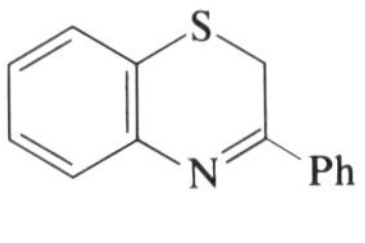

(10a) 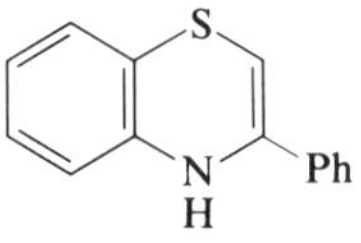   (10b)

After reexamination of the reaction product, using UV and ^{1}H NMR, two groups claimed that it was a mixture of the 2*H* (80%) and 4*H* (20%) tautomers.[13,14] However, more recently, Thomson *et al.* have shown that these results are also rather ambiguous.[15] Thus the 4'-methoxy and 4'-nitro derivatives of **10** could be obtained as monomers only by using a rapid workup procedure. The procedure of Unger and a later one described by Finar and Montgomery[16] both give only the dimer (**11**).

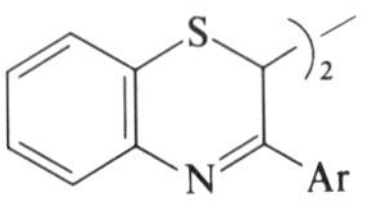

Ar = 4-CH$_3$OC$_6$H$_4$
Ar = 4-NO$_2$C$_6$H$_4$

(11)

---

[10] F. Chioccara, G. Prota, and R. H. Thomson, *Tetrahedron* **32**, 1407 (1976).
[11] O. Unger, *Ber. Dtsch. Chem. Ges.* **30**, 607 (1897).
[12] W. Friedrich, F. Kröhnke, and P. Schiller, *Chem. Ber.* **98**, 3804 (1965).
[13] P. Bottex, B. Sillion, G. de Gaudemaris, and J.-J. Basselier, *C. R. Hebd. Acad. Sci., Seances Ser. C* **267**, 186 (1968).
[14] M. Wilhelm and P. Schmidt, *J. Heterocycl. Chem.* **6**, 635 (1969).
[15] N. E. MacKenzie, R. H. Thomson, and C. W. Greenhalgh, *J.C.S. Perkin I*, 2923 (1980).
[16] I. L. Finar and A. J. Montgomery, *J. Chem. Soc.*, 367 (1961).

X-Ray studies of the system, using dimethyl 4-formyl-2,3-dihydro-1,4-benzothiazine-2,3-dicarboxylate,[17] 4H-1,4-benzothiazine 1,1-dioxide,[18] and the 3-methyl derivative,[19] showed these molecules to be essentially planar with only small distortions associated mainly with the heteroatoms in the thiazine ring. The various bond lengths and angles were related to those of indole in an attempt to rationalize their reactivity toward electrophiles, which has been reported[20] to be similar to that of indole. Molecular mechanics calculations similarly indicated essentially planar rings for *trans*-2-dimethyl-4-acetyldihydrobenzothiazine.[21]

Mass spectral fragmentation patterns of dihydro-1,4-benzothiazine have been studied and dissociation mechanisms discussed.[22] The neutral 1,4-benzothiazinyl radical **12** has recently been reported[23] and compared with phenothiazinyl radicals.[24–26]

(12)

The acidity of the NH group has been examined[27] in a range of heterocycles including 1,4-benzothiazines. As the heterocyclic ring becomes smaller, the NH acidity increases, and therefore the nitrogen center becomes less reactive with respect to electrophiles. Thus change in the pK is enhanced when sulfur and nitrogen are present in the ring.

II. Natural Occurrence

The 1,4-benzothiazine ring is the basic unit for the phaeomelanins and trichochrome melanin pigments found in mammals, particularly in red hair

---

[17] H. Ogura, K. Kikuchi, H. Takayanagi, K. Furuhata, Y. Iitaka, and R. M. Acheson, *J.C.S. Perkin I*, 2316 (1975).
[18] G. D. Andreetti, G. Bocelli, and P. Sgarabotto, *Cryst. Struct. Commun.* **3**, 305 (1974).
[19] G. D. Andreeti, G. Bocelli, L. Coghi, and P. Sgarabotto, *Cryst. Struct. Commun.* **5**, 315 (1976).
[20] G. Pagani and S. B. Pagani, *Tetrahedron Lett.*, 1041 (1968).
[21] R. M. Davidson, *Ph. D. Thesis*, University of Kent (1981).
[22] H. Budzikiewicz and U. Lenz, *Org. Mass Spectrom.* **10**, 992 (1975).
[23] F. Ciminale, G. Liso, and G. Trapani, *Tetrahedron Lett.* **22**, 1455 (1981).
[24] M. F. Chiu, B. C. Gilbert, and P. Hanson, *J. Chem. Soc. B*, 1700 (1970).
[25] Y. Tsujino, *Tetrahedron Lett.*, 411 (1968).
[26] C. Jackson and N. K. D. Patel, *Tetrahedron Lett.*, 2255 (1967).
[27] M. V. Gorelik, T. V. Levandovskaya, B. A. Korolev, M. I. Terekhova, E. S. Petrov, and A. I. Shatenshtein, *J. Org. Chem. USSR (Engl. Transl.)* **14**, 2202 (1978) [*CA* **90**, 10366 (1979)].

and feathers. The occurrence and biogenesis of pigments have been reviewed,[28-30] and so only a resumé will be given.

Phaeomelanins are reddish-brown, nitrogen and sulfur-containing macromolecular pigments, which are found in phaeomelanocytes. They are derived from tyrosinase oxidation of tyrosine and subsequent reaction with cysteine. Trichochromes are produced in a similar manner, but they are of low molecular weight and usually yellow, red, or violet. The commonly encountered trichochromes have the structures 13 and 14.

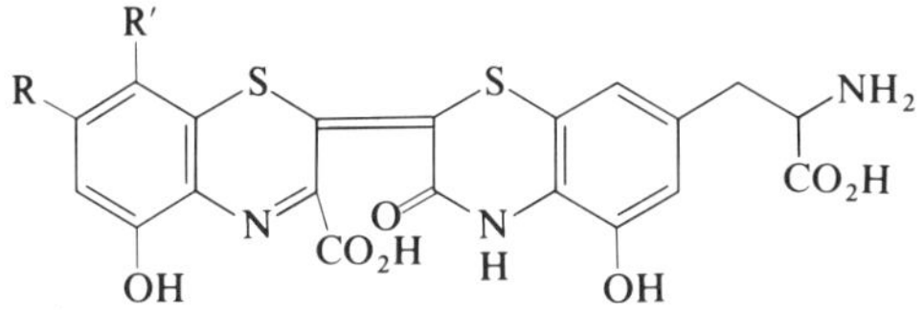

(13) Trichochrome B, R = H, R' = $CH_2CH(NH_2)CO_2H$
(14) Trichochrome C, R = $CH_2CH(NH_2)CO_2H$, R' = H

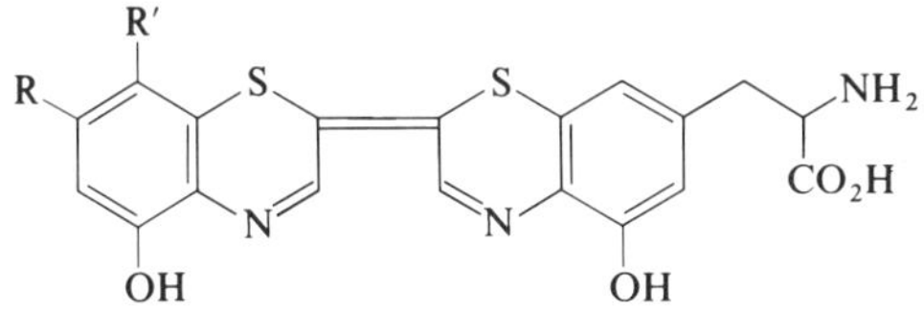

(15) Trichochrome E, R = H, R' = $CH_2CH(NH_2)CO_2H$
(16) Trichochrome F, R = $CH_2CH(NH_2)CO_2H$, R' = H

Trichochromes E and F are based on the 2H-1,4-benzothiazine ring, while trichochromes B and C are mixed systems, with both 1,4-benzothiazine and 1,4-benzothiazin-3-one fragments. These may be extracted from red hair and feathers under alkaline conditions.[31,32] Apparently[33] only trichochromes B and C occur in red feathers, and trichochromes E and F, reportedly isolated from this source, arise as artifacts. The chromophore in the trichochrome

---

[28] G. Prota and R. H. Thomson, *Endeavour* **35**, 32 (1976).
[29] R. H. Thomson, *Angew. Chem., Int. Ed. Engl.* **13**, 305 (1974).
[30] G. Prota, in "Pigmentation: Genesis and Biological Control" (V. Riley, ed.), p. 615. Appleton, New York, 1972.
[31] R. A. Nicolaus, G. Prota, C. Santacroce, G. Scherillo, and D. Sica, *Gazz. Chim Ital.* **99**, 323 (1969).
[32] G. Prota, A. Suarato, and R. A. Nicolaus, *Experientia* **27**, 1381 (1971).
[33] G. Agrup, C. Hansson, H. Rorsman, A-M. Rosengren, and E. Rosengren, *Acta Derm. Venereol.* **58**, 269 (1978) [*CA* **90**, 36495 (1979)].

system is analagous to that of indigo. Consequently, many recent studies have been aimed at reproducing the chromophore system **17** synthetically.

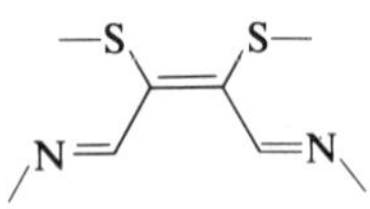

cis-(**17**) 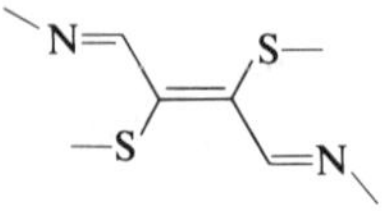   trans-(**17**)

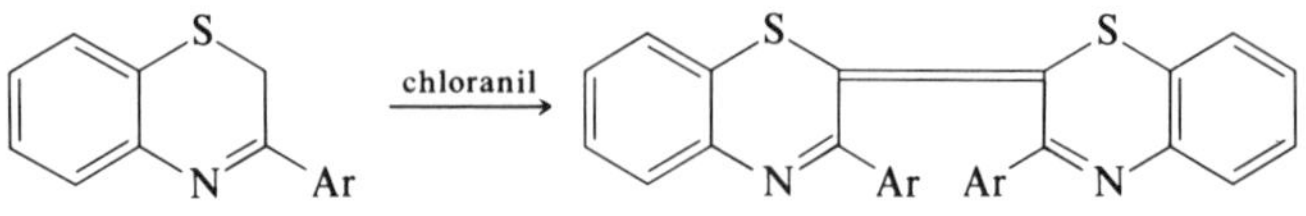

(**18**)

Prota and co-workers[34] prepared **18** by oxidative coupling of 3-aryl-2H-1,4-benzothiazines, and established the structures of the cis and trans isomers of the 3-p-bromophenyl derivatives by X-ray crystallography. The cis and trans isomers are readily interconverted in solution at room temperature, with the red, cis isomer predominating over the yellow, trans isomer, both isomers being stable in the solid state.

The dominance of the cis isomer can be attributed to Van der Waals forces, which maintain the two benzene rings at the 3,3'-positions in an almost fixed conformation. In the trans isomer, the benzene rings are relatively free. As the temperature is increased, thermal motion destabilizes the cis configuration and the less strained trans isomer becomes favored.

Kaul synthesized the chromophore based on 1,4-benzothiazin-3-one to obtain the skeleton **19** for trichochromes B and C.[35]

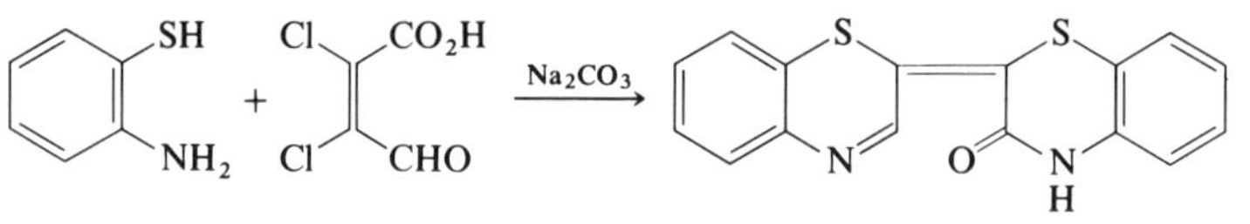

(**19**)

Prota's method cannot be employed in the synthesis of the trichochrome E and F skeleton because it would require the unstable 2H-1,4-benzothiazine **8** as starting material. However, the Italian group observed the spontaneous formation of compound **20** while attempting to prepare 2H-1,4-benzothiazine

---

[34] F. Giordano, L. Mazzarella, G. Prota, C. Santacroce, and D. Sica, *J. Chem. Sec. C*, 2610 (1971).
[35] B. L. Kaul, *Helv. Chim. Acta* **57**, 2664 (1974).

(8) and developed from this a simple method for the preparation of **20** in good yield.[36]

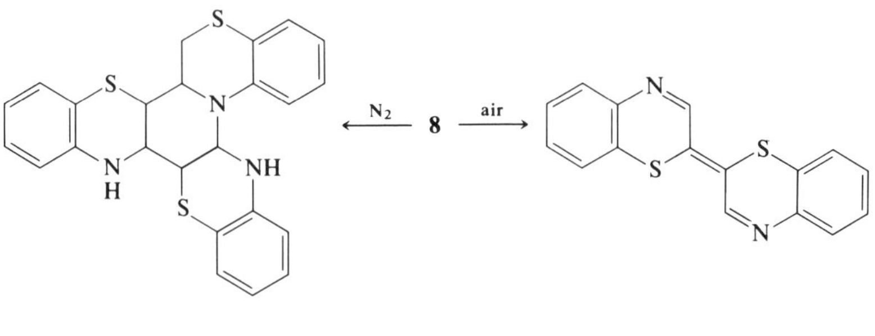

(21) (20)

Under anaerobic conditions, the trimer **21** was formed, presumably by an aldol reaction.[37] Treatment with hydrochloric acid in ethanol exposed to the air gave the dimer. This had led to the suggestion that under aerobic conditions, the trimer is converted to phaeomelanic pigments, which may be formed by aldol-type condensation and not by oxidative coupling as previously proposed.[38–40]

As mentioned, the amino acid cysteine is thought to be involved in the biosynthesis of trichochromes. Cysteine reacts with 4-methyl-*O*-benzoquinone to give **22**, which on treatment with mineral acid gives the trichochrome chromophore system,[43] and reactions of cysteine ethyl ester and of 2-aminoethanethiol with quinones were studied to elucidate and duplicate the natural synthesis.[41,42]

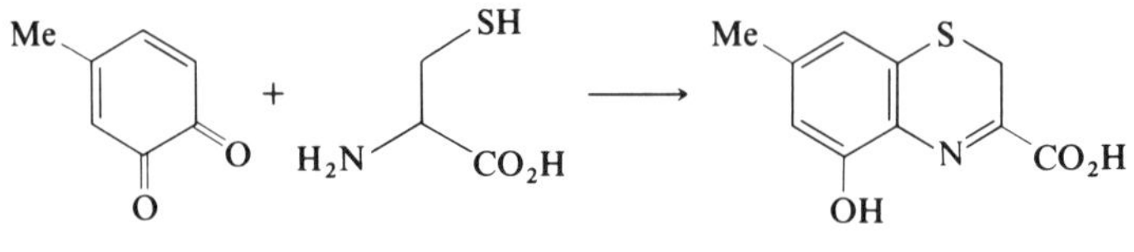

(22)

[36] G. Prota, E. Ponsiglione, and R. Ruggiero, *Tetrahedron* **30**, 2781 (1974).
[37] F. Chioccara, E. Novellino, G. Prota, and G. Sodano, *J.C.S. Chem. Commun.*, 50 (1977).
[38] G. Prota, G. Scherillo, E. Petrillo, and R. A. Nicolaus, *Gazz. Chim. Ital.* **99**, 1193 (1969).
[39] D. Sica, C. Santacroce, and G. Prota, *J. Heterocycl. Chem.* **7**, 1143 (1970).
[40] S. Crescenzi, G. Misuraca, E. Novellino, and G. Prota, *Chim. Ind. (Milan)* **57**, 392 (1975).
[41] G. Prota and E. Ponsiglione, *Tetrahedron Lett.*, 1327 (1972).
[42] G. Prota, O. Petrillo, C. Santacroce, and D. Sica, *J. Heterocycl. Chem.* **7**, 555 (1970).
[43] G. Prota, G. Scherillo, E. Napolano, and R. A. Nicolaus, *Gazz. Chim. Ital.* **97**, 1451 (1967).

The biosynthesis of luciferin (23) in the luminescent click beetle, *Pyrophorus pellucens*, is proposed to occur via the reaction of quinone with cysteine to give an intermediate benzothiazine, which on further reaction with cysteine gives the benzothiazole. But preliminary studies with ^{14}C-labeled cysteine indicated, however, that an alternative fate for the first-formed thiazine, involving oxidation and ring contraction, may have to be considered.[44]

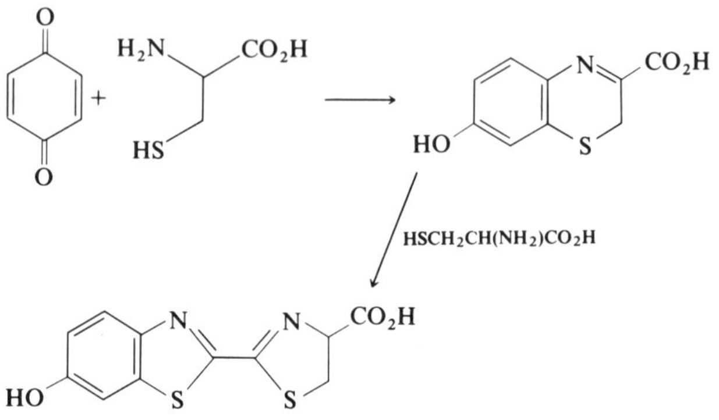

(23)

Rifamycin Verde from mutant strains of *Nocardia mediterranea* contains the 1,4-benzothiazine skeleton, the formation of which is proposed from cysteine of cysteamine derivatives and Rifamycin S.[45,46]

III. Pharmacology of 1,4-Benzothiazines and Their Derivatives

Much interest in the pharmacology of 1,4-benzothiazines is due to their similarity in structure with phenothiazine drugs, such as chlorpromazine (24),

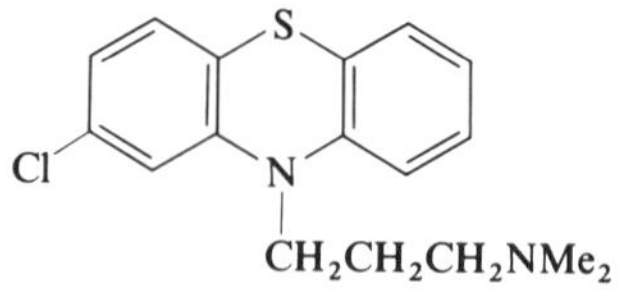

(24)

[44] F. McCapra and Z. Razavi *J.C.S. Chem. Commun.*, 153 (1976).
[45] R. Cricchio, P. Antonini, G. C. Lancini, G. Tamborini, R. J. White and E. Martinelli, *Tetrahedron* **36**, 1415 (1980).
[46] R. Cricchio, *Tetrahedron* **36**, 2009 (1980).

Sec. III.A] 1,4- AND DIHYDRO-1,4-BENZOTHIAZINES 143

which are widely used in many areas of medicine. The synthesis of phenothiazine drugs has been reviewed along with their chemistry.[47,48] 1,4-Benzothiazine derivatives display pharmacological activity in a number of areas, which are considered in the sections below.

A. ANTIINFLAMMATORY AND ANTIHISTAMINIC ACTIVITY

Substituted 1,4-benzothiazines were first reported to exhibit antihistaminic properties by Fujii.[49] 4-(Dimethylaminoethyl)-6-chloro-3,4-dihydro-1,4-benzothiazine (**25**) is particularly active.[49]

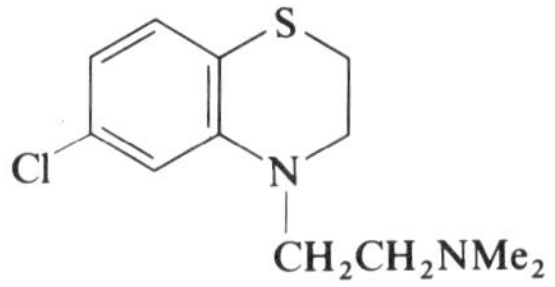

(**25**)

Further studies produced active compounds **26** with a carbonyl function in the side chain, and these display antifungal activity in addition to antiinflammatory properties.[50] Winthrop prepared compounds **27**, which display

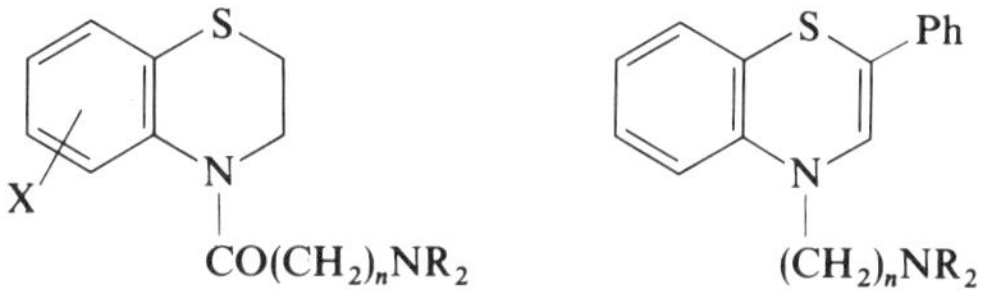

(**26**)   (**27**)

X = halogen or alkyl; R = alkyl

antihistaminic and spasmolytic activity.[51] The benzothiazines **28** also exhibit antiinflammatory properties.[52,53]

---

[47] D. Lednicer and L. A. Mitscher, "Organic Chemistry of Drug Synthesis," p. 372. Wiley, New York, 1977.
[48] C. Bodea and I. Silberg, *Adv. Heterocycl. Chem.* **9**, 321 (1968).
[49] K. Fujii, *Japanese Patent* 5241 (1958) [*CA* **53**, 17156 (1959)].
[50] H. S. Lowrie, *U.S. Patent* 2,947,744 (1960) [*CA* **55**, 583 (1961)].
[51] S. O. Winthrop and R. Gaudry, *U.S. Patent* 2,989,528 (1961) [*CA* **56**, 4777 (1962)].
[52] J. Krapcho and C. E. Turk, *J. Med. Chem.* **16**, 776 (1973).
[53] R. C. Millonig, M. B. Goldlust, W. E. Maguire, B. Rubin, E. Schulze, R. J. Wojnar, A. R. Turkheimer, W. F. Schreiber, and R. J. Brittain, *J. Med. Chem.* **16**, 780 (1973).

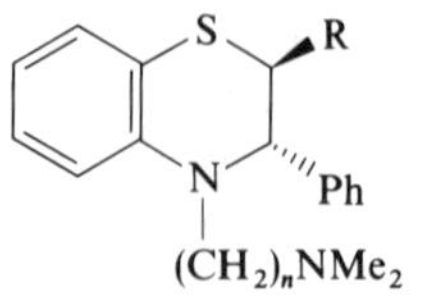

(28) R = CH(OH)Me; COMe

B. ATARACTIC ACTIVITY

1,4-Benzothiazines find extensive use as ataractic agents, i.e., tranquilizers and central nervous system (CNS) depressants. 4-(Hydroxyalkylcyclicamino)propyl-2-phenyl-3,4-dihydro-1,4-benzothiazines (**29**) show CNS depressant activity[54] without including the anorexia associated with the use of related 1,4-benzothiazine derivatives[55,56] such as **30**.

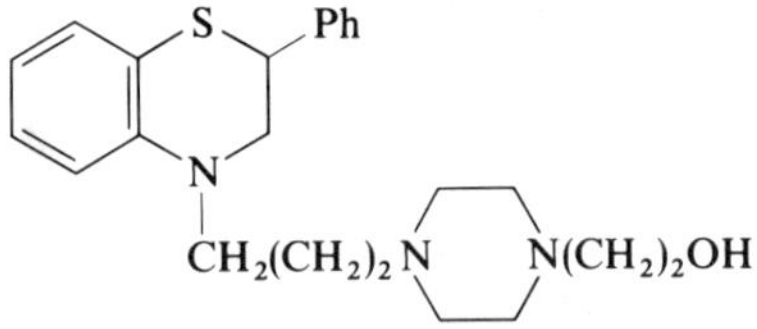

(29) 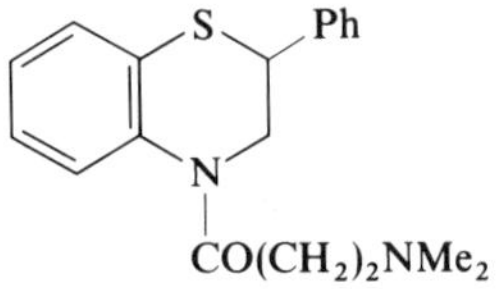   (30)

Krapcho[57] found the acid salts of N-aminoalkylbenzothiazine derivatives (**31**) to have ataractic and antispasmodic properties, and to be particularly useful in the treatment of parkinsonism. Substituted 2-phenyl-1,4-benzothiazin-3-ones can antagonize the activity of 1,4-dipyrrolidine-2-butyne, which induces tremors and is used as a guide to evaluate compounds

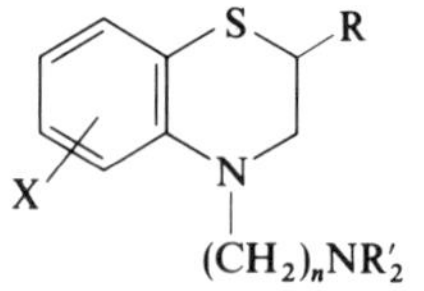

(31) 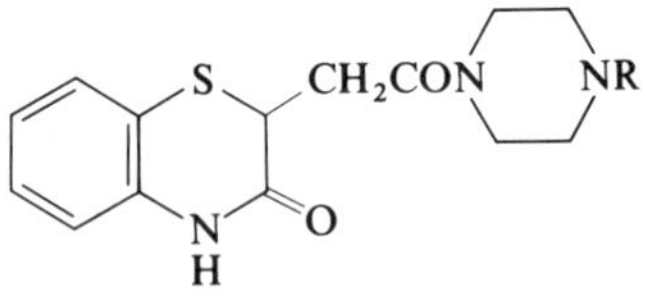   (32)

R = alkyl, aryl; R' = H, alkyl, hydroxyalkyl

[54] H. S. Lowrie, U.S. Patent 3,124,577 (1964) [CA **60**, 14515 (1964)].
[55] J. Krapcho, U.S. Patent 3,143,545 (1964) [CA **61**, 9507 (1964)].
[56] J. Krapcho, U.S. Patent 3,089,872 (1963) [CA **59**, 11541 (1963)].
[57] J. Krapcho and H. L. Yale, U.S. Patent 3,117,124 (1964) [CA **60**, 8048 (1964)].

in the treatment of parkinsonism.[58] Other derivatives display CNS depressant activity and, in some cases, antibacterial activity.[59] CNS depressant, antispasmodic, and antiulcer activities are exhibited[60] by 2-alkylcyclicaminoacetyl derivatives 32.

C. MISCELLANEOUS APPLICATIONS

Various derivatives are also active as anthelmintics,[61-63] muscle relaxants,[2,64] diuretics,[65] antimycotics,[66] antihypertensives,[67] and anticholesterics.[68] The metabolism of 4-[3-dimethylaminopropyl]-3,4-dihydro-2-(1-hydroxyethyl)-3-phenyl-2H-1,4-benzothiazine *in vivo* by rats, dogs, and monkeys has been studied, and at least five metabolites were isolated.[69]

IV. Agricultural Uses

The plant physiological activity of 1,4-benzothiazines has been little studied. Worley reports that compounds 33 destroy or prevent growth of some broad-leaved weeds, yet promote the growth of some dicotyledonous

(33) R = H, alkyl
R^1, R^2 = H, alkyl, alkylidene, aryl

(34) X = halogen

---

[58] J. Krapcho, A. Szabo, and J. Williams, *J. Med. Chem.* **6**, 214 (1963).
[59] J. Krapcho, U.S. Patent 3,401,166 (1968) [*CA* **69**, 106721 (1968)].
[60] J. F. Cavalla and J. Michael, British Patent 1,244,481 (1971) [*CA* **75**, 129824 (1971)].
[61] A. Mackie and J. Raeburn, *J. Chem. Soc.*, 787 (1952).
[62] A. Mackie, G. M. Stewart, A. A. Cutler, and A. L. Misra, *Br. J. Pharmacol. Chemother.* **10**, 7 (1955).
[63] A. Mackie and G. M. Stewart, *Arch. Int. Pharmacodyn. Ther.* **102**, 476 (1955).
[64] W. K. Hoya, U.S. Patent 3,148,188 (1964) [*CA* **61**, 13324 (1964)].
[65] H. Kano, S. Takahashi, Y. Ogawa, T. Yoshizaki, and T. Kitakaze, *Shionogi Kenkyusho Nempo* **11**, 1 (1961) [*CA* **56**, 4749 (1962)].
[66] H. Böshagen and M. Plempel, U.S. Patent 3,911,126 (1975) [*CA* **84**, 155700 (1976)].
[67] R. N. Prasad, *J. Med. Chem.* **12**, 290 (1969).
[68] K. Irmscher, J. Kraemer, G. Cimbollek, D. Orth, H. Novwack, and K. O. Freisberg, *Ger. Offen.* 1,809,454 (1970) [*CA* **73**, 66594 (1970)].
[69] S. J. Lan, T. J. Chando, A. I. Cohen, I. Weliky, and E. C. Schreiber, *Drug Metab. Dispos.* **1**, 619 (1973).

crops such as legumes.[70] Compounds **34** display herbicidal activity, but do not damage main crops and show low toxicity to fish and mammals.[71]

V. Industrial Uses

1,4-Benzothiazines have been used in industry as dyes and as antioxidants of rubber and natural elastomers. They have been used widely to prepare azo-dye compounds with specific properties, such as water insolubility, which is useful in the dyeing of polypropylene fibers, and increased stability to gas fading.[72,73] Several reports are concerned with the antioxidant properties of benzothiazines,[74–77] and their use as color photography developers.[78]

VI. Synthesis

A. PREPARATION FROM 2-AMINOBENZENETHIOLS

2-Aminobenzenethiol is the most commonly used starting material, reacting with a large variety of functional groups to give substituted 1,4-benzothiazines.

1. *Reaction with Unsaturated Carbon–Carbon Bonds*

2-Aminobenzenethiol reacts with acetylenes and olefins to give 1,4-benzothiazines. Reaction with acetylene dicarboxylic acid and esters gives the 2-alkylidene derivates (**35**).[79,80]

---

[70] J. W. Worley, *U.S. Patent* 3,923, 709 (1975) [*CA* **84**, 74279 (1976)].
[71] T. Yematsu, S. Hashimoto, and H. Oshio, *British Patent* 2,018,754 (1979) [*CA* **93**, 46693 (1980)].
[72] R. Tanaka, K. Teramura, and S. Yokoyama, *Japanese Patent* 41/5833 (1966) [*CA* **65**, 9062 (1966)].
[73] J. B. Dickey, W. H. Strain, and R. A. Corbitt, *U.S. Patent* 2,364,347 (1944) [*CA* **39**, 4233 (1945)].
[74] J. B. Dickey and J. G. McNally, *U.S. Patent* 2,374,181 (1945) [*CA* **40**, 1180 (1946)].
[75] L. Convert and R. Fabre, *French Patent* 1,344,437 (1963) [*CA* **61**, 5885 (1964)].
[76] Rhone-Poulenc S. A., *French Patent* 84,357 (1965) [*CA* **62**, 14914 (1965)].
[77] I. Ya Postovskii, *U.S.S.R. Patent* 707,915 (1980) [*CA* **92**, 148252 (1980)].
[78] W. H. Strain and J. B. Dickey, *U.S. Patent* 2,381,935 (1945) [*CA* **40**, 1889 (1964)].
[79] L. K. Mushkalo and V. A. Besemskaiya, *Ukr. Khim. Zh.* **18**, 163 (1952) [*CA* **48**, 13692 (1953)].
[80] Y. Maki and M. Suzuki, *Chem. Pharm. Bull.* **20**, 832 (1972).

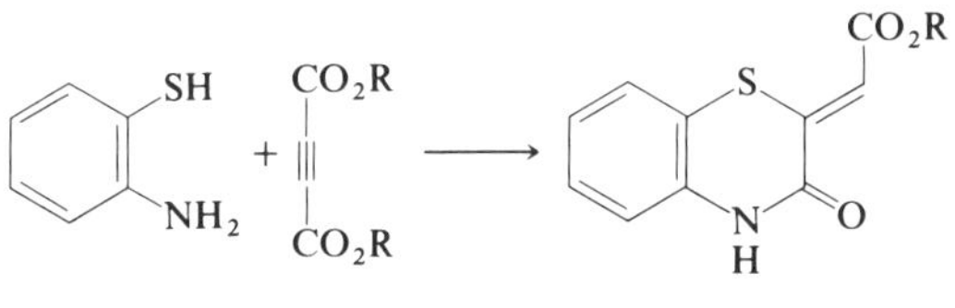

(35)

Similar reactions occur with α,β-unsaturated acids,[81] though earlier literature had suggested that 1,5-benzothiazepines (36) were produced.[82]

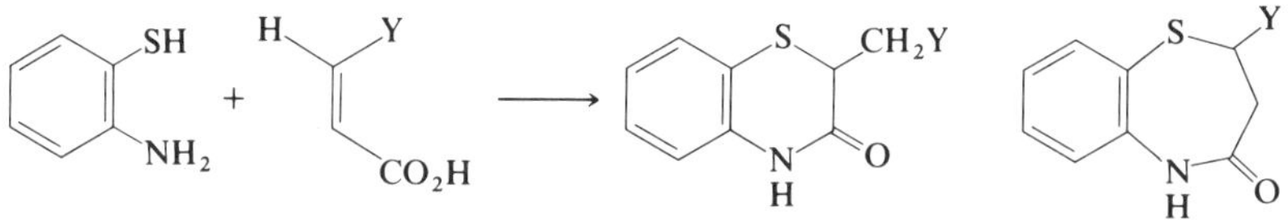

(36)

2. Reactions with Epoxides

One synthesis of benzothiazines via α-epoxy ketones[83] involves acid-catalyzed ring expansion of a benzothiazoline intermediate (37) to give the benzothiazine derivative 38.

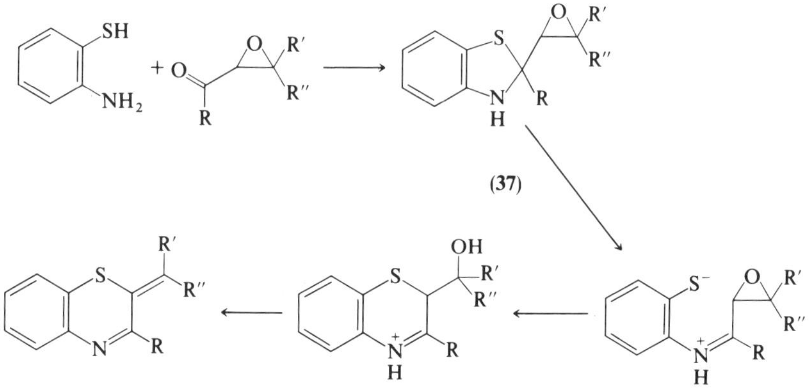

(37)

(38)

[81] F. K. Kirchner and E. J. Alexander, *J. Am. Chem. Soc.* **81**, 1721 (1959).
[82] W. H. Mills and J. B. Whitworth, *J. Chem. Soc.*, 2738 (1927).
[83] L. G. Gobalenko, L. K. Mushkalo, and V. A. Chiuguk, *Ukr. Khim. Zh.* **35**, 1278 (1969) [*CA* **72**, 78963 (1970)].

Early reports that 2-substituted dihydro-1,4-benzothiazines (**39**) could be prepared from 2-aminobenzenethiol with epoxides[84,85] were later shown incorrect.[86-88] The actual product is the amino alcohol **40**. Fujii[88] claimed, however, that heating the hydrochloride salt of **40** (R = H) did produce 3,4-dihydro-1,4-benzothiazine (**39**, R = H), and more recently it has been shown that benzothiazines can be obtained by reaction of perfluoroepoxides with 2-aminobenzenethiol.[89]

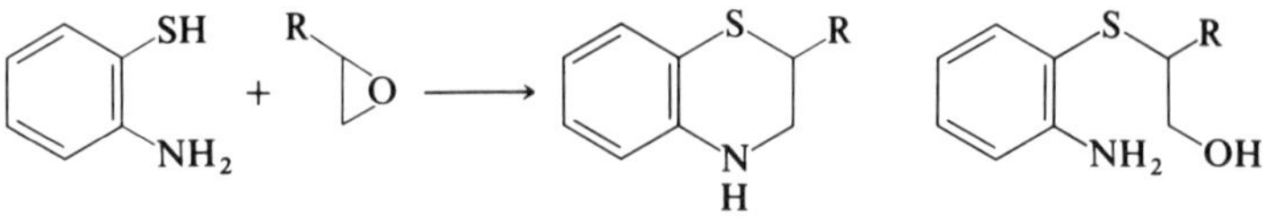

(39)   (40)

3. Reaction with α-Halo Acids and α-Halo Esters

Funke et al. used α-bromophenylacetic acid to prepare 2-phenyl-3,4-dihydro-1,4-benzothiazine (**41**) via a lithium aluminum hydride reduction of the 1,4-benzothiazin-3-one intermediate.[90] Similarly, chloroacetic acid gives products not substituted in the 2-position.[65]

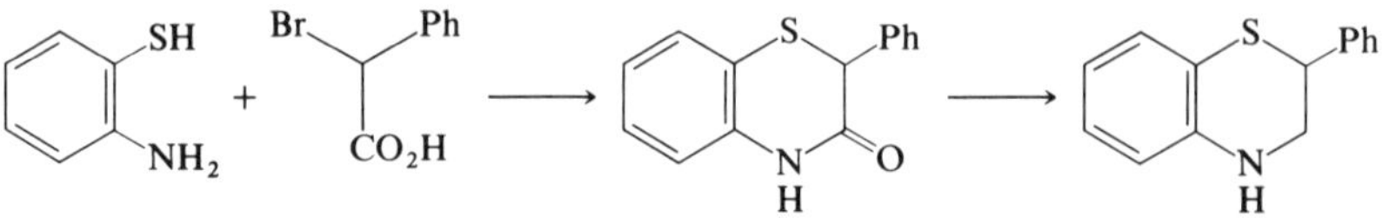

(41)

α-Halo-β-carbonyl esters give 4H-1,4-benzothiazines (**42**) with anhydrous diethyl ether or aqueous potassium hydroxide/methanol.[91,92] The chlorine

---

[84] C. C. J. Culvenor, W. Davies, and N. S. Heath, *J. Chem. Soc.*, 278 (1949).
[85] R. Fusco and G. Palazzo, *Gazz. Chim. Ital.* **81**, 735 (1951).
[86] J. F. Kerwin, J. E. McCarty, and C. A. VanderWerf, *J. Org. Chem.* **24**, 1719 (1959).
[87] O. Hromatka, J. Augl, A. Brazda, and W. Grünsteidl, *Monatsh. Chem.* **90**, 544 (1959).
[88] K. Fujii, *Yakugaku Zasshi* **77**, 355 (1957) [*CA* **51**, 12012 (1957)].
[89] N. Ishikawa and S. Sasaki, *Bull. Chem. Soc. Jpn* **50**, 2164'(1977).
[90] A. Funke, G. Funke, and B. Millet, *Bull. Soc. Chim. Fr.*, 1524 (1961).
[91] F. Duro, P. Condorelli, G. Scapini, and G. Pappalardo, *Ann. Chim. (Rome)* **60**, 383 (1970).
[92] F. Duro, P. Condorelli, and G. Ronsisvalle, *Ann. Chim. (Rome)* **63**, 45 (1973).

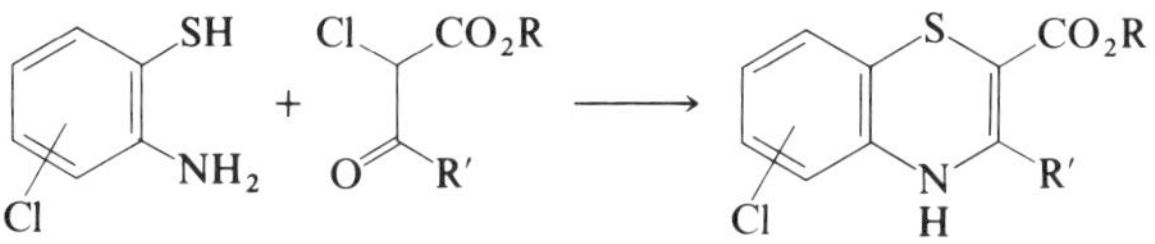

(42) R, R' = Me, Et

atom at the 6-position on the benzothiazine ring provides a considerable stabilizing influence, preventing dimerization on hydrolysis of the ester.

4. Reaction with α-Halo Ketones and Derivatives

α-Halo ketones, especially phenacyl bromides, react with 2-aminobenzenethiol as previously described,[10–14,16,93,94] although, as explained in Section I, these reactions are not always straightforward.[15] This method does, however, represent one of the simplest routes to benzothiazines. Using 2-hydroxyphenacyl halides, Curtze and Thomas prepared the benzothiazine **43**, which on treatment with acetic anhydride cyclized **44**.[95]

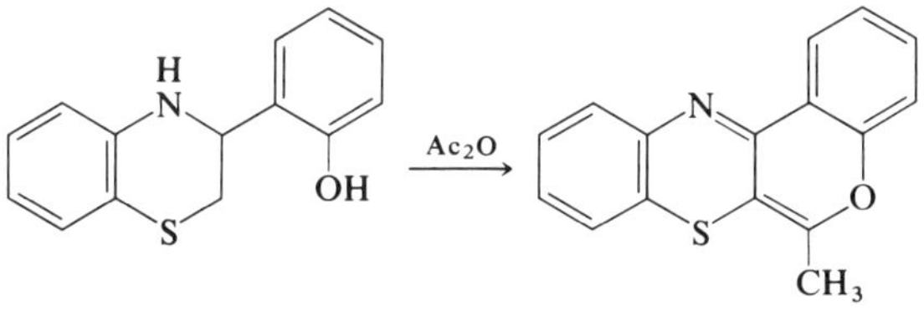

(43)   (44)

α-Halo carbonyl compounds, when used in the protected acetal form,[10,37,94,96] give the unsubstituted parent ring **8**, which cannot be isolated

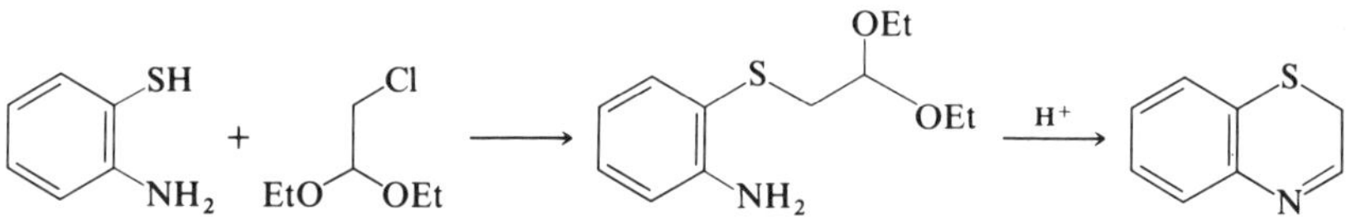

(8)

---

[93] K. Fujii, *Yakugaku Zasshi* **77**, 347 (1957) [*CA* **51**, 12100 (1957)].
[94] C. Santacroce, D. Sica, and R. A. Nicolaus, *Gazz. Chim. Ital.* **98**, 85 (1968).
[95] J. Curtze and K. Thomas, *Liebigs Ann. Chem.*, 328 (1974).
[96] F. Chioccara, G. Prota, and R. H. Thomson, *Tetrahedron Lett.*, 811 (1975).

owing to its instability. While this route, employing TFA as the catalyst, does lead to the formation of **8** in solution, the product decomposed easily to a variety of products, including 2-methylbenzothiazole and the dye **45**.[10] Use of aqueous HCl leads to the trimer **21**.[37]

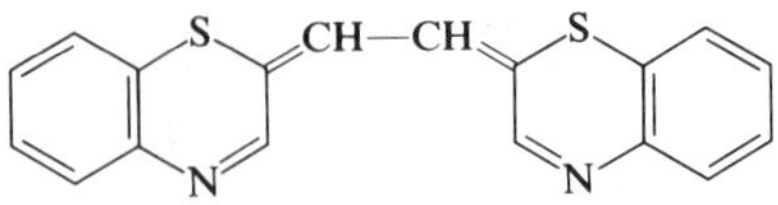

(45)

A thermally stable polymer (**47**) based on the 1,4-benzothiazine ring has been prepared, using a bis-α-halo ketone and the diphenyl derivative **46**. This polymer has a molecular weight of between 10,000 and 20,000 and is stable up to ~350°C in air.[97]

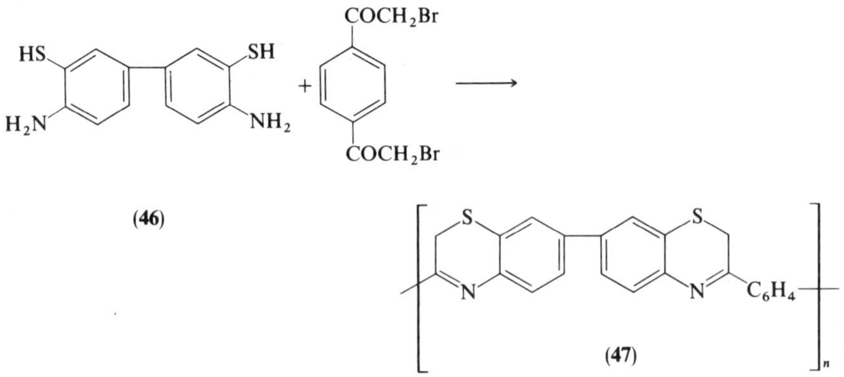

(46)

(47)

5. Reactions with 1,3-Diketones and β-Keto Esters

When DMSO is the solvent, 1,3-diketones and β-keto esters react with 2-aminothiophenol to give 4H-1,4-benzothiazines (**48**).[98] In benzene or methanol, however, benzothiazolines such as **49** are produced according to the work of Kiprianov and Portnyagina.[99] An oxidative cyclization mechanism

---

[97] P. Bottex, B. Sillion, and G. de Gaudemaris, *C. R. Hebd. Seances Acad. Sci., Ser. C* **267**, 711 (1968).
[98] S. Miyano, N. Abe, K. Sumoto, and K. Teramato, *J.C.S. Perkin I*, 1146 (1976).
[99] A. I. Kiprianov and V. A. Portnyagina, *Zh. Obshch. Khim.* **25**, 2257 (1955) [*CA* **50**, 9378 (1950)].

involving an intramolecular nucleophilic attack on an enamino ketone (**50**) was postulated to account for the results in DMSO.[98] This procedure has been used in the preparation of a series of 4H-1,4-benzothiazines **48** substituted in the aromatic ring or carrying fluorine in the side chains.[100,101]

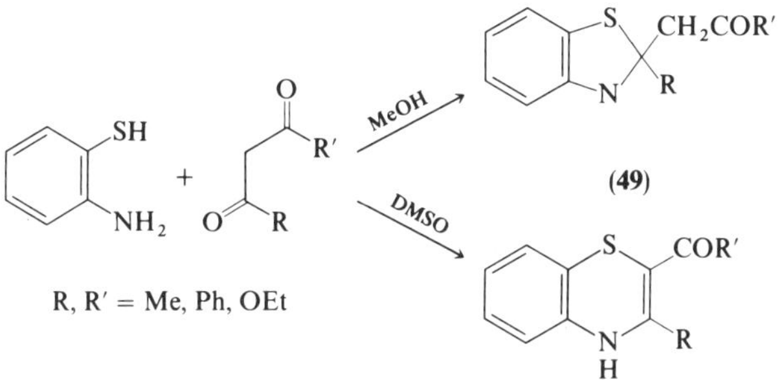

6. *Reaction with Other Compounds*

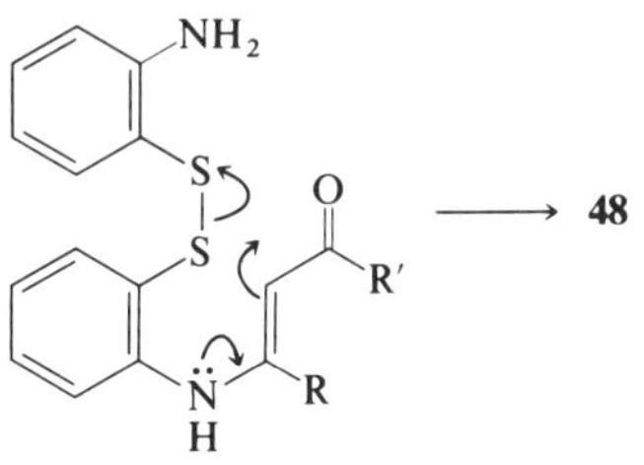

Reactions of 2-aminobenzenethiol with a variety of other compounds provide the 1,4-benzothiazine ring system. An early report by Langlet claimed that the reaction with ethylene dibromide gave dihydro-1,4-benzothiazine (**51**).[102] However, more recent work by Santacroce indicates the product is 2-methylbenzothiazole (**52**).[94] However, 1,2-dichloroalkenes have been employed successfully by Kaul in the preparation of heterocyclic dye **53** having the 1,4-benzothiazin-3-one ring.[35]

---

[100] R. R. Gupta, K. G. Ojha, G. S. Kalwania, and M. Jumar, *Heterocycles* **14**, 1145 (1980).
[101] R. P. Soni and M. L. Jain, *Tetrahedron Lett.* **21**, 3795 (1980).
[102] N. A. Langlet, *Bi. Sven. Vetenskaps akad. Handl.* **2211**, No. 1, 8 (1896) [*Beilstein* **27**, 34].

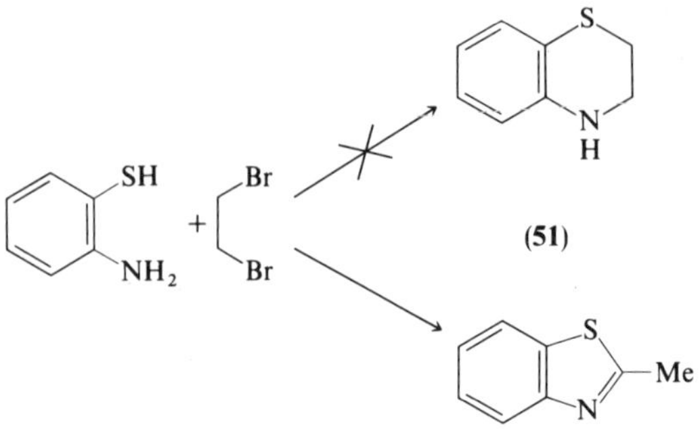

Hirano et al. prepared 2,3-dihydro-3-imino-4H-1,4-benzothiazine (**55**), using chloroacetonitrile under acid conditions[103] via the intermediate **54**, previously assigned structure **55** by Riolo.[104] These workers also reported the intriguing conversion of the benzothiazolium salt (**56**) to 4-methyl-3-iminodihydro-1,4-benzothiazine (**57**).[103] This reaction appears to involve

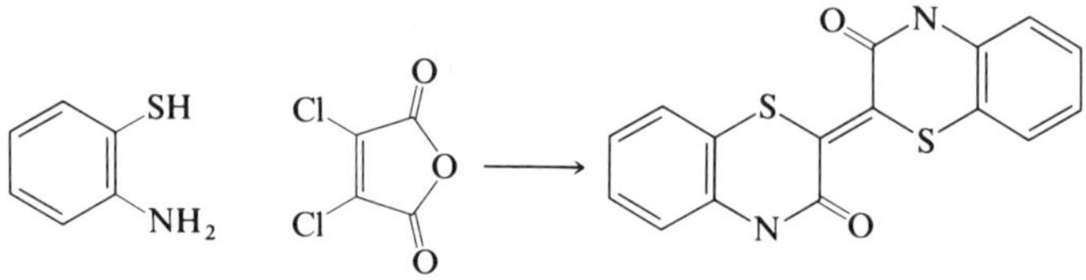

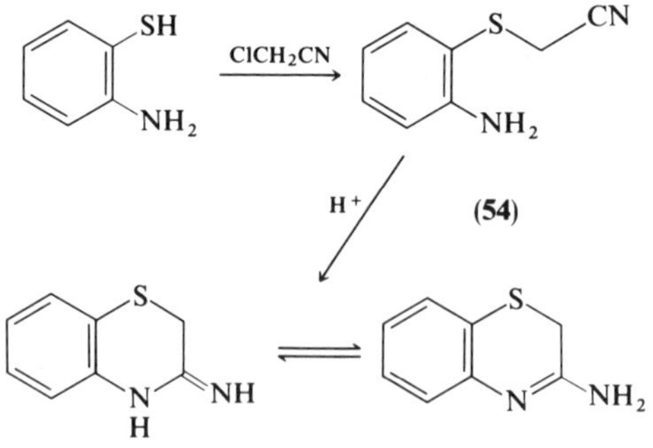

[103] H. Hirano, M. Takamatsu, K. Sugiyama, and T. Kurihara, *Chem. Pharm. Bull.* **27**, 374 (1979).
[104] C. B. Riolo, *Ann. Chim.* (*Rome*) **45**, 1174 (1955).

prior deformylation of the nitrogen, in contrast to the work previously reported by Friedrich[12] (see Section VI,B). 2,3-Diimino-4H-1,4-benzothiazine (58) has been prepared by Weidinger and Kranz, starting from dimethyl oxaldiimidate.[105]

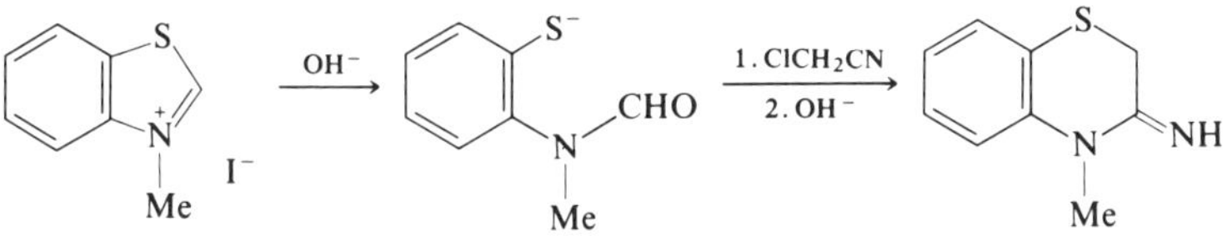

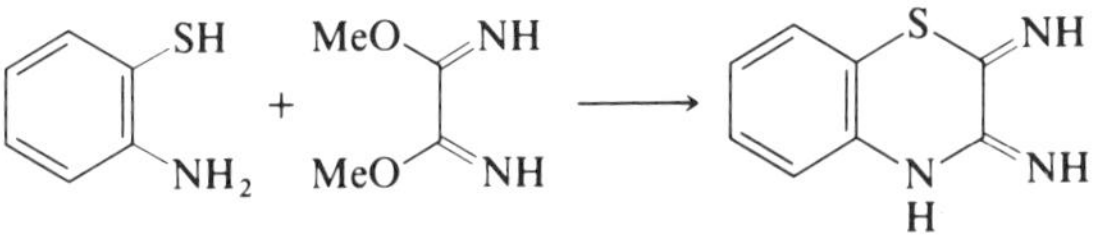

Miyano et al. observed an interesting cyclization of the vinylogous thiol ester (59), prepared from 2-aminobenzenethiol and 3-chloro-5,5-dimethylcyclohex-2-enone. On heating in DMSO, 59 gave benzothiazine 60. A

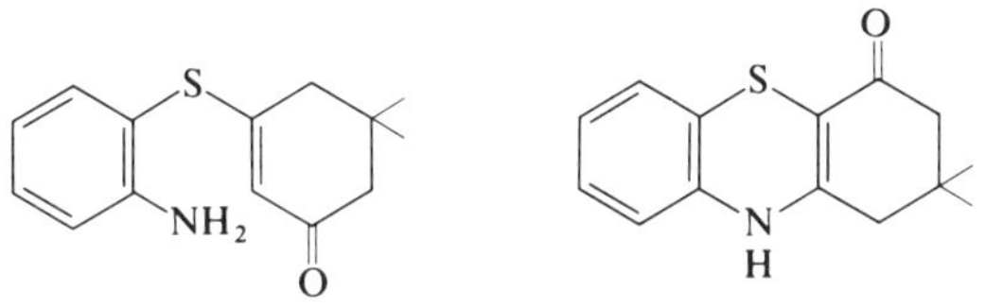

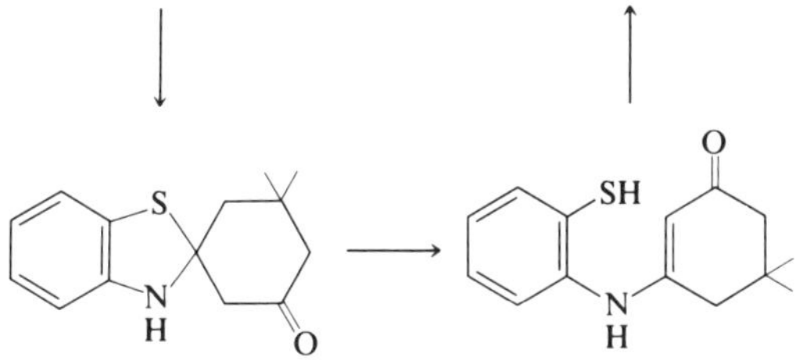

[105] H. Weidinger and J. Kranz, Chem. Ber. 97, 1599 (1964).

route involving the rearrangement of **59** to **62** via the spiro compound **61**, followed by oxidative cyclization of **62**, was proposed.[98]

B. Preparation by Ring Expansion

2-Aminobenzenethiol reacts readily with carbonyl compounds to give 1,3-benzothiazoles (**63**).[106] Certain derivatives of this ring system have been found to undergo ring expansion under specific conditions to give benzothiazines. The first report in the literature of this type of reaction involved the treatment of benzothiazolium salt **64** with base; the postulated mechanism[12] is shown in Scheme 1.

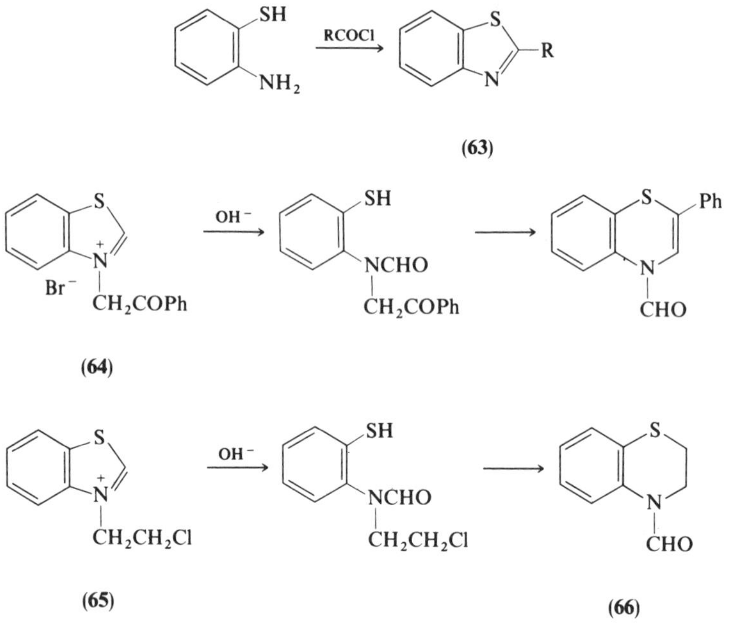

Scheme 1

On the other hand, treatment of *N*-(2-chloroethyl)benzothiazolium salt **65** with base affords the *N*-formyldihydro-1,4-benzothiazine **66** through ring

---

[106] R. C. Elderfield, ed., "Heterocyclic Compounds," Vol. 5, p. 506. Wiley, New York, 1957.

closure by S-alkylation.[107] Treatment of 3-methylbenzothiazolium with phenacyl bromide in the presence of triethylamine also results in ring expansion and formation of a 1,4-benzothiazine.[108]

Chioccara et al. reported ring expansion of benzothiazoline derivatives (67), using sulfuryl chloride in anhydrous dichloromethane to give N-acetylated 1,4-benzothiazines (69) accompanied by small quantities of the corresponding keto sulfides (68), which could also be cyclized in the presence of acid to the benzothiazines (69). The presence of keto sulfides was ascribed to ring opening of the intermediate benzothiazinyl cations by traces of water in the "dry" dichloromethane.[109] In the case of 67 (R = Ph, R' = Me), two isomeric thiazines (70 and 71) in the ratio 3:1 were obtained, elimination of the benzylic proton evidently being little favored over elimination of the methyl proton in

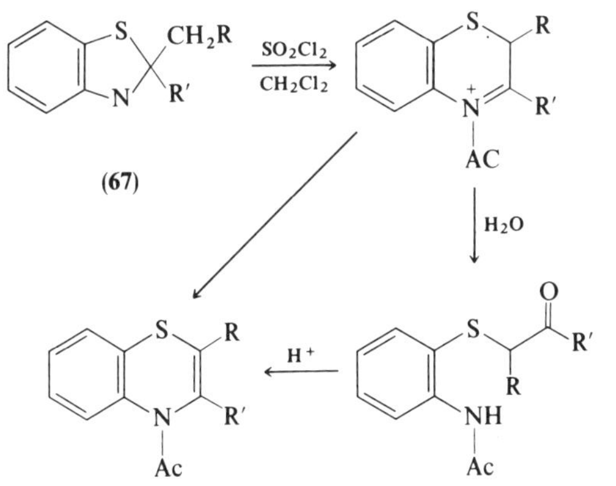

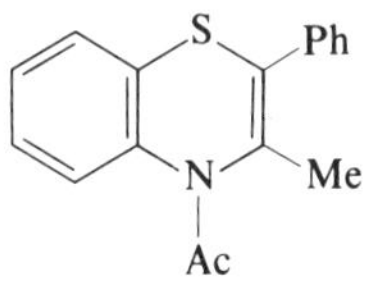

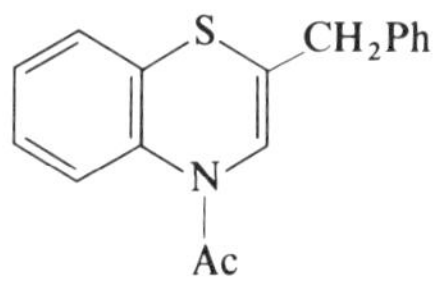

[107] H.-J. Federsel and J. Bergman, Tetrahedron Lett. 21, 2429 (1980).
[108] J. A. Van Allen, J. D. Mee, C. A. Magguilli, and R. S. Menion, J. Heterocycl. Chem. 12, 1005 (1975).
[109] F. Chioccara, R. A. Nicolaus, E. Novellino, and G. Prota, Chim. Ind. (Milan) 58, 546 (1976).

formation of the intermediate thiazinyl cations.[110] Similar reactions in dry diethyl ether give comparable yields.[111,112]

It was subsequently observed that when an excess of sulfuryl chloride was employed instead of 1 equivalent, the product was the dimer 2,2'-(ethane-1,2-diylidene)-bis-1,4-benzothiazine (72), with the cyanine dye linkage.[113]

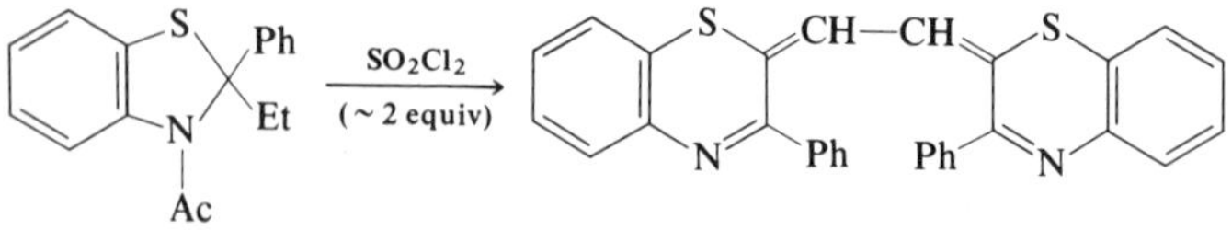

(72)

Benzothiazolines undergo ring expansion in boiling DMSO to give 4H-1,4-benzothiazines (73) in low to moderate yield.[114] Similar treatment of N-acetylated benzothiazolines results in an N → C acyl migration to give the benzothiazoles (74), rather than N-acylbenzothiazines.[115] It is possible to transform in low yields benzothiazolines to 2,3-dihydro-4H-1,4-benzothiazin-3-ones, using chloroacetyl chloride.[116]

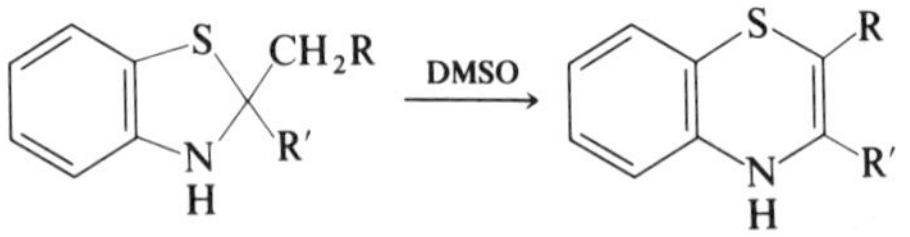

(73)

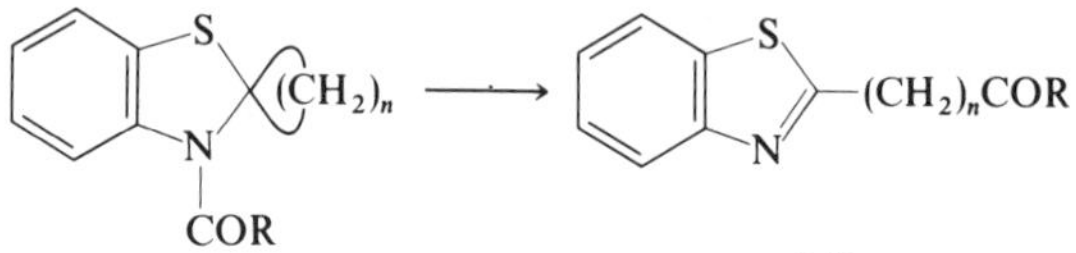

(74)

Prota et al. and a group in Japan independently reported ring expansion reactions of benzothiazoline S-oxides, which are mechanistically similar to the

---

[110] F. Chioccara, G. Prota, R. A. Nicolaus, and E. Novellino, Synthesis, 876 (1977).
[111] F. De Simone, A. Dini, E. Ramundo, and G. La Bella, Rend. Accad. Sci. Fis. Mat., Naples 44, 387 (1977) [CA 90, 54891 (1979)].
[112] F. De Simone, A. Dini, R. A. Nicolaus, E. Ramundo, M. Di Rosa, and P. Persico, Farmaco, Ed. Sci. 35, 333 (1980) [CA 93, 132429 (1980)].
[113] F. Chioccara, V. Manciacapra, E. Novellino, and G. Prota, J.C.S. Chem. Commun., 863 (1977).
[114] G. Liso, G. Trapani, A. Latrofa, and P. Marchini, J. Heterocycl. Chem. 18, 279 (1981).
[115] G. Liso, G. Trapani, A. Reho, and A. Latrofa, Tetrahedron Lett. 1641 (1981).
[116] M. Hori, T. Kataoska, M. Shimizu, and Y. Imai, Heterocycles 9, 1413 (1978).

sulfuryl chloride reactions described above.[117-120] These Italian workers reported fair yields of N-acetyl-4H-1,4-benzothiazines (**69**) from the N-acetylbenzothiazolines **67** on treatment with 3-chloroperbenzoic acid in chloroform.[117] The reaction appears to work equally well for either the cis or trans isomers of the initially formed sulfoxides but, apart from one case, does not appear to offer any significant advantage over the sulfuryl chloride procedure.

Hori et al. reported the ring expansion of 3-acetylbenzothiazoline 1-oxide (**75**) in refluxing acetic anhydride[118-120] to give 4-acetylbenzothiazines (**76**) via a sulfenic anhydride intermediate; this reaction is nonstereospecific, supporting the proposed nonconcerted mechanism.

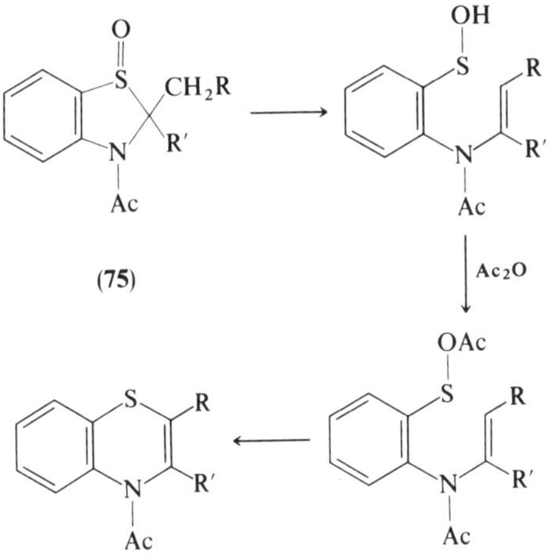

The reaction is unusual in that sulfoxides processing an α-hydrogen atom normally undergo a Pummerer reaction under these conditions to form the acetoxy sulfide. The reaction can be compared with the Morin rearrangement of penicillin sulfoxide to cephalosporins.[121]

Ring expansion of benzothiazoles (**77**) with dimethylacetylenedicarboxylate (DMAD) in aqueous methanol requires water as an essential agent to

---

[117] F. Chioccara, L. Oliva, G. Prota, and E. Novellino, *Synthesis*, 744 (1978).
[118] M. Hori, T. Kataoka, H. Shimizu, and Y. Imai, *Chem. Pharm. Bull.* **27**, 1982 (1979).
[119] M. Hori, T. Kataoka, H. Shimizu, Y. Imai, and N. Ueda, *Fukusokan Kagaku Toronkai Koen Yoshishu, 12th*, 61 (1979) [*CA* **93**, 71673 (1980)].
[120] M. Hori, T. Kataoka, H. Shimizu, and N. Ueda, *Tetrahedron Lett.* **22**, 1701 (1981).
[121] P. G. Sammes, *Chem. Rev.* **76**, 113 (1976).

trap the benzothiazolium zwitterion intermediate **78** in contrast to the routes already described, which rely on anhydrous conditions.[122]

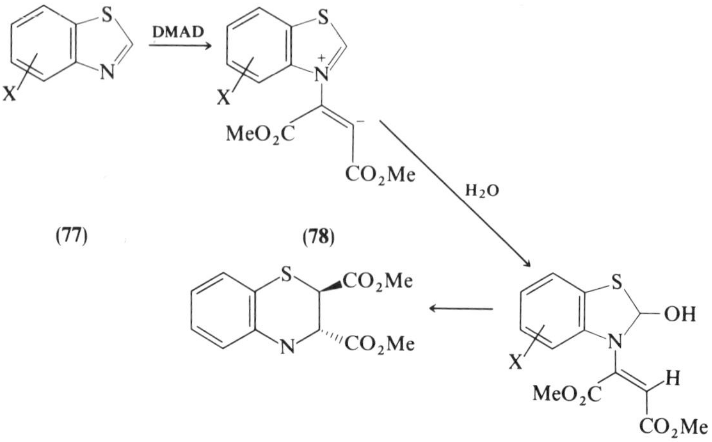

Preparation by ring expansion can be contrasted with that of ring contraction recently reported, whereby 1,6-benzothiazocine derivatives (**79**) undergo ring contraction in the presence of lead tetraacetate to give the 2*H*-1,4-benzothiazine derivatives **80** and **81**. The proposed mechanism for the reaction involves a transannular sulfur migration. The structure of **81** was confirmed by X-ray crystallography.[123]

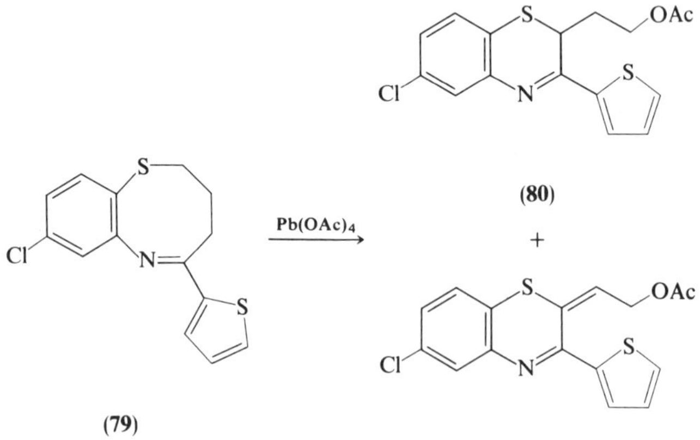

---

[122] A. McKillop, T. S. B. Sayer, and G. C. A. Bellinger, *J. Org. Chem.* **41**, 1328 (1976).
[123] J. B. Press, N. H. Eudy, F. M. Lovell, and N. A. Perkinson, *Tetrahedron Lett.* **21**, 1705 (1980).

C. Preparation from 2-Nitrobenzenethiols

Syntheses of the heterocycle from 2-nitrobenzenethiol involving reductive cyclization methods have been employed by a number of research groups.

1. Reaction with Unsaturated Carbon–Carbon Bonds

2-Nitrobenzenethiol will add across the double bond of maleic anhydride[124]; the adduct can then be cyclized in the presence of zinc and acetic acid to give 1,4-benzothiazin-3-one-2-acetic acid (**82**). Similar studies with citraconic acid and the acid anhydride, where the thiol added predominantly to the position β to the methyl in contrast to an earlier report,[125] gave the citraconic anhydride and acid adducts. When treated with zinc and acetic acid, the adducts formed diastereoisomers (**83a** and **83b**) of 1,4-benzothiazin-3-onepropanoic acid.

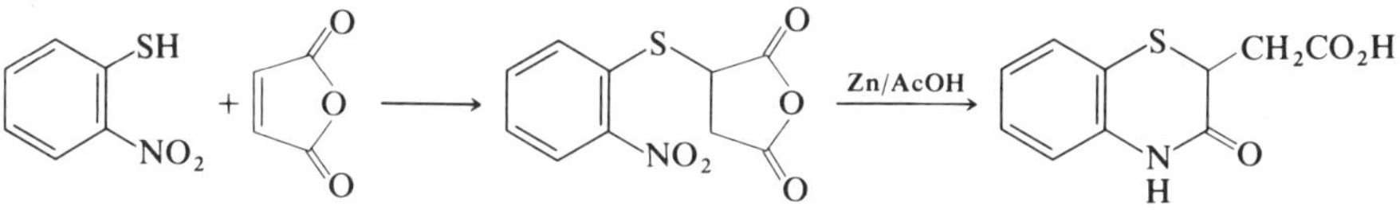

(82)

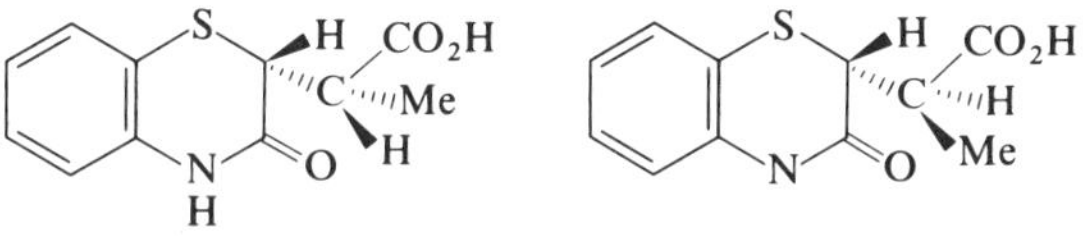

(83a)　　(83b)

2. Reaction with α-Halo Acids and Esters

As early as 1897, Friedlander treated 2-nitrobenzenethiol with chloroacetic acid to obtain the thioacetic acid adduct, which then was cyclized with zinc in acetic acid to dihydro-1,4-benzothiazin-3-one.[126] A similar synthesis by

---

[124] J. Bouroais, *C. R. Hebd. Seances Acad. Sci. Ser. C* **262**, 1701 (1966).
[125] F. B. Zienty, B. D. Vineyard, and A. A. Schleppnik, *J. Org. Chem.* **27**, 3140 (1962).
[126] H. Friedlander and C. Chwala, *Monatsh. Chem.* **28**, 271 (1897).

Claasz generated the mercaptan *in situ* from the disulfide. Claasz also treated α-bromo esters with the thiol and subjected the adducts to reductive cyclization, which produced the benzothiazinones.[127]

A different reduction was employed by Coutts *et al.* on β-(2-nitrophenylthio) esters to give hydroxamic acid derivatives (84), sodium borohydride with palladium–charcoal being used to effect the cyclization.[128]

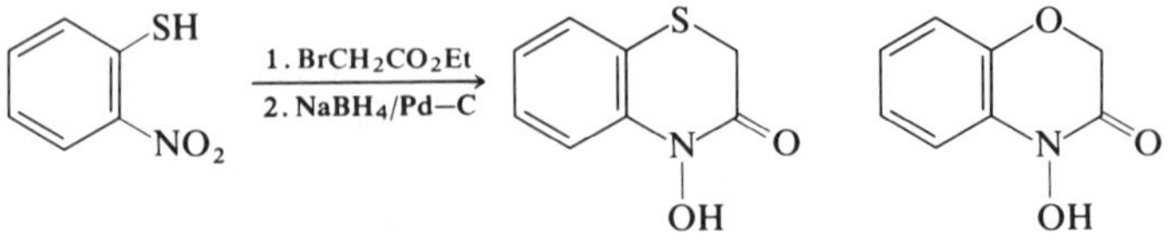

(84)   (85)

Angelini described the preparation of dihydro-1,4-benzothiazine and related compounds from analogous esters by a two-stage reduction using tin and HCl or lithium aluminum hydride.[129] It is interesting to compare the reaction of 2-nitrophenol with similar reagents to produce benzooxazine derivatives. 2-Nitrophenol reacts with α-bromoethyl acetate, and the adduct is cyclized with zinc and ammonium chloride to the hydroxamic derivative 85.[130] The adduct with phenacyl chloride has been similarly cyclized under a variety of conditions.[131,132]

3. Reaction with Halides

4H-1,4-Benzothiazine 1,1-dioxide (87) has been prepared by the action of allyl bromide on the sulfinic acid 86, followed by ozonolysis and catalytic hydrogenation.[20]

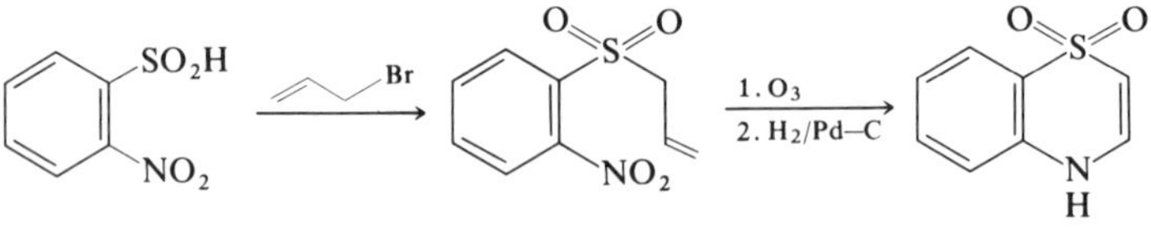

(86)   (87)

---

[127] M. Claasz, *Chem. Ber.* **45**, 2424 (1912).
[128] R. T. Coutts, H. W. Peel, and E. M. Smith, *Can. J. Chem.* **43**, 3221 (1965).
[129] C. Angelini and G. Grandolini, *Ann. Chim.* (*Rome*) **46**, 235 (1956).
[130] E. Honkanen and A. Virtanen, *Acta Chem. Scand.* **14**, 1214 (1960).
[131] P. Battistoni, P. Bruni, and G. Fava, *Synthesis*, 220 (1979).
[132] P. Battistoni, P. Bruni, and G. Fava, *Tetrahedron* **35**, 1771 (1979).

Sec. VI.D]   1,4- AND DIHYDRO-1,4-BENZOTHIAZINES   161

Fusco reported to have prepared 3,4-dihydro-1,4-benzothiazine by treating 2-nitrobenzenethiol with β-chloroethanol and cyclizing the intermediate with tin and hydrochloric acid.[85] Later work, however, showed the product to be 2-(2-aminophenylthio)ethanol.[86–89,93]

4. Other Reactions with the SH Group

In addition to the 2-amino and 2-nitro derivatives, various other substituted benzenethiols have been used as precursors for the benzothiazine system. 6,7-Dimethoxy-1,4-benzothiazin-3-one (**88**) has been prepared

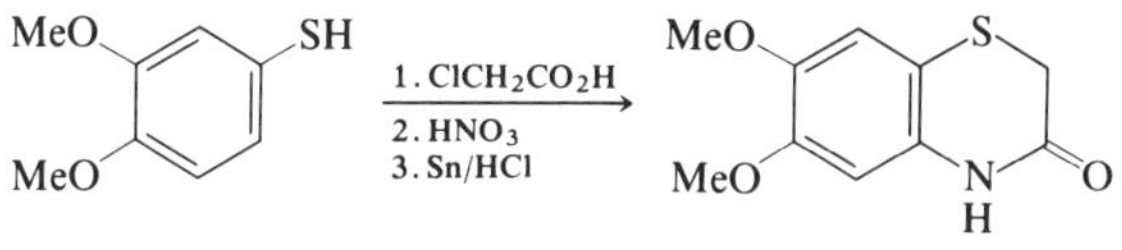

(**88**)

from 3,4-dimethoxybenzenethiol.[133] 2-Chlorobenzenesulfinic acid, α-bromoacetone, and ammonia give 3-methyl-4H-1,4-benzothiazine 1,1-dioxide (**89**).[134]

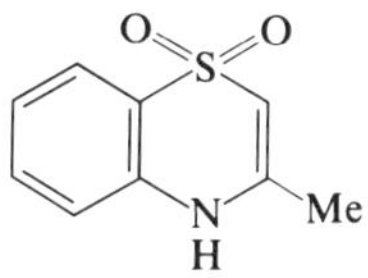

(**89**)

D. PREPARATION FROM 2-NITROBENZENESULFENYL CHLORIDES

All of the synthetic routes reviewed so far have primarily involved the use of sulfur as a nucleophile. Sulfur can also act as an electrophile in the formation of precursors of the benzothiazine ring system. Reaction of sulfenyl chloride with ketones gives β-keto sulfides. The acetophenone adduct **90** has been cyclized with stannous chloride and hydrogen chloride in acetic acid to give 3-phenyl-6-chloro-2H-1,4-benzothiazine (**91**).[135]

[133] K. J. Baldick and F. Lions, *J. Proc. R. Soc. N.S.W.* **71**, 112 (1938) [*CA* **32**, 34006 (1938)].
[134] G. Pagani and S. Maiorana, *Chim. Ind.* (*Milan*) **53**, 468 (1971).
[135] T. Zinke and J. Baeumer, *Justus Liebigs Ann. Chem.* **416**, 86 (1918).

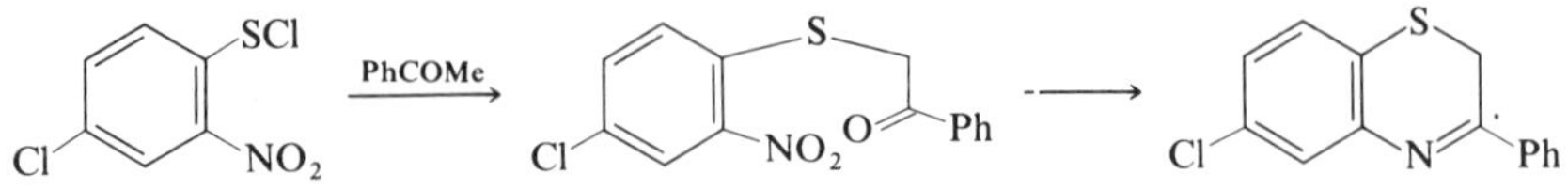

(90) (91)

Dihydro-1,4-benzothiazines have been prepared in good yield by catalytic reduction of β-keto sulfides at elevated temperatures and pressures using a variety of catalysts.[78] Coutts et al. unsuccessfully attempted to prepare benzothiazines by reductive cyclization of β-keto sulfides with sodium borohydride or palladium–charcoal. They isolated only the β-(2-nitrophenylthio) alcohols,[136] in contrast to the successful synthesis of the desired ring by the same reduction of β-(2-nitrophenylthio) esters.[128] A wide range of β-keto sulfides has been successfully reduced to dihydro-1,4-benzothiazines, using tin and hydrochloric acid, in either glacial acetic acid or ether solution.[21]

The sulfide 92, prepared by treatment of N-methyl-3-methylindole with 2-nitrobenzenesulfenyl chloride, was converted to the rearranged benzothiazine 93 by treatment with triethyl phosphite under nitrogen.[137]

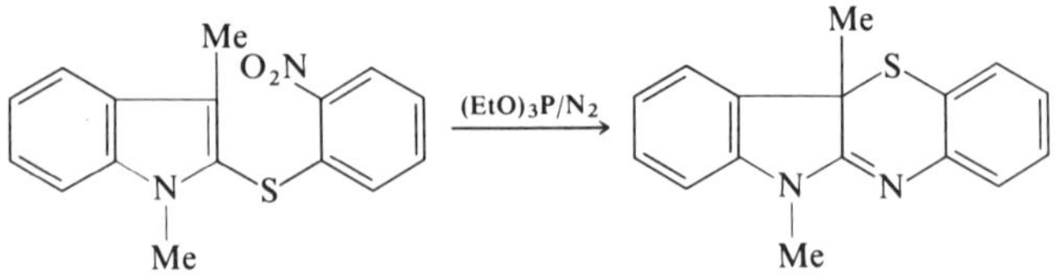

(92) (93)

1,3-Diketones are also readily sulfenylated in the 2-position by benzenesulfenyl halides, and Shvedov et al. have reported the reductive cyclization of products such as 94 with zinc in acetic acid to benzothiazines

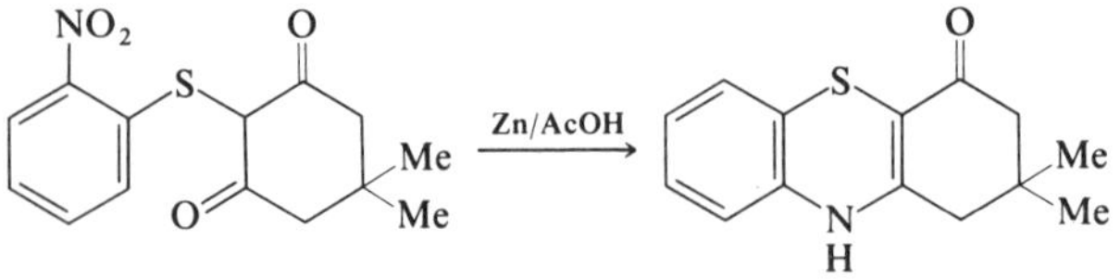

(94) (95)

[136] R. T. Coutts, S. J. Matthias, and H. W. Peel, Can. J. Chem. 48, 2448 (1970).
[137] A. H. Jackson, D. N. Johnston, and P. V. R. Shannon, J.S.C. Chem. Commun., 911 (1975).

95.[138] However, attempts to reduce the analogous product from 2,4-pentanedione gave 3-methyl-3,4-dihydro-1,4-benzothiazine[21] together with 2-methylbenzothiazole. The latter had previously been reported to be the product of this reduction, using stannous chloride.[16]

E. PREPARATION FROM DISULFIDES, CYSTEINE DERIVATIVES, AND OTHER REAGENTS

It has been found possible to construct the 1,4-benzothiazine ring by direct interaction of bis-(2-aminobenzene) disulfide (**96**) with carbonyl compounds.[139,140] The reaction is most efficient when conducted under a nitrogen atmosphere with a 1:1 ratio of reactants; otherwise, the principle products are benzothiazoles. While reduction of the benzothiazine **97** with sodium borohydride gives a stable dihydro derivative, the unsaturated benzothiazines themselves were prone to autoxidation, giving rise to benzothiazoles and benzothiazine sulfoxides.[141]

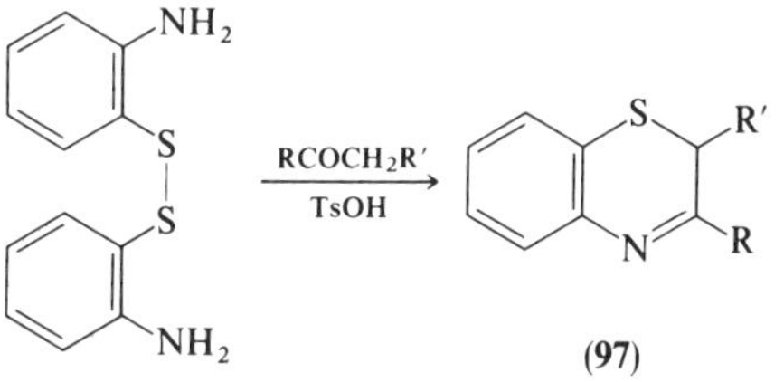

The postulated mechanism for the formation of the 1,4-benzothiazine ring involves the formation of the imine **98**, followed by cleavage of the disulfide bond (see Section VI,A,5). This reaction is general for keto compounds, β-keto esters,[142] and acetylenic ketones and esters[143]; the heterocycle is produced in yields of 70–80%.

[138] V. I. Shvedov, L. B. Altukhova, V. M. Lyubchanskaya, and A. N. Grinev, *Khim. Geterosikl. Soedin.*, 1509 (1972) [*CA* **78**, 43395 (1973)].
[139] V. Carelli, P. Marchini, F. Micheletti-Moracci, and G. Liso, *Tetrahedron Lett.*, 3421 (1969).
[140] V. Carelli, P. Marchini, M. Caroellini, F. Micheletti-Moracci, G. Liso, and M. G. Lucarelli, *Tetrahedron Lett.*, 4619 (1969).
[141] V. Carelli, F. Micheletti-Moracci, F. Liberatore, M. Cardellini, M. G. Lucarelli, P. Marchini, G. Liso, and A. Reho, *Int. J. Sulf. Chem.* **8**, 267 (1973).
[142] P. Marchini, G. Trapani, G. Liso, and V. Berardi, *Phosphorus and Sulphur* **3**, 309 (1977).
[143] G. Liso, G. Trapani, V. Bernardi, and P. Marchini, *J. Heterocycl. Chem.* **17**, 377 (1980).

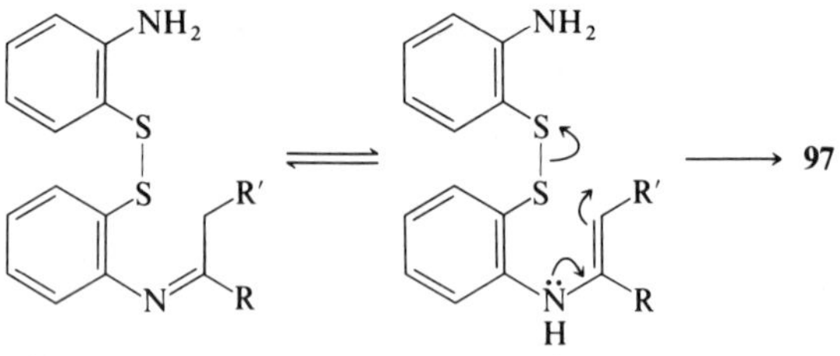

(98)

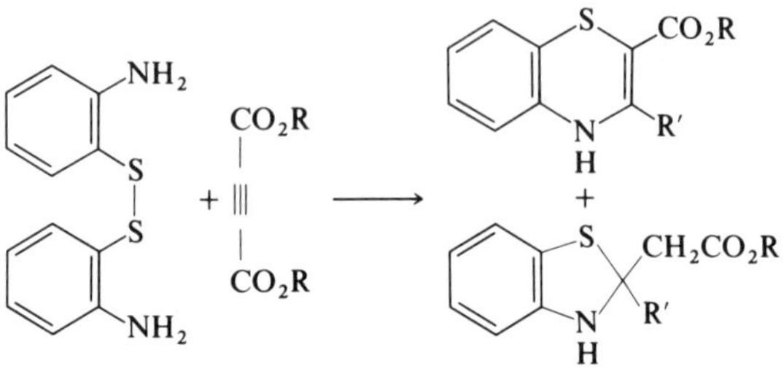

The involvement of cysteine in the biosynthesis of phaeomelanins and trichochromes, natural pigments in mammals, has prompted work on the reaction of cysteine with a variety of reagents to produce the 1,4-benzothiazine ring system.

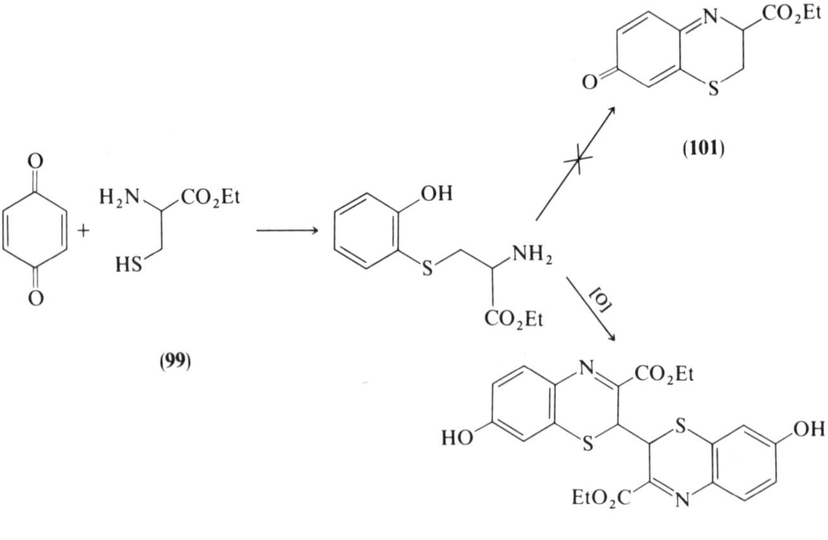

Cysteine ethyl ester (**99**) reacts with *p*-benzoquinone[41] to give a dimeric product (**100**) and not the quinonoid structured product (**101**) as reported by Kuhn and Hammer.[144]

The reaction of "epoxydon" (**102**) with cysteine hydrochloride has been found to give a 1,4-benzothiazine derivative (**103**).[145]

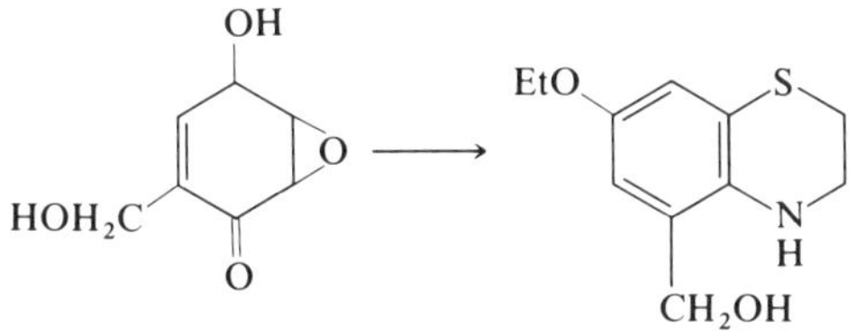

The formation of the 1,4-benzothiazin-3-one (**104**) has been observed in the course of sequential analysis of proteins, using 2-methyl-*N*-benzenesulfonyl-*N'*-bromoacetylquinone diimide.[146] The protein chain is cleaved at the cysteine amino acid unit, with cysteine forming a benzothiazinone with the diimide reagent.

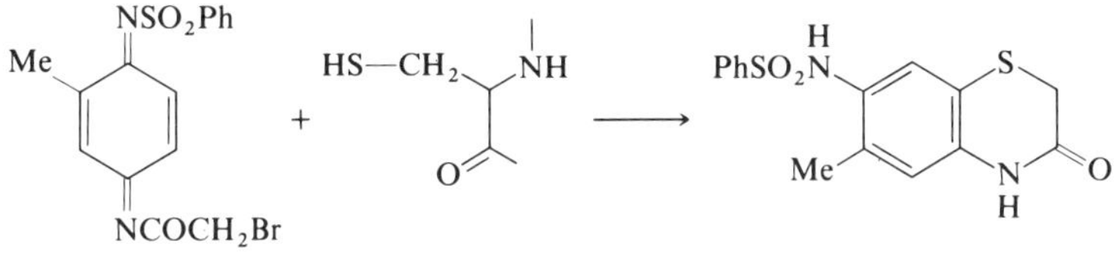

F. OTHER ROUTES

Various other routes have been reported to give the 1,4-benzothiazine ring, some of which are summarized below. Synthesis of the 1,4-benzothiazine 1,1-dioxides (**105**) from benzaldimine and methylene sulfone ylide has been reported.[147]

---

[144] R. Kuhn and I. Hammer, *Chem. Ber.* **84**, 91 (1951).
[145] K. Nabeta, A. Ichihara, R. Sakai, and S. Sakumara, *Agric. Biol. Chem.* **36**, 2261 (1972).
[146] T. J. Holmes and R. G. Lawton, *J. Am. Chem. Soc.* **99**, 1984 (1977).
[147] M. Rai, S. Kumar, K. Krishan, and A. Singh, *Chem. Ind. (London)*, 26 (1979).

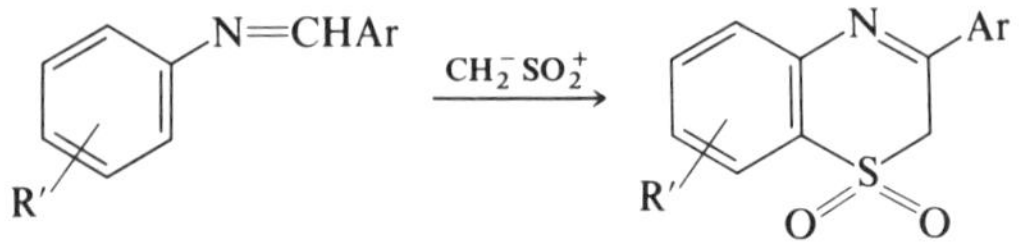

(105)

An attempt to prepare triorganoantimony compounds containing tridentate ligands failed as a result of their spontaneous decomposition to triorganostibenes R_3Sb and 2-acetyl-3-methyl-4H-1,4-benzothiazine (106).[148]

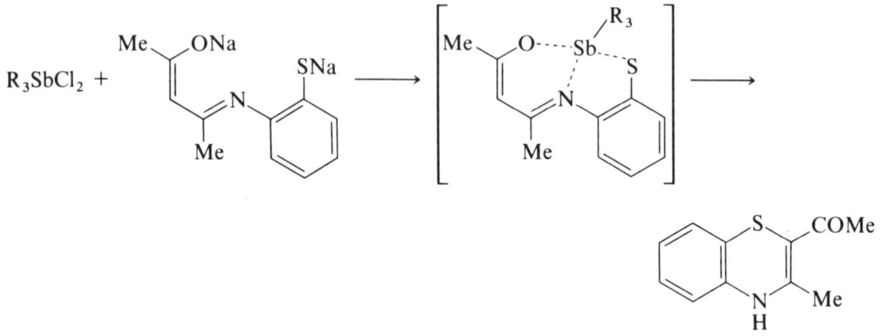

(106)

The formation of thiazine was explained in terms of the previously observed instability of triorganoantimony dimercaptides, which spontaneously decompose with the formation of R_3Sb and RSSR.[149,150]

Secondary amine derivatives containing sulfur have been cyclized to give 1,4-benzothiazine rings. Cyclization under base conditions have been studied by Knyazev et al., using 1-(β-N-aminoethylthio)-2,4,6-trinitrobenzene (107). On cyclization, this produced 6,8- (108) and/or 5,7-dinitro-4-methyl-2,3-dihydro-1,4-benzothiazine (109), the ratio of isomers produced depending on the base and solvent used to effect cyclization.[151] The secondary amides 110 can also be cyclized to give the 1,4-benzothiazine ring. Heating at 180–200°C produced 1,4-benzothiazin-3-one, which on treatment with lithium aluminum hydride gave dihydro-1,4-benzothiazine.[152]

---

[148] F. Di Bianca, E. Rivarola, A. L. Spek, H. A. Meinema, and J. G. Noltes, *J. Organomet. Chem.* **63**, 293 (1973).

[149] Y. Matsumara, M. Shindo, and R. Okawara, *Inorg. Nucl. Chem. Lett.* **3**, 219 (1967).

[150] H. Schmidbaur and K. Mitschke, *Chem. Ber.* **104**, 1824 (1971).

[151] V. N. Knyazev, V. N. Drozd, and V. M. Minov, *Tetrahedron Lett.*, 4825 (1976).

[152] R. N. Prasad and K. Tietje, *Can. J. Chem.* **44**, 1247 (1966).

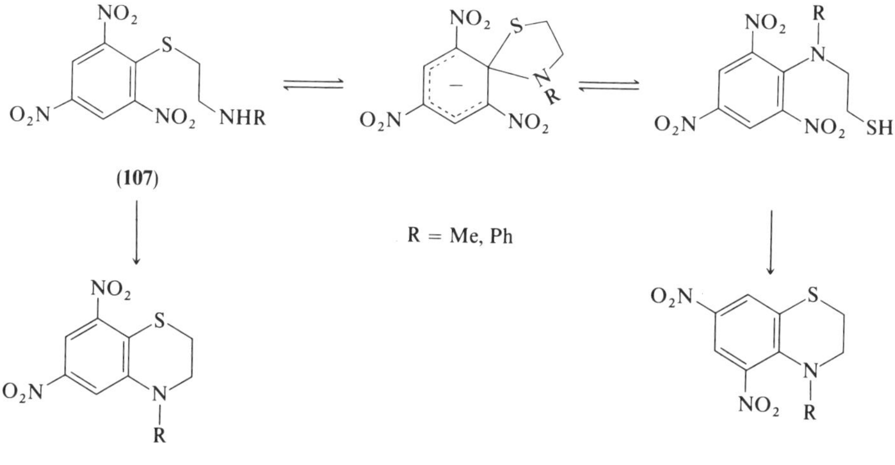

Sec. VII.A] 1,4- AND DIHYDRO-1,4-BENZOTHIAZINES

(107) ⇌ [R = Me, Ph] ⇌

(108)   (109)

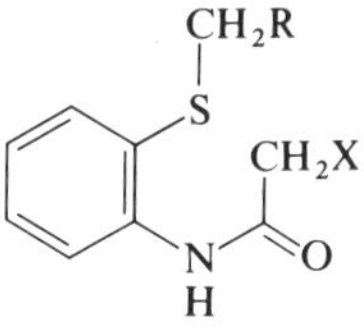

(110) R = H, Me, Ph; X = Cl, Br

As outlined in Section II, the reaction between benzoquinones, or their phenolic precursors, and 2-aminoethanethiols leads to a variety of benzothiazines, depending on reaction conditions.[41–43]

VII. Reactions of 1,4-Benzothiazines

The 1,4-benzothiazines do not have a particularly well studied chemistry; with certain exceptions (see below), most efforts have been expended on their preparation. However, some reactions have received some attention, and these are summarized in this section.

A. N-Alkylation and N-Acylation

N-Alkylation of the 4H-1,4-benzothiazine 95 has been reported.[138] Kiprianov et al. described the quaternary salts of some 2H-1,4-benzothiazines which are less conveniently prepared by direct N-alkylation than by synthesis from the appropriate N-substituted 2-aminobenzenethiols and α-halo ketones.[93,153]

Dihydro-1,4-benzothiazines are generally more reactive, and direct N-alkylation has been reported, but this reaction is not always straightforward.[21] Funke et al. could not alkylate 2-phenyldihydro-1,4-benzothiazine with ω-chloramines, even under forcing conditions.[90] However, later work showed that this reaction was possible in toluene solution,[57] and other workers have also reported direct alkylations.[143] The 1,4-benzothiazin-3-ones are, however, more easily alkylated, and reduction of the N-alkyl derivatives of these compounds, usually with lithium aluminum hydride, affords the corresponding N-alkyldihydro-1,4-benzothiazines.[52,56,70,90,154] These products can also be prepared in one step from the corresponding 1,4-benzothiazines, e.g., **111** → **112**, presumably via intermediate dihydro-1,4-benzothiazines, by sodium borohydride in the presence of a carboxylic acid. Boron derivatives, such as $Na[(RCOO)_3BH]$ and $Na[(RCOO)_4B]$ are suggested as the species responsible for N—C bond formation.[155]

(111) $\xrightarrow{CH_3CO_2H, NaBH_4}$ (112)

N-Acylation of dihydro-1,4-benzothiazines is, however, a smooth process, and most of the usual reagents and reaction conditions have been successfully employed.[21,50,55,64,152] Reduction of the acyl derivatives with lithium aluminum hydride affords the corresponding N-alkyl compounds.[54,152]

Hirano[156] observed that when attempting to prepare the acetyl derivatives of 2,3-dihydro-3-imino-4H-1,4-benzothiazine (**55**) with acetic anhydride, the product obtained was bis-[1,4]-benzothiazino [3,2-b; 2'3'-e]pyridine (**113**). A similar reaction[103] with 4-methyl-3-iminodihydro-1,4-benzothiazine (**114**) produced 6,7,9,16-tetramethyl-8,15-diazophenothiazino[8,7-h]-phenothiazine (**115**).

---

[153] A. I. Kiprianov and Z. N. Pazenko, *Zh. Obshch. Khim.* **21**, 156 (1951) [*CA* **45**, 7574 (1951)].

[154] J. Piechaczeck and H. Bojarska-Dahlig, *Acta Pol. Pharm.* **24**, 244 (1967) [*CA* **69**, 19098 (1968)]. (1968)].

[155] P. Marchini, G. Liso, and A. Reho, *J. Org. Chem.* **40**, 3453 (1976).

[156] H. Hirano, K. Sugiyama, S. Nagata, and T. Kurihara, *Chem. Pharm. Bull.* **27**, 2488 (1979).

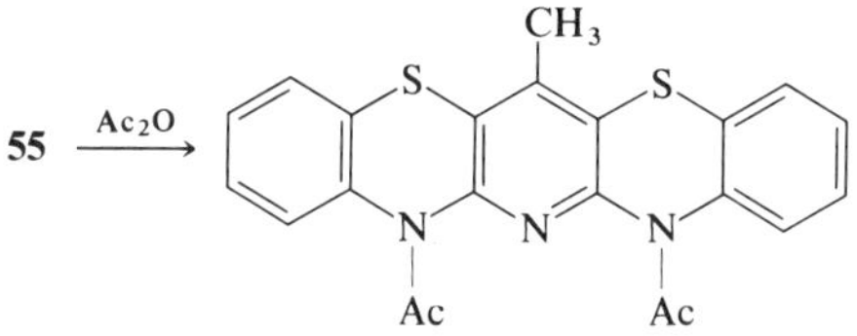

Sec. VII.B] 1,4- AND DIHYDRO-1,4-BENZOTHIAZINES 169

(113)

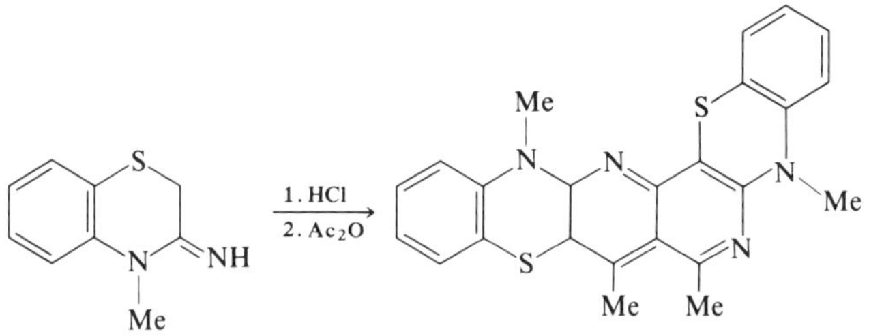

(114)   (115)

N-Acyl-2,3-dihydro-1,4-benzothiazines and the related phenothiazines react with both organolithium and organomagnesium reagents to give the β-dicarbonyl derivative 116, indicating that the N-acyl derivatives may have a role to play as useful acylating agents.[157] Mechanistic studies have provided novel explanations for this behavior and that of the related 1-oxides and 1,1-dioxides (117).[158]

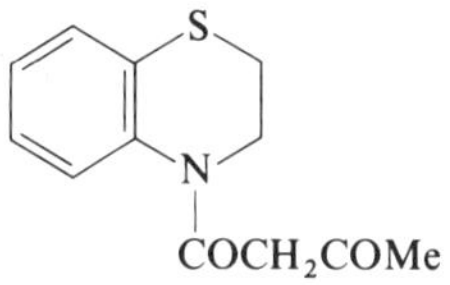

(116) 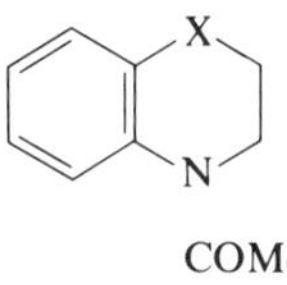   (117) X = SO, SO_2

B. Electrophilic Substitution

There have been relatively few reports of electrophilic substitution reactions with this system. One example is the reaction of the indolobenzothiazine 118

[157] F. Ciminale, L. DiNunno, and S. Florio, *Tetrahedron Lett.*, 3001 (1980).
[158] S. Florida, J. L. Leng, and C. J. M. Stirling, *J. Heterocycl. Chem.* **18**, 857 (1981).

with oxygen to give the spirooxindole **119**. This may proceed via the electrophilic attack of oxygen on the activated 2-position of **118** to give the hydroperoxide **120** and hence **119** via the hydroxy species **121**.[137] The analogous autoxidation reaction of simpler benzothiazines to give 2-acyl-benzothiazolines has been shown to be a radical chain process.[141]

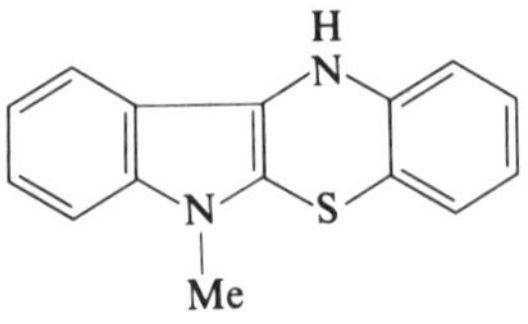

(118) 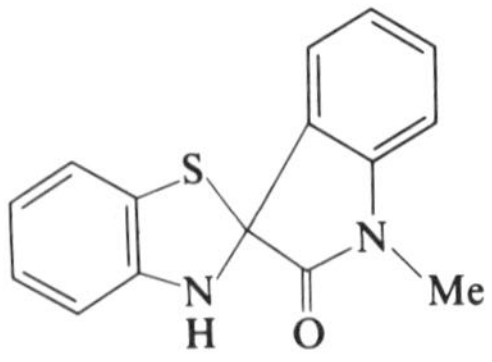 (119) 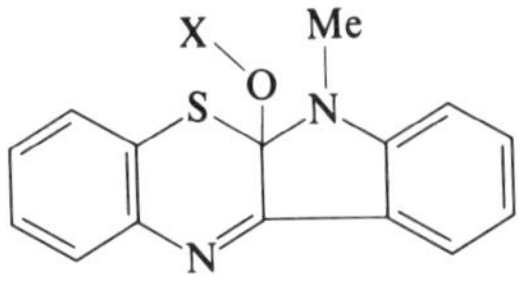 (120) X = OH
              (121) X = H

4H-1,4-Benzothiazine 1,1-dioxide undergoes a variety of electrophilic substitutions, including nitration, bromination, Vilsmeier formylation, and the Mannich reaction. All substitutions occur at the 2-position, and an analogy with indole was drawn by the authors.[20]

3,4,-Dihydro-2H-1,4-benzothiazin-3-one and N-alkyl derivatives have been chlorinated, also in the 2-position, with sulfuryl chloride in dichloromethane.[159] Reaction of the first-mentioned of these with Vilsmeier reagent yields the chlorinated product **122**,[160] which hydrolyzes to give a variety of benzothiazine products, depending on the reaction conditions.[161]

2-Alkylidene derivatives **123** have been prepared by condensation of benzothiazin-3-ones with aldehydes[52] and by reductive cyclization of the nitro acids **124**.[162] Quaternary salts of benzothiazines also condense readily

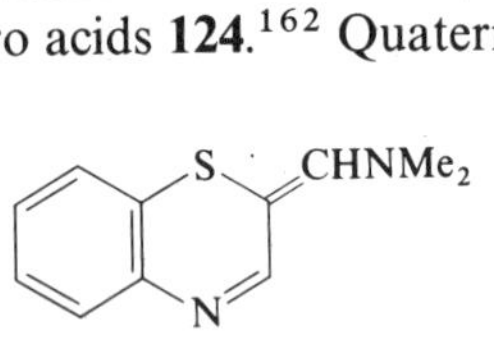

(122) 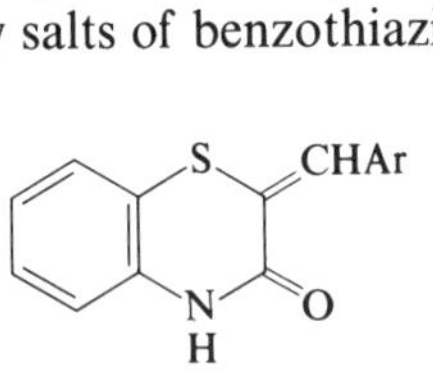 (123) 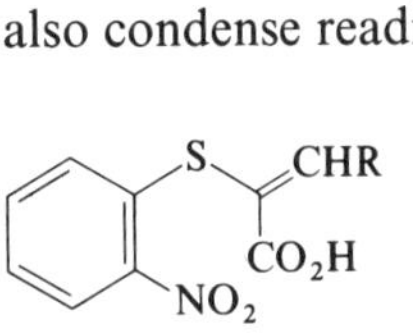 (124)

[159] J. W. Worley, K. W. Ratts, and K. L. Cammack, *J. Org. Chem.* **40**, 1731 (1975).
[160] M. R. Chandramohan, M. S. Sardessai, S. R. Shah, and S. Seshadri, *Indian J. Chem.* **7**, 1008 (1969).
[161] S. R. Shah and S. Seshadri, *Indian J. Chem.* **10**, 820 (1972).
[162] V. Baliah and T. Rangarajan, *J. Chem. Soc.*, 4703 (1960).

with *p*-dimethylaminobenzaldehyde to give intensely colored styryl products such as **125**.[163]

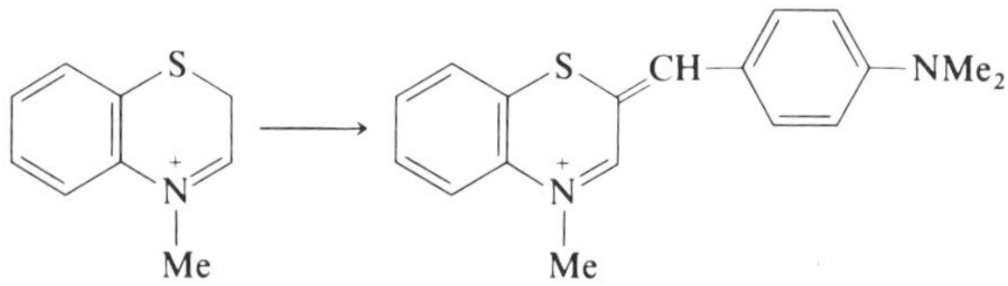

(125)

C. NUCLEOPHILIC SUBSTITUTION

Nucleophilic attack on the 2-chlorine of **126** by triethyl phosphite accounts for the formation of the phosphonates **127**, which are useful in the synthesis of 2-alkylidene derivatives analogous to **123** via the Wadsworth–Emmons reaction.[159]

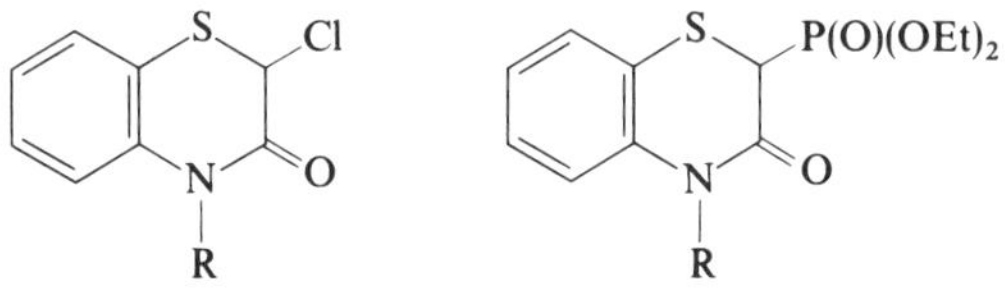

(126)   (127)

The Vilsmeier product **122** also undergoes nucleophilic displacement of the 3-chlorine when treated with amines in acetic acid containing pyridine. The ultimate product is the aldehyde **128** via the pyridinium salt **129**.[164]

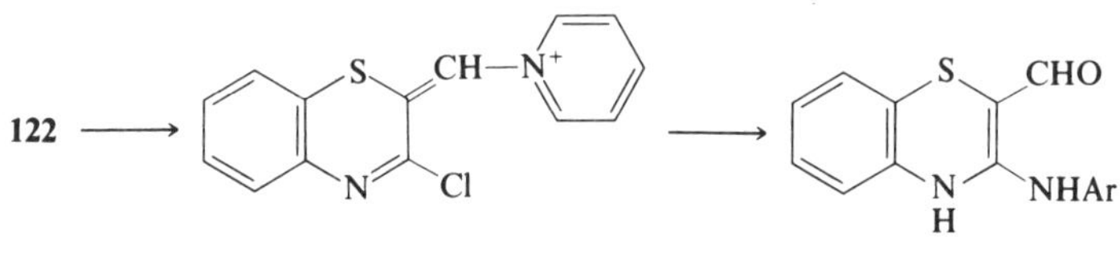

122 ⟶

(129)   (128)

Nucleophilic substitution of hydrogen by chlorine in the 7-position occurs when the *N*-hydroxy derivative **130** is treated with hydrochloric acid,

[163] A. I. Kiprianov and Z. N. Pazenko, *Zh. Obshch. Khim.* **21**, 170 (1951) [*CA* **45**, 7576 (1951)].
[164] S. R. Shah and S. Seshadri, *Indian J. Chem.* **10**, 977 (1972).

probably via **131**, to give the product **132** as shown below.[165] When the 7-position is blocked, as in **133**, the reaction does not occur.[166]

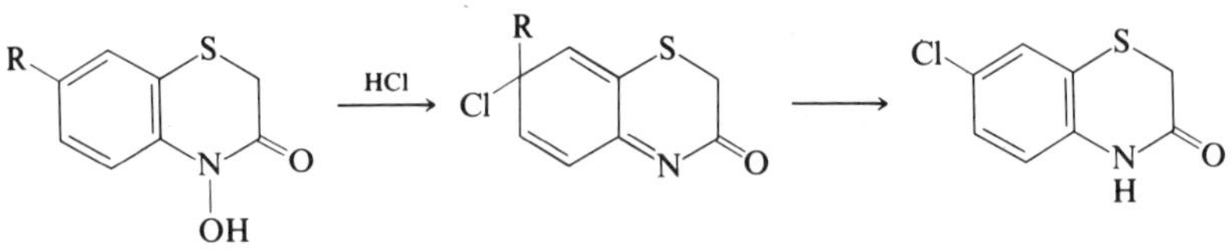

(130) R = H (131) R = H (132)
(133) R = Me

6-Amino-2,3-dihydro-1,4-benzothiazin-3-one (**134**) is a useful source of a wide variety of 6-substituted derivatives via diazotization. The amino group has been replaced by halogens and pseudohalogens, by NO, and by NO_2, as well as by HgCl, H_2AsO_3, and H_2SbO_3.[61]

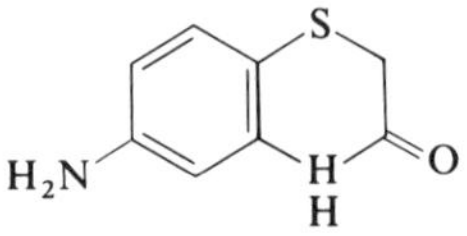

(134)

D. REARRANGEMENT REACTIONS

Benzothiazines have been reported to undergo a number of rearrangement reactions, one of the most commonly encountered involving ring contraction. Thus among the products of autoxidation of 1,4-benzothiazines are benzothiazoline derivatives.[137,141,167] For example, the benzothiazine **135** on treatment with oxygen in acidified cyclohexane gave the acyl derivative **136**.

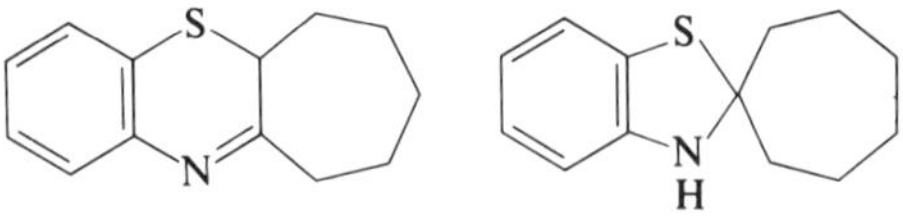

(135) (136)

---

[165] R. T. Coutts and N. J. Pound, *Can. J. Chem.* **48**, 1859 (1970).
[166] R. T. Coutts, S. J. Matthias, E. Mah, and N. J. Pound, *Can. J. Chem.* **48**, 3727 (1970).
[167] G. Liso, P. Marchini, A. Reho, and F. Micheletti-Moracci, *Phosphorus and Sulphur* **2**, 117 (1976).

The intermediacy of hemithioketals was indicated by the isolation of the analogous species **137**, which could be converted to the thiazoline **138**.

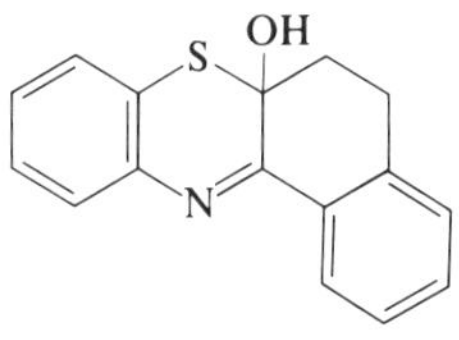

(137) 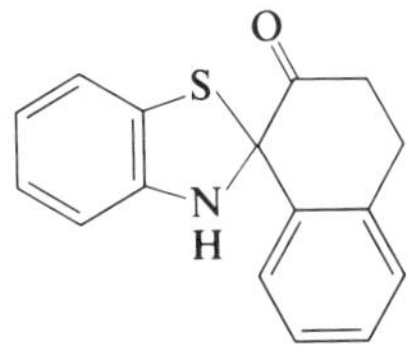   (138)

As might be anticipated, derivatives fully substituted at the 2-position are not prone to such autooxidation, even when the 3-substituent has a β-hydrogen, as in **139**, thus supporting the contention that a radical mechanism via a species such as **140** operates.[141]

Ring contraction has also been observed in hydrogenation reactions of benzothiazines. Catalytic hydrogenation of 3-phenyl-2H-1,4-benzothiazine (**141**) gave 2-methyl-2-phenylbenzothiazoline (**142**), and not the dihydro-1,4-benzothiazine as might have been expected.[14]

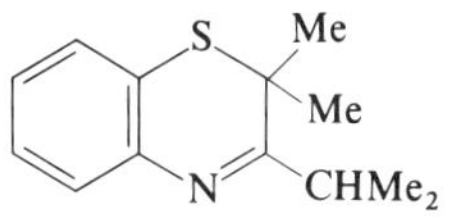

(139) 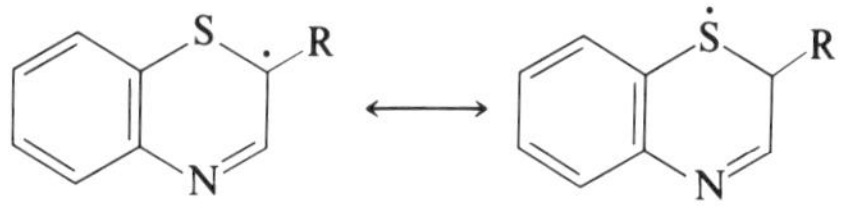   (140)

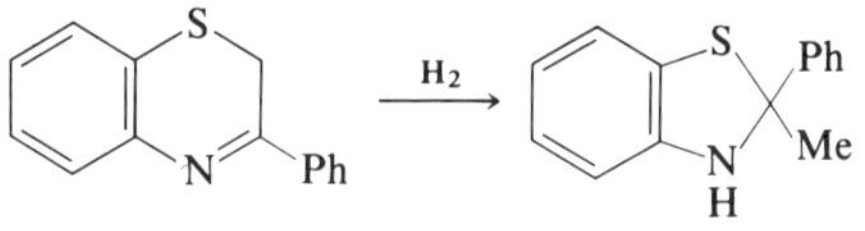

(141)   (142)

A thiophenol intermediate is assumed to participate in the formation of the benzothiazole ring, this being consistent with the mechanism postulated for hydrogenolytic thiazine–thiazole rearrangement.[168]

Benzothiazines (**143**) prepared from β-keto esters and disulfide will react further under acid-catalyzed, thermal conditions to give an isomeric product (**144**) derived via a [1,3]-sulfur shift.[142]

---

[168] D. Sica, C. Santacroce, and R. A. Nicolaus, *Gazz. Chim. Ital.* **98**, 488 (1968).

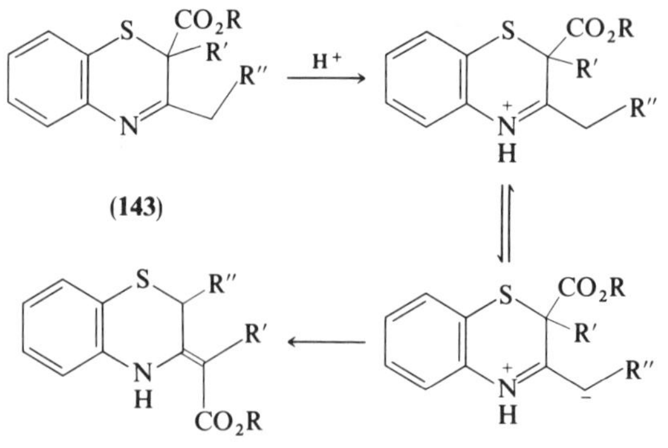

(143)

(144)

The last reaction can be compared with the Grignard reagent-induced conversion of the 1,4-benzothiazin-3-one **145** to the dihydro-1,4-benzothiazine **146**. The α,β-unsaturated ketone **147** was postulated as an intermediate to account for this rearrangement.[52]

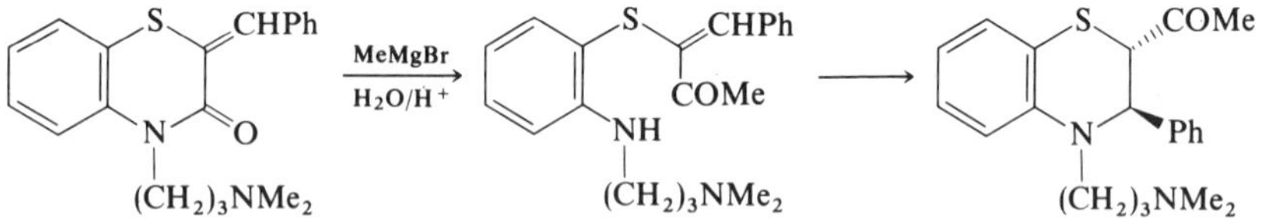

(145)   (147)   (146)

Coutts and Pound observed the isomerization of **148** to **149** with glacial acetic acid, presumably by the route indicated.[169]

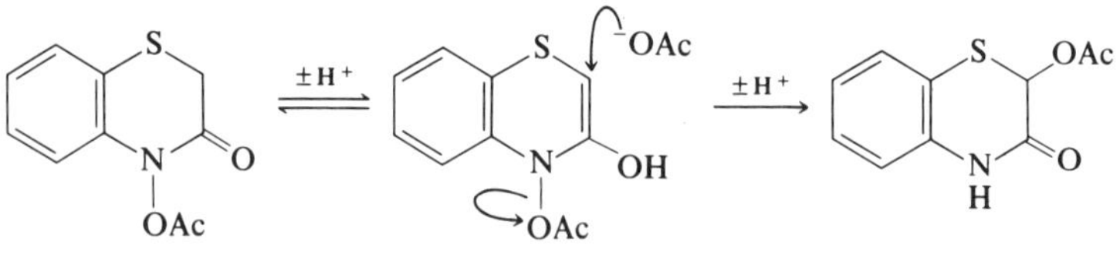

(148)   (149)

Configurational isomerism in 2-methylene derivatives can be induced photochemically as demonstrated by the conversion of Z isomer **150** to its E isomer **151**. The E isomer undergoes a facile thermal dimerization to form the cyclobutane derivative **152**.[80,170]

[169] R. T. Coutts and N. J. Pound, *J. Chem. Soc. C*, 2696 (1971).
[170] Y. Maki and M. Sako, *Chem. Pharm. Bull.* **24**, 2250 (1976).

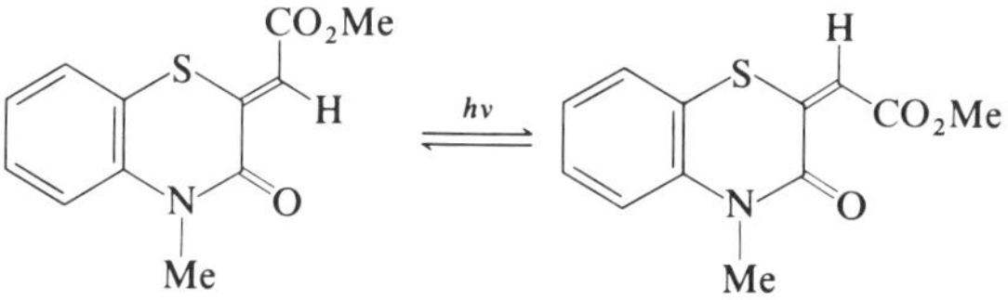

E. Oxidative Coupling of 1,4-Benzothiazines

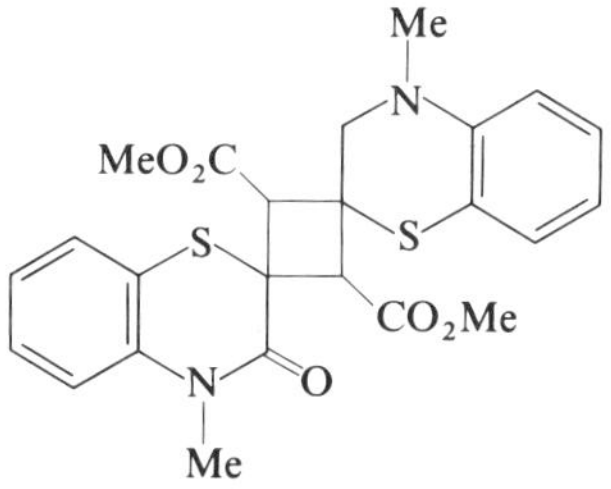

Difficulties are often encountered in the preparation of even quite simple 1,4-benzothiazines, and it is now clear that these can often be explained in terms of readily occurring oxidative coupling reactions of the ring. Indeed, many early methods for the synthesis of 1,4-benzothiazines do, in fact, yield the dimers.[15] Dehydrodimers are important as the basis of the naturally occurring pigments, the trichochromes discussed in Section II.

Such dimers were first reported by Fujii, who, on attempting the preparation of the picrate from 3-phenyl-2H-1,4-benzothiazine, isolated a dimeric species which he formulated as **153** (X = H, Ar = Ph) with unspecified stereochemistry. He also reported the 6-chloro analogue **153** (X = Cl, Ar = Ph), which formed spontaneously from the monomeric thiazine.[93]

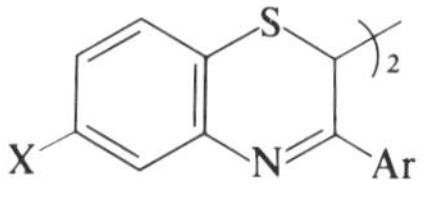

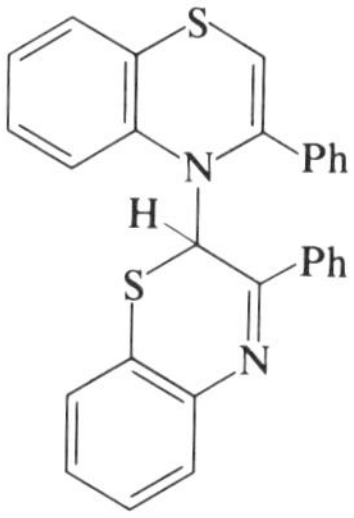

Some time later, Bottex et al., observed a similar phenomenon: the dimer **153** (X = H, Ar = Ph) formed readily from solutions of the monomer in ethanol, pyridine, or ether in the presence of air, or from heating in benzene with diethyl azodicarboxylate. Ether solutions protected from air did not yield the dimer. On the basis of limited NMR evidence, these authors formulated their dimer as **154**.[13] Wilhelm and Schmidt also obtained this dimer, which they formulated[14] as the symmetrical structure **153**, subsequently confirmed by NMR and X-ray studies.[39] These studies also indicated the meso stereochemistry for **153** (X = H, Ar = 4-BrC$_6$H$_4$). With the exception of Bottex *et al.*, all workers assign the 2*H*-1,4-benzothiazine structure to both monomeric and dimeric products. However, Duro *et al.* have shown that the 4*H*-1,4-benzothiazine **155** also forms a 2*H* dimer (**156**).[92]

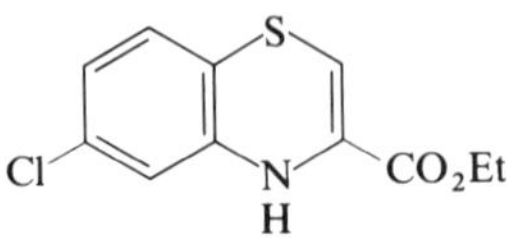

(155) 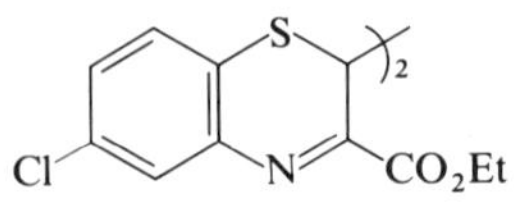   (156)

With more powerful oxidizing agents, such as chloranil, benzothiazines can be converted to the dehydro dimers $\Delta^{2,2'}$-bis-(2*H*-1,4-benzothiazines) (**157** and **158**). These isomeric species interconvert readily even at ambient temperatures.[36] The structures were confirmed by X-ray analysis, and the cis isomer is shown to predominate at room temperature.

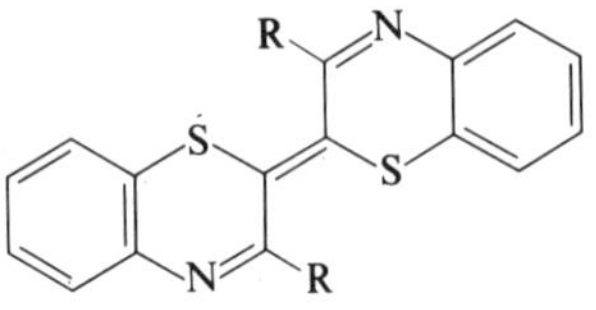

(157) R = Ar   (158) R = Ar
(159) R = H   (160) R = H

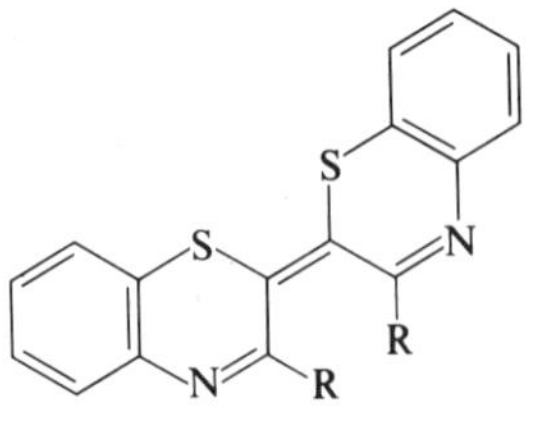

Subsequently, it was discovered that even exposure to air was sufficient to convert the parent compound, 2*H*-1,4-benzothiazine **8**, to the trans dimer **159**, a yellow-orange solid. Solutions of this isomer are photochromic, **159** interconverting readily with the cis isomer **160** on exposure to sunlight.[36] Such photochromism had already been observed with related natural pigments based on the $\Delta^{2,2'}$-bi-(2*H*-1,4-benzothiazine) system.[171]

---

[171] P. Boldt and E. Hermstedt, *Z. Naturforsch. Anorg. Chem., Org. Chem., Biochem., Biophys., Biol.* **22B**, 718 (1967).

ADVANCES IN HETEROCYCLIC CHEMISTRY, VOL. 38

The Chemistry of Hydantoins

C. AVENDAÑO LÓPEZ AND G. GONZÁLEZ TRIGO

*Departamento de Química Orgánica y Farmacéutica, Facultad de Farmacia,
Universidad Complutense, Madrid, Spain*

I. Introduction 178
II. Methods of Synthesis 178
   A. Introduction 178
   B. Amino Acids or Related Compounds and Cyanate or Thiocyanate Salts:
      The Read Synthesis. 178
   C. Amino Acids and Alkyl or Aryl Isocyanates and Isothiocyanates 180
   D. Amino Acids and Ureas 182
   E. α-Dicarbonyl Compounds and Urea Derivatives 183
   F. Bucherer–Bergs Synthesis from Carbonyl Compounds. 184
      Stereochemistry of the Bucherer–Bergs and Read Reactions. 187
   G. Conversion of Other Heterocyclic Compounds to Hydantoins 191
      1. Pyrimidine Rearrangement Ring Contraction 191
      2. From Other Heterocyclic Compounds 194
   H. Cycloaddition Reactions with Heterocumulenes 194
      1. Reaction with 3-Dimethylamino-2H-azirines 194
      2. Reaction with Nucleophilic Carbenoids or Ylides 197
      3. Reaction with 1-Oxa-4-aza- and 1,4-Diazabutadienes 198
      4. Reaction with Phenylacetylene 198
   I. Miscellaneous Methods Starting from Amino Acid
      Derivatives or Related Substrates 198
   J. Other Methods 200
      1. Condensation of Imines with Isonitriles and Thiocyanic Acid 200
      2. Reaction of N-Cyanoamines with 1-*tert*-Butyl-3,3-diphenylaziridinone . . . 201
   K. Mutual Transformations between Hydantoins and Thiohydantoins 201
III. Physicochemical Studies on Hydantoins 203
   A. Ionization 203
   B. Ultraviolet Spectroscopy 204
   C. Mass Spectrometry 204
   D. Infrared Spectroscopy 205
   E. Nuclear Magnetic Resonance Spectroscopy 206
   F. Crystal Structure Determinations 207
   G. Quantum Mechanical Calculations 209
IV. Reactivity of Hydantoins and Their Derivatives 209
   A. Electrophilic Substitution 209
      1. Protonation 209
      2. Reactions with Lewis Acids 209
      3. N-Alkylation 210

177

Copyright © 1985 by Academic Press, Inc.
All rights of reproduction in any form reserved.
ISBN 0-12-020638-2

4. N-Acylation 214
5. Halogenation 215
6. Hydantoins as Heterocyclic Compounds Containing
   Active Methylene Groups. 217
B. Reactions with Nucleophiles. 219
   1. Hydrolysis 219
   2. Ammonolysis 221
C. Reactions with Reducing Agents 222
D. Other Significant Reactions 224
V. Uses and Applications 224
A. Hydantoins as Medicinal Products 224
B. Other Uses. 227

I. Introduction

Reviews of the chemistry of hydantoins (2,4-imidazolidinediones) appeared in 1950[1] and in 1957[2]; the latest in 1966[3] deals only with thiohydantoins. Ware's 1950 review[1] is particularly exhaustive.

II. Methods of Synthesis

A. INTRODUCTION

Mechanistic and preparative aspects of the formation of the hydantoin ring show considerable development since Ware's review. Some synthetic routes previously reviewed continue to be of prime importance, especially the Bucherer–Bergs method.[4]

B. AMINO ACIDS OR RELATED COMPOUNDS AND CYANATE OR THIOCYANATE SALTS: THE READ SYNTHESIS

A large number of hydantoins and 2-thiohydantoins wtih substituents in the C-5 position as well as a limited number of N-1 substituted examples have been prepared by this route. The ring closure of the hydantoic acid intermediates (1) to hydantoins (2) is reversible, the forward reaction being

---

[1] E. Ware, *Chem. Rev.* **46**, 403 (1950).
[2] E. S. Schipper and A. R. Day, *in* "Heterocyclic Compounds" (R. C. Elderfield, ed.), Vol. 5, p. 254. Wiley, New York, 1957.
[3] J. T. Edward, *Chem. Org. Sulfur Comp.* **2**, 287 (1966).
[4] H. T. Bucherer and W. Steiner, *J. Prakt. Chem.* **140**, 291 (1934); H. T. Bucherer and V. A. Lieb, *ibid.* **141**, 5 (1934).

Sec. II.B]   HYDANTOINS   179

favored by acid medium and the reverse by alkali.[1] Several kinetic studies have also been reported.[5-7]

$$\text{R—CH}\begin{array}{c}\text{CO}_2\text{H}\\\text{NH}_2\end{array} + \text{KCNX} \xrightarrow{\text{H}^+} \text{R—CH}\begin{array}{c}\text{CO}_2\text{H}\\\text{NHCXNH}_2\end{array} \rightleftharpoons \begin{array}{c}\text{O}\\\|\\\text{R}\end{array}\!\!\begin{array}{c}\text{NH}\\\diagup\diagdown\\\diagdown\diagup\\\text{N}\text{X}\\\text{H}\end{array}$$

(1)   (2)

A number of reagents convert an optically pure 2-phenylhydantoic acid to an optically active 5-phenylhydantoin. Trifluoroacetic anhydride and dilute hydrochloric acid are the reagents of choice. The temperature dependence of the kinetics of racemization of 5-phenylhydantoin in dilute hydrochloric acid supports the conclusion that the resultant products are nearly, if not entirely, optically uniform.[8]

In a modification of the Read reaction, instead of the free amino acids, their esters or amides, and especially their nitriles, are used.[9] This procedure has also been termed the Strecker method on the basis of the synthesis of the starting amino nitriles.

Practically quantitative yields of 5-monsubstituted 2-thiohydantoins (7) are reported when ammonium thiocyanate is used instead of potassium thiocyanate in acetic anhydride. The amino acid is first acetylated to give 3 which is then successively converted to the mixed anhydride 4, the acyl isothiocyanate 5, and finally the 1-acetyl-2-thiohydantoin 6. The acetyl group may be

$$\text{R—CH}\begin{array}{c}\text{NHCOR}^1\\\text{CO}_2\text{H}\end{array} \xrightarrow{\text{Ac}_2\text{O}} \text{R—CH}\begin{array}{c}\text{NHCOR}^1\\\text{COOCOCH}_3\end{array} \xrightarrow{\text{CNS}^-}$$

(3)   (4)

$$\text{R—CH}\begin{array}{c}\text{NHCOR}^1\\\text{CO—N}\end{array}\!\!\!{=}\!\text{C}{=}\text{S} \longrightarrow \text{R}\!\!\begin{array}{c}\text{O}\\\|\\\text{NH}\\\text{N}\text{—S}\\\text{COR}^1\end{array} \longrightarrow \text{R}\!\!\begin{array}{c}\text{O}\\\|\\\text{NH}\\\text{N}\text{—S}\\\text{H}\end{array} + \text{R}^1\text{CO}_2\text{H}$$

(5)   (6)   (7)

[5] V. Stella aand T. Higuchi, *J. Org. Chem.* **38**, 1527 (1973).
[6] I. B. Blagoeva, I. G. Pojarlieff, and V. S. Dimitrov, *J.C.S. Perkin II*, 887 (1978).
[7] F. Güler and R. B. Moodie, *J.C.S. Perkin II*, 1752 (1980).
[8] K. H. Dudley and D. L. Bries, *J. Heterocycl. Chem.* **10**(2), 173 (1973).
[9] W. T. Read, *J. Am. Chem. Soc.* **44**, 1766 (1922).

removed by very mild acid or alkaline hydrolysis.[3] This reaction sequence has been applied by Schlack and Kumpf[10] to remove the C-terminal amino acid of a polypeptide.[3,11]

C. AMINO ACIDS AND ALKYL OR ARYL ISOCYANATES AND ISOTHIOCYANATES

N^3-Substituted hydantoins and 2-thiohydantoins (9) are prepared by acid-catalyzed cyclization of the substituted ureido or thioureido acids (8) obtained from a reaction of α-amino acids with alkyl or aryl isocyanates and isothiocyanates.[1,3,12] This method has been applied to prepare 2-selenohydantoins,[13] and also in the stereospecific synthesis of (S)-(+)-1,5-cyclotrimethylene-3-phenyl-2-thiohydantoin.[14]

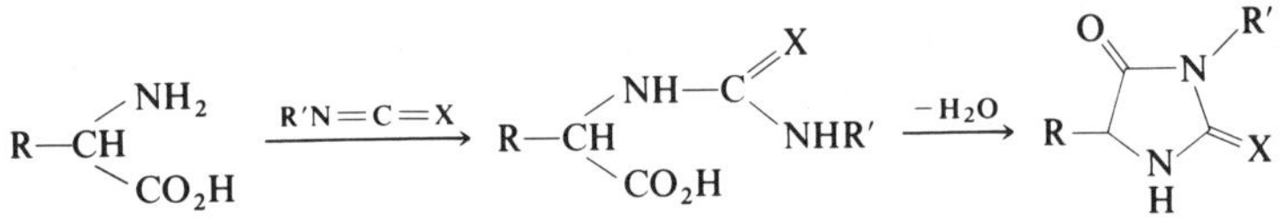

X = O, S, Se   (8)   (9)

Acid treatment of the substituted thioureas resulting from the treatment of esters, amides, or peptides of α-amino acids (10) with isothiocyanates (usually phenyl isothiocyanate) also affords 2-thiohydantoins (11). These reactions furnish the basis of the Edman's method for the stepwise degradation of peptides.[3]

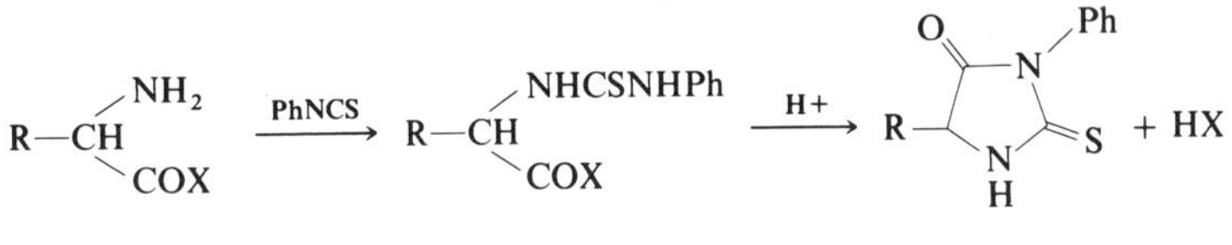

(10)   (11)

X = OR', NHR'

The rate constants for 3,5-disubstituted 2-thiohydantoins prepared by Edman's procedure have been reported.[15] 3-Phenyl-2-thiohydantoins (PTH)

[10] P. Schlack and W. Kumpf, *Hoppe-Seyler's Z. Physiol. Chem.* **154**, 125 (1926).
[11] H. Kubo, T. Nakajima, and Z. Tamura, *Chem. Pharm. Bull.* **19**(1), 210 (1971).
[12] H. Fujiwara, A. K. Bose, M. S. Manhas, and J. M. Vander Veen, *J.C.S. Perkin II*, **5**, 653 (1979).
[13] M. J. Korohoda, *Pol. J. Chem.* **54**(4), 683 (1980).
[14] J. H. Poupaert and G. Lhoest, *Bull. Soc. Chim. Belg.* **88**(5), 339 (1979).
[15] V. Knoppova, J. Kovac, and I. Basnak, *Collect. Czech. Commun.* **39**(3), 773 (1974).

are obtained by cyclization of phenylthiocarbamoyl amino acids in an acetic–hydrochloric acid medium, following the Sjöquist modification of Edman's method.[16] This modification fails in the case of L-azetidine-2-carboxylic acid and **14a** is not obtained. However, the resultant phenylthiocarbamoyl derivative (**12a**) with dicyclohexylcarbodiimide (DCC) gives 2-phenylimino-5-thiazolidinone (**13a**) that thermally rearranges to racemic **14a**. If the *p*-nitrophenyl ester of **12a** is used, then **14a** is obtained with retained optical activity.[17] In the oxo series, only the last procedure through the *p*-nitrophenyl ester gives **14b**.

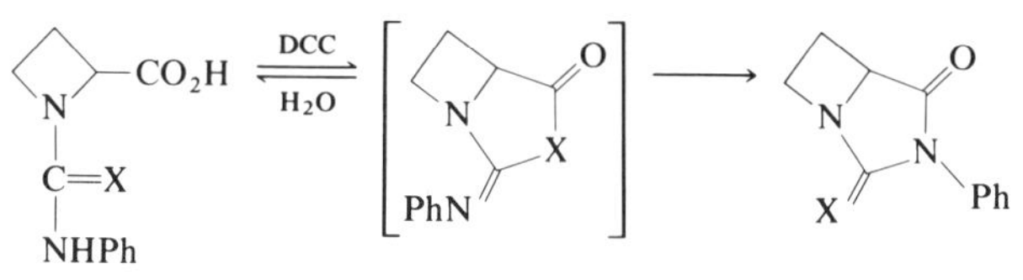

(12a,b)   (13a,b)   (14a,b)

a X = S
b X = O

Amino acids with protected side chains have also been characterized as phenylthiohydantoins by reaction with phenyl isothiocyanate. This procedure has been applied to monitor solid-phase peptide synthesis.[18]

Bridgehead nitrogen hydantoins have been obtained from imidazole-, pyrrolidine-, and piperidine-2-carboxylates,[19] as well as spirohydantoins, such as **15**, derived from the aminolysis products of isatincarboxamides.[20] Amino nitriles react in a similar manner.[1,3,21,22]

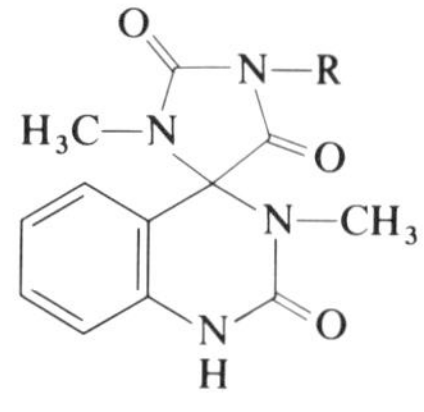

(15)

---

[16] J. Sjöquist, *Ark. Kemi* **11**, 129 (1957).
[17] H. T. Nagasawa, P. S. Fraser, and J. A. Elberling, *J. Org. Chem.* **37**(3), 516 (1972).
[18] G. Milhand, D.J.P. Raulais, P. C. Rivaille, B. Daguilhanes, and G. J. G. Lefevre, *J. Chem. Res. Synop.* **1**, 11 (1978).
[19] L. Capuano, M. Welter, and R. Zander, *Chem. Ber.* **103**(8), 2394 (1970).
[20] L. Capuano and K. Benz, *Chem. Ber.* **110**(12), 3849 (1977).
[21] W. Oldfield and C. H. Cashin, *J. Med. Chem.* **8**(2), 239 (1965).
[22] U. Krueger and G. Zinner, *Arch. Pharm. (Weinheim, Ger.)* **311**(1), 39 (1978).

Chlorosulfonyl isocyanate is uniquely useful for the formation of hydantoins from sterically hindered acid- and base-labile amino nitriles, e.g., the synthesis of optically active spirohydantoins such as the biologically active (4S)-2,3-dihydro-6-fluorospiro(4H-1-benzopyran-4,4'-imidazolidine)-2',5'-dione.[23]

Condensation of α-amino esters, a primary amine, and diphenyl carbonate is the modification of choice for derivatives containing carboxyl or hydroxyl groups.[24]

D. AMINO ACIDS AND UREAS

This reaction has been known since Ware's report,[1] giving excellent yields in the case of amino acetamides and acetates with nitrourea.[25]

The reaction of urea or phosgene with amino acids gives ^{11}C-labeled hydantoins with potential utility as radiopharmaceuticals.[26]

The α-halogen acids are often used in reactions with primary amines and urea,[27,28] as well as in cyclocondensations of chloroacetates and ureas. Some 3-(benzothiazol-2-yl)-2-thiohydantoins derivatives have been obtained by this fashion.[29]

Similarly N^1-(4'-arylthiazole-2'-yl)thioureas, prepared from 2-amino-4-arylthiazoles with benzoyl thiocyanate and subsequent hydrolysis of the product with alkali, condense with monochloroacetic acid in pyridine to give 2-thiohydantoins.[30]

Other amino acid derivatives, such as α-hydroxy acids and cyanohydrins, also react with urea to give hydantoins,[1] e.g., the synthesis of 5-(4-hydroxyphenyl)hydantoin.[31]

Urea, methylurea, and dimethylureas react with glyoxylic acid and its methyl ester to give α-substituted hydantoic acid derivatives (16) and substituted allantoic acid derivatives (17), which can be cyclized to 5-substituted hydantoins (18). Although allantoin (18) formation from urea and

---

[23] R. Sarges, H. R. Howard, and P. R. Kelbaugh, *J. Org. Chem.* **47**, 4081 (1982).
[24] K. Iwata and S. Hara, *J. Heterocycl. Chem.* **15**(7), 1231 (1978).
[25] E. Schipper and E. Chinery, *J. Org. Chem.* **26**, 3597 (1961).
[26] D. Roeda and G. Westera, *Int. J. Appl. Radiat. Isot.* **32**(11), 843 (1981).
[27] R. O. Kochkanyan, Yu. A. Israelyan, and A. N. Zaritovskii, *Khim. Geterotsikl. Soedin.* **1**, 87 (1978).
[28] S. Icli and L. D. Colebrook, *J. Pure Appl. Sci.* **9**(1), 39 (1976).
[29] M. Behera, P. N. Dhal, and A. Nayak, *J. Indian Chem. Soc.* **52**(11), 1067 (1975).
[30] P. N. Dhal, T. E. Achary, A. Nayak, and M. K. Rout, *J. Indian Chem. Soc.* **50**(10), 680 (1973).
[31] T. Ohashi, S. Takahashi, T. Nagamachi, K. Yoneda, and H. Yamada, *Agric. Biol. Chem.* **45**(4), 831 (1981).

glyoxylic acid was described in 1876,[32] this reaction is particularly interesting because the reverse sequence **18** → **16** is part of the biological degradation of uric acid to urea and glyoxylic acid.

$$\begin{array}{c} \text{HO} \\ \diagdown \\ \text{CHCO}_2\text{H} \\ | \\ \text{NHCONH}_2 \end{array} \underset{}{\overset{\text{H}_2\text{NCONH}_2}{\rightleftharpoons}} (\text{H}_2\text{NCONH})_2\text{CHCO}_2\text{H} \rightleftharpoons \begin{array}{c} \text{H}_2\text{NCONH} \end{array} \begin{array}{c} \text{O} \\ \diagup\diagdown \\ \text{NH} \\ \diagdown\diagup \\ \text{N} \\ | \\ \text{H} \end{array} \text{O}$$

(16)   (17)   (18)

Syntheses of relatively stable adducts of ureas and glyoxylic acid have been reinvestigated. Very good yields of 5-hydroxy-, 5-methoxy-, and 5-isopropylthiohydantoins have been achieved.[33]

Some 1-aryl-substituted hydantoins have been synthesized by acid-catalyzed condensation of monosubstituted ureas and glyoxal with moderate yields.[34]

E. α-DICARBONYL COMPOUNDS AND UREA DERIVATIVES

5,5-Disubstituted hydantoins and 2-thiohydantoins, especially their 5,5-diaryl derivatives, have been prepared by means of a condensation that involves a rearrangement similar to that in which benzilic acid is formed from benzil.[1,3] Examples include the synthesis of fluorinated hydantoins,[35] spirofluorene-,[36] and spiroacenaphthylenehydantoins,[37] 2-thiohydantoins,[30,38,39] and 5,5-diphenylhydantoin-2,4,5-[13]C_3.[40]

1-Substituted biguanides (**19**) also react with benzil on heating in alcohol in the absence of any catalyst to give 2-substituted guanilydene-5,5-diphenylhydantoins (**20**), whose formation involves an anionotropic migration of a phenyl group.[41] N-Amidino-O-alkylisoureas and 1-aryl- or 1-alkyl-3-amidino-2-thioureas behave similarly.[42]

---

[32] E. Grimaux, *Hebd. Seances C. R. Acad. Sci.* **83**, 62 (1876).
[33] D. Ben-Ishai, J. Altman, and Z. Bernstein, *Tetrahedron* **33**(10), 1191 (1977).
[34] J. Nematollahi, Z. Mehta, and J. Langstron, *J. Pharm. Sci.* **62**(2), 340 (1973).
[35] K. C. Joshi, V. N. Pathak, and M. K. Goyal, *J. Heterocycl. Chem.* **18**(8), 165 (1981).
[36] W. R. Dunnavant, *J. Org. Chem.* **22**, 991 (1957).
[37] P. A. Crooks, T. Deeks, and F. de Simone, *Gazz. Chim. Ital.* **107** (5–6), 353 (1977).
[38] H. C. Chiang and J. R. Ko, *Heterocycles* **19**, 529 (1982).
[39] H. P. Das and G. N. Mahapatra, *Indian J. Chem.* **18B**(3), 257 (1979).
[40] J. A. Kepler, J. W. Lytle, and G. F. Taylor, *J. Labelled Compd.* **10**(4), 683 (1974).
[41] M. Furukawa, Y. Fijimo, Y. Kojima, and S. Hayashi, *Chem. Pharm. Bull.* **20**, 521 (1972).
[42] S. Hayashi, M. Furukawa, K. Matsuoka, T. Yoshida, Y. Kojima, and S. Ayashi, *Chem. Pharm. Bull.* **22**(1), 1 (1974).

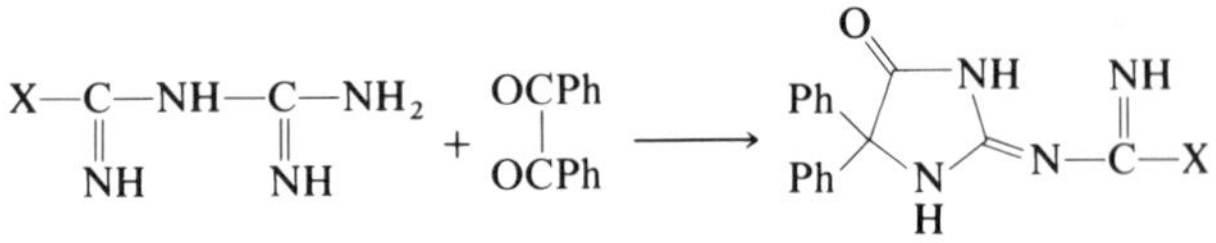

(19)   (20)

X = NHR, OR

1,2-Cyclohexanedione specifically reacts with the guanidino group of arginine and arginyl residues of proteins in 0.2 M alkali to form hydantoin (21).[43]

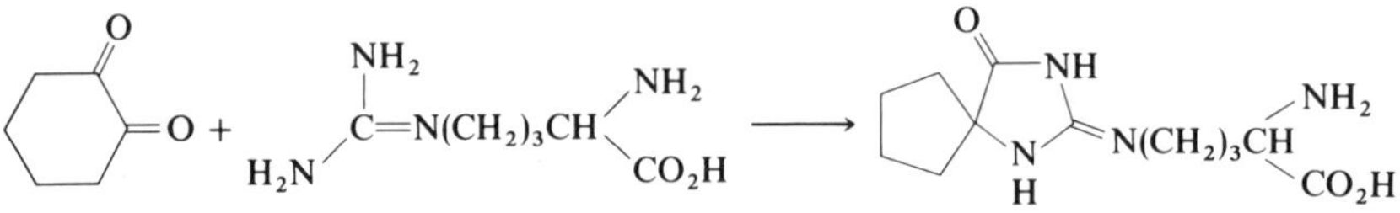

(21)

F. Bucherer–Bergs Synthesis from Carbonyl Compounds

The Bucherer–Bergs synthesis is of general application to carbonyl compounds, employing potassium cyanide and ammonium carbonate.[1,4] Carbonyl derivatives such as semicarbazones, thiosemicarbazones, oximes, azines, phenylhydrazones, imidazolidines, and azomethines also are readily converted directly to the corresponding hydantoins.[44] The extent to which the reaction occurs appears unrelated to the hydrolytic stability of the starting material. The proposed mechanism is given in Scheme 1.

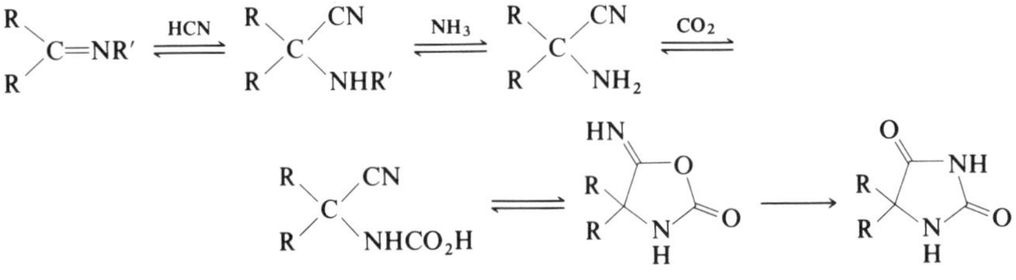

Scheme 1

[43] K. Toi, E. Bynum, E. Norris, and H. A. Itano, *J. Biol. Chem.* **242**(5), 1036 (1967).
[44] J. N. Coker, W. L. Kohlase, T. F. Martens, A. O. Rogers, and G. G. Allan, *J. Org. Chem.* **27**(9), 3201 (1962).

Substitution of $(NH_4)_2CO_3$ by $H_2NCOSNH_4$ (prepared from COS and alcoholic NH_3) in the presence of NaCN gives 4-thiohydantoins. Preformed α-amino nitriles also give 4-thiohydantoins, employing carbonyl sulfide.[45]

The Carrington modification, using carbon disulfide, ammonium chloride, and sodium cyanide, affords 2,4-dithiohydantoins.[46] If preformed α-amino nitriles react with carbon disulfide, 2,4-dithiohydantoins are also obtained. 4,4-Dialkyl-5-iminothiazolidine-2-thiones are probably first formed and then isomerize by the mechanism shown in Scheme 2.[3]

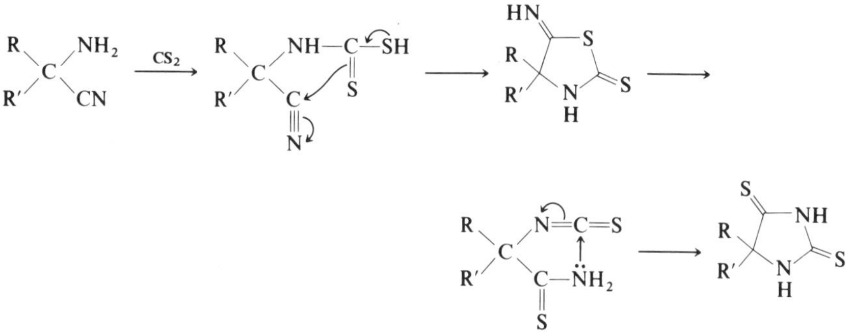

SCHEME 2

The mechanism is supported by the observation that α-methylamino nitriles (**22**), which cannot isomerize by this mechanism, afford 4,4-dialkyl-5-imino-3-methylthiazolidine-2-thiones (**23**), which on prolonged reaction are converted to 5-methyliminothiazolidine-2-thiones (**24**) and 5,5-dialkyl-2,4-dithiohydantoins (**25**).[47] The apparent methyl transfer shown in this scheme takes place by the interchange of ammonia and methylamine.[48]

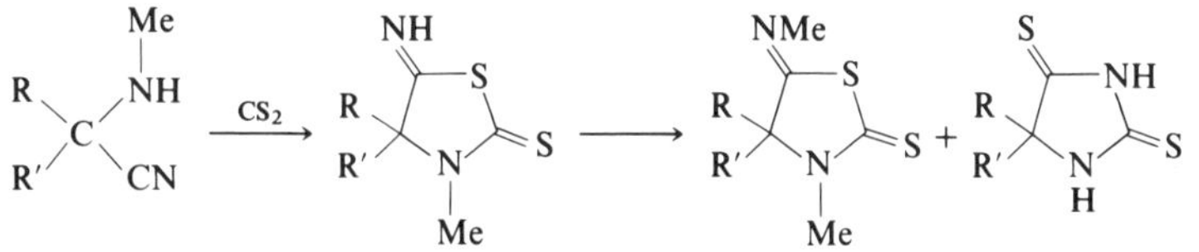

(22)   (23)   (24)   (25)

When the Carrington modification is applied to some sterically hindered piperidones such as tropinone or pseudopelletierine, the corresponding

---

[45] H. C. Carrington, C. H. Vasey, and W. S. Waring, *J. Chem. Soc.*, 396 (1959).
[46] H. C. Carrington, *J. Chem. Soc.*, 681 (1947).
[47] F. L. Chubb and J. T. Edward, *Can. J. Chem.* **59**(18), 2724 (1981).
[48] J. Taillades and A. Commeyras, *Tetrahedron* **30**, 127, 2493, 3407 (1974).

thiocyanates (26) are the main reaction products.[49] During the Bucherer–Bergs reaction, ester groups are saponified.[50]

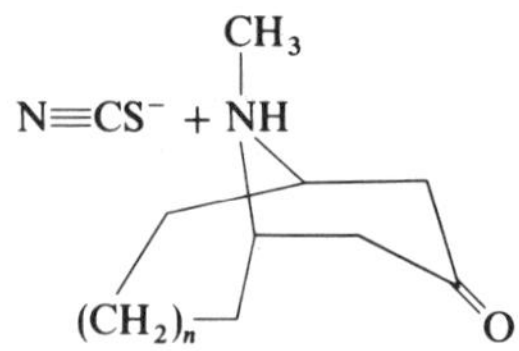

(26)

$n = 0, 1$

The Bucherer–Bergs synthesis has been successfully applied to a great variety of cyclic ketones, ranging from cyclopropanone to steroidal ketones. However, the difficulty of preparing small-ring ketones has led to the development of alternative methods to obtain the corresponding spirohydantoins. Isocyanate intermediate 27 may be prepared in several steps from malonic ester; typical Bucherer–Bergs chemistry follows.[51]

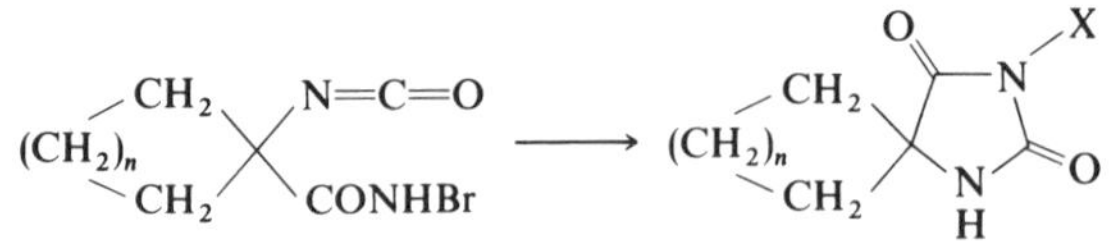

(27) $X = Br, H$
       $n = 0, 1$

The synthesis of 5-(phenyl-$^{13}C_6$)-5-phenylhydantoin is valuable in biomedical research.[52] The Bucherer–Bergs reaction is recognized to be suitable for characterization of 3-oxoalkylpyrroles[53] as well as 1-(2-furyl)-2-alkanones.[54] A kinetic study of the Bucherer–Bergs synthesis of p-hydroxy-5-phenylhydantoin has been reported.[55]

---

[49] G. G. Trigo, C. Avendaño, and E. Santos, *An. Quim.* **75**(9–10), 761 (1979).
[50] G. G. Trigo, C. Avendaño, P. Ballesteros, and A. Sastre, *J. Heterocycl. Chem.* **17**(1), 103 (1980).
[51] W. Dvonch, H. Fletcher, and H. E. Alburn, *J. Org. Chem.* **29**(9), 2764 (1964).
[52] J. H. Poupaert, M. Winand, and P. Dumont, *J. Labelled Compd. Radiopharm.* **18**, 1827 (1981).
[53] A. Treibs, R. Wilhelm, and E. Herrmann, *Liebigs Ann. Chem.* **5**, 849 (1981).
[54] E. H. Sund and D. S. Hunter, *J. Heterocycl. Chem.* **11**(6), 1123 (1974).
[55] A. P. Khardin, A. Gutmanis, A. I. Val'dman, B. I. Panfilov, D. I. Val'dman, S. Yu Sizoo, L. V. Semenova, I. Leucbergs, A. Ozoes, and M. Dziena, *Latv. PSR Zinat. Akad. Vestis, Kim. Ser.* **4**, 430 (1981).

Finally, yields from the Bucherer–Bergs reaction have been improved by several optimization procedures.[56,57]

Stereochemistry of the Bucherer–Bergs and Read Reactions

When the Bucherer–Bergs and Read methods are applied to some substituted cyclanones, two different epimeric products, designated α and β, are found. After a long-standing controversy about the configurations of the hydantoins and their corresponding amino acids derived by hydrolysis,[45,46,58–63] a mechanism proposed by Edward gives a coherent explanation.[64] The key to the problem is that the Bucherer–Bergs reaction is under thermodynamic control, whereas Read's is kinetically controlled. Edward concludes that the main epimer obtained by the Bucherer–Bergs procedure (α) has the C-4 carbonyl group in the less-hindered position, while the Read procedure yields the spirohydantoin of the opposite configuration (β epimer). The modification of the Bucherer–Bergs reaction, using COS or CS_2 to obtain 4-thio- or 2,4-dithiohydantoins, gives β (with the C-4 thiocarbonyl group in the more crowded position). Finally, the use of thiocyanate instead of cyanate in the Read method yields 2-thiohydantoins with the β configuration (C-4 carbonyl group in the more crowded position).

The mechanistic considerations are based mainly on the fact that amino nitriles **29** and **29′** rapidly equilibrate in alkaline solution through the intermediacy of imine **28** but not in acid solution. The Read reaction is conducted under acidic conditions so that the interconversion of **29′** to (**29**) is

---

[56] F. L. Chubb, J. T. Edward, and S. Ch. Wong, *J. Org. Chem.* **45**(12), 2315 (1980).
[57] C. Pedregal, G. G. Trigo, M. Espada, D. Mathieu, R. Phan Tan Luu, C. Barceló, J. Lamarca, and J. Elguero, *J. Heterocycl. Chem.* **21**, 1527 (1984).
[58] H. L. Hoyer, *Chem. Ber.* **83**, 491 (1950).
[59] H. C. Brimelow and C. H. Vasey, British Patent 807,678 (1959) [*CA* **53**, 12303 (1959)].
[60] L. Munday, *J. Chem. Soc.*, 4372 (1961).
[61] H. C. Brimelow, H. C. Carrington, C. H. Vasey, and W. S. Waring, *J. Chem. Soc.*, 2789 (1962).
[62] R. J. Cremlyn and M. Chisholm, *J. Chem. Soc. C*, 2269 (1967).
[63] Y. Maki and T. Masugi, *Chem. Pharm. Bull.* **21**, 685 (1973).
[64] J. T. Edward and C. Jitrangsri, *Can. J. Chem.* **53**, 3339 (1975).

prevented. The amino nitrile (mostly **29′**) reacts with cyanic acid to form a urea (**30′**), which can then cyclize to β-hydantoin (**31′**).

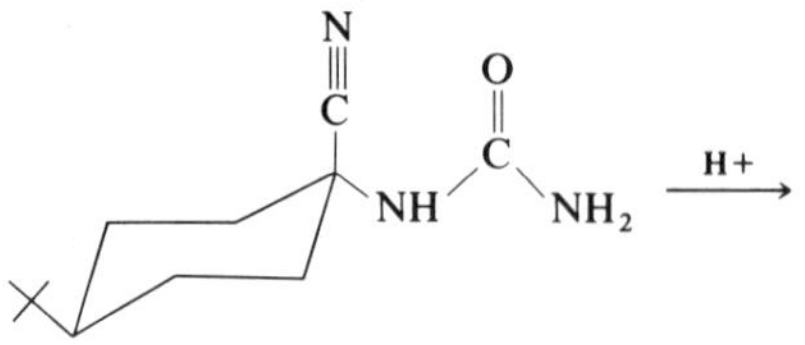

(**30′**)

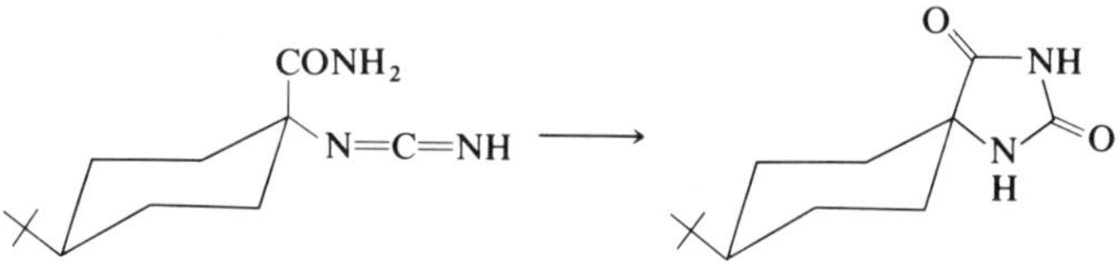

(**31′**)

On the other hand, the same amino nitrile treated with carbon dioxide in aqueous ethanol gives the α-hydantoin (**31**). The elaborated Edward's mechanism (shown only for the β-series in Scheme 3) assumes the rapid interconversion **29 ⇌ 29′** and that the rate-determining step in the path to **31′** is **29′a → 29′b**. The path leading to the α compound should be less favored in its earlier, preequilibrium steps, but the overall energy barrier must be lower for steric reasons.

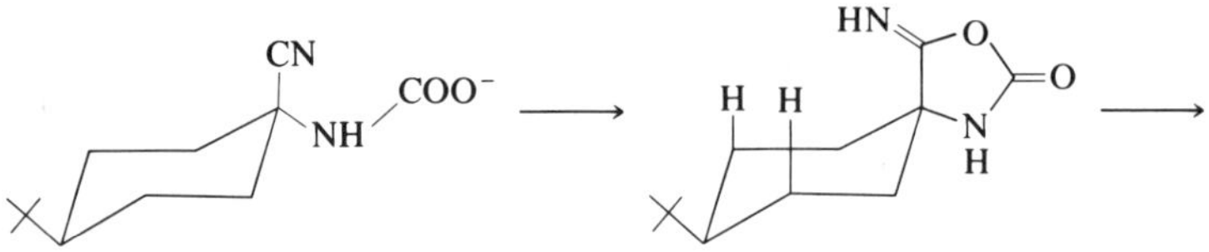

(**29′a**)   (**29′b**)

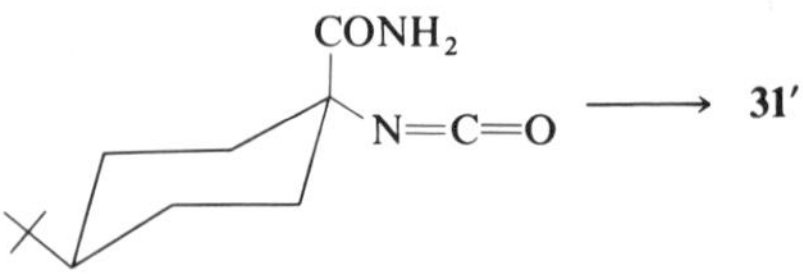

(**29′c**)

SCHEME 3

When the carbon dioxide is replaced by carbon disulfide, the reaction takes the β path because the much greater nucleophilicity of the sulfur atom makes the reaction **29'a** → **29'b** exothermic and, therefore, its transition step should resemble **29'a** rather than **29'b**. Consequently, the steric barrier impeding the β path to the hydantoin is missing on the β path to the 2,4-dithiohydantoin, whereas the various steric factors impeding the α path are still present. The observation that, when carbon dioxide is replaced by carbon oxysulfide, the β- rather than the α-4-thiohydantoin is obtained can be explained by similar reasoning.

The reaction in the presence of CO_2 and a primary amine, in place of ammonium carbonate, gives **32** instead of the expected derivative (**33**) because the Bucherer–Bergs reaction requires N-unsubstituted α-amino nitriles, able to give the isocyanate intermediate **29'c**.[65]

(32)   (33)

Kinetic and thermodynamic studies[66] show the stability of the α-amino nitrile carbamate **29a** versus pH to be determined mainly by the concentration of dissolved CO_2 and by the equilibrated formation of products from degradation of the α-amino nitrile, i.e., the amino dinitrile and the cyanohydrin.

Among many studies which confirm the Edward's rule is that dealing with tropane-3-spiro-5'-hydantoins **34–36** in Scheme 4.[49,67]

N-Alkylgranatanine-3-spiro-5'-hydantoins obtained by the Bucherer–Bergs methods have the α-configuration.[68–71] The stereoselectivity of the Bucherer–Bergs and Read reactions of bicyclo[3.2.1]octan-3-one[72] and N-methyl-10-azabicyclo[4.3.1]decan-8-one (homogranatanine) has also been

[65] G. G. Trigo, C. Avendaño, and F. Marti, *An. Quim.* **73**(3), 426 (1977).
[66] A. Rousset, M. Lasperas, J. Taillades, and A. Commeyras, *Tetrahedron* **36**(18), 2649 (1980).
[67] G. G. Trigo, C. Avendaño, E. Santos, J. T. Edward, and S. Ch. Wong, *Can. J. Chem.* **57**(12), 1456 (1979).
[68] F. Florencio, P. Smith-Verdier, and S. García-Blanco, *Acta Crystallogr., Sect. B* **B34**(4), 1317 (1978).
[69] F. Florencio, P. Smith-Verdier, and S. García-Blanco, *Acta Crystallogr., Sect. B* **B34**(7), 2220 (1978).
[70] F. Florencio, P. Smith-Verdier, and S. Garcia-Blanco, *Cryst. Struct. Commun.* **9**(3), 687 (1980).
[71] G. G. Trigo, C. Avendaño, P. Ballesteros, and A. González, *J. Heterocycl. Chem.* **15**, 833 (1978).
[72] H. N. Christensen, M. E. Handlogten, J. V. Vagdama, E. de la Cuesta, P. Ballesteros, G. G. Trigo, and C. Avendaño, *J. Med. Chem.* **26**, 1374 (1983).

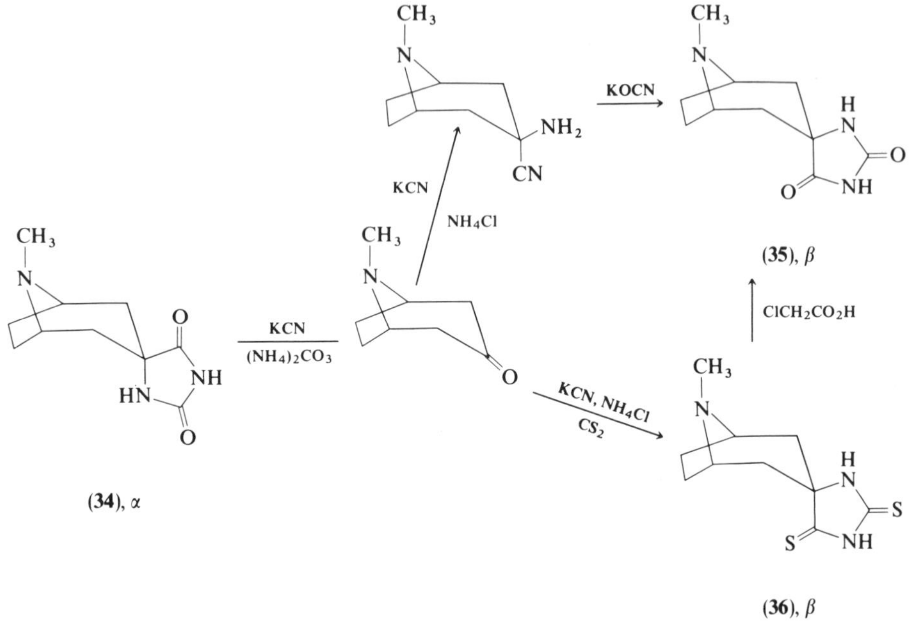

SCHEME 4

confirmed.[73] Other azabicyclo ketones have also been studied.[74-78] In all cases the configuration of the hydantoins agrees with the stereochemical predictions based on the Edward mechanism and in some cases with the stereochemistry of other reactions, such as reduction[79] and the addition of organometallic reagents to carbonyl groups.[80]

The results of the Bucherer–Bergs and Read reactions on cyclopentanones such as cis-bicyclo[3.3.0]octan-3-one and cis-3,4-dimethylcyclopentanone show a stereochemical course related to the preferred conformation of the cyclopentane rings.[67] The stereochemical preferences of biased cyclopentanones are much less pronounced than those of biased cyclohexanones, but Edward's rule indicates trends.

[73] G. G. Trigo, E. Martínez, and E. Llama-Hurtado, *J. Heterocycl. Chem.* (in press).
[74] G. G. Trigo, E. Gálvez, M. Espada, and C. Bernal, *J. Heterocycl. Chem.* **16**, 977 (1979).
[75] P. Smith-Verdier, F. Florencio, and S. García-Blanco, *Acta Crystallogr., Sect. B* **B35**(8), 1911 (1979).
[76] F. Florencio, P. Smith-Verdier, and S. García-Blanco, *Acta Crystallogr., Sect. B* **B35**(10), 2422 (1979).
[77] G. G. Trigo, E. Gálvez, and C. Avendaño, *J. Heterocycl. Chem.* **15**, 907 (1978).
[78] E. Martínez, C. del Campo, and G. G. Trigo, *Helv. Chim. Acta* **66**, 338 (1983).
[79] H. O. House, H. C. Muller, C. G. Pitt, and P. P. Wickham, *J. Org. Chem.* **28**, 2407 (1963).
[80] H. O House and W. H. Bryant, *J. Org. Chem.* **30**, 3634 (1965).

Other rigid cyclopentanones thus far studied have been (±)-camphor and (±)-norcamphor. The stereochemistry of the reactions confirms Edward's rule. While (±)-norcamphor gives, by the Bucherer–Bergs method, the hydantoin with the C-4 carbonyl group in the exo position,[81] (±)-camphor gives the hydantoin with an endo C-4 carbonyl group.[82] These are, respectively, the more accessible sides.[83,84]

G. Conversion of Other Heterocyclic Compounds to Hydantoins

Hydantoins have been prepared from cyclic derivatives of urea by mechanisms that are not very reliable in most cases.[1] Some new ring transformations have clarified some mechanistic aspects, however.

1. *Pyrimidine Rearrangement Ring Contraction*

Reaction of several 5-bromo-3-alkyl-6-methyl-1-aryluracils (**37**) with monoalkylamines causes ring contraction to give hydantoin derivatives

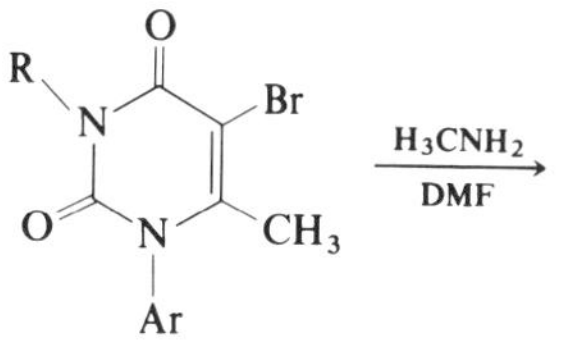

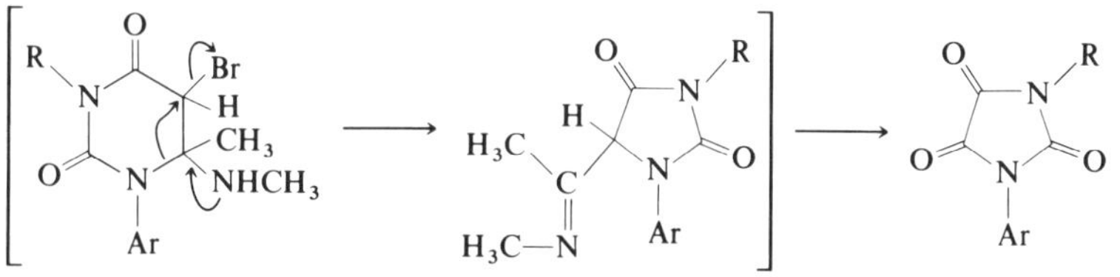

[81] H. S. Tager and H. N. Christensen, *J. Am. Chem. Soc.* **94**, 968 (1972).
[82] Y. Maki, T. Masugi, and K. Ozeki, *Chem. Pharm. Bull.* **21**(11), 2466 (1973).
[83] H. C. Brown, *Chem. Br.* **2**, 199(1966).
[84] E. L. Eliel, *in* "Stereochemistry of Carbon Compounds," p. 303. McGraw-Hill, New York, 1962.

besides the normal substitution products. A reasonable reaction mechanism implies that intermediate **38** tautomerizes and hydrolyzes to give **39**.[85]

Treatment of 5-acetoxy-6-(acetoxymethyl)uracils (**40**) with very dilute sodium hydroxide solution affords hydantoins (**42**) via the exocyclic methylene intermediate **41**.[86]

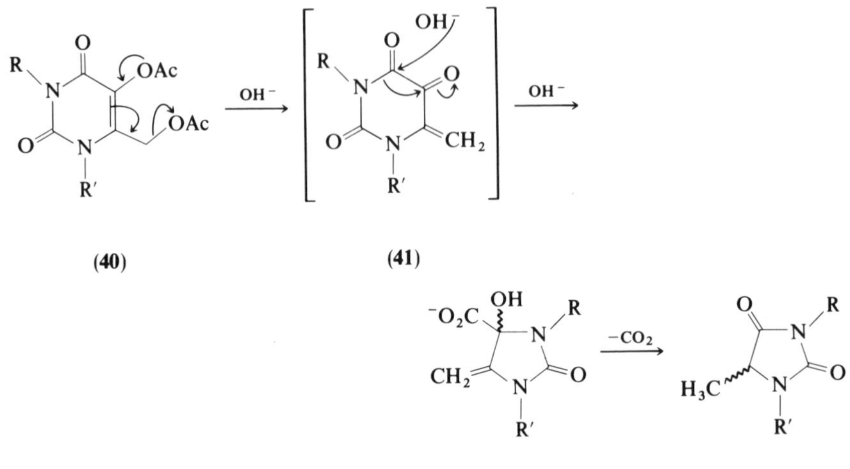

(**40**) (**41**)

(**42**)

N^5,N^{10}-Dialkylisoalloxazines react with molecular oxygen to give 4a-hydroperoxides (**43a**), which provide, on spontaneous decomposition, 4a-pseudobases (**43b**). These intermediates undergo ring contraction in basic medium to yield spirohydantoins (**44**).[87]

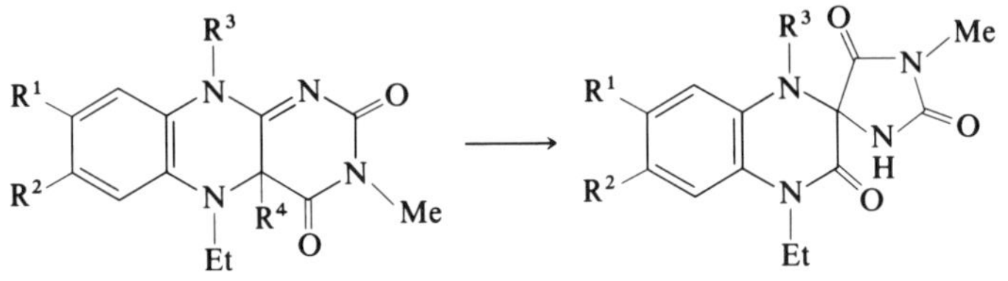

(**43a**) $R^4 = OOH$
(**43b**) $R^4 = OH$

(**44**)

The autooxidation of 1,3,5,6,7-pentamethyl-5,6,7,8-tetrahydrolumazine also yields spirohydantoins identified by X-ray analysis.[88] Early results

---

[85] Sh. Senda, K. Hiroka, and K. Banno, *Tetrahedron Lett.* **35**, 3087 (1974).
[86] B. A. Otter, I. M. Sasson, and R. P. Gagnier, *J. Org. Chem.* **47**(3), 508 (1982).
[87] M. Iwata, T. C. Bruice, H. L. Carrell, and J. P. Glusker, *J. Am. Chem. Soc.* **102**(15), 5036 (1980).
[88] H. Van Koningsveld, *Tetrahedron* **32**(17), 2121 (1976).

Sec. II.G] HYDANTOINS 193

concerning the rearrangement of uric acid and its derivatives probably involve a similar mechanism.[1]

Hydrolysis of isoalloxazines (flavins, **45**) generally occurs via nucleophilic addition of OH^- to the 10a-position to provide hydantoins (**46**), which by intramolecular dehydration in acidic medium give **45**.[89,90]

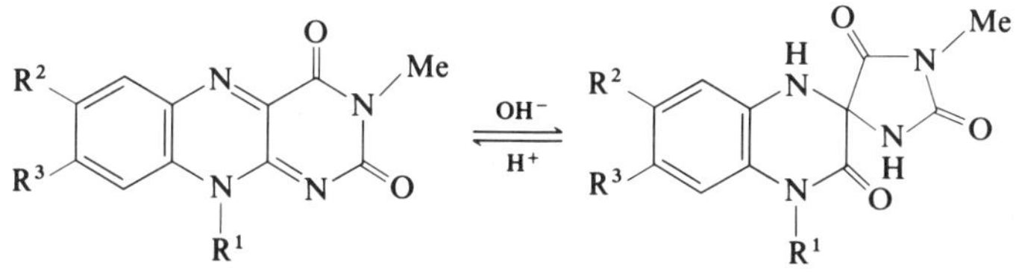

(**45**)   (**46**)

Spirohydantoins have also been found to result from reactions of several nucleophiles with 1,3,10-trimethylflavinium perchlorate.[91] In the same context is described the pyrimidine ring contraction of the flavinium cation **47** with OH^- through the 4a-adduct **48** or its isomer 10a-adduct **49** to the spirohydantoins **50** and **51**.[92]

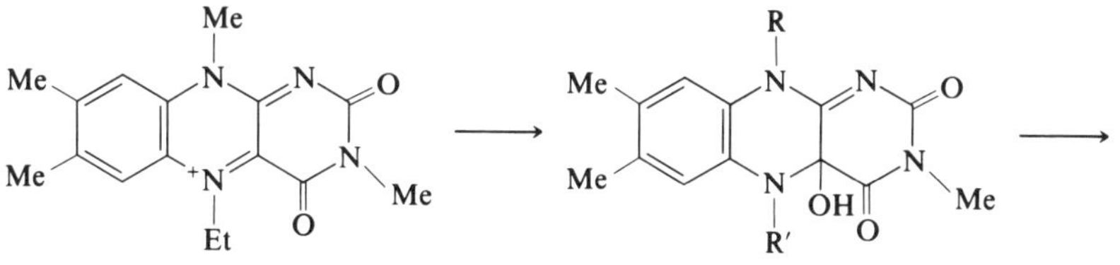

(**47**)

(**48**) R = Me, R' = Et
(**49**) R = Et, R' = Me

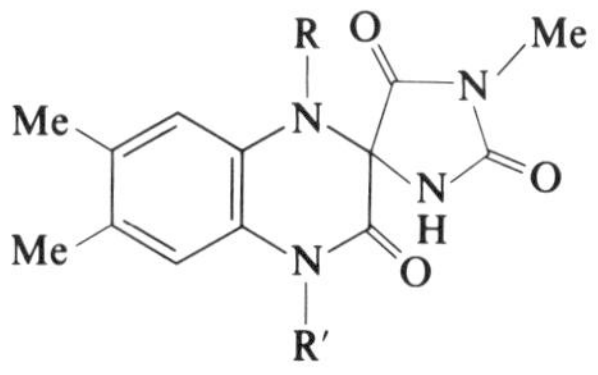

(**50**) R = Me, R' = Et
(**51**) R = Et, R' = Me

---

[89] T. C. Bruice, T. W. Chan, J. P. Taulane, I. Yokoc, D. L. Elliot, R. F. Williams, and M. Novak, *J. Am. Chem. Soc.* **99**(20), 6713 (1977).
[90] F. Yoneda, Y. Sakuma, and K. Shinozuka, *J.C.S. Chem. Commun.* **6**, 175 (1977).
[91] K. H. Dudley and P. Hemmerich, *J. Org. Chem.* **32**(10), 3049 (1967).
[92] H. I. Mager, *Tetrahedron Lett.* **37**, 3549 (1979).

Some pyrimidine derivatives, such as alloxan, react in the absence of mineral acid with 3,4-diaminopyridines to give hydantoins through a ureide[93] in analogy to the cyclization of 1,4,5-triazanaphthalenecarboxy ureides.[94]

2. From Other Heterocyclic Compounds

4-Arylmethylene-2-benzylthio-2-thiazolin-5-ones form 3-aryl-5-arylmethylene-2-thiohydantoins by treatment with ammonia or primary aromatic amines in acetic acid.[95,96] The mechanism probably involves, in a first stage, ring opening of the thiazolone ring at C-5. Treatment with diborane of the bicyclic hydantoin derivative 2,3-dihydro-6,6-diphenylimidazo(2,1-*b*)oxazol-5(6*H*)-one[97] gives 5,5-diphenyl-3-ethylhydantoin.[98]

H. CYCLOADDITION REACTIONS WITH HETEROCUMULENES

1. Reaction with 3-Dimethylamino-2H-azirines

Azirine 52 (Scheme 5) reacts with carbon disulfide and isothiocyanates with splitting of the azirine N-1=C-3 double bond to give dipolar five-membered heterocyclic 1:1 adducts. In some cases, these products can undergo secondary reactions to yield 1:2 and 1:3 adducts. Hydrolysis of the adducts formed with isothiocyanates affords thioureas, which easily cyclize to yield 2-thiohydantoins (53).[99] Amines trap the adducts with carbon disulfide and give dithiohydantoins (54) with thiourea intermediates.[100] Similarly, hydrolysis of the 1:1, 1:2, and 1:3 adducts with isocyanates yields hydantoins (55).[101] With arylsulfonyl isocyanates, the adducts lead to amidines, which readily rearrange with a shift of the sulfonyl group to give imidazolinones, which afford 1-arylsulfonylhydantoins (56).[102] Some 2-monosubstituted azirines also afford hydantoins by treatment with aryl isocyanates and subsequent hydrolysis.[103]

---

[93] J. W. Clark-Lewis and R. P. Singh, *J. Chem. Soc.*, 3162 (1962).
[94] J. W. Clark-Lewis and M. J. Thompson, *J. Chem. Soc.*, 430 (1957).
[95] A. M. Khalil, I. I. Abd El-Gawad, and M. Hammouda, *Aust. J. Chem.* **27**(9), 2035 (1974).
[96] A. H. Harhash, M. H. Elnagdi, and Ch. A. S. Elsannib, *J. Prakt. Chem.* **315**(2), 211 (1973).
[97] V. E. Marquez, L. Twanmoh, H. B. Wood, and J. S. Driscoll, *J. Org. Chem.* **37**, 2558 (1972).
[98] V. E. Marquez, T. Hirata, L. Twanmoh, B. Harry, and J. S. Driscoll, *J. Heterocycl. Chem.* **9**(5), 1145 (1972).
[99] E. Schaumann, E. Kausch, and W. Walter, *Chem. Ber.* **110**, 820 (1977).
[100] E. Schaumann, E. Kausch, S. Grabley, and H. Behr, *Chem. Ber.* **111**(4), 1486 (1978).
[101] G. Mukherjee-Müller, H. Heimgartner, and H. Schmid, *Helv. Chim. Acta* **62**(5), 1429 (1979).
[102] E. Schaumann and S. Grabley, *Chem. Ber.* **113**, 934 (1980).
[103] E. Schaumann, S. Grabley, and G. Adiwidjaja, *Liebigs Ann. Chem.* **2**, 264 (1981).

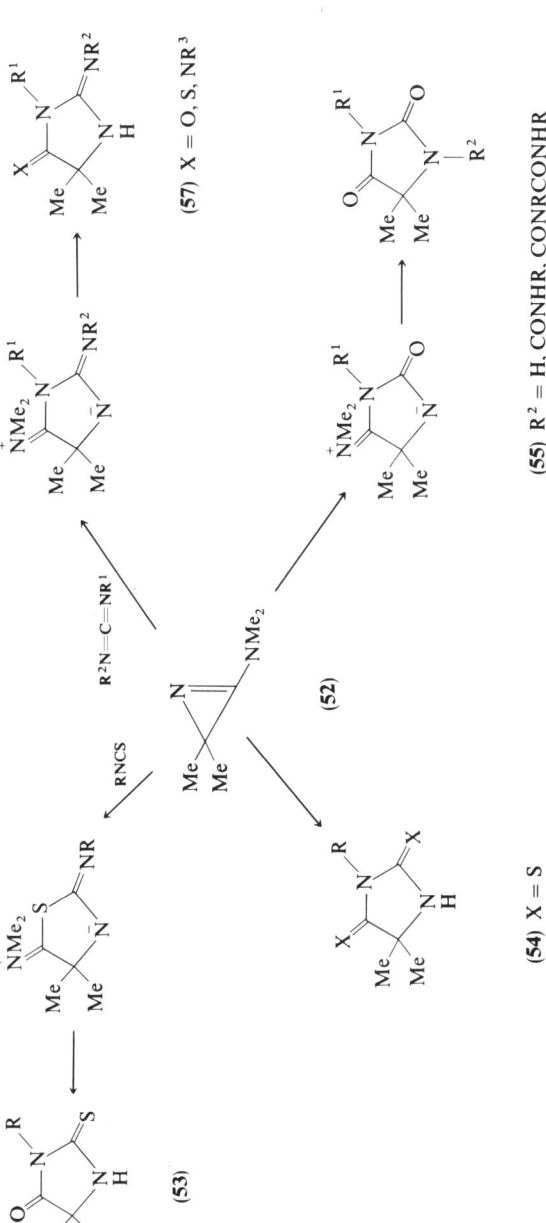

SCHEME 5

Some carbodiimide adducts with **52** give, by hydrolysis, thiolysis, or aminolysis, 2-iminohydantoins, 2-imino-4-thiohydantoins and 2,4-diiminohydantoins, respectively (**57**).[103] The carbon diselenide adducts afford 2,4-diselenohydantoins (**58**) with amines.[104]

Dipolar adducts, formed by the mechanism shown in Scheme 6, have been proposed to explain the unusual hydrolysis of the α-thiocarbamoyl-substituted carbodiimide **59** to a thiourea, which then eliminates dimethylamine and closes to form the 2-thiohydantoin (**60**).[105]

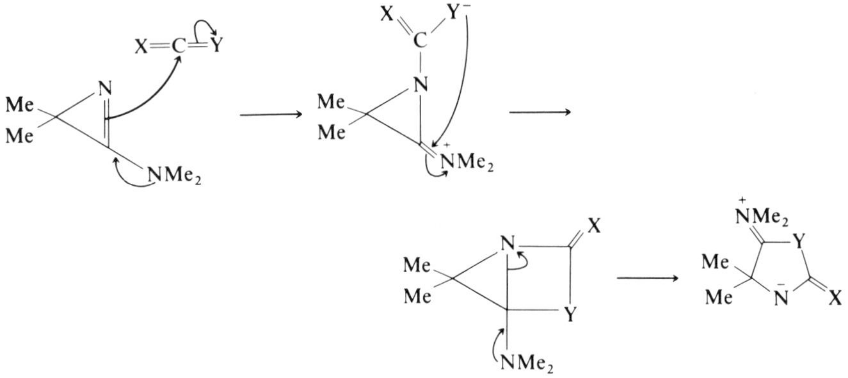

SCHEME 6

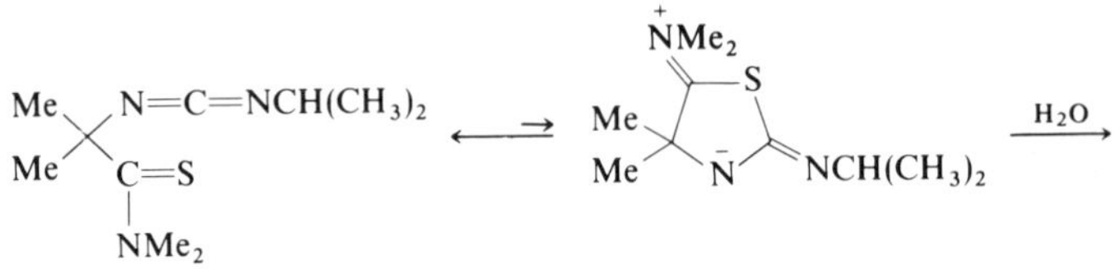

(59)

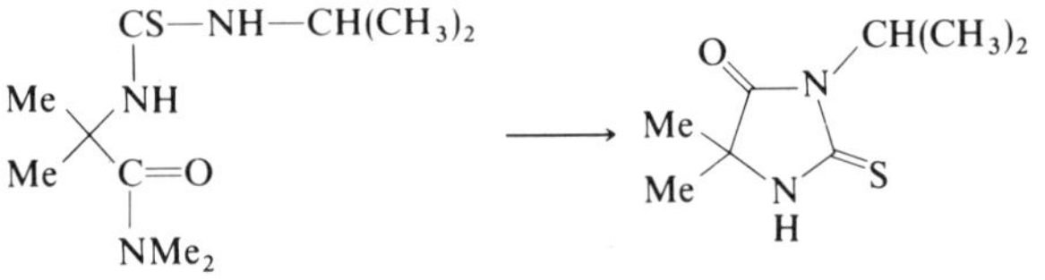

(60)

Diazomethylcarbonyl compounds react similarly with tosyl isocyanate to give hydantoin derivatives.[106]

[104] E. Schaumann, H. Nimmesgern, C. Adiwidjaja, and L. Carlsen, *Chem. Ber.* **115**, 2516 (1982).
[105] E. Schaumann, E. Kausch, K.H. Klaska, and B. Metz, *Naturwissenschaften* **64**(10), 528 (1977).
[106] M. Regitz, B. Weber, and A. Heydt, *Liebigs Ann. Chem.* **2**, 305 (1980).

2. Reaction with Nucleophilic Carbenoids or Ylides

Nucleophilic carbenoids (also called ylides by some authors) react with heterocumulenes, such as isocyanates and isothiocyanates, to yield hydantoins, 2-thiohydantoins, and 2,4-dithiohydantoins via dipolar intermediates. Thus dimethoxycarbene adds to aryl isocyanates and isothiocyanates to form 5,5-dimethoxyhydantoins and dithiohydantoins.[107] Thiazolium ylides (61) (resulting from thiazolium salts and NEt_3 in DMF) give dipolar 1:1 adducts (62) or 1:2-cycloadducts (63) with isocyanates or isothiocyanates.[108-110] Other substrates are imidazolidine[111] or 2-imidazoline derivatives.[112]

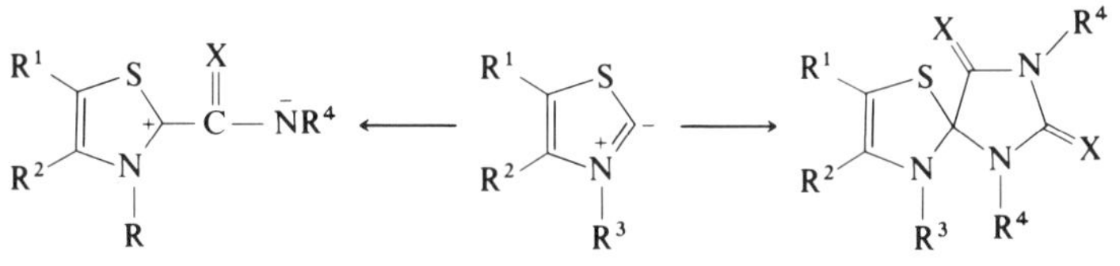

(62)   (61)   (63)

A related reaction uses cyclic peraminoethylenes, particularly $\Delta^{2,2'}$-bis(1,3-disubstituted imidazolidines) (64), as starting materials. These powerful bases are extremely reactive and, in the presence of isocyanates or isothiocyanates, yield inner salts (65) with a 1,3-dipolar structure, which undergo cycloaddition with suitable dipolarophiles, such as isocyanates or isothiocyanates, to give the corresponding spirohydantoins. In many cases the spiro derivative is obtained in one step.[113]

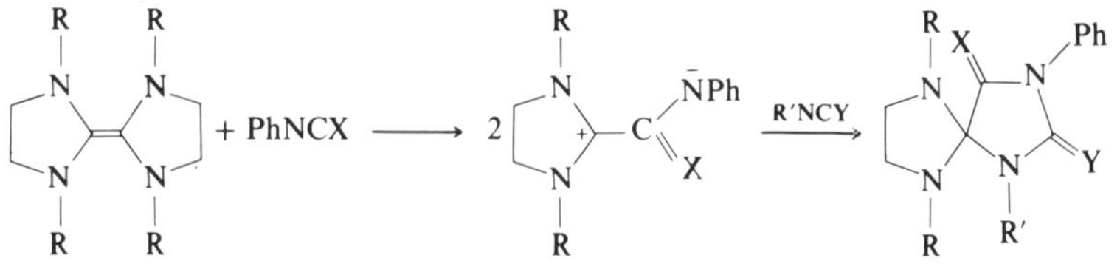

(64)   (65)

---

[107] R. W. Hoffmann, K. Steinbach, and B. Dittrich, *Chem. Ber.* **106**(7), 2174 (1973).
[108] A. Takamizawa, K. Hirai, and S. Matsumoto, *Tetrahedron Lett.* **37**, 4027 (1968).
[109] J. Hocker and R. Merten, *Liebigs Ann. Chem.* **751**, 145 (1971).
[110] A. Takamizawa, S. Matsumoto, and Sh. Sakai, *Chem. Pharm. Bull.* **22**(2), 293 (1974).
[111] R. W. Hoffmann, B. Hagenbruch, and D. M. Smith, *Chem. Ber.* **110**(1), 23 (1977).
[112] H. Giesecke and J. Hocker, *Synthesis* **11**, 806 (1977).
[113] H. E. Winberg and D. D. Coffman, *J. Am. Chem. Soc.* **87**, 2776 (1965).

3. Reaction with 1-Oxa-4-aza- and 1,4-Diazabutadienes

1-Oxa-4-azabutadienes and 1,4-diazabutadienes give, on reaction with heterocumulenes in 1,3-cycloaddition reactions, various 5,5-disubstituted derivatives of 1,3-diarylhydantoins (66).[114,115] Relatively high yields, mild conditions, and a very weak effect of solvent polarity on the reaction rate suggest a synchronous mechanism involving 1,2-migration of a substituent.

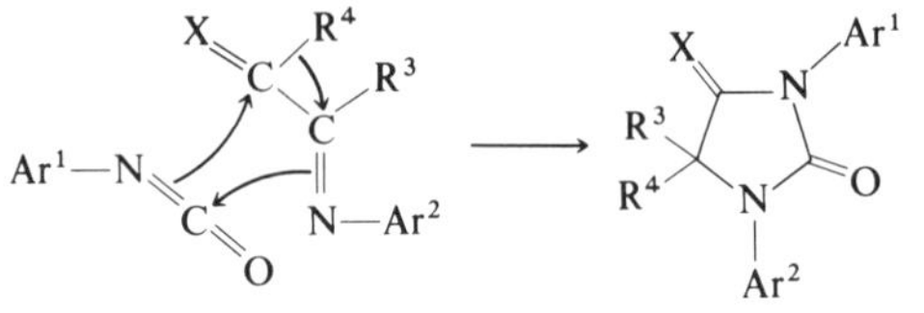

$X = O, ArN$
$R^3 = COR, CONHAr$

(66)

If $R^3 = CONHR$ and $R^4 = Ar$, then heating the cycloaddition product above its melting point splits off aryl isocyanate and gives 1,3,5-triarylhydantoin.[116]

4. Reaction with Phenylacetylene

In the reaction of phenyl isocyanate with phenylacetylene in the presence of $Fe(CO)_5$, 5-benzylidene-1,3-diphenylhydantoin was obtained in 85% yield by an addition reaction and hydrogen shift.[117] Other phenylacetylenes behave similarly.[118]

I. MISCELLANEOUS METHODS STARTING FROM AMINO ACID DERIVATIVES OR RELATED SUBSTRATES

As discussed in Sections II,B–D, these starting materials have been applied with success to prepare hydantoins.[1] Recently, new and sophisticated methods have been developed to improve and expand the utility of amino acids. Thus,

---

[114] J. Moskal, A. Moskal and P. Milart, *Tetrahedron* **38**(12), 1787 (1982).
[115] J. Moskal, J. Bronowski, and A. Rogosvki, *Monatsh. Chem.* **112**, 1405 (1981).
[116] J. Moskal and A. Moskal, *Synthesis* **10**, 794 (1979).
[117] I. Oshiro, K. Kinugasa, T. Minami, and T. Ogawa, *J. Org. Chem.* **35**, 2136 (1970).
[118] A. Baba, Y. Ohshiro, and T. Agawa, *J. Organomet. Chem.* **87**(2), 247 (1975).

N-(1-benzotriazolylcarbonyl)amino acids are especially useful to prevent racemization.[119]

A modification of the stepwise degradation of peptides involves their N-terminal cyclization with N,N'-carbonyl- or N,N'-thiocarbonyldiimidazole.[120] Acetylation of N-hydroxyphenylurethane (67) yields the diester 68, which by simply heating in refluxing ethanol furnishes the N-acetoxyhydantoin 69.[121]

$$\text{ArNHCOCH}_2-\overset{\text{OH}}{\underset{|}{\text{N}}}-\text{CO}_2\text{C}_6\text{H}_5 \xrightarrow{\text{Ac}_2\text{O}} \text{ArNHCOCH}_2\text{N}\genfrac{}{}{0pt}{}{\text{OCOCH}_3}{\text{CO}_2\text{C}_6\text{H}_5} \longrightarrow$$

(67)   (68)   (69)

Normal substitution products of potassium N-cyanoanilide and nitriles or esters yield hydantoins by acid hydrolysis. Hydrogen sulfide treatment of 70 gives 2-thiohydantoins in excellent yield.[122]

$$\underset{\text{PhNC}\equiv\text{N}}{\overset{\text{Ph}_2\text{CX}}{|}}$$

(70)

X = CN, CO$_2$Et

Racemic and optically active hydantoins have been prepared through isocyanate derivatives, starting from chiral disubstituted cyanoacetic acids 71.[123] Similarly, N^2,N^2-disubstituted thiocyanatocarboxylic hydrazides give 3-amino-2-thiohydantoin derivatives 72,[124] and 1-phenyl-3-aminohydantoin is obtained in excellent yield in the reaction of ethyl N-phenyl-N-carbethoxyglycinate and hydrazine hydrate.[125]

Isomerization of thiocyanatofumaric and -maleic esters affords a mixture of the corresponding isothiocyanate esters, which react with primary amines to give 2-thiohydantoins.[126]

---

[119] B. Zorc and I. Butula, *Croat. Chem. Acta* **54**, 441 (1981).
[120] F. Esser and O. Roos, *Angew. Chem.* **90**(6), 495 (1978).
[121] S. C. Bell, G. Conklin, and R. J. McCaully, *J. Heterocycl. Chem.* **13**(1), 51 (1976).
[122] G. Simig, K. Lempert, and J. Tamas, *Tetrahedron* **29**(22), 3571 (1973).
[123] J. Knabe and W. Wunn, *Arch. Pharm. (Weinheim, Ger.)* **313**(6), 538 (1980).
[124] H. Boehme, F. Martín, and J. Strahl, *Arch. Pharm. (Weinheim, Ger.)* **313**(1), 10 (1980).
[125] B. T. Gillis and J. G. Dain, *J. Heterocycl. Chem.* **8**(2), 339 (1971).
[126] P. C. Thieme and E. Haedicke, *Liebigs Ann Chem.* **2**, 227 (1978).

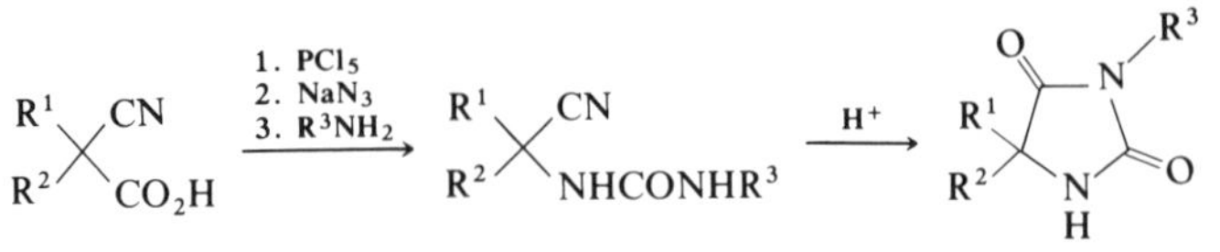

(71)

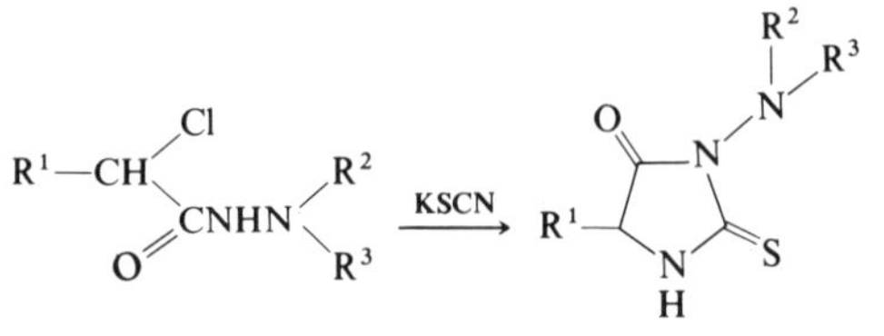

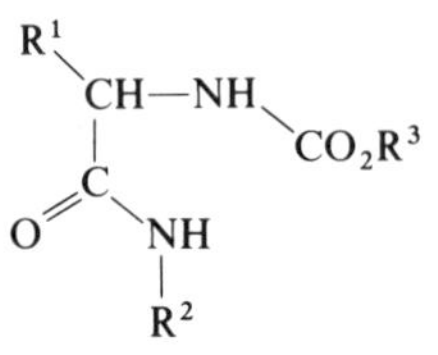

(72) (73)

The old method of reaction between urea and unsaturated acids[1] has been reported for the synthesis of 5-aroylmethylthiohydantoins by reaction of β-aroylacrylic acids and thiourea.[127] N-Carbethoxyamino acid amides (73) easily cyclize in the presence of tetrabutylammonium fluoride to give substituted hydantoins in excellent yields.[128]

Mild aqueous alkaline treatment of peptides containing 2,2-dibenzyloxy-carbonylamidopropionic acid as the N-terminal residue easily gives 5-amino-5-methylhydantoins, which are not available from 5-halohydantoins.[129,130] Other urethanes react similarly.[131] The 5-amino group reacts very easily with nucleophiles such as water to give 5-hydroxyhydantoins.[132]

J. OTHER METHODS

1. Condensation of Imines with Isonitriles and Thiocyanic Acid

4-Iminohydantoins have been prepared according to Scheme 7.[133,134] The hydrolysis of the alkylimino groups affords hydantoins and 2-thiohydantoins.

[127] R. D. Khachikyan, S. M. Atashyan, and S. G. Agbalyan, *Ann. Khim. Zh.* **34**(9), 775 (1981).
[128] J. Pless, *J. Org. Chem.* **39**(17), 2644 (1974).
[129] E. J. McMullen, H. R. Henze, and B. W. Wyatt, *J. Am. Chem. Soc.* **76**, 5636 (1954).
[130] A. B. Evnim, A. Lam, and J. Blyskal, *J. Org. Chem.* **35**, 3097 (1970).
[131] G. Zanotti and F. Pinnen, *J. Heterocycl. Chem.* **18**, 1629 (1981).
[132] G. Lucente and G. Zanotti, *J. Heterocycl. Chem.* **8**(6), 1027 (1971).
[133] I. Ugi, *Angew. Chem.* **74**, 9 (1962).
[134] I. Ugi, K. Risendahll, and F. Bodesheim, *Justus Liebigs Ann. Chem.* **666**, 54 (1963).

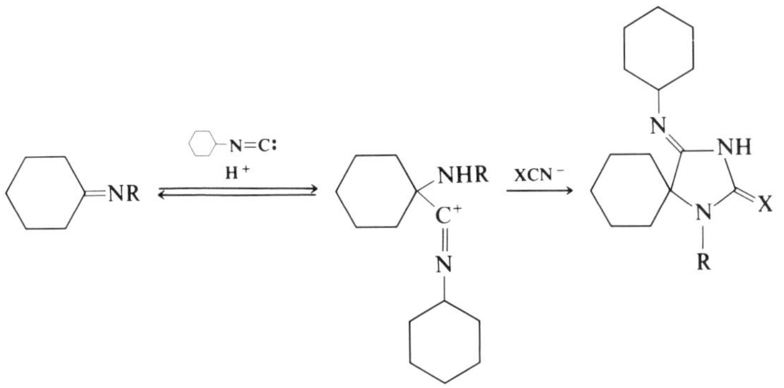

Sec. II.K]   HYDANTOINS   201

SCHEME 7

2. *Reaction of* N-*Cyanoamines with 1*-tert-*Butyl-3,3- diphenylaziridinone*

The reaction of N-cyanoamines with 1-*tert*-butyl-3,3-diphenylaziridinones is a general method for the synthesis of 1-alkyl-, 1-aralkyl-, and 1-aryl-5,5-diphenylhydantoins and glycocyamidines and is a new approach for the synthesis of hydantoins in which the aziridinone ring constitutes the C-5—C-4—N-3 unit of the hydantoin ring and the cyano amine the N-1—C-2 unit.[135] In a similar reaction scheme, α-chloro-α,α-diphenylacetamides and potassium N-cyanoanilide give hydantoins.[136]

K. MUTUAL TRANSFORMATIONS BETWEEN HYDANTOINS AND THIOHYDANTOINS

Most 5,5-disubstituted hydantoins (**74**) give 5,5-disubstituted 2,4-dithiohydantoins (**75**) by reaction with phosphorus trisulfide[137] or pentasulfide[138] (Scheme 8).

5,5-Disubstituted 2,4-dithiohydantoins (**75**) react with ammonia or amines preferentially at the 4-position to give imino derivatives (**76**) which can be hydrolyzed by acid to 2-thiohydantoins (**77**). Carrington[46] found ethanolamine convenient for this purpose. Acid hydrolysis of the methylthio

[135] G. Simig, K. Lempert, J. Tamas, and G. Czira, *Tetrahedron* **31**(9), 1195 (1975).
[136] G. Simig and K. Lempert, *Tetrahedron* **31**, 983 (1975).
[137] H. R. Henze and P. E. Smith, *J. Am. Chem. Soc.* **65**, 1090 (1943).
[138] H. C. Carrington and W. S. Waring, *J. Chem. Soc.*, 354 (1950).

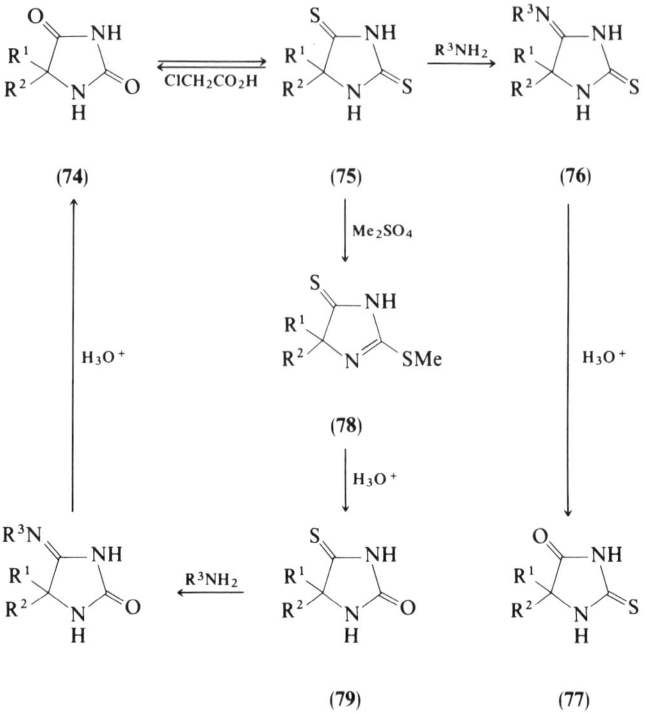

SCHEME 8

derivatives (78) obtained by methylation of the 2,4-dithiohydantoins affords 4-thiohydantoins (79). Boiling 20% aqueous chloroacetic acid completely removes the sulfur from 2,4-dithiohydantoins to give hydantoin derivatives. In some compounds, such as 80, the reaction with phosphorus pentasulfide gives bicyclic derivatives (81).[139]

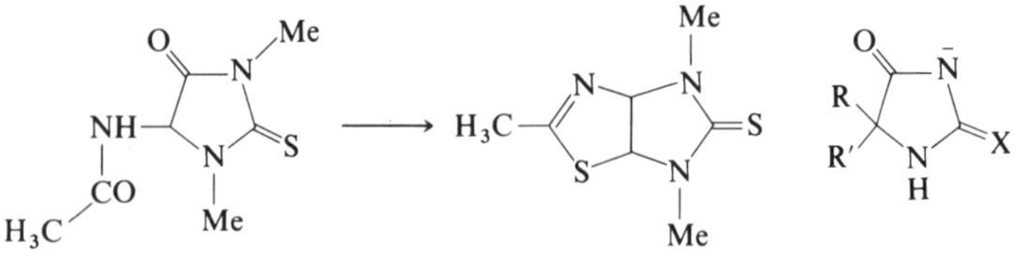

[139] Z. I. Moskalenko and G. P. Shumelyak, *Khim. Geterosikl. Soedin.* 7, 932 (1974).

III. Physicochemical Studies on Hydantoins

^{1}H-NMR, IR, and UV techniques have been used to study ionization and tautomerism in hydantoins.[140]

A. IONIZATION

Hydantoins are weak acids which owe their acidic character to dissociation of the proton bonded to the 3-nitrogen atom since this allows for maximum delocalization of charge in anion **82**.[1] For unsaturated hydantoin derivatives, such as 5-arylidene- or 5-ethoxycarbonylmethylidenehydantoins, IR, UV, and NMR data show that the N-3-substituted derivatives lose the proton from the 1-nitrogen atom in alkaline solution, forming an anion which can delocalize through the 5-substituent.[141,142]

Because the thiocarbonyl group is more strongly electron attracting than the carbonyl, 2-thiohydantoin (pK_a 8.5)[3] is a slightly stronger acid than hydantoin (pK_a 9.12).[143,144] 4-Thiohydantoins appear to be weaker acids than 2-thiohydantoins,[3] and 2,4-dithiohydantoins are stronger.[145,146]

Ionization constants are affected only to a minor extent by alkyl substitution at the 1- and 5-positions, although some N-1 substituents greatly affect the pK_a values. For instance, 1-benzenesulfonylphenytoin has pK_a 4.89, while phenytoin has a pK_a value of 8.31.[147]

3-Substituted hydantoins show no measurable acidity in aqueous alkali,[148] and 3-substituted 2-thiohydantoins are very weak acids ($pK_a \sim 11$).[149] Reported pK_a values of hydantoins have covered a relatively narrow range of acid strengths (pK_a 8.33–9.19).[143,144,150] Remarkably low pK_a values (7–7.5) have been found for granataninespirohydantoins.[151]

---

[140] B. I. Ivin, G. V. Rutkovskii, N. A. Smorygo, A. I. D'yachkov, G. M. Frolova, and E. G. Sochilin, *Zh. Org. Khim.* **9**(11), 2405 (1973).
[141] B. A. Ivin, G. V. Rutkovskii, N. A. Smorygo, and E. G. Sochilin, *Zh. Org. Khim.* **6**(12), 2601 (1970).
[142] G. V. Rutkovskii, B. A. Ivin, V. A. Kirillova, N. A. Smorygo, and E. G. Sochilin, *Zh. Obshch. Khim.* **40**(7), 1583 (1970).
[143] M. Zief and J. T. Edsall, *J. Am. Chem. Soc.* **59**, 2245 (1937).
[144] L. W. Pickety and M. McLean, *J. Am. Chem. Soc.* **61**, 423 (1939).
[145] J. T. Edward and O. J. Chin, *Can. J. Chem.* **41**, 1650 (1963).
[146] E. Santos, I. Rosillo, B. del Castillo, and C. Avendaño, *J. Chem. Res., Synop.*, 131 (1982).
[147] H. Fujioka and T. Tari, *J. Pharm. Dyn.* **5**, 475 (1982).
[148] R. E. Stuckey, *J. Chem. Soc.*, 331 (1947).
[149] J. T. Edward and S. Nielsen, *J. Chem. Soc.*, 5075 (1957).
[150] S. P. Agarwal and M. I. Blake, *J. Pharm. Sci.* **57**, 1434 (1968).
[151] E. Santos, C. Avendaño, I. Rosillo, and G. G. Trigo, *J. Perkin II*, 389 (1982).

B. Ultraviolet Spectroscopy

The ultraviolet absorptions of thioamides have been studied by Janssen.[152] The band of low intensity and at the longest wavelength (Band I) that is assigned with reasonable certainty to $n \to \pi^*$ transitions of the thiocarbonyl group often appears as a shoulder or inflection of the slope of the more intense type II bands at shorter wavelengths. These bands have been assigned to $n \to \sigma^*$ transitions of the thiocarbonyl group.

UV spectra have proved that potentially tautomeric thiohydantoins have thione and not mercapto structures, since their spectra are profoundly affected by S-methylation and only slightly by N-methylation.[138,153]

Ionization of 2-thiohydantoins in alkaline solution leads to characteristic changes in absorption. It results in a slight hypsochromic shift but a big increase in intensity of the main peak, and a hypsochromic shift and decrease in intensity of the subsidiary peak. Several pK_a values of thiohydantoins have been determined by spectrophotometry.[145,146,149,154]

The pK_a value for 5,5-diphenylhydantoin has also been determined by spectrophotometric measurements,[150] although weak optical absorption values have been found for other hydantoins.[146]

C. Mass Spectrometry

The mass spectral decomposition of hydantoin derivatives containing alkyl and phenyl substituents have been investigated, using labeling techniques. The mass spectra of hydantoin,[155] thiohydantoins,[156–158] 5-substituted hydantoins,[155,159–161] and 3-arylhydantoins[162] have shown that cleavage of the hydantoin ring occurs by α-fission at the C-4 carbonyl group with loss of CO and RNCO. Aryl isocyanate ion formation occurs through several pathways, and the substituent effects on its formation obey the Hammett equation.[163]

---

[152] M. J. Janssen, Recl. Trav. Chim. Pays-Bas 79, 454 (1960).
[153] J. T. Edward and J. K. Liu, Can. J. Chem. 50, 2423 (1972).
[154] J. T. Edward and J. K. Liu, Can. J. Chem. 47, 1117 (1969).
[155] R. A. Corral, O. O. Orazi, A. M. Duffield, and C. Djerassi, Org. Mass Spectrom. 5, 551 (1971).
[156] R. E. Ardrey and A. Darbre, J. Chromatogr. 87, 499 (1973).
[157] T. Sizuki, K. D. Song, and K. Tuzimura, Org. Mass Spectrom. 11, 557 (1976).
[158] T. Suzuki, S. Matsui, and K. Tuzimura, Agric. Biol. Chem. 36, 1061 (1972).
[159] R. A. Locock and R. T. Coutts, Org. Mass Spectrom. 3, 735 (1970).
[160] A. J. Atkinson, J. MacGee, J. Strong, D. Gartiez, and T. E. Gaffney, Biochem. Pharmacol. 19, 2483 (1970).
[161] H. M. Fales, G. W. A. Milne, and N. C. Law, Arch. Mass Spectral Data 2(4), 650 (1971).
[162] G. Ruecker, P. N. Natarajan, and A. F. Fell, Arch. Pharm. (Weinheim, Ger.) 304, 833 (1971).
[163] B. M. Kwon and S. Ch. Kim, J. S. Perkin II, 761 (1983).

The heterocyclic rings of 1,3,5-triarylhydantoins cleave via the aziridinone intermediates (83) with ejection of aryl isocyanate. The hydantoins containing substituents possessing carbonyl groups joined directly to the heterocyclic ring in position 5 reveal a great tendency to eliminate them under electron impact.[164]

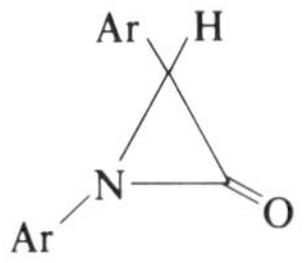

(83)

The mass spectra of 1,5,5- and 3,5,5-trisubstituted hydantoins have also been discussed.[122,135,165]
Mass spectra have been used in the analysis of amino acid phenylthiohydantoin derivatives formed during Edman degradation of proteins.[166] Gas chromatographic–mass spectrometric methods to detect and quantify very low levels of biologically active hydantoins have been developed.[167]

D. INFRARED SPECTROSCOPY

Hydantoins show two bands in the carbonyl region at about 1720 and 1780 cm^{-1}. However, their interpretation has been the subject of controversy. The low-frequency band has been attributed both to the C-4[168] and the C-2 carbonyl group.[169] Infrared and Raman spectra have also been interpreted in terms of carbonyl coupling similar to that found in imides.[170] Other studies have confirmed this coupling.[171]
Many studies concerning the origin and localization of the characteristic IR frequencies of the CO and NH groups in hydantoins have been

---

[164] J. Moskal, K. Nagraba, and A. Moskal, *Org. Mass Spectrom.* 15(5), 257(1980).
[165] K. Lempert, K. Zaer, J. Møller, and G. Schroll, *Acta Chem. Scand.* 26, 1542 (1972).
[166] H. M. Fales, Y. Nagai, G. W. A. Milne, H. B. Brewer, T. Brouzert, and J. J. Pisano, *Anal. Biochem.* 43(1), 288 (1971).
[167] J. Plowman, D. B. Lakings, E. S. Owens, and R. H. Adamson, *Pharmacology* 15(4), 359 (1977).
[168] J. Derkosch, *Monatsh. Chem.* 92, 361 (1961).
[169] P. S. A. Demoen, *Bull. Soc. Chim. Belg.* 75, 524 (1966).
[170] Ch. Fayat and A. Foucaud, *Bull. Soc. Chim. Fr.* 3, 987 (1971).
[171] J. Bellanato, C. Avendaño, P. Ballesteros, and M. Martinez, *Spectrochim. Acta, Part A* 35A, 807 (1979).

reported.[172-174] In 2-thiohydantoins, strong coupling occurs between C—S and C—N vibration, resulting in numerous bands in the 1550–1200 cm^{-1} region.[175,176] The identification in the literature is often based on the "thioureide band."[175,177] Antisymmetric stretching modes of NCS bonds complete the characterization.[178]

Besides N—H···O hydrogen bonds, IR results have revealed in some granatanine- and nortropane-3-spiro-5′-hydantoins other strong hydrogen bonds between the N-3—H group and the piperidine nitrogen atom.[171] Some of these compounds exist as zwitterion structures in the solid state.[179] Infrared results also support the idea that hydrogen bonding is the main factor in the formation of molecular compounds between 5,5-disubstituted hydantoins and some pyrazolone derivatives such as antipyrine.[147]

Urey–Bradley force constants have been calculated, using the vibrational frequencies of hydantoin or 2-thiohydantoin and its N-1,N-3-dideutero derivatives.[180,181]

E. Nuclear Magnetic Resonance Spectroscopy

^{1}H-NMR spectroscopy is particularly useful for localization of substituents in the hydantoin ring, considering the chemical shifts of protons at 1,3- and 5-positions and the characteristic changes of the N-1—H and N-3—H signals on addition of bases.[182]

From the NMR study of 5,5-diphenylhydantoin, its thioxo analogues, and their possible N- and S-methyl derivatives, correlations have been found between the chemical shifts of the resonance signals and the chemical structures of the compounds.[183]

---

[172] J. H. Elliot and P. N. Natarajan, *J. Pharm. Pharmacol.* **19**, 209 (1967).
[173] A. R. Katritzky and P. J. Taylor, in "Physical Methods in Heterocyclic Chemistry" (A. R. Katritzky, ed.), Vol. 4, p. 265. Academic Press, New York, 1971.
[174] S. Goenechea, *Mikrochim. Acta* **3**, 276 (1972).
[175] D. T. Elmore, *J. Chem. Soc.*, 3489 (1958).
[176] C. Cogrossi, *Spectrochim. Acta, Part A* **28A**(5), 855 (1972).
[177] K. A. Jenssen and P. H. Nielsen, *Acta Chem. Scand.* **20**, 597 (1966).
[178] J. Poupaert and R. Bouche, *J. Pharm. Sci.* **65**(8), 1258 (1976).
[179] J. Bellanato, C. Avendaño, P. Ballesteros, E. Santos, and G. G. Trigo, *Spectrochim. Acta, Part A* **36A**, 879 (1980).
[180] K. Geetharani and D. N. Sathyanarayana, *Indian J. Pure Appl. Phys.* **16**(2), 88 (1978).
[181] Y. Saito and K. Machida, *Bull. Chem. Soc. Jpn.* **51**(1), 108 (1978).
[182] R. A. Corral and O. O. Orazi, *Spectrochim. Acta* **21**(12), 2119 (1965).
[183] P. Sohár, J. Nyitrai, K. Zaner, and K. Lempert, *Acta Chim. Acad. Sci. Hung.* **65**(2), 189 (1970).

^{1}H- and ^{13}C-NMR spectroscopy have helped to solve several configurational and conformational problems in hydantoins,[71,79,184–186] and the energy barrier for free rotation about the N—C bond in 3-aryl-2-thiohydantoins has been calculated by ^{1}H-NMR experiments.[187] Diastereomeric solute–solvent interactions of 1- and 3-arylhydantoins in chiral solvents have been studied.[188]

^{1}H- and ^{13}C-NMR data have been reported for diagnostic purposes in direct analysis of phenylthiohydantoin amino acid derivatives (PTH) produced in the Edman degradation of peptides and proteins.[189–193] The insensitivity of ^{1}H-NMR spectroscopy constitutes a major hurdle for its application in the sequence study of peptides.[194,195] Alternatively, identification of the cleaved amino acids in the automated Edman degradation has been solved in some cases by using IR,[196–198] mass,[199] and gas chromatographic techniques.[200]

F. CRYSTAL STRUCTURE DETERMINATIONS

Most X-ray diffraction studies have been undertaken to confirm the structure of hydantoins and thiohydantoins or their derivatives obtained by unusual methods.[87,88,111,126,201,202] Many others have emphasized con-

---

[184] G. G. Trigo, M. Martinez, and E. Gálvez, *J. Pharm. Sci.* **70**(1), 87 (1981).
[185] L. D. Colebrook and M. A. Khadin, *Org. Magn. Reson.* **19**(1), 27 (1982).
[186] I. Attia and I. Z. Siemion, *Rocz. Chem.* **50**(12), 2063 (1976).
[187] I. A. Attia, T. Glowiak, and I. Z. Siemion, *Bull. Acad. Pol. Sci. Chim.* **24**(10), 781 (1976).
[188] L. D. Colebrook, S. Icli, and F. H. Hund, *Can. J. Chem.* **53**(11), 1556 (1975).
[189] K. Izumi, *Anal. Biochem.* **85**(1), 306 (1978).
[190] I. Z. Siemion, I. Attia, and K. Nowak, *Bull. Akad. Pol. Sci., Ser. Sci. Chim.* **23**(7), 575 (1975).
[191] I. Z. Siemion, K. Sobczyk, and B. Picur, *Pol. J. Chem.* **52**(3), 667 (1978).
[192] K. Sobczyk and I. Z. Siemion, *Pol. J. Chem.* **54**(9), 1833 (1980).
[193] M. L. Bouguerra and Y. Leraux, *C. R. Hebd. Seances Acad. Sci., Ser. C* **273**(16), 991 (1971).
[194] C. S. Tsai, N. L. Fraser, H. Avdovich, and J. P. Farrant, *Can. J. Biochem.* **53**(9), 1005 (1975).
[195] T. Suzuki, T. Tomioka, and K. Tuzimura, *Can. J. Biochem.* **55**(5), 521 (1977).
[196] J. Ramachandran, *Nature (London)* **206**, 927 (1965).
[197] C. Djerassi, K. Undheim, R. C. Sheppard, W. G. Terry, and B. Shoberg, *Acta Chem. Scand.* **15**, 903 (1961).
[198] J. A. Poupaert, M. Claesen, J. Delegaen, and P. Dumont, *Bull. Soc. Chim. Belg.* **86**(6), 465 (1977).
[199] N. S. Vul'fson, V. M. Stepanov, V. A. Puchkov, and A. M. Zyakun, *Iz. Akad. Nauk SSRR, Ser. Khim.* **8**, 1524 (1963).
[200] H. Tschesche, R. Obermeier, and S. Kupfer, *Angew. Chem., Int. Ed. Engl.* **9**(11), 893 (1970).
[201] T. Kinoshita, S. Sato, and Ch. Tamura, *Tetrahedron Lett.* **40**, 3695 (1971).
[202] A. F. Cameron, I. R. Cameron, and F. D. Cuncanson, *J.C.S. Perkin II* **5**, 789 (1981).

figurational and conformational studies,[203–206] especially those of García-Blanco and co-workers.[68–70,75,76,207–212]

The separate existence of two individual tautomeric forms of the S-methyl derivative of 5,5-diphenyl-2-thiohydantoin (**84** and **85**), which had been originally claimed on the basis of IR evidence, has been proved by X-ray structure determination.[213] Although the hydantoin ring is essentially planar, in some cases the deviation from planarity is not negligible.[214] A great variety of N—H···N and especially N—H···O hydrogen bond systems have been found.[215] The first type is particularly strong in some hydantoins containing basic centers. The solid structure of 2-thiohydantoin does not show N—H···S=C hydrogen bonds.[216] However, some 2-thiohydantoin derivatives have shown N^1—H···S=C hydrogen bonds.[217]

(84)   (85)

---

[203] A. Camerman and N. Camerman, *Acta Crystallogr., Sect. B* **B27**(11), 2205 (1971).
[204] P. H. Bird, L. D. Colebrook, A. R. Fraser, and H. G. Giles, *J.C.S., Chem. Commun.* **6**, 225 (1974).
[205] R. W. Miller and A. T. McPhail, *J. Chem. Res., Synop.* **10**, 330 (1979).
[206] M. H. Koch, G. Germain, J. P. Declerq, and Y. Dusansoy, *Acta Crystallogr., Sect. B* **B31**(10), 2547 (1975).
[207] P. Smith-Verdier, F. Florencio, and S. García-Blanco, *Acta Crystallogr., Sect. B* **B33**(11), 3381 (1977).
[208] P. Smith-Verdier, F. Florencio, and S. García-Blanco, *Acta Crystallogr., Sect. B* **B35**(1), 216 (1979).
[209] F. Florencio, P. Smith-Verdier, and S. García-Blanco, *Acta Crystallogr., Sect. B* **B38**, 2089 (1982).
[210] J. Vilches, F. Florencio, P. Smith-Verdier, and S. García-Blanco, *Acta Crystallogr., Sect. B* **B37**(1), 201 (1981).
[211] J. Vilches, F. Florencio, P. Smith-Verdier, and S. García-Blanco, *Acta Crystallogr., Sect. B* **B37**(11), 2076 (1981).
[212] J. Vilches, F. Florencio, and S. García-Blanco, *Acta Crystallogr., Sect. B* **B37**(2), 361 (1981).
[213] K. Lempert, J. Nyitrai, K. Zaner, A. Kalman, Gy. Argay, A. J. M. Duisemberg, and P. Sohar, *Tetrahedron* **29**(22), 3565 (1973).
[214] E. Arte, B. Tinant, J. P. Declerq, G. Germain, and M. Van Meerssche, *Bull. Soc. Chim. Belg.* **89**(5), 379 (1980).
[215] R. E. Cassady and S. W. Hawkinson, *Acta Crystallogr., Sect. B* **B38**, 164 (1982).
[216] L. A. Walker, K. Folting, and L. L. Merritt, *Acta Crystallogr., Sect. B* **B25**, 88 (1969).
[217] H. Fujiwara and J. M. Van der Veen, *J.C.S. Perkin II* **5**, 659 (1979).

G. QUANTUM MECHANICAL CALCULATIONS

Extensive MO calculations for bond lengths and angles of 1-phenyl-2-thio-5-carbomethoxymethylhydantoin have been reported.[218] He(I) excited photoelectron spectra of hydantoin and 1-methyl-hydantoin are assigned by comparison with the spectrum of succinimide and by INDO/S calculations.[219]

IV. Reactivity of Hydantoins and Their Derivatives

As would be expected, hydantoins and thiohydantoins can react with nucleophilic and electrophilic as well as with other types of reagents.

A. ELECTROPHILIC SUBSTITUTION

1. Protonation

On protonation in strongly acidic solution, hydantoin cations (**86** or **87**) are formed.[220] UV spectral changes indicate that 2,4-dithiohydantoins are first protonated on the sulfur atom at the 2-position, while 4-hydantoins are protonated on oxygen at the 2-position.[154,221] Both O- and S-protonation are separately observed in 2-thiohydantoins.[222]

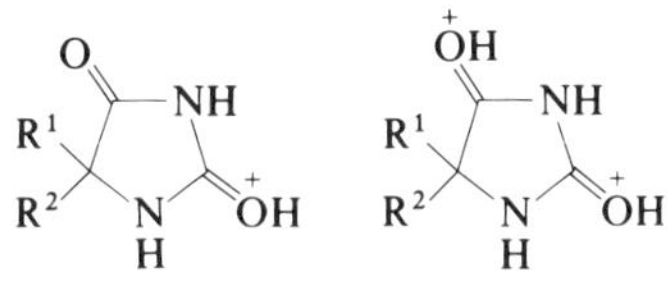

(86)   (87)

2. Reactions with Lewis Acids

Oxidation and the benzilic acid rearrangement occur during the reaction leading from benzoin to 5,5-diphenyldithiohydantoin (**88**) (Scheme 9). The

---

[218] R. Baltrusis, Z. Beresnevicius, I. Vizgaitis, and Yu. V. Gatilov, *Khim. Geterotsikl. Soedin.* **12**, 1669 (1981).
[219] D. Ajo, M. Casarin, G. Granozzi, A. Poli, and E. Tondello, *J. Mol. Struct.* **82**, 277 (1982).
[220] W. I. Congdon and J. T. Edward, *Can. J. Chem.* **50**, 3767 (1972).
[221] J. T. Edward and J. K. Liu, *Can. J. Chem.* **47**(7), 1123 (1969).
[222] J. T. Edward and S. Ch. Wong, *Can. J. Chem.* **57**(15), 1980 (1979).

product on treatment with $AlCl_3$ undergoes a reductive retrobenzilic acid rearrangement (Section II,E). However, in the benzilic acid rearrangement the phenyl group migrates anionically; the migration of the phenyl group in this case is most likely accomplished cationically.[223–225]

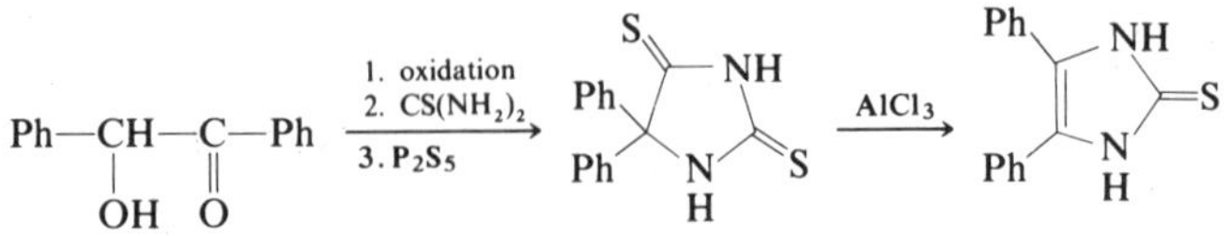
Ph—CH—C—Ph
OH O
1. oxidation
2. CS(NH2)2
3. P2S5
S
NH
Ph
Ph
N
H
S
AlCl3
Ph
NH
S
Ph
N
H

(88)

SCHEME 9

5,5-Diaryl-2,4-dithiohydantoins, when refluxed with BF_3 dimethyl etherate or mixtures of the former reagent with BF_3 in toluene or chlorobenzene, are selectively methylated at the S2 atom and/or undergo the above-mentioned rearrangement to yield imidazole derivatives (**89**).[226]

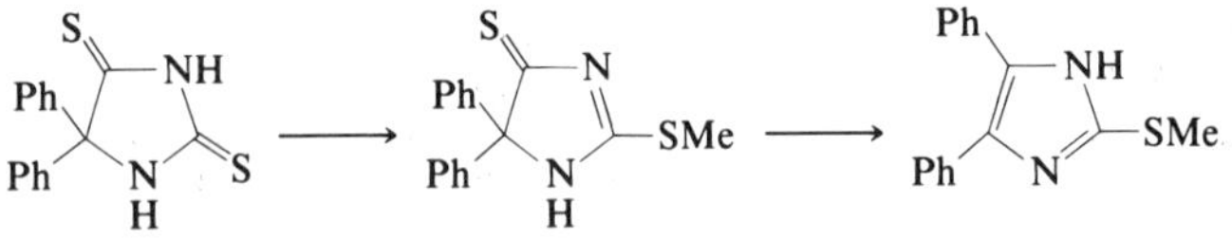
S
NH
Ph
Ph
N
H
S
S
N
Ph
SMe
Ph
N
H
Ph
NH
SMe
Ph
N

(89)

3. *N-Alkylation*

Hydantoins can easily be alkylated in the N-3 position by treatment with alkyl halides in alkaline solution in protic or aprotic solvents. Other alkylating agents include dimethyl sulfate and diazomethane. Amide nitrogen (N-1) alkylations are well-known in 3-substituted hydantoins, but they occur under more severe conditions than the simple N-3 imide alkylations. For instance, N-3 substituted hydantoins are N-1 alkylated with good yields in DMF and sodium hydride.[227] Mono-N-1-alkylated hydantoins can be obtained by protecting the imide nitrogen with an aminomethyl group, followed by alkylation of the amide nitrogen and the removal of the protecting group by

[223] J. Nyitrai, *Acta Chim. Acad. Sci. Hung.* **51**, 95 (1967).
[224] E. Koltai, J. Nyitrai, K. Lempert, and L. Bursics, *Chem. Ber.* **104**, 290 (1971).
[225] R. Markovits-Komis, J. Nyitrai, and K. Lempert, *Chem. Ber.* **104**, 3080 (1971).
[226] J. Fetter, J. Nyitrai, and K. Lempert, *Tetrahedron* **27**(23), 5933 (1971).
[227] O. O. Orazi, R. A. Corral, and H. Schttenberg, *J.C.S. Perkin Trans. II*, 219 (1974).

mild alkaline hydrolysis.[228] However, in intramolecular processes, such as the conversion of **90** to **91**, only amide cyclized monomers have been obtained.[229] If strongly alkaline conditions or phase-transfer catalysis is used, N-1,N-3-disubstituted hydantoins are obtained.[230,231]

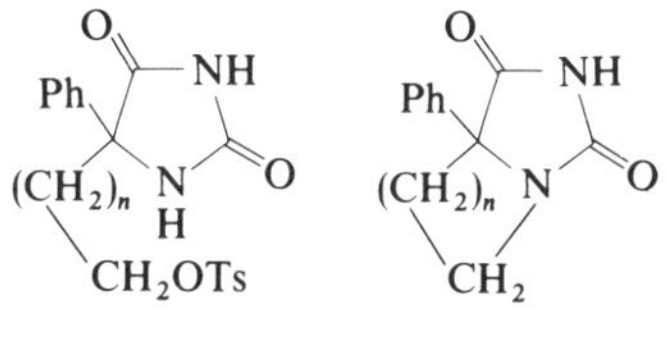

(90) (91)

$n = 2, 3$ $n = 2, 3$

The products of alkylation of thiohydantoins are kinetically controlled and hence are frequently the less stable S-alkyl compounds.[138,232] Methylation products of 2-thio- and 2,4-dithiohydantoins exist almost entirely in one tautomeric form (**92**) or the other (**93**) depending on the solvent.[153] 5-Monosubstituted 2,4-dithiohydantoins exist as the tautomer (**94**), and alkylation takes place on sulfur.[221]

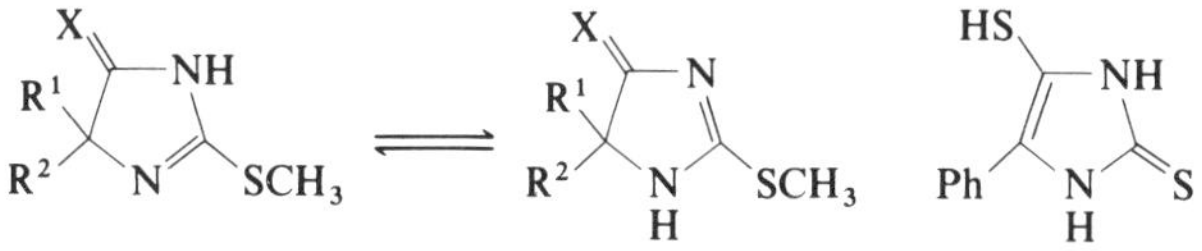

(92) (93) (94)

S-Methyl derivatives of 5,5-diphenylmono and dithiohydantoins may be demethylated by the hydrogen sulfide anion, thiolate anions, or phosphorus pentasulfide. The latter simultaneously converts carbonyl to thiocarbonyl groups. The S-benzyl groups may also be removed by briefly boiling with benzene in the presence of aluminum chloride. The removal of N-3-benzyl groups needs more vigorous conditions.[233]

---

[228] O. O. Orazi and R. A. Corral, *Experientia* **21**, 508 (1965).
[229] E. E. Smissman, P. L. Chien, and R. A. Robinson, *J. Org. Chem.* **35** (11), 3818 (1970).
[230] G. G. Trigo, C. Avendaño, I. Varela, and M. Martínez, *An. Quim.* **74**(7–8), 1090 (1978).
[231] C. Pedregal, G. G. Trigo, M. Espada, and J. Elguero, *J. Heterocycl. Chem.* **21**, 477 (1984).
[232] N. Kornblum, R. A. Smiley, R. K. Blackwood, and D. C. Iffland, *J. Am. Chem. Soc.* **77**, 6261 (1955).
[233] G. Domany, J. Nyitrai, K. Zaner, K. Lempert, and S. Bekassy, *Acta Chim. Acad. Sci. Hung.* **80**(1), 101 (1974).

Direct N-1 and S-alkylation of 5,5-disubstituted 2-thiohydantoins has been reported.[38]

Some N-3-substituted hydantoins, such as 95, lead to N-bridgehead bicycles (96), which are oxygen analogues of levamisole and are powerful alkylating agents.[234] The intramolecular process is controlled by the thermodynamic stabilities of the two possible bicyclic products resulting from N- or O-alkylation.[97]

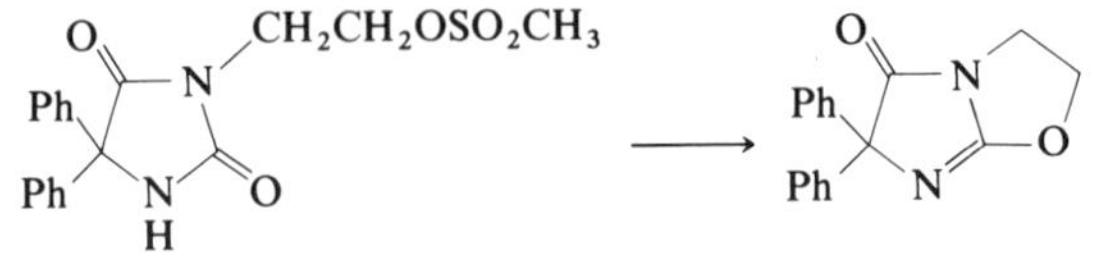

(95) (96)

In contrast, reaction of 5,5-diphenylthiohydantoin with 1-bromo-2-chloroethane generates two isomeric imidazo(2,1-b)thiazole derivatives (97 and 98) through intramolecular S,N-dialkylation.[235] The formation of imidazo(2,1-b)thiazoles by this kind of process is rather common.[236] Similarly, reaction with 1,3-dibromopropane gives two isomeric diphenylimidazothiazines.[237,238]

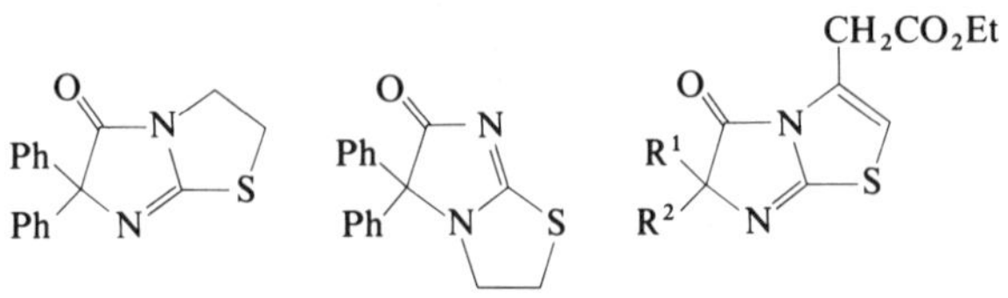

(97) (98) (99)

5-Substituted 2-thiohydantoins and ethyl 4-bromacetoacetate give 5,6-dihydroimidazo(2,2-b)thiazole derivatives (99).[239]

N-Alkoxymethyl derivatives are formed from the hydantoin alkali salt with chloromethyl alkyl ether[240] but the alkoxymethylation of alcohols, using

[234] K. Okada, J. A. Kelley, and J. S. Driscoll, *J. Heterocycl. Chem.* **14**(3), 511 (1977).
[235] K. Okada, J. A. Kelley, and J. S. Driscoll, *J. Org. Chem.* **42**(15), 2594 (1977).
[236] E. G. Delegan, I. V. Smolanka, and Yu. V. Melika, *Khim. Geterotsikl. Soedin.* **11**, 1572 (1974).
[237] J. Karolak-Wojciechowska, M. W. Wieczorek, K. Kiec-Kononowicz, A. Zejc, M. Mikolajezyk, and A. Zatorskii, *Tetrahedron* **36**(8), 1079 (1980).
[238] J. Karolak-Wojciechowska, M. W. Wieczorek, K. Kiec-Kononowicz, A. Zejc, M. Mikolajezyk, and A. Zatorskii, *Tetrahedron* **37**(2), 409 (1981).
[239] R. B. Blackshire and C. J. Sharpe, *J. Chem. Soc. C* **21**, 3602 (1971).
[240] J. A. Vida, M. H. O'Dea, and C. M. Samour, *J. Med. Chem.* **18**, 383 (1975).

dialkoxymethane and phosphorus pentoxide, has also been applied.[241] N-Alkylation with chloroformate esters has been described.[242]

The Mannich reaction gives N-3-aminomethyl derivatives.[243-246] These products have been proposed as prodrugs for a series of drugs containing a primary aromatic amino group to prevent metabolic inactivation by N-acetylation (Scheme 10).[247]

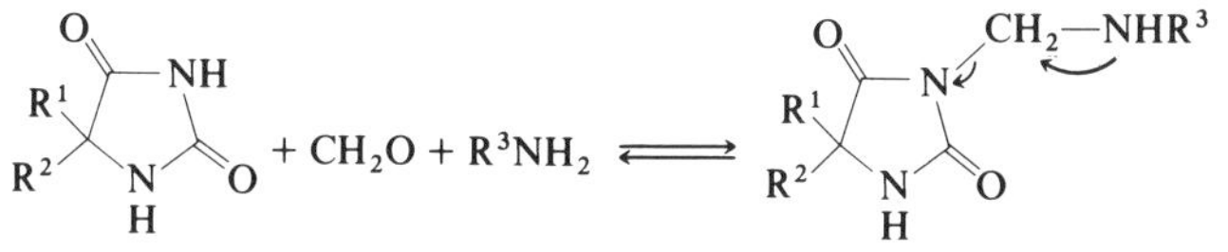

SCHEME 10

Reduction with $NaBH_4$ in Me_2SO of the N-3-aminomethyl derivatives has been also proposed as a method for N-methylation of aromatic primary amines.[248] Alternatively, N-1- or N-3-aminomethyl derivatives can be obtained by reaction of N-1- or N-3-halohydantoins with diazomethane and the amine.[249]

Aminoethylations with ethylenimine[250] and Michael reactions have also demonstrated a preference for the acidic imide function. Thus the N^3—H group adds to acetylene to give N-3-vinylhydantoins[251] and to activated ethylenes, such as acrylonitrile or 4-vinylpyridine, to give N-3-substituted hydantoins.[25,252,253] Under some conditions 1,3-disubstituted hydantoins are obtained.[251]

2-Thiohydantoins behave as S-nucleophiles and, with acrylic acid, give compounds **100**, which cyclize by acetic anhydride to 6,7-dihydro-3,3-disubstituted $5H$-imidazo(2,1-b)(1,3)thiazine-2(3H)-5-diones (**101**).[254]

---

[241] J. Gal, *J. Pharm. Sci.* **68**(12), 1562 (1979).
[242] J. A. Vida, *Tetrahedron Lett.* **37**, 3921 (1972).
[243] O. O. Orazi and R. A. Corral, Tetrahedron **15**, 93 (1961).
[244] M. B. Winstead, D. E. Barr, C. R. Hamel, D. J. Reun, H. I. Parker, and R. M. Newmann, *J. Med. Chem.* **8**(1), 117 (1965).
[245] O. O. Orazi and R. A. Corral, *An. Asoc. Quim. Argent.* **51**(2), 180 (1963).
[246] B. Sucka-Sobstel and A. Zejc, *Diss. Pharm. Pharmacol.* **22**(1), 13 (1970).
[247] H. Bundgaard and M. Johansen, *Int. J. Pharm.* **8**(3), 183 (1981).
[248] K. Horiki, *Heterocycles* **5**(1), 203 (1976).
[249] R. A. Corral and O. O. Orazi, *Tetrahedron Lett.* **25**, 1693 (1964).
[250] J. W. Shaffer, R. Scheasley, and M. B. Winstead, *J. Med. Chem.* **10**, 739 (1967).
[251] Wn. O. Jones, British Patent 846, 601 (1960) [*CA* **55**, 10471 (1961)].
[252] Ch. Chia and Ch. Hsing, *Hua Hsueh Hsueh Pao* **29**(6), 433 (1963).
[253] J. W. Shaffer, E. Steinberg, V. Krimsley, and M. B. Winstead, *J. Med. Chem.* **11**(3), 462 (1968).
[254] J. Fetter, K. Harsanyi, J. Nyitrai, and K. Lempert, *Acta Chim. Acad. Sci. Hung.* **78**(3), 325 (1973).

(100)   (101)   (102)

4. N-Acylation

When hydantoins are treated with acetic anhydride, sodium hypochlorite, or nitiric acid, substitution takes place at the N-1 position, followed by the formation of a 1,3-disubstituted hydantoin.[1] Intramolecular processes yield only the amide cyclized products (102).[255,256]

Thiohydantoins could form S-acyl compounds, which would speedily rearrange to the more stable N-acyl compounds.[3] 2,4-Dithiohydantoin is converted by acetic anhydride to 1,3-diacetyl-2,4-dithiohydantoin.[257]

Brief fusion of a mixture of isopropenyl esters and an acid catalyst with hydantoins gives high yields of N-acylated products.[258] Treatment of hydantoins with trihalomethylsulfenyl chlorides yields N-3-substituted and N-1,N-3 disubstituted compounds, which exhibit antifungal activity.[259] 3-Aryl- or 3-alkylsulfonylhydantoins are prepared by direct acylation with RSO_2Cl in C_5H_5N or Et_3N.[260]

Acid or basic hydrolysis of 1,3-diacyl and 1,3-disulfonyl derivatives gives N-1 derivatives.[1,261] 3-Acetyl-1,5,5-trimethylhydantoin exhibits a remarkable regioselectivity in acylation reactions of amines or phenols bearing alcoholic groups (Scheme 11).[262] Nitrosation in an aqueous medium gives satisfactory yields in only a few examples.[263] The reaction with sodium nitrite and methanesulfonic acid in dichloromethane affords 1-nitrosohydantoins.[264] 1,3-

---

[255] J. L. Szabo, F. R. Scholer, and K. H. Wildrick, *J. Med. Chem.* **9**, 142 (1966).
[256] H. R. Henze and R. J. Speer, *J. Am. Chem. Soc.* **64**, 522 (1942).
[257] A. H. Cook, I. Heilbron, and A. L. Levy, *J. Chem. Soc.*, 201 (1948).
[258] E. S. Rothman, S. Serota, and D. Swern, *J. Org. Chem.* **29**(3), 646 (1964).
[259] R. J. W. Cremlyn, *J. Chem. Soc.* **2**, 6240 (1964).
[260] T. Nakamura and K. Nakamura, Japanese Patent 70/24, 777 (1970) [*CA* **74**, 76419 (1971)].
[261] S. Umemoto, T. Nakamura, and K. Nakamura, Japanese Patent 70/24, 778 (1970) [*CA* **74**, 76416 (1971)].
[262] O. O. Orazi and R. A. Corral, *J. Am. Chem. Soc.* **91**(8), 2162 (1969).
[263] E. H. White, *J. Am. Chem. Soc.* **77**, 6008 (1955).
[264] O. O. Orazi, R. A. Corral, and A. O. Sanchez, *Synthesis* **4**, 205 (1972).

Bisdithiocarboxylic acid derivatives of hydantoins are obtained by treatment with alkali and CS_2.[265]

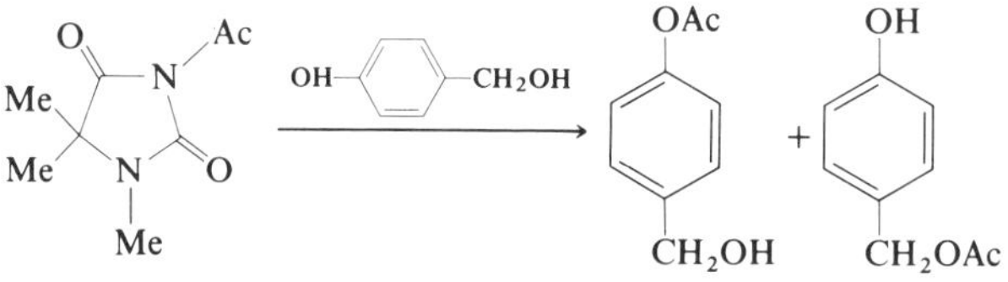

SCHEME 11

5. Halogenation

Although halogenation of the hydantoin ring with several agents occurs preferentially at position 3, monohalogenation of nitrogen-unsubstituted hydantoins leads in all cases to the 1-halo derivative as a result of the intermediate formation of the 1,3-dihalo compound and later transfer of the halogen atom at the N-3 position to another hydantoin molecule.

N-Chloro derivatives have been prepared, using various chlorinating agents[1] and N-bromo derivatives by employing bromine in alkaline solution. N-Monoiodo compounds are obtained by treatment of the silver derivative of the hydantoin with iodine while N-1,N-3-diiodo derivatives are prepared by reaction of hydantoins with iodine monochloride.[266]

Several halohydantoins have been patented as bleaching agents, antiseptics, and fungicides (Section V), and as N-halosuccinimides; these halo derivatives are potent halogenating and oxidizing reagents.[266–269]

1,3-Diiodo-5,5-dimethylhydantoin shows general applicability for nuclear iodination of aromatic and heteroaromatic compounds activated by electron-donating substituents, and, like N-iodosuccinimide, reacts with enol acetates, affording α-iodo ketones in good yields.[270] 3-Bromo-5,5-dimethylhydantoin is particularly useful in regio- and stereo-controlled conversion of unsaturated acids, such as **103**, to bromo lactones, such as **104**.[270a] Allylic

---

[265] T. Takeshima, M. Ikeda, M. Yokoyama, N. Fukada, and M. Muraoka, *J.C.S. Perkin I* **3**, 692 (1979).
[266] R. A. Corral and O. O. Orazi, *J. Org. Chem.* **28**, 1100 (1963).
[267] Y. Ishii, T. Ito, and Sh. Kato, *Kogyo Kagaku Zasshi* **61**, 1254 (1958).
[268] M. J. Fumarola and O. A. Orio, *An. Asoc. Quim. Argent.* **60**(1), 61 (1972).
[269] O. D. Madoery and O. A. Orio, *An. Asoc. Quim. Argent.* **58**(4), 277 (1970).
[270] O. O. Orazi, R. A. Corral, and H. E. Bertorello, *J. Org. Chem.* **30**, 1101 (1965).
[270a] Ch. Cook and Y. Chung, *Arch. Pharm. Res.* **4**(2), 133 (1981).

halogenation with 1-bromo-3,5,5-trimethylhydantoin has been studied,[271] and significant reactivity differences with phenylpropenes under Ziegler conditions have been found.[272] 1,3-Dibromo-5,5-dimethylhydantoin has been used for stereospecific bromohydroperoxylation of 4,4-dimethyl-3,5-diphenyl-4H-pyrazole.[273]

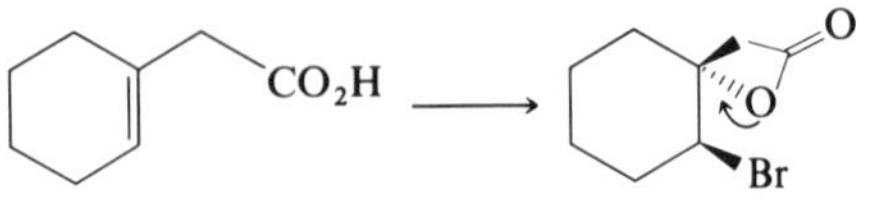

(103) (104)

1-Chlorohydantoins react with methyl alkyl sulfides to give N-sulfonium salts (105), which undergo with tertiary amines a Stevens rearrangement with formation of unstable imidates (106), which isomerize to 107 by heating.[274] Similarly, 1,3-dichloro-5,5-dimethylhydantoin and methyl sulfides gives disulfonium salts, which undergo Stevens rearrangement in the presence of tertiary amines.[275]

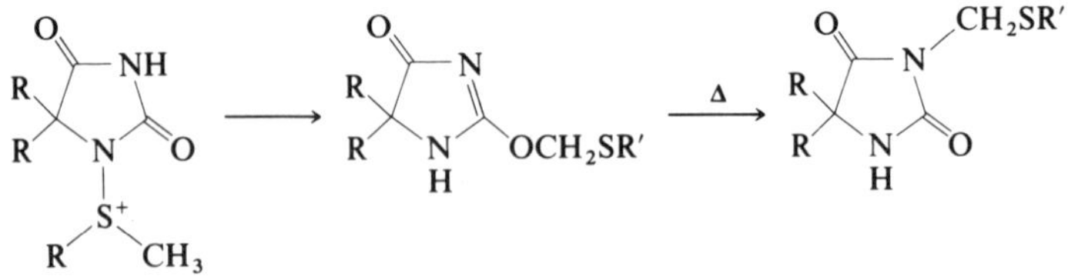

(105) (106) (107)

5-Unsubstituted 2-thiohydantoins are brominated with bromine in acetic acid to give 5-bromo derivatives, which are intermediates in several reactions. Thus by refluxing with either thiourea or thioacetamides, thiazolothiohydantoin derivatives (108) are obtained.[276]

The electrophilic reactivity of 5-halohydantoins with several aromatic systems has been studied.[277] 5-Halohydantoins undergo the Arbuzov reaction, giving bidentate ligands (109).[278]

[271] A. R. Suarez and O. A. Orio, *An. Asoc. Quim. Argent.* **65**, 163 (1977).
[272] V. T. Balzaretti, A. R. Suarez, and O. A. Orio, *An. Asoc. Quim. Argent.* **68**, 163 (1980).
[273] M. E. Landis, R. L. Lindsay, W. H. Watson, and V. Zabel, *J. Org. Chem.* **45**(3), 525 (1980).
[274] E. Vilsmaier, R. Bayer, I. Laengenfelder, and U. Welz, *Chem. Ber.* **111**(3), 1136 (1978).
[275] E. Vilsmaier, R. Bayer, U. Welz, and K. Dittrich, *Chem. Ber.* **111**(3), 1147 (1978).
[276] B. Dash and S. K. Mahapatra, *Curr. Sci.* **39**(24), 559 (1970).
[277] H. E. Zangg, J. E. Leonard, and D. L. Arendsen, *J. Heterocycl. Chem.* **11**(5), 833 (1974).
[278] M. Willson, T. Bouisson, and F. Mathis, *C. R. Hebd. Seances Acad. Sci.* **295**, 567 (1982).

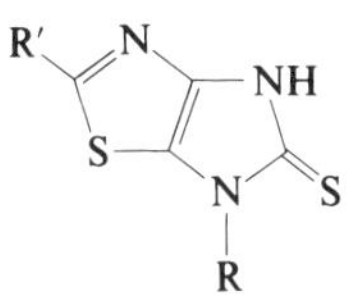

Sec. IV.A]   HYDANTOINS   217

(108) 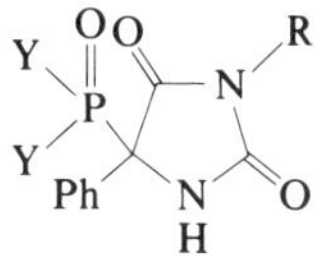   (109)

6. Hydantoins as Heterocyclic Compounds Containing Active Methylene Groups

Besides the previously mentioned halogenation in the C-5 position, hydantoins having a free methylene group condense with aldehydes (especially aromatic aldehydes) or their precursors.[279] 2-Thiohydantoins condense more readily than do the corresponding hydantoins.[1] These C-5-unsaturated hydantoin derivatives may be reduced by any common reducing agents as well as by catalytic methods[1] and are versatile synthetic intermediates for the preparation of amino acids[280] and for other reactions.[281]

Treatment of the 5-arylidene derivatives of 1,3-diphenyl-2-thiohydantoin with malononitrile affords tetrahydropyrano(2,3-d)imidazoles (110), while 3-phenyl- or unsubstituted 2-thiohydantoins originate pyrrolo(1,2-c)imidazoles (111). These products are also obtained in a Michael-type reaction of 2-thiohydantoins with activated nitriles.[282]

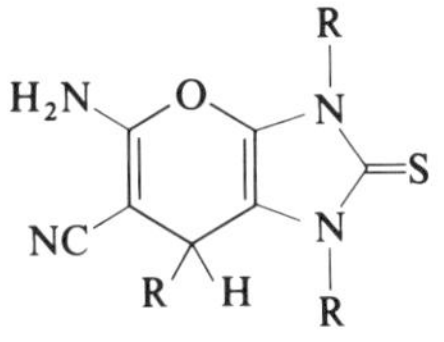

(110) 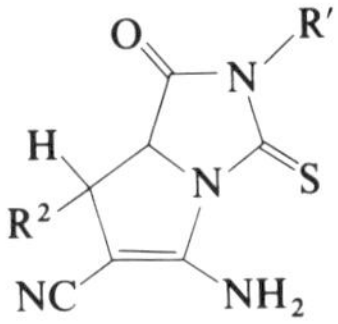   (111)

Some 5-benzal-2-thiohydantoins, especially the 5-anisal derivative, have been selected and proposed for spectrophotometric determination of palladium,[283] and several condensation products of aldehydes and hydantoin

[279] T. Moriya, K. Hagio, and N. Yoneda, *Chem. Pharm. Bull.* **28**(6), 1891 (1980).
[280] U. Jacoby and F. Zymalkowski, *Arch. Pharm. (Weinheim, Ger.)* **304**(4), 271 (1971).
[281] B. A. Ibin, G. V. Rutkovskii, E. G. Sochilin, and I. Y. Tsereli, *Zh. Org. Khim.* **8**(3), 640 (1972).
[282] H. A. F. Daboun, S. E. Abdou., M. M. Hussein, and M. H. Elnagdi, *Synthesis*, 502 (1982).
[283] M. T. Montana González and J. L. Gómez Ariza, *Microchem. J.* **25**(3), 360 (1980).

or 2-thiohydantoin have been prepared as possible antituberculosis substances.[284]

Condensation products of aromatic aldehydes and 4-thiohydantoins react with Grignard reagents to give the colorless products **112**. Compounds **113** and **114** are formed by alkylation and diazomethane addition reactions.[285] These condensation products also react as dienes with acrylonitrile, ethyl acrylate, and other dienophiles giving adducts **115**.[286]

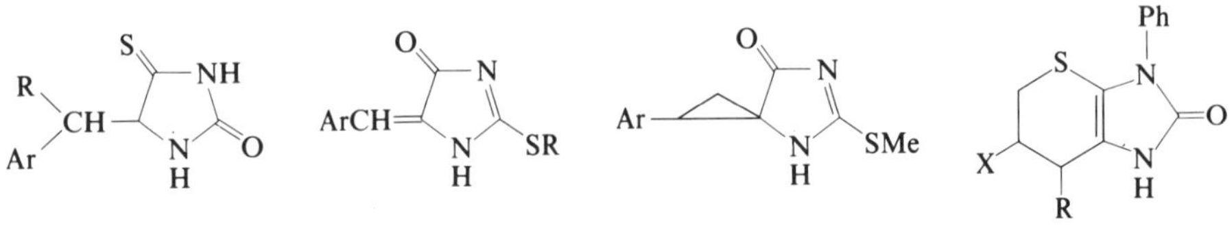

(112)   (113)   (114)   (115)

$X = CN, CO_2Et$

Treatment of 2-thiohydantoins, 1-phenyl-2-thiohydantoins, and 1,2-diphenyl-2-thiohydantoins with aryldiazonium salts yields 5-arylazo-2-thiohydantoins (**116**) and with $p\text{-Me}_2NC_6H_4NO$ the anil **117**, in which the exocyclic azomethine linkage can be cleaved by $PhNHNH_2$ and/or AcOH–HCl to give **118**.[287]

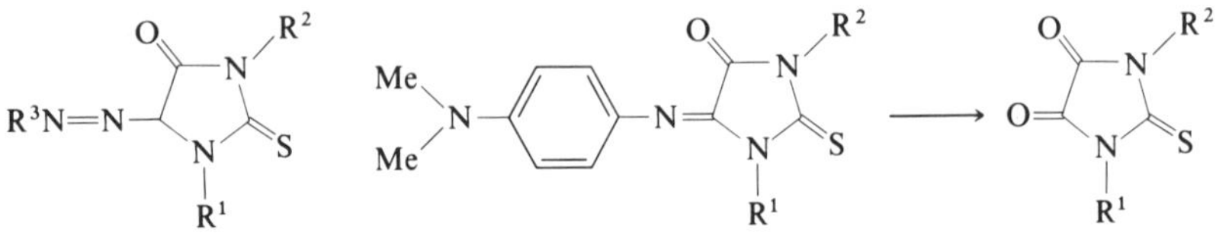

(116)   (117)   (118)

The tautomerism of some 5-arylazo-2-thiohydantoin derivatives as a criterion of their acidity has been studied, and the apparent pK_a values have been discussed in terms of Hammett correlations.[288]

---

[284] H. Tielemann, *Sci. Pharm.* **39**(1), 8 (1971).
[285] A. F. A. Shalaby, H. A. Daboun, and S. S. M. Bghdadi, *Z. Naturforsch., B: Anorg. Chem., Org. Chem.* **B29**(1–2), 99 [1974].
[286] H. A. Daboun, M. A. Abdel Aziz, and F. A. Abdel Aal, *Heterocycles* **19**, 677 (1982).
[287] A. F. A. Shalaby, H. A. Daboun, and M. A. Abdel Aziz, *Indian J. Chem.* **12**(6), 577 (1974).
[288] H. M. Fahmy, M. A. A. Aziz, M. Abortabl, and M. A. Azzeu, *Indian J. Chem.* **20A**(6), 593 (1981).

Sec. IV.B]   HYDANTOINS   219

1-Phenyl-4-thiohydantoin also couples with aryldiazonium salts to give colored 5-arylazo derivatives. The introduction of the electron-withdrawing arylazo group activates the hydantoin ring as well as the thione group.[289]

B. REACTIONS WITH NUCLEOPHILES

1. *Hydrolysis*

Hydantoins and thiohydantoins are hydrolyzed by alkaline media to give the corresponding ureido or thioureido acids, respectively (Section II.A), which under very vigorous conditions give amino acids.[1] Several detailed kinetic investigations on the mechanism of hydrolysis of hydantoin derivatives have been published since the first report of Ingold and co-workers.[290] Two parallel reactions are proposed: attack of OH^- on either the free hydantoin ring or on its N-3 anion.[291–295] Blagoeva *et al.*[6] have established the general validity of the mechanism shown in Eq. (1).

$$\text{(1)}$$

The alkaline hydrolysis of hydantoins appears to be faster than that of the corresponding 2-thiohydantoins, even when account is taken of the greater degree of ionization of the latter compounds and its probable effect on rates.[3]

Studies on alkaline hydrolysis of 2-thio-5-arylidenehydantoins and some of their derivatives have been reported.[296,297] However, very little attention has

---

[289] A. F. A. Shalaby, H. A. Daboun, and S. S. M. Boghdadi, *Z. Naturforsch., B: Anorg. Chem., Org. Chem.* **31B**(6), 865 (1976).
[290] C. K. Ingold, S. Sako, and J. F. Thorpe, *J. Chem. Soc.*, 1177 (1922).
[291] B. A. Ivin, G. V. Rutkowskii, E. G. Sochilin, and I. Yu. Tsereteli, *Zhr. Org. Khim.* **8**, 840, 1951 (1972).
[292] B. A. Ivin, G. V. Rutkovskii, T. N. Rusavskaya, and E. G. Sochilin, *Zhr. Org. Khim.* **11**, 2188 (1975).
[293] B. A. Ivin, G. V. Rutkovskii, and E. G. Sochilin, *Zhr. Org. Khim.* **9**, 179 (1973).
[294] B. A. Ivin, G. V. Rutkovskii, S. A. Andreev, and E. G. Sochilin, *Zhr. Org. Khim.* **9**, 420, 2194 (1973).
[295] G. D. Vogels, F. E. de Windt, and W. Bassi, *Recl. Trav. Chim. Pays. Bas* **88**, 940 (1969).
[296] B. A. Ivin, T. N. Rusavskaya, A. I. D'yachkov, G. U. Rutkovskii, and Yu. A. Ignat'ev, *Khim. Geterotsikl. Soedin.* **3**, 406 (1979).
[297] B. A. Ivin, T. N. Rusavskaya, A. I. D'yachkov, and G. V. Rutkovskii, *Khim. Geterotsikl. Soedin.* **6**, 813 (1979).

been paid to the alkaline hydrolysis of derivatives with electron-withdrawing groups attached to C-5, as in the 5-arylazo-2-thiohydantoin derivatives. These compounds are cleaved by aqueous sodium hydroxide and rearrange to different products, depending on the N-1 substituents.[298]

In 1-acyl-2-thiohydantoins, the acyl group is easily attacked by hydroxide ion, even in very dilute alkali, making possible the stepwise degradation of peptides in Schlack and Kumpf's procedure.[10] 2,4-Dithiohydantoins appear to be resistant to alkaline hydrolysis, perhaps because they are so completely converted to the conjugate base that hydrolysis rates become neglibible.[3] Certain N-3-aminoalkylhydantoins undergo hydrolytic ring opening under very mild conditions, suggesting the intramolecular mechanism shown in Scheme 12.[299]

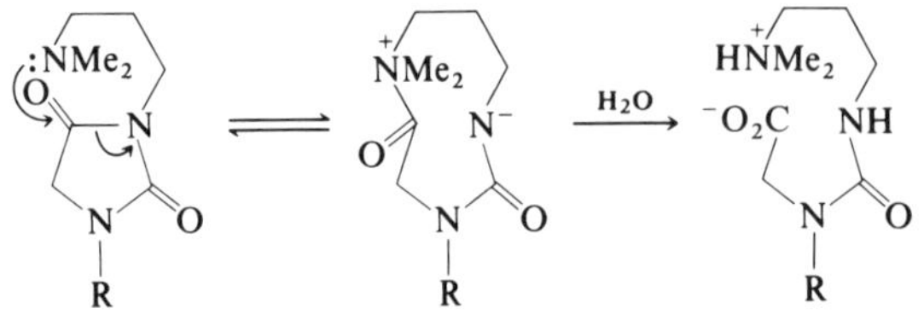

SCHEME 12

Since the discovery of hydantoin in 1861, when Baeyer isolated it in his uric acid studies, that system has been an important precursor of α-amino acids owing to its lability toward alkali, especially for those acids that are difficult to prepare by other methods.[300] Furthermore, the stereochemical courses of the Bucherer–Bergs and Read methods of synthesis for hydantoins (Section II,E), permit the preparation of epimeric amino acids.[301–305] Some of these amino acids have been tested as possible tumor growth inhibitors,[306,307] as metabolism-resistant amino acid analogues for transport system studies,[72,308]

---

[298] A. F. A. Shalaby, H. A. Daboun, and M. A. Abdel Aziz, *Z. Naturforsch., B: Anorg. Chem., Org. Chem.* **33B**, 937 (1978).
[299] C. F. Spencer, *J. Heterocycl. Chem.* **10**(4), 455 (1973).
[300] E. B. Henson, P. M. Gallop, and P. V. Hanschka, *Tetrahedron* **37**(15), 2561 (1981).
[301] R. J. W. Cremlyn, *J. Chem. Soc.*, 3977 (1962).
[302] G. Nathansohn, G. F. Odasso, C. R. Pasqualucci, and E. Testa, *Steroids* **5**(3), 263 (1965).
[303] Y. Maki, M. Sato, and K. Obata, *Chem. Pharm. Bull.* **13**(2), 1377 (1965).
[304] R. J. W. Cremlyn, R. M. Ellam, and T. K. Mitra, *Indian J. Chem.* **8**(3), 218 (1970).
[305] F. F. Knapp, *J. Org. Chem.* **44**(6), 1007 (1979).
[306] T. A. Connors and W. C. J. Ross, *J. Chem. Soc.*, 2119 (1960).
[307] L. Nicole and L. Berlinguet, *Can. J. Chem.* **40**, 353 (1962).
[308] G. G. Trigo, C. Avendaño, E. Santos, H. N. Christensen, and M. Handlogten, *Can. J. Chem.* **58**(21), 2295 (1980).

and as intermediates in the synthesis of spiro-4′-oxazolidines (**119**),[309] pyrrolines (**120**),[310] and analgesic-type compounds (**121**).[311]

(**119**)   (**120**)   (**121**)

D-Amino acids, used as intermediates for semisynthetic penicillins and cephalosporins, have been obtained by stereospecific enzyme-catalyzed hydrolysis of DL-hydantoins.[312] The enzymes have been described as hydantoinase and N-carbamoyl-D-amino acid amidohydrolase,[313] although some authors identify the former with dihydropyrimidinase.[314]

Amino acids have also been obtained by acid hydrolysis of hydantoins and 2-thiohydantoins under relatively drastic conditions.[1,3] 4-Thiohydantoin is hydrolyzed with hot concentrated hydrochloric acid to hydantoin, and boiling 20% aqueous chloroacetic acid completely removes sulfur from 2,4-dithiohydantoin to form hydantoin derivatives. (Scheme 8). These reactions reflect the intrinsic stability of the hydantoin ring.

2. Ammonolysis

4-Thio- and 2,4-dithiohydantoins react with ammonia or amines at the 4-position to give 4-imino compounds (Scheme 8). Some 2,4-dithiohydantoins react with ethanolamine at the 2-position to give imino compounds, probably because the 4-position is too hindered.[3]

An unusual base-catalyzed cleavage reaction of the fused spirothiazolidine ring of the hydantoin **122** to **123** occurs by reaction with liquid ammonia.[315]

2-Thiohydantoins, like hydantoins,[1,316] react with hydrazine, and probably with amines, to give hydrazides and amides of the thioureido acids.[317]

---

[309] P. Michon and A. Rassat, *J. Org. Chem.* **39**(14), 2121 (1974).
[310] K. E. Schulte, J. Reisch, and M. Plener, *Arch. Pharm.* (*Weinheim, Ger.*) **303**(5), 435 (1970).
[311] P. W. Feit and H. H. Freit, British Patent 1,115,817 (1968) [*CA* **70**, 28834 (1969)].
[312] D. Dinelli, W. Marconi, F. Cecer, G. Galli, and F. Morisi, *Enzyme Eng.* **3**, 477 (1978).
[313] R. Olivieri, E. Fascetti, L. Angelini, and L. Degan, *Biotechnol. Bioeng.* **23**(10), 2173 (1981).
[314] H. Yamada, S. Takakasshi, Y. Kii, and H. Kumagau, *J. Ferment. Technol.* **56**(5), 484, 492 (1978).
[315] A. Takamizawa and S. Matsumoto, *Chem. Pharm. Bull.* **21**(6), 1300 (1973).
[316] S. R. F. Kagaruki, M. R. Khan, and H. Wevers, *Pak. J. Sci. Ind. Res.* **24**, 99 (1981).
[317] J. T. Edward and S. Nielsen, *J. Chem. Soc.*, 5084 (1957).

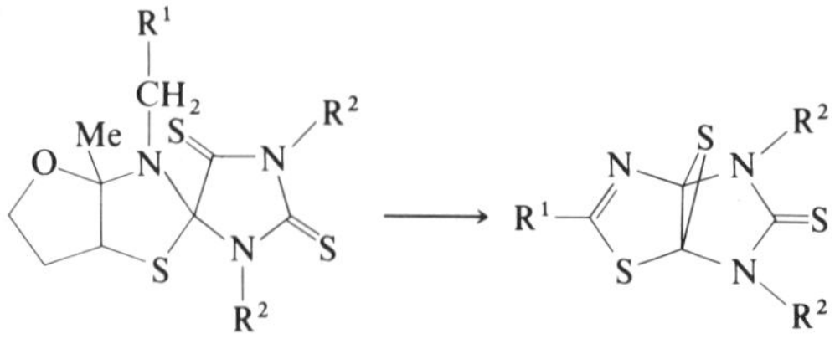

(122) (123)

R^1 = Ar
R^2 = Me

Formation of 3-aminohydantoin derivatives in some cases implies ring opening–ring closure of the hydantoin ring.[318,319] In some of these processes, such as the reaction of the methylmercaptohydantoin **124** to form the glycocyamidine (**125**), rearrangements are involved.[320]

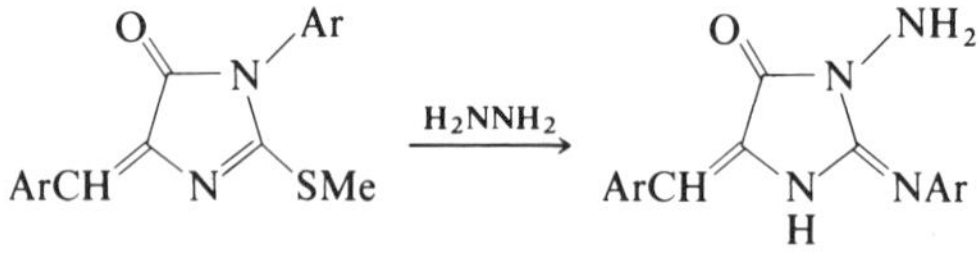

(124) (125)

C. REACTIONS WITH REDUCING AGENTS

Lithium aluminum hydride ($LiAlH_4$) reductions of hydantoins have been reported to yield a variety of products depending on the C-5, N-1, and N-3 substituents. Imidazolones (**126**)[321] and 4-hydroxy-2-imidazolidinones (**127**)[97,322] are formed from room temperature reductions, while slightly more vigorous conditions (reflux, ether or THF) give **127**,[97] 2-imidazolidinones (**128**),[323,324] imidazoles (**129**),[321] and imidazolidines (**130**).[321,323,324] Re-

---

[318] R. A. Wildonger and M. B. Winstead, *J. Med. Chem.* **10**(5), 981 (1967).
[319] K. Lempert, *Chem. Ber.* **96**, 2246 (1963).
[320] H. A. Daboun and Y. A. Ibrahim, *J. Heterocycl. Chem.* **19**, 41 (1982).
[321] I. J. Wilk and W. Close, *J. Org. Chem.* **15**, 1020 (1950).
[322] S. Cortes and H. Kohn, *J. Org. Chem.* **48**(13), 2247 (1983).
[323] F. J. Marshall, *J. Am. Chem. Soc.* **78**, 3696 (1956).
[324] E. de la Cuesta, P. Ballesteros, and G. G. Trigo, *Heterocycles* **16**(10), 1647 (1981).

Sec. IV.C] HYDANTOINS 223

cently, selective ring opening of 3-substituted hydantoins to produce N-methylethylenediamines (131) has been reported.[322]

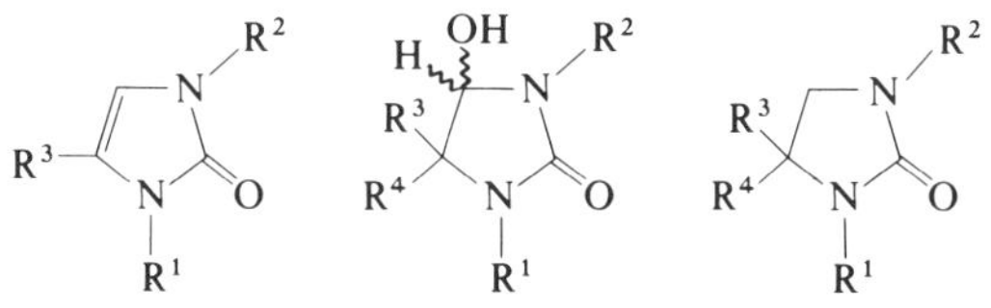

(126)   (127)   (128)

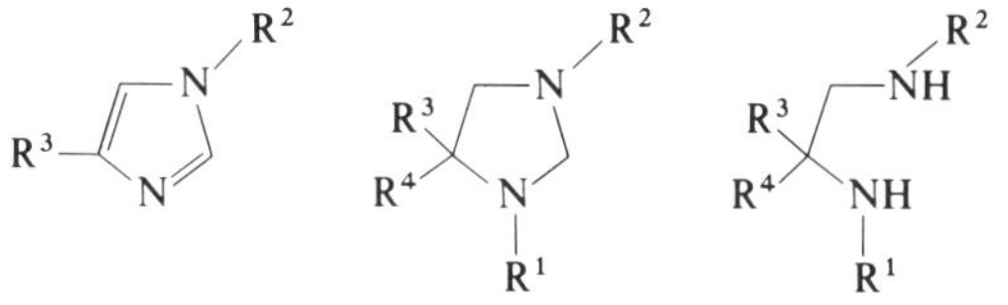

(129)   (130)   (131)

Little is known about the behavior of hydantoins in dissolving metal reductions. The reaction of hydantoin (132) with 5 equivalents of lithium in *tert*-butyl alcohol and liquid ammonia gives the 4-imidazolin-2-one 133. The use of methanol instead of *tert*-butyl alcohol gives a mixture of reduced products, probably originated by ring cleavage as well as by reduction.[325]

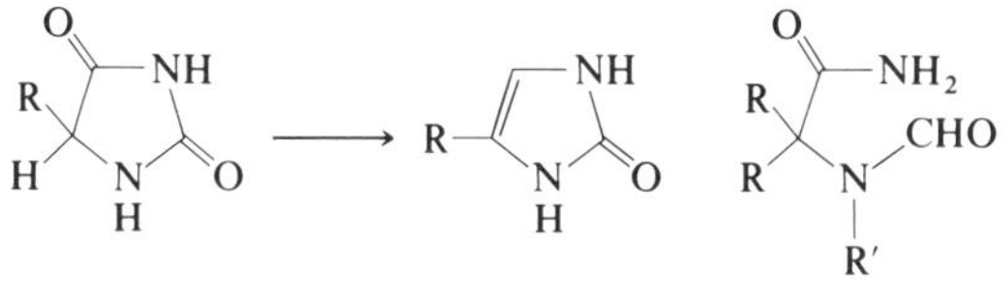

(132)   (133)   (134)

Because of its greater polarization, the thiocarbonyl group would be expected to be more easily reduced than would the carbonyl group. 2-Thiohydantoins are reduced by sodium in ethanol to give 4-imidazolidinones under conditions that do not affect hydantoins. Reduction with Raney nickel yields the same products, although the less reduced 4-imidazolones are obtained under some conditions. If the 5-substituents are alkyl instead of phenyl, α-N-formylamino amides (134) become the major products of Raney nickel reduction. 4-Thiohydantoins are reduced by this reagent to 2-imidazolidinones and 2,4-dithiohydantoins to imidazoles.[3]

[325] H. R. Divanfard, Y. A. Ibrahim, and M. M. Jouillié, *J. Heterocycl. Chem.* **15**, 691 (1978).

D. OTHER SIGNIFICANT REACTIONS

As predicted in 1965, some hydantoin derivatives are precursors of carbonium–immonium ions, which are powerful electrophiles.[326] These include the 4-hydroxy-2-imidazolidinones (Section IV,C). 5-Alkoxyhydantoins (135) undergo successful inter- and intramolecular amidoalkylations through cation intermediates (136).[327]

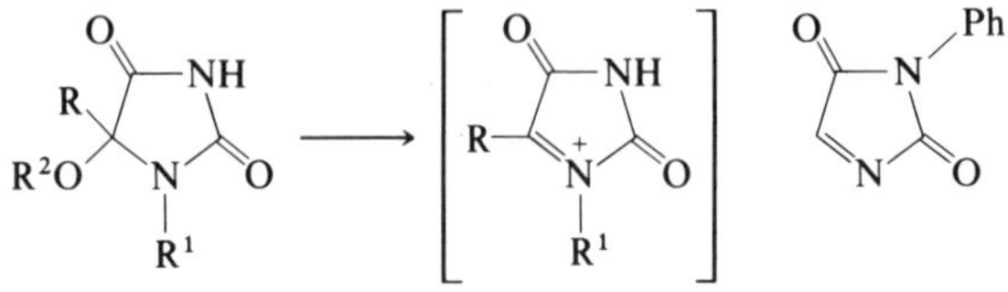

(135)   (136)   (137)

3-Phenyl-5-methoxyhydantoins are also precursors of dieneophiles, such as 137, reacting thermally or under acid catalysis with dienes to give Diels–Alder adducts. The acid-catalyzed reactions are more stereospecific than the thermal reactions.[328] In the total synthesis of the antibiotic streptonigrin,[329] the pyridine ring has been constructed in this fashion. 5-Alkyl-5-aminohydantoins are also suitable intermediates to obtain 5-alkyl-5-substituted derivatives.[129] Finally, optically active 3-hydroxyhydantoins have been designed as acyl activating reagents for asymmetrically selective peptide synthesis.[330,331]

V. Uses and Applications

A. HYDANTOINS AS MEDICINAL PRODUCTS

Although the hydantoin ring itself does not present any medicinal activity, the 5,5-disubstituted hydantoins have found wide use in medicine. They have mainly been considered as anticonvulsant agents, having been reviewed in

---

[326] H. E. Zangg and W. B. Martin, *Org. React. (N.Y.)* **14**, 60 (1965).
[327] D. Ben-Ishai, G. Ben-Et, and A. Warshawsky, *J. Heterocycl. Chem.* **7**, 1289 (1970).
[328] D. Ben-Ishai and E. Goldstein, *Tetrahedron* **27**, 3119 (1971).
[329] S. M. Weinreb, F. Z. Basha, S. Hibino, N. A. Khatu, D. Kim, W. E. Pye, and T. Wu, *J. Am. Chem. Soc.* **104**(2), 536 (1982).
[330] T. Teramoto, T. Kurosaki, and M. Okawara, *Tetrahedron Lett.* **18**, 1523 (1977).
[331] T. Teramoto, T. Kurosaki, and M. Okawara, *Pept. Chem.* **15**, 43 (1978).

1950[1] and 1963.[332] The hydantoin derivative which has achieved the greatest prominence in this field is 5,5-diphenylhydantoin (Phenitoin, Dilantin), recommended since 1938 for the treatment of epilepsy by Merritt and Putman.[333] Because of its regulating effect on the bioelectric activity of the nervous system, it is used widely as an anticonvulsant and cardiac antiarrhythmic.[334] Despite important toxic side effects, it is still the anticonvulsant of choice for epilepsy treatment.[335] Several analytical, spectrophotometric, and, more recently, GLC techniques have been developed for the determination of quantitative plasma phenytoin levels.[336] Other 5,5- and/or N-1,N-3-disubstituted hydantoin derivatives have been prepared and tested in order to improve and expand the anticonvulsant activity as well as to prevent side effects.[21,337–340]

Significant antiarrythmic activity has been found in certain piperidinespirohydantoins.[341–343]

Recently, antidiabetic activity has been reported for some spirohydantoins. Thus compounds of type **138–145** are potent inhibitors of aldose reductase activity and sorbitol accumulation, being particularly useful in treatment of chronic diabetic complications.[344–352]

---

[332] A. Spinks and W. S. Waring, *Prog. Med. Chem.* **3**, 313 (1963).
[333] H. H. Merrit and T. J. Putman, *JAMA, J. Am. Med. Assoc.* **111**, 1068 (1938).
[334] S. Bogoch and J. Dreyfus, "The Broad Range of Use of Diphenylhydantoin." Dreyfus Medical Foundation, New York, 1970.
[335] J. K. Perry and M. E. Newmark, *Ann. Intern. Med.* **89**, 207 (1979).
[336] P. A. Schwartz, C. T. Rodes, and J. W. Cooper, *J. Pharm. Sci.* **66**(7), 994 (1977).
[337] R. Eberhad, W. Persch, and A. Schmidt, *Arzneim.-Forsch.* **5**, 357 (1955).
[338] W. Stumpf and K. Rombusch, German Patent 1,173,102 (1964) [*CA* **61**, 9504 (1964)].
[339] M. A. Davis, S. O. Winthrop, R. A. Thomas, F. Hen, M. P. Charest, and R. Gandry, *J. Med. Chem.* **7**(4), 439 (1964).
[340] H. Arnold, E. Kuehas, and N. Brock, German Patent 1,135,915 (1962) [*CA* **58**, 3440 (1963)].
[341] C. Casagrande, A. Galli, R. Ferrini, G. Miragoli, and G. Ferrari, *Farmaco, Ed. Sci.* **29**, 757 (1974).
[342] A. García-Sacristán, M. Illera, and F. Sanz, *Arch. Pharmacol. Toxicol.* **3**(1), 57 (1977).
[343] M. R. Martínez-Larrañaga, A. Anadon, and F. Sanz, *Arch. Pharmacol. Toxicol.* **3**(3), 247 (1977).
[344] R. Sarges, U.S. Patent 4,130,714 (1978) [*CA* **92**, 94401 (1980)].
[345] M. J. Peterson, R. Sarges, C. E. Aldinger, and D. P. MacDonald, *Metab., Clin. Exp.* **28**, 456 (1979).
[346] R. Sarges, U.S. Patent 4,286,098 (1981) [*CA* **95**, 203958 (1981)].
[347] R. Sarges and R. C. Schnur, U.S. Patent 4,127,665 (1978) [*CA* **90**, 87464 (1979)].
[348] R. P. Kelbaugh and R. Sarges, U.S. Patent 4,147,797 (1978) [*CA* **91**, 20511 (1979)].
[349] D. R. Brittain and R. Wood, European Patent Appl. 28,906 (1981) [*CA* **95**, 150660 (1981)].
[350] R. Sarges and R. C. Schmur, U.S. Patent 4,248,882 (1981) [*CA* **94**, 192339 (1981)].
[351] R. Sarges, U.S. Patent 4,235,911 (1980) [*CA* **94**, 122540 (1981)].
[352] R. C. Schnur, U.S. Patent 4,176,185 (1979) [*CA* **92**, 111015 (1980)].

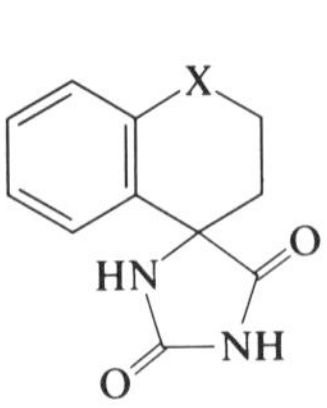

(138) 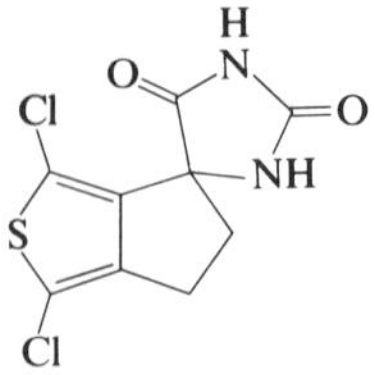   (139) 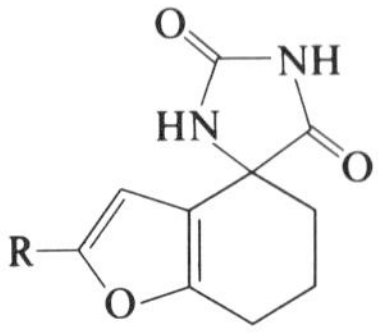   (140)

X = O, S

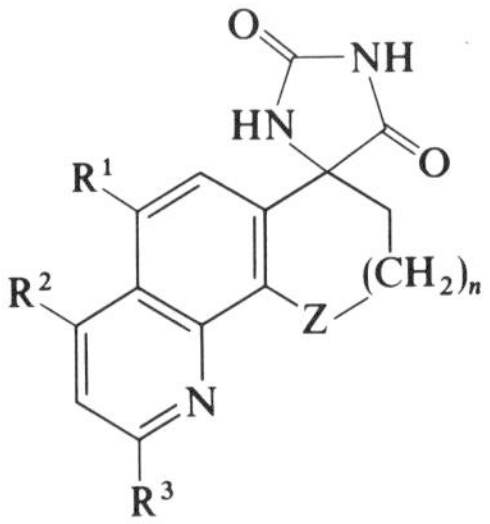

(141) 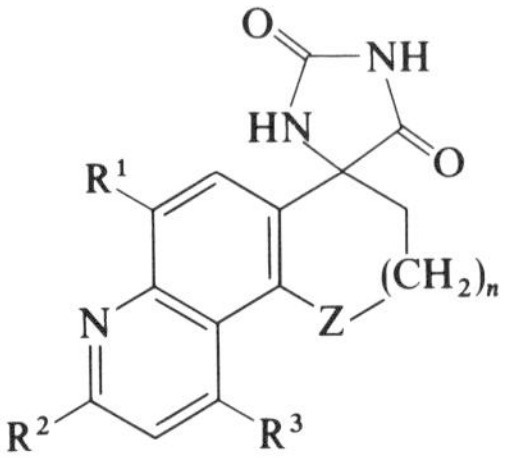   (142) 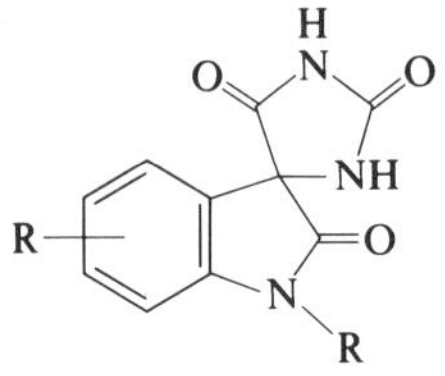   (143)

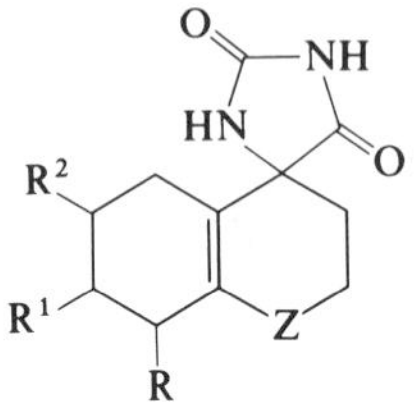

(144) 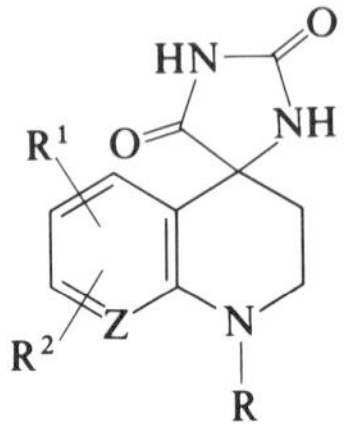   (145)

Some hydantoins are very useful carriers of the nitrogen mustard bis(β-chloroethyl)amine, being active on central nervous system tumors, B16 melanocarcinoma, leukemia L1210, leukemia P388, Lewis lung carcinoma, and ependymoblastoma.[353] The response of 9L tumor cells *in vitro* to *N*-3-bis(β-chloroethyl)aminoethylcyclohexanespiro-5'-hydantoin has suggested its utility in multiagent therapy regimens.[354]

[353] G. Peng, V. Marquez, and J. S. Driscoll, *J. Med. Chem.* **18**(8), 846 (1975).
[354] F. D. Deen, T. Hoshino, M. E. Williams, K. Nomura, and P. M. Bartle, *Cancer Res.* **39**(11), 4336 (1979).

Hydantoins and other imidazole analogues of prostaglandins of type **146** and **147** have been synthesized as racemic compounds.[355] The less polar diastereoisomer of **146b** is a potent inhibitor of platelet aggregation in human platelet-rich plasma.

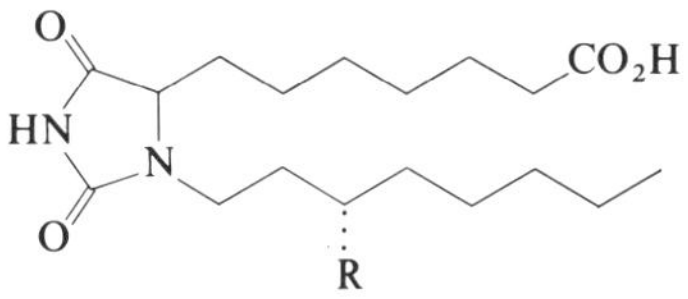

(146) 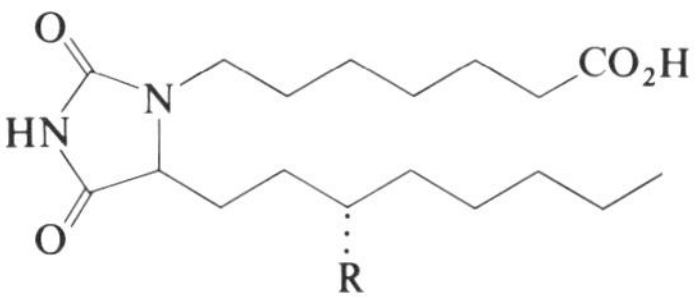   (147)

a R = H
b R = OH
c R = O

B. Other Uses

Hydantoins, thiohydantoins, and their substituted products are particularly useful as catalysts and as stabilizer agents in polymer chemistry and also in the preparation of epoxy resins.[356,357] Thus hydantoins **148** are readily hardened, giving suitable polymers for the preparation of molding and lacquers.

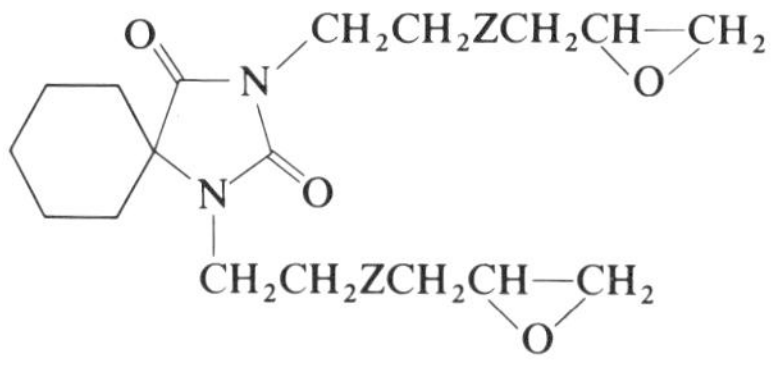

(148)

Z = COO, O

N-Trihalomethylsulfenylhydantoins exhibit fungicidal activity.[358,359] Other derivatives are herbicides.[360]

---

[355] A. G. Caldwell, C. J. Harris, R. Stepney, and N. Whittaker, *J.C.S. Perkin I*, 495 (1980).
[356] D. Porret, *Makromol. Chem.* **108**, 73 (1967).
[357] E. H. Catsiff, R. E. Coulehan, J. F. DiPrima, D. A. Gordon, and R. Seltzer, *Org. Coat. Plast. Chem.* **39**, 139 (1978).
[358] C. J. Mappes, E.-H. Pommer, C. Rentzea, and B. Zeeh, U.S. Patent 4,198,423 (1980) [*CA* **93**, 71784 (1980)].
[359] E. Klauke, E. Kuehle and F. Gravem, Belgian Patent 631,731 (1963) [*CA* **61**, 1873 (1964)].
[360] H. Ohta, T. Jikihara, Ko. Wakabayashi, and T. Fujita, *Pestic. Biochem. Physiol.* **14**, 153 (1980).

Hydantoins and thiohydantoins have been used as analytical reagents for heavy metals because of their capacity for complex formation.[361-363]

ACKNOWLEDGMENT

We express sincere thanks to Dr. J. Elguero for his advice and guidelines and to Dr. P. Ballesteros for her comments and criticism.

---

[361] M. J. Blais, O. Enea, and G. Berthon, *Thermochim. Acta* **30**, 45 (1979).
[362] F. Barragan, M. T. Montaña, and J. L. Gómez-Ariza, *Microchem. J.* **25**, 524 (1980).
[363] M. T. Montaña and J. L. Gómez-Ariza, *Microchem. J.* **25**, 360 (1980).

Recent Progress in Barbituric Acid Chemistry

JACEK T. BOJARSKI, JERZY L. MOKROSZ, HENRYK J. BARTOŃ,
AND MARIA H. PALUCHOWSKA

*Department of Organic Chemistry, Nicolaus Copernicus
Academy of Medicine, Kraków, Poland*

I. Introduction 229
II. Physicochemical Properties of Barbiturates 231
   A. Tautomerism and Solvation of the Barbituric Acid Ring. 231
   B. Spectral Properties. 236
      1. UV Spectra 236
      2. Infrared and Raman Spectroscopy 240
      3. ^{1}H-NMR Spectroscopy 241
      4. ^{13}C-NMR Spectroscopy. 243
      5. Mass Spectrometry 246
   C. Structure and Conformation 252
      1. Structure and Conformation in the Solid State 252
         a. Hydrogen Bonding and Polymorphism 252
         b. Ring Conformations 255
         c. Orientation of Substituents. 255
      2. Conformation in Solution 257
   D. Optical Isomers. 259
III. Reactivity of Barbiturates. 263
   A. Reactions at the C-5 Atom 263
   B. Substitution at Nitrogen 269
   C. Reactions of Carbonyl Groups 273
   D. Stability of the Pyrimidine Ring 276
   E. Photochemical Reactions 282
   F. Other Reactions 283
IV. Analytical Methods for Barbiturates 288
V. Correlation Analysis in Barbituric Acid Chemistry 292
VI. Closing Remarks. 295
VII. Addendum 297

I. Introduction

The chemistry of barbituric acid and its derivatives has been studied for over 100 years if we consider 1864, the year the parent compound of this class was prepared by von Baeyer,[1] as its starting point. In 1903 the therapeutic

---
[1] A. von Baeyer, *Ann. Chem. Pharm.* **130**, 129 (1864).

value of 5,5-diethylbarbituric acid (barbital) as a hypnotic agent was reported by Fisher and von Mering[2] and since that time constant progress in chemical and pharmacological studies on barbiturates has been observed and reported.

Several earlier review articles are cited in the fundamental work of Doran in 1959.[3] This book provides general information on the chemistry and pharmacology of barbiturates, but its main value lies in a compilation of an extensive bibliography on the preparation and pharmacological activity of these compounds.

More details on synthesis, properties, and reactivity are provided in the review by Levina and Velichko in 1960.[4] Since that time further studies dealing with structure and conformation, spectral properties, and structure–activity relationships have been among the main advances in barbiturate chemistry. These topics were reviewed by Bobrański.[5] This review presents the most important recent information on barbituric acids with special emphasis on their physicochemical properties.

Although barbituric acid [2,4,6-(1H,3H,5H)-pyrimidinetrione] (1) and its derivatives now are listed in *Chemical Abstracts* by their IUPAC names, there is a well-established tradition to use the name "barbituric acid" along with prefixes denoting appropriate substitution. This system was used in *Chemical Abstracts* until 1972; we adopt it for convenience.

The terms barbituric acid and barbiturates are used interchangeably for a general description of derivatives of 1, while the subclasses and individual compounds are named according to the mode of substitution, e.g., 5,5-dialkylbarbituric acids (2) or 5-ethyl-1-methyl-5-phenylbarbituric acid (3).

Several barbiturates of therapeutic importance have common names or names adopted by manufacturers but we will use only their recommended international nonproprietary names [e.g., methylphenobarbital (3)].[6]

[2] E. Fisher and J. R. von Mering, *Ther. Ggw.* **44**, 97 (1903).
[3] W. J. Doran, *Med. Chem.* **4**, 1 (1959).
[4] R. Ya. Levina and F. K. Velichko, *Russ. Chem. Rev.* (*Engl. Transl.*) **29**, 437 (1960).
[5] B. Bobrański, *Wiad. Chem.* **31**, 231 (1977).
[6] M. Negwer, "Organic-chemical Drugs and their Synonyms." Akademie-Verlag, Berlin, 1978.

II. Physicochemical Properties of Barbiturates

A. TAUTOMERISM AND SOLVATION OF THE BARBITURIC ACID RING

Barbituric acid exists in the solid state in the trioxo structure as shown by X-ray[7,8] and ^{14}N-NQR[9] methods. NMR investigation of the oxo–hydroxy equilibrium also indicates that only the oxo form is present in a solution in anhydrous DMSO.[10,11] The hydroxy form (**4b**) appears in a water-free acid but the oxo form (**4a**) is still predominant.[12,13] Relaxation methods[12] and UV measurements[13] give $pK_T = 1.9$ and 1.3, respectively, for **4a/4b** in water, and substitution at the nitrogen atoms (N-methyl- and N,N'-dimethylbarbituric acid) has no influence on pK_T (1.65 and 1.83, respectively). Introduction of substituents in the 5-position results in the predominance of the hydroxy form, pK_T being 0.88, 0.76, and 0.21, respectively, for Me, Cl, and Br; results can be explained by inductive effects.[13] The oxo form dominates in aprotic solvents.

In amphiprotic solvents, the amount of the oxo form decreases, and in proton-donating solvents it increases again. IR spectra of barbituric and 5-chlorobarbituric acids in the solid state and in solution are consistent with those results.[14]

X-ray and IR analyses show that 1,3-diethylbarbituric acid exists in the trioxo form, while its sulfur analogue is in the monohydroxy form in the solid.[15]

Generally, 5,5-di- and 1,5,5-trialkyl, -alkenyl or -aryl substituted barbiturates have the trioxo form in the crystalline state. Included are

---

[7] G. A. Jeffrey, S. Ghose, and J. O. Warwicker, *Acta Crystallogr.* **14**, 881 (1961).
[8] W. Bolton, *Acta Crystallogr.* **16**, 166 (1963).
[9] T. Maruizumi, Y. Hiyama, and E. Niki, *Bull. Chem. Soc. Jpn.* **53**, 1443 (1980).
[10] G. A. Neville and D. Cook, *Can. J. Chem.* **47**, 743 (1969).
[11] J. A. Glasel, *Org. Magn. Reson.* **1**, 481 (1969).
[12] M. Eigen, G. Ilgenfritz, and W. Kruse, *Chem. Ber.* **98**, 1623 (1965).
[13] W. I. Slesarev and B. A. Ivin, *Zh. Org. Khim.* **10**, 113 (1974).
[14] N. A. Smorygo and B. A. Ivin, *Khim. Geterotsikl. Soedin.*, 1411 (1975).
[15] J.-P. Bideau, P. V. Huong, and T. Toure, *Acta Crystallogr., Sect. B* **B32**, 481 (1976).

5,5-diethyl-,[16] 5-ethyl-5-(1′-methylbutenyl)-,[17] 5-methyl-5-phenyl-,[18] 1-methyl-5-ethyl-5-phenyl-,[19] or 1-(p-bromophenyl)-5,5-diallylbarbituric acid[20] (see also references in Section II,C,1). In the solid state, however, monopotassium salts exist in the dioxo structure, which is confirmed by X-ray[21] and IR investigations.[22] Moreover, there are no indications of the presence of the hydroxy form in the ^{1}H-NMR spectra of the 5,5-disubstituted barbiturates in CF_3COOH, p-dioxane, or DMSO.[10] IR results show that for 5-alkyl-5-α- and 5-β-alkoxyethylbarbiturates in solution the oxo–hydroxy equilibrium depends on the inductive effect of substituents, i.e., 5-α-alkoxyethyl substituents shift the equilibrium toward the hydroxy form.[23]

In the solid state and in aprotic solvents, 5-benzylidenebarbituric acid and its meta- and para-substituted derivatives (5) also have the trioxo structure, according to their ^{1}H-NMR, IR, and UV spectra.[24]

(5)

1,5,5-Trisubstituted barbituric acids (6) exist in the hydrogen-bonded, monohydroxy form in the solid state, as revealed by a comparison of IR spectra. 1,3,5,5-Tetrasubstituted derivative 7, which cannot form an intramolecular hydrogen bond, was used as a model.[25]

(6)   (7)

[16] B. M. Craven, E. A. Vizzini, and M. M. Rodrigues, *Acta Crystallogr., Sect. B* **B25**, 1978 (1969).
[17] B. M. Craven and C. Cusatis, *Acta Crystallogr., Sect. B* **B25**, 2291 (1969).
[18] G. Bravic, J. Housty, and J. P. Bideau, *C. R. Hebd. Seances Acad. Sci., Ser. C* **266**, 969 (1968).
[19] J.-P. Bideau, L. Marly, and J. Housty, *C. R. Hebd. Seances Acad. Sci., Ser. C* **269**, 549 (1969).
[20] D. Pyżalska, R. Pyżalski, and T. Borowiak, *Acta Crystallogr., Sect. B* **B36**, 1672 (1980).
[21] J. Berthon, B. Rérat, and C. Rérat, *Acta Crystallogr.* **18**, 768 (1965).
[22] C. F. Gavrilin, V. E. Chistyakov, and G. A. Kononenko, *Zh. Obshch. Khim.* **40**, 669 (1970).
[23] A. V. Bogatskii, G. Y. Glinskaya, and A. I. Gren, *Zh. Obshch. Khim.* **39**, 2568 (1969).
[24] B. A. Ivin, A. I. Dyachkov, I. M. Vishnyakov, N. A. Smorygo, and E. G. Sochilin, *Zh. Org. Khim.* **11**, 1337 (1975).
[25] H. J. Roth, K. Jäger, and R. Brandes, *Arch. Pharm. (Weinheim, Ger.)* **298**, 885 (1965).

1,3-Disubstituted-5-acyl derivatives of barbituric acid (**8a**) enolize with the aid of the 5-β carbonyl group. Formation of an intramolecular hydrogen bond with the carbonyl group at C-4 or C-6 gives **8b**, and then proton transfer leads to the monohydroxy form **8c**, as demonstrated by ^{1}H-NMR and UV studies.[26]

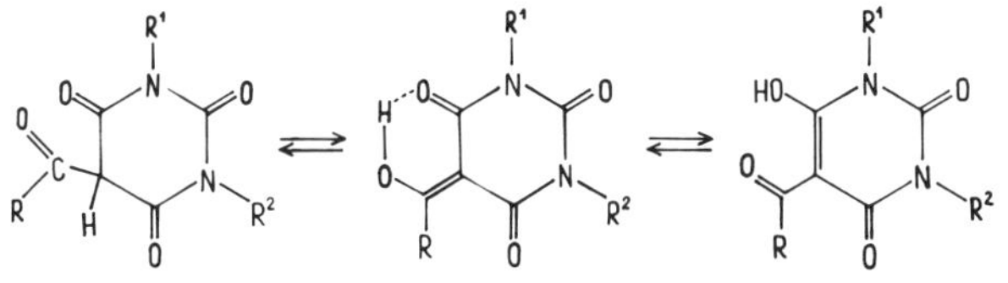

(**8a**)   (**8b**)   (**8c**)

Anhydrous 5-nitrobarbituric acid (**9**) exists in the monohydroxy form (**8a**),[27] but the trihydrate is in the *aci*-nitro form (**9b**)[28] in the crystalline state as shown by X-ray analysis. IR and UV investigations[29] and dipole moment measurements[30] are consistent with those conclusions. The results of MO calculations show that the monohydroxy structure (**9a**) is the most stable. UV spectra[31] as well as ^{14}N-NQR data[32] also support tautomer **9a** in water.

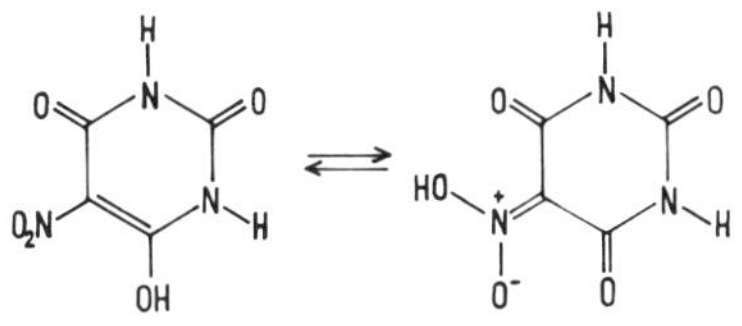

(**9a**)   (**9b**)

Crystal data show that dialuric acid (**10**) exists in the dihydroxydioxo form (**10b**),[33,34] but ^{1}H-NMR studies of DMSO solutions suggest the existence of a monohydroxytrioxo structure (**10a**).[11] No evidence of tautomerism is

---

[26] K. Rehse and W.-D. Kapp, *Arch. Pharm. (Weinheim, Ger.)* **315**, 502 (1982).
[27] W. Bolton, *Acta Crystallogr.* **16**, 950 (1963).
[28] B. M. Craven, S. Martinez-Carrera, and G. A. Jeffrey, *Acta Crystallogr.* **17**, 891 (1964).
[29] F. Mihai and R. Nutiu, *Rev. Roum. Chim.* **13**, 39 (1968).
[30] R. Nutiu, L. Kurunczi, and Z. Simon, *Rev. Roum. Chim.* **14**, 1435 (1969).
[31] Z. Simon, F. Mihai, and R. Nutiu, *Rev. Roum. Chim.* **13**, 147 (1968).
[32] S. N. Subbarao and P. J. Bray, *J. Chem. Phys.* **67**, 1085 (1977).
[33] W. Bolton, *Acta Crystallogr.* **19**, 1051 (1965).
[34] B. M. Craven and T. M. Sabine, *Acta Crystallogr., Sect. B* **B25**, 1970 (1969).

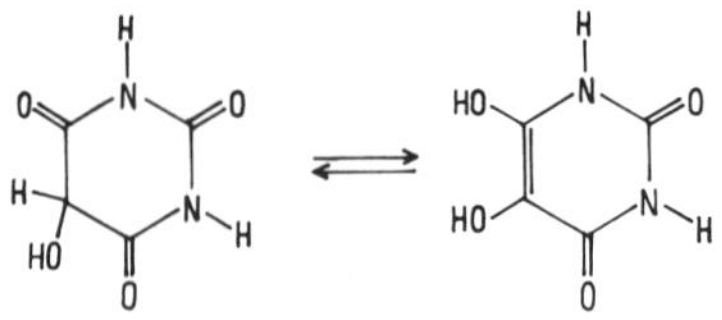

(10a)   (10b)

observed for 5,5-dihydroxybarbituric acid (11) from X-ray,[35,36] ^{1}H-NMR,[11] and ^{14}N-NQR[9] investigations.

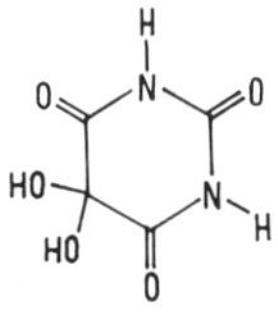

(11)

In aqueous solution the hydroxy forms of barbituric acids are in equilibrium with their mono- and dianionic conjugate bases (Scheme 1). The acid–base equilibria are strongly dependent on the site of substitution and the nature of the substituents[37–45] (Table I).

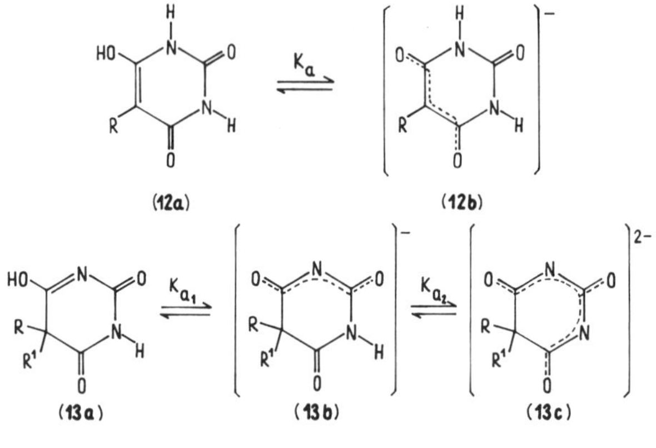

(12a)   (12b)

(13a)   (13b)   (13c)

SCHEME 1

Resonance and large inductive effects of substituents easily account for changes in pK_a values, but small changes are more difficult to interpret.[13,40,43] Monoanions 12b and 13b are 10^4–10^8 times weaker acids than their neutral conjugate acids. The influence of electronic effects of substituents on the

TABLE I
INFLUENCE OF SUBSTITUTION ON pK_{a_1} AND pK_{a_2} OF BARBITURATES

Barbituric acid	pK_{a_1}	References	pK_{a_2}	References
5-Chloro	0.00	13	—	—
5-Bromo	0.26	13	—	—
5-Phenyl	2.54	38	—	—
5-Alkyl	3.4–4.9	37,38	—	—
Unsubstituted	4.10	13,37	12.60	13
1-Methyl	4.35	13	12.90	13
5-Ethyl-5-phenyl	7.41	13,39	12.14	40
5-Ethyl-5-alkenyl	7.4–8.0	11,38,40	12.2–12.7	40
5,5-Dialkyl	7.8–8.5	39–43	12.4–12.8	40
1-Methyl-5,5-disubstituted	7.9–8.8	40,41	—	—
5,5-Alkylene	8.7–8.9	44,45	—	—

second step of ionization is less clear, owing to fewer studies. Acid strength is almost independent of the length of the straight-chain alkyl group attached to the C-5 atom of the ring, e.g., $\Delta pK_{a_1} = 0.07$ for 5-ethyl-5-n-pentyl- and 5,5-diethylbarbituric acids. Introduction of a methyl group in the α position of an alkyl chain at C-5 increases the pK_{a_1} by about 0.2 unit, while the introduction of a methyl group at the β or γ carbon atom causes either a very slight decrease in the pK_{a_1} (~0.06 unit) or has no significant effect.[39–42] Replacement of the ethyl substituent at C-5 by a methyl group results in a small acid-weakening effect and pK_{a_1} values are higher by about 0.3.[40,41,43–46]

Large alkyl groups at the C-5 position shield the C-4 and C-6 carbonyl groups and hinder solvation.[44–46] On the other hand, an increase in the ionic strength of the solvent results in a decrease of pK_{a_1}, followed by stronger solvation, for 5,5-disubstituted barbiturates.[39]

A comparison of the pK_{a_1} values for 5,5-diethylbarbituric acid (8.00 ± 0.02) and its spirocyclic derivatives (14) ($n = 4$ and 5; $pK_{a_1} = 8.83$ and 8.88,

---

[35] D. Mootz and G. A. Jeffrey, *Acta Crystallogr.* **19**, 717 (1965).
[36] C. Singh, *Acta Crystallogr.* **19**, 759 (1965).
[37] W. F. Smyth, T. Jenkins, J. Siekiera, and A. Baydar, *Anal. Chim. Acta* **80**, 233 (1975).
[38] H. Koffer, *J.C.S. Perkin II*, 1429 (1974).
[39] M. E. Krahl, *J. Phys. Chem.* **44**, 449 (1940).
[40] D. A. Doornbos and R. A. de Zeeuw, *Pharm. Weekbl.* **104**, 233 (1969).
[41] D. A. Doornbos and R. A. de Zeeuw, *Pharm. Weekbl.* **106** 134 (1971).
[42] J. M. A. Sitsen and J. A. Fresen, *Pharm. Weekbl.* **108**, 1053 (1973).
[43] R. H. McKeown, *J.C.S. Perkin II*, 504 (1980).
[44] R. H. McKeown and R. J. Prankerd, *J.C.S. Perkin II*, 481 (1981).
[45] R. H. McKeown, *J.C.S. Perkin II*, 515 (1980).
[46] J. Mokrosz and J. Bojarski, *Pol. J. Chem.* **56**, 491 (1982).

respectively) shows significant influence of the substituent with almost the same electronic nature but different shielding effect on solvation and acidity of these compounds. However, 5,5-dimethylbarbituric acid, where the steric effect of methyl groups should be similar to that in compounds **14**, has a $pK_{a_1} = 8.51$. This acid-strengthening effect is explained by the hyperconjugative effect of methyl groups.[44,46]

The acidity of 5-methyl- and 5-ethylbarbituric acids is slightly greater than that of barbituric acid ($pK_{a_1} = 3.39, 3.69$, and 4.02, respectively), a conclusion

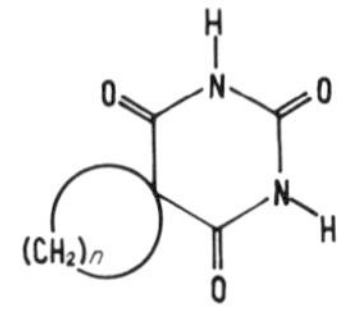

(**14**)

not expected from a consideration of the usual inductive electron-donating effect of alkyl groups.[38] Relaxation methods show that the influence of substituents on individual reaction rate constants among oxo, hydroxy, and anion forms is responsible for the unusual acidity of these compounds.[47]

Further studies on the acid–base equilibria of 5,5-biscarboxymethyl-,[48] 5-azo-,[49] and 5-ylidenebarbiturates[50,51] are also available.

B. SPECTRAL PROPERTIES

1. *UV Spectra*

Near-UV absorption spectra of barbiturates depend on the type of substitution and the state of ionization of the molecule.[52–54] Absorption bands at ~210 nm for undissociated compounds are shifted to 240–270 nm

---

[47] H. Koffer, *J.C.S. Perkin II*, 819 (1975).
[48] J. Mirek, M. Adamczyk, and P. Król, *Chem. Anal.* **24**, 739 (1979).
[49] V. Madajová and J. Zelenský, *Collect. Czech. Chem. Commun.* **46**, 987 (1981).
[50] P. Schuster and O. E. Polansky, *Monatsh. Chem.* **99**, 1234 (1968).
[51] R. Bednar, O. E. Polansky, and P. Wolschann, *Z. Naturforsch., B: Anorg. Chem., Org. Chem.* **30B**, 582 (1975).
[52] J. J. Fox and D. Shugar, *Bull. Soc. Chim. Belg.* **61**, 44 (1952).
[53] L. A. Gifford, W. P. Hayes, L. A. King, J. N. Miller, D. T. Burns, and J. W. Bridges, *Anal. Chem.* **46**, 94 (1974).
[54] A. J. Berdnikov and S. P. Bystrov, *Farmatsiya* **25**, 32 (1976).

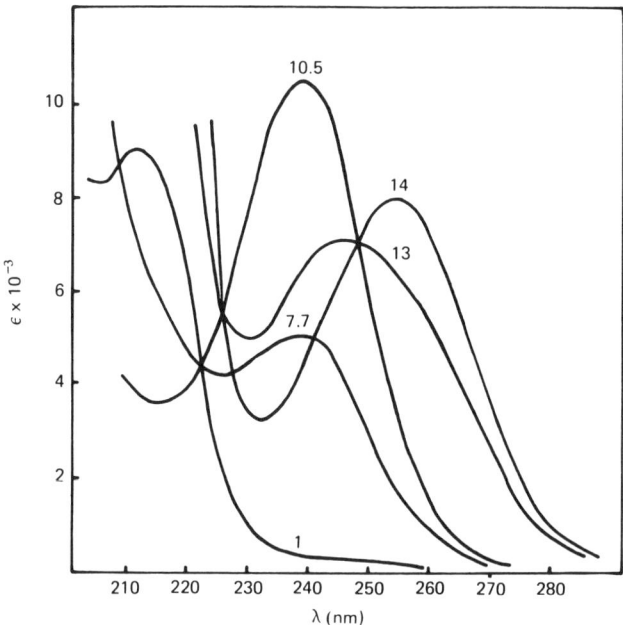

FIG. 1. UV spectra of 5,5-diethylbarbituric acid at various pH values.

for ionized forms (Fig. 1). Monoionized forms (Scheme 1) exhibit the highest absorption intensity ($\varepsilon \sim 10^4$). Substitution at the nitrogen and C-5 atoms results in a slight bathochromic shift and a decrease in intensity.[52] UV-absorption spectra of barbiturates with the same substitution and ionization modes are independent of the kind of substituents[55-58] (Table II). Calculations by the PCILO method agreed with that conclusion and indicated that neither the chemical structure nor the conformational aspects of the substituents attached to C-5 exert a significant perturbation on the ground-state electronic charge distributions in the barbituric acid ring.[59] On the contrary, for 2-thio-[60] and 5-ylidenebarbiturates[24] a significant bathochromic shift and an increase in intensity of absorption bands are observed (Table II).

---

[55] G. Milch and I. Vagi, *Lumin.-Spektroszk. Kut. Alkamazasi Probl. Magy. Tud. Akad. Spektroszk. Albizottsaga*, 173 (1979) [*CA* **95**, 15339g (1981)].
[56] C. I. Miles and G. H. Schenk, *Anal. Chem.* **45**, 130 (1973).
[57] H. Barton and J. Bojarski, *Pharmazie* **38**, 630 (1983).
[58] J. Mokrosz, M. Klimczak, H. Barton, and J. Bojarski, *Pharmazie* **35**, 205 (1980).
[59] B. Pullman, J. L. Coubeils, and P. Courriere, *J. Theor. Biol.* **35**, 375 (1972).
[60] Z. Kazimierczuk, A. Psoda, and D. Shugar, *Acta Biochim. Pol.* **20**, 83 (1973).

TABLE II
Typical UV Parameters of Barbiturates in Buffer Solutions for Undissociated (A), Monoanionic (B), and Dianionic (C) Forms[a]

Position of substitution	A		B		C	
	λ_{max} (nm)	$\varepsilon \times 10^{-3}$	λ_{max} (nm)	$\varepsilon \times 10^{-3}$	λ_{max} (nm)	$\varepsilon \times 10^{-3}$
None	205	10.4	257.5	21.5[b] 19.8[c]	260	15.2[b] 15.1[c]
5	—	—	263–271	17.1–21.6	—	—
5,5	212.5	9.15	239–241	9.3–10.6	255–256	6.8–7.95
1,5,5	222.5	7.95	245–246	8.5–8.7	—	—
1,3,5,5	228–230	6.3–10.0	—	—	—	—
1	219.5	8.7	258.5	20.3	260	15.0
1,3	226	7.9	260	18.9	—	—
5,5-, 2-Thio	285	24	305	26	306	28
5-Benzylidene[d]	230–250 315–452	8.0–15.0 17.0–40.5	—	—	—	—

[a] Data from refs. 24, 52, 53, 57, 58, 60.
[b] From reference 52.
[c] From reference 53.
[d] Dioxane solution.

Luminescence spectra of barbiturates have been studied, usually for analytical purposes by room-[53,61–65] or low-temperature (77 K) techniques.[53,56,65–67] The structure–luminescence relationships[53,68] and the influence of substituents on the fluorescence characteristics were also investigated.[53,69–71] Only 5,5-disubstituted barbiturates, as the dianion species, show significant fluorescence at 420 nm. However, the dianions of other barbiturates exhibit very weak emissions or none at all, except for 5-phenylbarbituric acid.[53] Barbiturates also exhibit low-temperature flu-

[61] S. Udenfriend, D. E. Duggan, B. M. Vasta, and B. B. Brodie, *J. Pharmacol. Exp. Ther.* **120**, 26 (1957).
[62] J. E. Swagzdis and T. L. Flanagan, *Anal. Biochem.* **7**, 147 (1964).
[63] P. G. Dayton, J. M. Perel, M. A. Langrau, L. Brand, and L. C. Mark, *Biochem. Pharmacol.* **16**, 2321 (1967).
[64] C. I. Miles and G. H. Schenk, *Anal. Lett.* **4**, 61 (1971).
[65] L. A. King, J. N. Miller, and D. T. Burns, *Anal. Chim. Acta* **68**, 205 (1974).
[66] L. A. Gifford, W. P. Hayes, L. A. King, J. N. Miller, D. T. Burns, and J. W. Bridges, *Anal. Chim. Acta* **62**, 214 (1972).
[67] J. D. Winefordner and M. Tin, *Anal. Chim. Acta* **32**, 64 (1965).
[68] L. A. King and L. A. Gifford, *Anal. Chem.* **47**, 17 (1975).
[69] L. J. Cline-Love and L. M. Upton, *Spectrochim. Acta, Part A* **37A**, 879 (1981).
[70] L. A. King, *J.C.S. Perkin II*, 844 (1976).
[71] L. A. King, *J.C.S. Perkin II*, 1725 (1976).

orescence at 77 K, but slight blue shifts were observed.[53] Those with unsaturated substituents at C-5 show very weak fluorescence, but 5-phenyl derivatives exhibit strong phosphorescence near 400 nm with lifetimes of several seconds.[53,56,65,66] This long-lived emission is maintained even in acidic media, except for 5-phenyl- and 5-benzylbarbituric acids. King proposed that phosphorescence is derived from phenyl because of intramolecular energy transfer,[72] whereas the pyrimidine ring is responsible for fluorescence.[53] The effect of C-5 substituents on the relative fluorescence quantum yields was explained by King in terms of the inductive effect.[68]

Cline-Love and Upton found microsecond natural lifetimes for 5,5-disubstituted barbiturates.[69] This suggests that the fluorescence transition is of the $n \to \pi^*$ type, although the absorption should be of the $\pi \to \pi^*$ type. Moreover, natural lifetimes of the excited states for barbiturates with unsaturated substituents are longer than those for barbiturates with alkyl groups. A small mixing of electronic levels of the fluorophore with the levels of the electronic environment provided by the unsaturated substituents have been proposed to explain this phenomenon.[69]

The electronic structure of barbiturates has also been investigated by spectropolarimetric methods.[73–77] Circular dichroism (CD) spectra of a series of (S)-5-alkyl-5-(2′-pentyl)barbituric acids show three Cotton effects centered around 212, 240, and 260 nm, and the short-wavelength band is positive and has an opposite sign in relation to the longer wavelengths. However, in the case of (S)-5-(2′-pentyl)barbituric acid, the signs of all three Cotton effects are opposite to those of all the other derivatives. The solvent studies indicate that the 212- and 260-nm bands arise from the $\pi \to \pi^*$ and $n \to \pi^*$ transitions, respectively, but the 240-nm band is due to $n \to \sigma^*$ or the second $n \to \pi^*$ transitions.[74] An influence of concentration and ionization mode has not been observed. Similar assignment of three Cotton effects has been reported for (S)-5-alkyl-5-(2′-pentyl)-2-thiobarbituric acids.[75]

Both experimental data[73,75] and theoretical calculations[78] show that differences in the sign of the Cotton effects of 5-mono- and 5,5-disubstituted barbiturates are sensitive to chiral distortions within the barbituric acid ring. The induced CD spectra of barbiturates as inclusion complexes with β-cyclodextrin support the foregoing conclusion.[79]

[72] L. A. King, *Spectrochim. Acta, Part A* **31A**, 1933 (1975).
[73] F. I. Carroll and R. Meck, *J. Org. Chem.* **34**, 2676 (1969).
[74] F. I. Carroll and A. Sobti, *J. Am. Chem. Soc.* **95**, 8512 (1973).
[75] F. I. Carroll, A. Philip, and C. G. Moreland, *J. Med. Chem.* **19**, 521 (1976).
[76] F. I. Carroll, D. Smith, G. N. Mitchell, and A. Sobti, *J.C.S. Perkin II*, 983 (1977).
[77] J. Knabe and W. Wunn, *Arch. Pharm. (Weinheim, Ger.)* **315**, 977 (1982).
[78] C. Y. Yeh and F. S. Richardson, *Theor. Chim. Acta* **39**, 197 (1975).
[79] M. Otagiri, T. Miyaji, K. Uekama, and K. Ikeda, *Chem. Pharm. Bull.* **24**, 1146 (1976).

A number of the CD spectra of barbituric acid derivatives were also reported by Knabe and co-workers[80] (see also references in Section II,D).

2. Infrared and Raman Spectroscopy

The main purpose of the IR and Raman analysis has been the detection and identification of particular derivatives as the free acid, their salts, or metal complexes.[81-91] Although detailed interpretation of the spectra largely concerned only the N—H and C=O stretching modes, even for these regions there were some controversies about the origin of the peaks.[81-87] Two bands of the N—H stretching vibration (at 3200 and 3090 cm^{-1}) are observed in the IR spectra of barbituric acid and its derivatives in the solid state.[82,92-94] The position and intensity of these bands depend on the degree of the hydrogen bonding.[88,92-96] However, in highly dilute solutions and in argon matrices, a strong vibration at ~ 3400 cm^{-1} and two broad bands in the region 3250–3100 cm^{-1} are observed, which correspond to the monomeric and dimeric forms, respectively.[92-94,97,98]

In the IR spectra between 1770 and 1680 cm^{-1}, three bands of the C=O modes are observed. Analysis of IR and Raman spectra of four barbiturates in low-temperature matrices showed that the highest frequency band is attributable to the 4,6-CO symmetric vibration, the middle to the 4,6-CO antisymmetric stretch, and the lowest frequency band to the 2-CO mode[92-94]

---

[80] J. Knabe, H. Junginger, and W. Geismar, *Justus Leibigs Ann. Chem.* **739**, 15 (1970).
[81] W. C. Price, J. E. S. Bradley, R. D. B. Fraser, and J. P. Quilliam, *J. Pharm. Pharmacol.* **6**, 522 (1954).
[82] L. Levi and C. E. Hubley, *Anal. Chem.* **28**, 1591 (1956).
[83] J. M. Manson and J. A. R. Cloutier, *Appl. Spectrosc.* **15**, 77 (1961).
[84] S. Goenechea, *Z. Anal. Chem.* **218**, 416 (1966).
[85] S. Pinchas, *Spectrochim. Acta* **22**, 1889 (1966).
[86] R. Bouché, L. Coclers, R. Delahaut, and J. Muquardt, *J. Pharm. Belg.*, 282 (1970).
[87] A. Sucharda-Sobczyk, *Rocz. Chem.* **41**, 1435 (1970).
[88] M. Kuhnert-Brandstätter and F. Bachleitner-Hofmann. *Arch. Pharm. (Weinheim, Ger.)* **304**, 580 (1971).
[89] J. N. Willis, R. B. Cook, and R. Jankow, *Anal. Chem.* **44**, 1228 (1972).
[90] A. Sucharda-Sobczyk, *Rocz. Chem.* **46**, 517 (1972).
[91] D. J. Moffatt and G. A. Neville, *Can. J. Spectrosc.* **26**, 14 (1981).
[92] A. J. Barnes, M. A. Stuckey, W. J. Orville-Thomas, L. Le Gall, and J. Lauransan, *J. Mol. Struct.* **56**, 1 (1979).
[93] A. J. Barnes, L. Le Gall, and J. Lauransan, *J. Mol. Struct.* **56**, 15 (1979).
[94] A. J. Barnes, L. Le Gall, and J. Lauransan, *J. Mol. Struct.* **56**, 29 (1979).
[95] R. J. Mesley, *Spectrochim. Acta, Part A* **26A**, 1427 (1970).
[96] K. Hollenbach, J. Mezösi, K. Pintye-Hodi, and G. Kedvessy, *Pharmazie* **34**, 240 (1979).
[97] N. A. Smorygo and B. A. Ivin, *Khim. Geterotsikl. Soedin.*, 1402 (1975).
[98] M. Guérin, J. M. Dumas, and C. Sandorfy, *Can. J. Chem.* **58**, 2080 (1980).

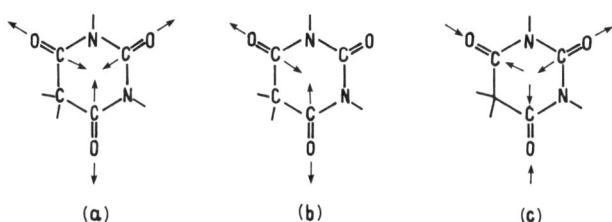

FIG. 2. Carbonyl stretching vibrations of barbituric acid ring: (a) 4,6-CO symmetric; (b) 4,6-CO antisymmetric, (c) 2-CO.

(Fig. 2). Band b is the most sensitive of the carbonyl stretches for conjugation within the pyrimidine ring.[87]

Raman spectra of barbituric acid could not be obtained because of its high luminescent backround,[99,100] but Willis et al. gave interpretations of Raman spectra for eight barbiturates,[89] and Barnes et al. reported a detailed analysis of IR and Raman spectra for the whole frequency range for 1-methyl-, 1,3-dimethyl-, and 5,5-diethylbarbituric acids.[92–94] Similar IR and Raman spectra analyses were described for 5-ethyl-5-phenylbarbituric acid and its N-mono- and N,N-disubstituted derivatives.[101] A detailed interpretation of IR spectra of the 1-acyl and 1-(4'-aminobutynyl) derivatives was also reported.[102,103]

3. ^{1}H-NMR Spectroscopy

The ^{1}H-NMR spectroscopy of common, medicinal 5,5-di- and 1,5,5-trisubstituted barbiturates is very well described in the literature.[10,24,26,104–110]

Chemical shifts of substituents attached to the nitrogen atoms or to the C-5 atom of the barbituric acid ring, together with their multiplicity, completely characterize particular barbituric acid derivatives and serve for identification.

[99] J. R. Nestor and E. R. Lippincott, *J. Mol. Spectrosc.* **51**, 351 (1974).
[100] L. Le Gall and A. J. Barnes, *Ber. Bunsenges. Phys. Chem.* **82**, 52 (1978).
[101] R. Buchet, H. Bill, and B. Siegfried, *Spectrochim. Acta, Part A* **38A**, 239 (1982).
[102] A. Sucharda-Sobczyk and J. Bojarski, *Rocz. Chem.* **44**, 2333 (1970).
[103] B. Danielsson, *Acta Pharm. Suec.* **2**, 47 (1965).
[104] G. Rücker, *Arch. Pharm. (Weinheim, Ger.)* **299**, 688 (1966).
[105] H. W. Avdovich and G. A. Neville, *Can. J. Pharm. Sci.* **4**, 51 (1969).
[106] H. Lackner and G. Doring, *Arch. Toxicol.* **26**, 237 (1970).
[107] G. A. Neville, H. W. Avdovich, and A. W. By, *Can. J. Chem.* **48**, 2274 (1970).
[108] E. L. Frochaux-Zeidler and B. Testa, *Pharm. Acta Helv.* **54**, 229 (1979).
[109] P. R. Andrews, A. J. Jones, G. P. Jones, A. Marker, and E. A. Owen, *Eur. J. Med. Chem.—Chim. Ther.* **16**, 145 (1981).
[110] M. Yogo, K. Hirota, and S. Senda, *Chem. Pharm. Bull.* **30**, 1333 (1982).

TABLE III
CHEMICAL SHIFTS OF THE NH PROTONS FOR 5,5-DI- AND 1,5,5-TRISUBSTITUTED BARBITURATES

Barbituric acid derivative	δ_{NH} (ppm)^a in different solvents		
	$CDCl_3$	p-Dioxane	DMSO-d_6
5-Allyl-5-cyclopentenyl	—	10.0	11.4
5-Allyl-5-phenyl	9.1	9.7	11.6
1-Methyl-5-(2'-bromallyl)-5-isopropyl	9.4	10.2	11.7
5,5-Diethyl	—	10.1	11.4
1-Methyl-5,5-diethyl	9.4	10.1	11.6
5-Ethyl-5-isopropyl	—	10.0	11.3
5-Ethyl-5-phenyl	8.7	10.2	11.7
1-Methyl-5-ethyl-5-phenyl	8.8	10.2	11.9

^{a} Chemical shift δ in ppm versus TMS.

Electronic effects of substituents at the C-5 position influence the chemical shift of the NH protons in 5,5-di- and 1,5,5-trisubstituted derivatives, and the possibility of an interaction of the barbituric acid nucleus with a solvent, usually through a hydrogen bond, was confirmed by NH chemical shifts as shown in Table III.

One of the imide protons (NH) of the barbituric acid ring engages in intermolecular hydrogen bonding with DMSO or N,N-dimethylacetamide.[10,107,111] The formation of the intramolecular hydrogen bond in 5-hydroxy-1,3,5-triethylbarbituric acid (**15**) causes different chemical shifts of the N-1 and N-3 ethyl groups owing to the anisotropic effect of the C-4 and C-6 carbonyl groups.[107]

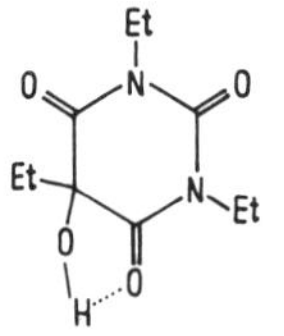

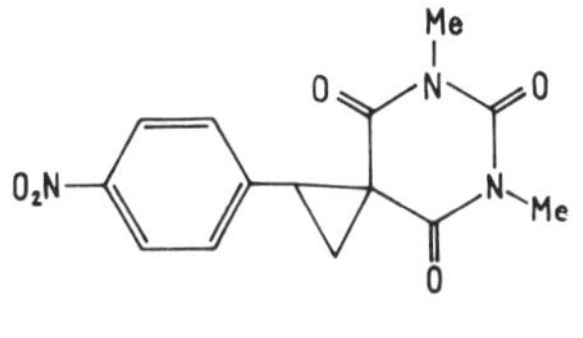

Two resonance signals for N-methyl groups in the spirobarbiturate **16** are also observed as a result of the long-range magnetic effect of the p-nitrophenyl moiety.[112]

[111] K. C. Tewari, F. K. Schweighardt, J. Lee, and N. C. Li, *J. Magn. Reson.* **5**, 238 (1971).
[112] R. Bednar, U. Herzig. I. Schuster, P. Schuster, and P. Wolschann, *Org. Magn. Reson.* **8**, 301 (1976).

The influence of solvent on the chemical shift of substituents at the 5-position was also investigated.[104,107,113] Proton-donating solvents result in significant downfield shifts of the resonance of 5-alkyl groups; for example, differences in the chemical shifts of the 1′-CH_2 protons in trifluoroacetic acid and dioxane is 0.3–0.5 ppm.[104,107] Bobrański explained this phenomenon by suggesting protonation of the barbiturate ring in strong acidic media, which results in the deformation of this ring from planarity and then in deshielding of the 5-substituent protons.[113] On the other hand, Neville and Cook claim that solute–solvent hydrogen-bond interactions are responsible.[107] In fact, both may be important.

The ^{1}H-NMR spectra of 5-arylidenebarbiturates (17) in DMSO-d_6 show, besides aromatic proton signals, two singlets for NH or NMe protons and a singlet for the olefinic proton (Table IV).[24,112,114,115]

(17)

The differences in chemical shifts of protons a and b are caused by the anisotropy of the C-4 and C-6 carbonyl groups owing to the geometry of 17. But an influence of substitution in the aromatic ring on the chemical shift of the NMe protons is less visible in relation to that for the NH protons (Table IV).

^{1}H-NMR spectra of some barbiturates complexed with lanthanide ions were also investigated.[116–118]

4. ^{13}C-NMR Spectroscopy

The ^{13}C-NMR spectra of over 100 barbiturates, thiobarbiturates, and arylidenebarbiturates are reported in the literature. Ranges in the chemical shifts of carbon atoms in the barbituric acid nucleus and in NMe groups for different classes of derivatives (18 and 19) are shown in Fig. 3.

---

[113] B. Bobrański, *Rocz. Chem.* **43**, 1971 (1969).
[114] R. Bednar, E. Haslinger, U. Herzig, O. E. Polansky, and P. Wolschann, *Monatsh. Chem.* **107**, 1115 (1976).
[115] E. Haslinger and P. Wolschann, *Org. Magn. Reson.* **9**, 1 (1977).
[116] J. Knabe and V. Gradmann, *Arch. Pharm. (Weinheim, Ger.)* **310**, 468 (1977).
[117] J. Triepel and H.-H. Otto, *Monatsh. Chem.* **108**, 1085 (1977).
[118] J. Ascenso, M. Candida, T. A. Vaz, and J. J. R. Frausto Da Silva, *J. Inorg. Nucl. Chem.* **43**, 1255 (1981).

TABLE IV
¹H-NMR CHEMICAL SHIFTS OF THE NH, NCH₃, AND =CH
PROTONS FOR 5-ARYLIDENEBARBITURATES (17)[a]

	R' = H			R' = CH₃		
R	(a)	(b)	=CH	(a)	(b)	=CH
m-OMe	11.52	11.69	8.50	3.38	3.49	—
p-Br	11.38	11.53	8.34	3.36	3.44	—
p-NO₂	11.35	11.45	8.40	3.36	3.45	—
H	11.29	11.39	8.42	3.22	3.28	—
o-NO₂	11.24	11.49	8.62	3.24	3.42	8.86
p-OMe	11.22	11.32	8.42	3.38	3.40	8.47
p-NMe₂	11.03	11.16	8.25	3.35	3.36	—

[a] Chemical shift δ in ppm versus TMS.

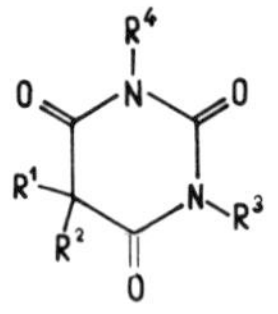

(18)

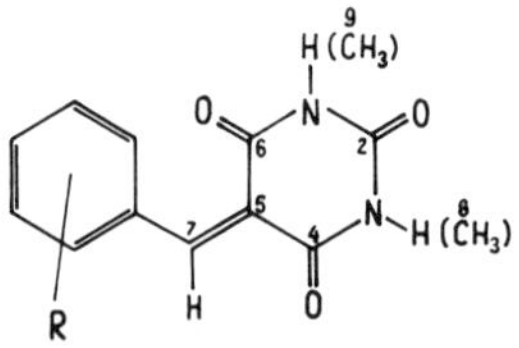

(19)

Differences in the ¹³C chemical shifts of the C-5 atom for compounds of type **18** are dependent on the nature of substituents R^1 and R^2 and amount to ~6–11 ppm, while those for C-2, C-4, and C-6 atoms are only 1.5–2.5 ppm.[42,75,119–127]

In the ¹³C-NMR spectra of sodium salts of 5,5-disubstituted derivatives (**18**) the signals of C-2 and C-4(6) carbonyl groups are shifted downfield by ~13 and 10 ppm, respectively, in relation to their undissociated forms.[75,109,120,123,126]

For meta- and para-substituted arylidenebarbiturates (**19**) the conjugation in the molecule extends from the aromatic moiety over the C-5=C-7 bond to

[119] K. Rehse, *Dtsch. Apoth.-Ztg.* **112**, 1185 (1972).
[120] A. Fratiello, M. Mardirossian, and E. Chavez, *J. Magn. Reson.* **12**, 221 (1973).
[121] J. Okada and T. Esaki, *Yakugaku Zasshi* **93**, 1014 (1973).
[122] J. Okada and T. Esaki, *Chem. Pharm. Bull.* **22**, 1580 (1974).
[123] R. C. Long and J. H. Goldstein, *J. Magn. Reson.* **16**, 228 (1974).
[124] F. I. Carroll and C. G. Moreland, *J.C.S. Perkin II*, 374 (1974).
[125] S. Asada and J. Nishijo, *Bull. Chem. Soc. Jpn.* **51**, 3379 (1978).
[126] H. W. Avdovich and G. A. Neville, *Can. J. Pharm. Sci.* **15**, 75 (1980).
[127] M. Rautio and E. Rahkamaa, *Org. Magn. Reson.* **15**, 53 (1981).

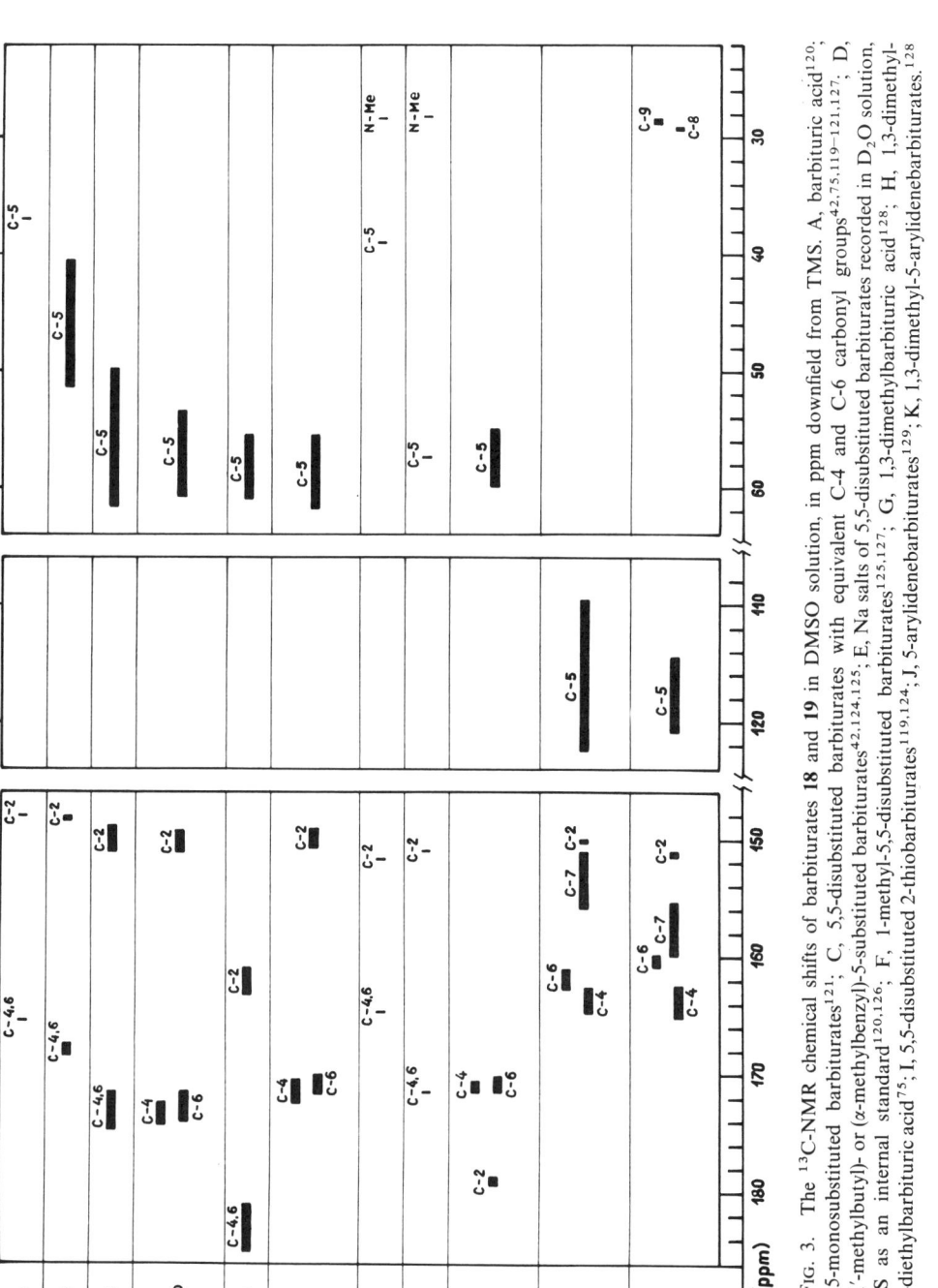

FIG. 3. The ^{13}C-NMR chemical shifts of barbiturates 18 and 19 in DMSO solution, in ppm downfield from TMS. A, barbituric acid[120]; B, 5-monosubstituted barbiturates[121]; C, 5,5-disubstituted barbiturates with equivalent C-4 and C-6 carbonyl groups[42,75,119–121,127]; D, 5-(1'-methylbutyl)- or (α-methylbenzyl)-5-substituted barbiturates[42,124,125]; E, Na salts of 5,5-disubstituted barbiturates recorded in D$_2$O solution, DSS as an internal standard[120,126]; F, 1-methyl-5,5-disubstituted barbiturates[125,127]; G, 1,3-dimethylbarbituric acid[128]; H, 1,3-dimethyl-5,5-diethylbarbituric acid[75]; I, 5,5-disubstituted 2-thiobarbiturates[119,124]; J, 5-arylidenebarbiturates[129]; K, 1,3-dimethyl-5-arylidenebarbiturates.[128]

the C-4 and C-6 carbonyl groups but is not observed at the C-2 carbonyl group, as evidenced by changes in chemical shifts. The exocyclic C=C bond is strongly polarized with an electron deficiency at C-7. The substitution of an electron-donating substituent R in **19** by an electron-attracting one results in an upfield shift of the C-7 atom by 4.5 ppm and in a longer downfield shift of the C-5 atom by ~13 ppm.[128,129]

The chemical shifts of the NMe carbon atoms are almost independent of substituents at C-5.

The C-4 and C-6 atoms can show separate resonances as a result of a chiral center in the side chain for compounds **18** ($R^3 = R^4 = $ H or Me),[75,120,124,126] E,Z isomerism around the exocyclic C=C bond for ylidene derivatives (**19**),[128,129] or for N-monomethylated barbiturates (**18**) ($R^3 = $ H, $R^4 = $ Me)[125] (Fig. 3). Differences in the chemical shifts of the C-4 and C-6 carbonyl groups for **18** ($R^1 = $ 1'-methylbutyl or 1'-phenylethyl, $R^3 = R^4 = $ H) are 0.1–1.1 ppm and the degree of nonequivalence depends on the R^2 substituent, while 5-ethyl-5-isopropylbarbituric acid, without a center of chirality, shows only one signal for these carbon atoms. However, two C-4- and C-6-atom signals of N-methylbarbiturates (**18**) are differentiated by 0.5–1.5 ppm.[124] The nonequivalence of the C-4 and C-6 atoms in 5-arylidenebarbiturates (**19**) is dependent on the nature of the R substituent, and for p-NO_2 and p-NMe_2 groups equals 1.4 and 2.0 ppm, respectively. The NMe carbon atoms (C-8, C-9) of 1,3-dimethyl-5-arylidenebarbiturates (**19**) are also nonequivalent, and the differences in their chemical shifts amount to 0.6–0.8 ppm.[128]

According to Haslinger and Wolschann[128] the C-6 atom in 5-arylidenebarbiturates (**19**) is more shielded than C-4 because of the steric interaction of the carbonyl oxygen with the aromatic ring, but Craik *et al.*, introducing the idea of the MLF (molecular lines of forces), suggest interactions through bonds.[130]

Okada and Esaki carried out a detailed analysis of the ^{13}C-NMR chemical shifts of all sp^3 carbon atoms for 5-mono- and 5,5-dialkylbarbiturates and calculated equations based on additivity for chemical shifts of these atoms at position 5 and in the side chain.[121,122]

5. *Mass Spectrometry*

Electron-impact (EI) mass spectra of numerous barbiturates[108,131–138] reflect the high stability of the barbituric acid ring. According to Costopanagiotis and Budzikiewicz,[131] unsubstituted barbituric acid has the molecular ion m/z 128. The base peak corresponds to CH_2=C=$\overset{+}{O}$ and a cleavage of the ring occurs by the elimination of the CO and HNCO molecules. The formation of O=$\overset{\cdot}{C}$—CH_2—C=O^+ also is observed, and loss of a hydrogen atom yields O=C=CH—C=O^+. However, barbiturates

Sec. II.B] BARBITURIC ACID 247

containing alkyl or alkenyl substituents at the C-5 position do not show observable molecular ions (20). Owing to an easy initial fragmentation, there are three types: (i) elimination of an olefin from the molecular ion via McLafferty rearrangement for compounds where ethyl or longer substituents are present, (ii) elimination of the 5-substituent as a radical, and (iii) elimination of an HX molecule or X radical for barbiturates substituted in the side chain. Because of these three processes, intense ions 21–23 are observed (Scheme 2). The elimination of HNCO from the molecular ion does not occur.[134] Fragments 26–29 are formed from 21–23 by a further cleavage of C-5 substituents. The elimination of HNCO from a barbituric acid ring produces

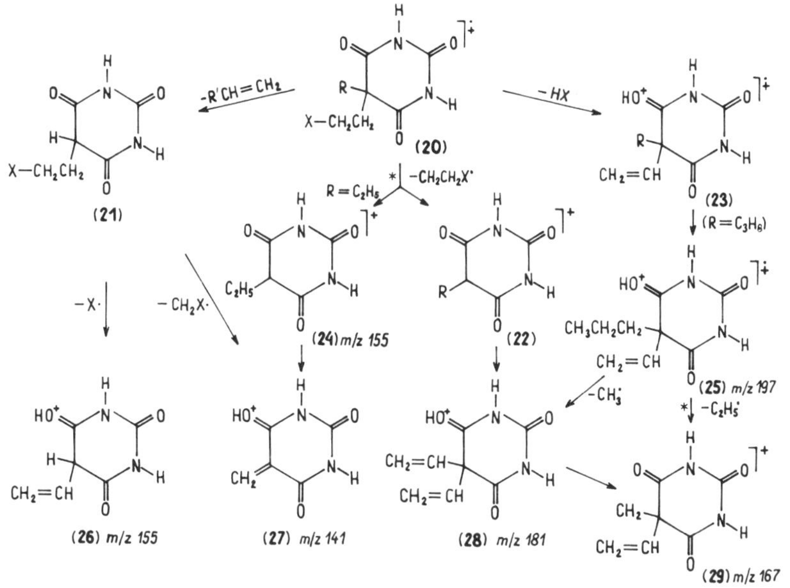

SCHEME 2. X = H, Cl, Br, I, SCN, NMe$_2$

---

[128] E. Haslinger and P. Wolschann, *Bull. Soc. Chim. Belg.* **86**, 907 (1977).
[129] C. N. Robinson and C. C. Irving, *J. Heterocycl. Chem.* **16**, 921 (1979).
[130] D. J. Craik, R. T. C. Brownlee, and M. Sadek, *J. Org. Chem.* **47**, 657 (1982).
[131] A. Costopanagiotis and H. Budzikiewicz, *Monatsh. Chem.* **96**, 1800 (1965).
[132] H. F. Gruetzmacher and W. Arnold, *Tetrahedron Lett.*, 1365 (1966).
[133] R. T. Coutts and R. Locock, *J. Pharm. Sci.* **57**, 2096 (1968).
[134] R. T. Coutts and R. Locock, *J. Pharm. Sci.* **58**, 775 (1969)
[135] J. N. T. Gilbert, B. J. Millard, and J. W. Powell, *J. Pharm. Pharmacol.* **22**, 897 (1970).
[136] M. Mizugaki, Y. Suzuki, M. Uchiyama, and H. Abe, *Eisei Kagaku* **17**, 241 (1971) [*CA* **76**, 68,814 (1972)].
[137] M. Klein, *Recent Dev. Mass Spectrom. Biochem. Med.* **1**, 471 (1978).
[138] S. Dilli and D. N. Pillai, *Aust. J. Chem.* **29**, 1769 (1976).

ions **30** and **32–35**. Fragmentation patterns presented in Scheme 2 were confirmed by accurate mass measurements and metastable ions.[137] Fragmentation of **24** to open-chain ion **31** is also postulated.[108]

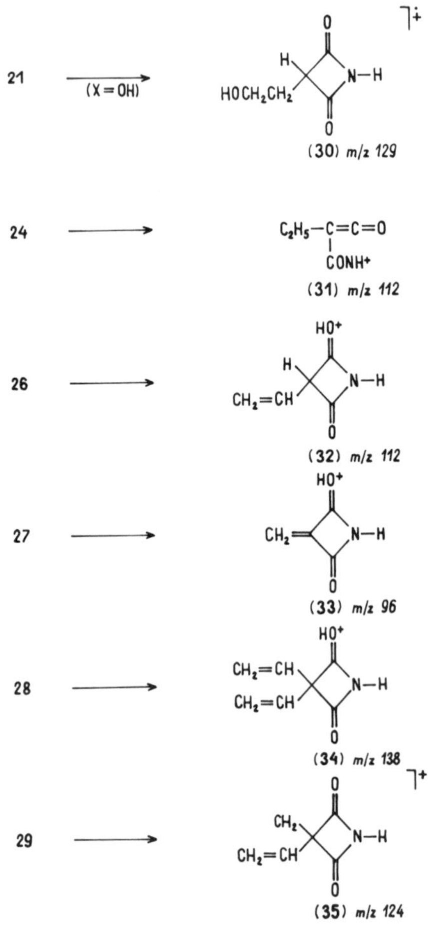

21 $\xrightarrow{(X=OH)}$ (**30**) m/z 129

24 $\longrightarrow$ $C_2H_5-C=C=O$ | $\overset{|}{C}ONH^+$ (**31**) m/z 112

26 $\longrightarrow$ (**32**) m/z 112

27 $\longrightarrow$ (**33**) m/z 96

28 $\longrightarrow$ (**34**) m/z 138

29 $\longrightarrow$ (**35**) m/z 124

Watson and Falkner[139,140] investigated EI-MS of 1,3-bis(trimethylsilyl)-5,5-disubstituted barbiturates (**36**). They found that these compounds have a molecular ion and give the characteristic fragment $M - 15$ (**37**) formed by the elimination of a CH_3 radical from one of the trimethylsilyl moieties. This fragmentation pattern was confirmed by the observation that trimethyl-d_9-silyl derivative **36** loses only 18 (not 15) mass units from the molecular ion.

[139] J. T. Watson and F. C. Falkner, *Org. Mass Spectrom.* **7**, 1227 (1973).
[140] F. C. Falkner and J. T. Watson, *Org. Mass Spectrom.* **8**, 257 (1974).

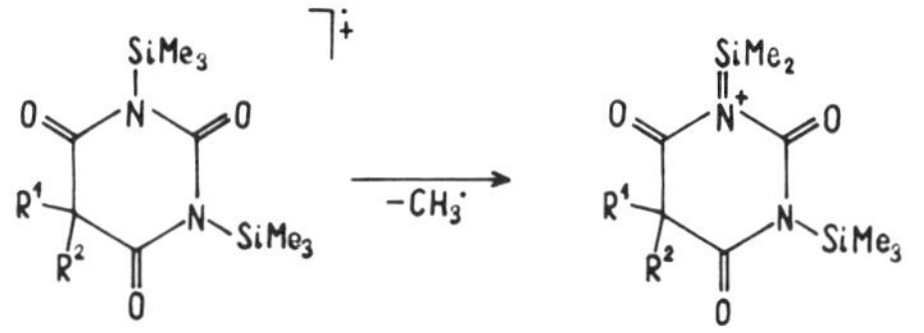

Sec. II.B] BARBITURIC ACID 249

(36) (37)

A strong interaction of the aromatic substituent with the pyrimidine ring in 5-arylidenebarbituric acid derivatives (38) results in the stabilization of the whole molecule. This stabilization is reflected in their EI mass spectra. In most cases this shows the molecular ion M^+ or quasi-molecular ions $[M - H]^+$ or $[M - R]^+$, where R is a substituent on the aromatic ring.[137,141] Klein[137] suggests further ring fragmentation by the elimination of HNCO and CO molecules and proposes, for 5-benzylidenebarbituric acid, the mechanism shown in Scheme 3. However, Ramana and Viswanadham[141] claim that the

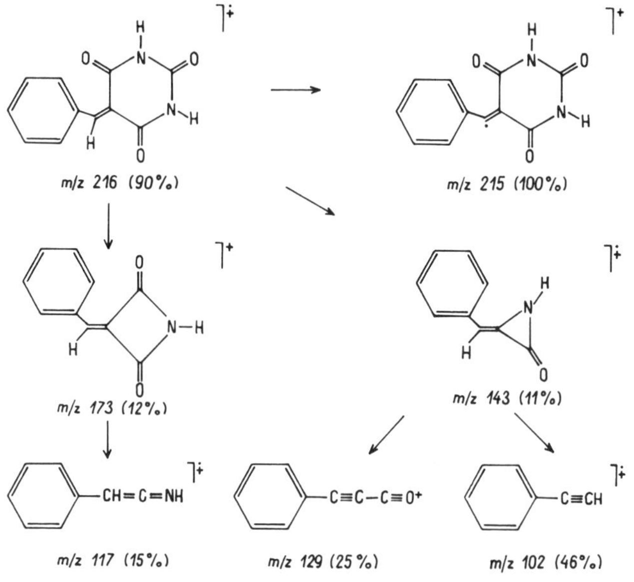

m/z 216 (90%)   m/z 215 (100%)

m/z 173 (12%)   m/z 143 (11%)

m/z 117 (15%)   m/z 129 (25%)   m/z 102 (46%)

SCHEME 3

molecular ion 38 loses $H_2NCONHCO$ or H· radicals to form fragments 39 and a cyclic cation 40, respectively (Scheme 4). The H atom is eliminated from the aromatic ring during cyclization. This was confirmed by the mass spectrum

[141] D. V. Ramana and S. K. Viswanadham, *Bull. Chem. Soc. Jpn* 53, 3004 (1980).

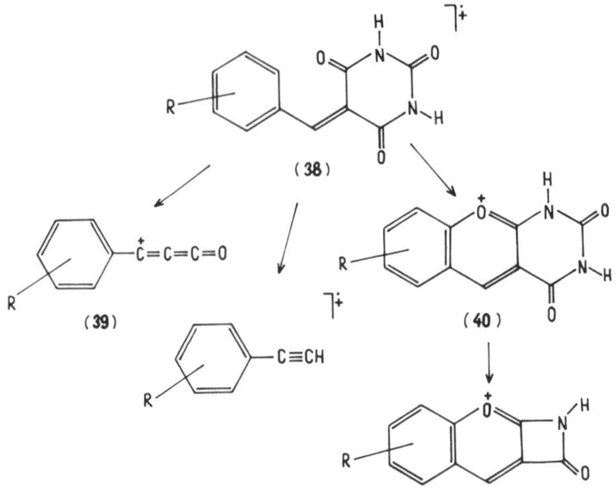

SCHEME 4

of 5-benzylidenebarbituric acid perdeuterated in the aromatic ring. This compound gives $[M - D]^+$ ion, and the formation of $[M - H]^+$ ion was not observed. Additional evidence for the formation of **40** is the fragmentation of 5-arylidene derivatives substituted in the aromatic ring. The ortho-substituted derivative **38** shows the presence of $[M - R]^+$ ion, but for meta- and para-substituted compounds this ion was not observed. Thus the fragmentation pattern of 5-arylidenebarbiturates proposed by Ramana and Viswanadham seems to be more feasible.

EI mass spectroscopy gives valuable information about the fragmentation of barbiturates, but the molecular ion is not usually observed. The presence of the molecular ion is very important for the identification of particular derivatives. Therefore, other methods of ionization of a molecule are also employed.

In field desorption (FD) and positive-ion chemical ionization (PICI) mass spectrometry of barbiturates, the molecular ion M^+ and/or quasimolecular ion $[M + 1]^+$ (**41**) are always observed.[142-144] Fales et al.[142] show that the

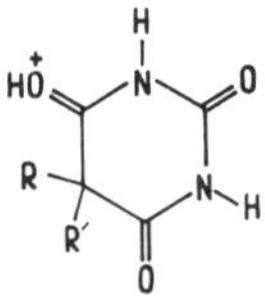

(41)

[142] M. M. Fales, G. W. A. Milne, and T. Axenrod, *Anal. Chem.* **42**, 1432 (1970).

PICI MS of barbiturates substituted at the C-5 position by alkyl, alkenyl, or aromatic substituents give $[M + 1]^+$ (41) ion, which is the base peak. Barbiturates containing only aliphatic substituents at that position give in their FD MS the $[M + 1]^+$ ion that is 2–6 times more intense in relation to the molecular ion M^+. Elongation of the carbon chain at that position results in an increase of the relative abundance of the $[M + 1]^+$ ion. On the other hand, a cyclic substituent at the 5-position causes a decrease of the relative intensity of the $[M + 1]^+$ ion, and molecular ion M^+ is the base peak.[143,144]

The negative ion chemical ionization (NICI) of some barbiturates is also described.[145–148] The molecular ion $M^{\bar{}}$ (42) is formed by electron capture, and this is followed by the elimination of the H atom or R radical, yielding stable anions 43–45 (Scheme 5). Anion 46, present in low abundance, may arise from a McLafferty rearrangement. Ring cleavage is manifested by the presence

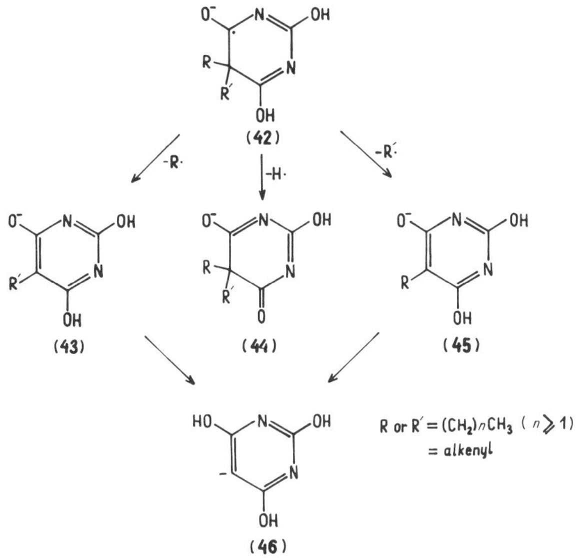

SCHEME 5

[143] D. E. Games, A. H. Jackson, K. T. Taylor, and N. J. Haskins, *Adv. Mass Spectrom. Biochem. Med.* **1**, 383 (1976).
[144] H.-R Schulten and D. Kümmler, *Anal. Chim. Acta* **113**, 253 (1980).
[145] D. F. Hunt, G. C. Stafford, F. W. Crow, and J. W. Russell, *Anal. Chem.* **48**, 2098 (1976).
[146] I. Dzidzic, D. I. Carroll, R. N. Stillwell, M. G. Hornig, and E. C. Hornig, *Adv. Mass Spectrom.* **A7**, 359 (1978).
[147] D. Frangi-Schnyder and H. Branderberger, *Fresenius'Z. Anal. Chem.* **290**, 153 (1978).
[148] L. V. Jones and M. J. Whitehouse, *Biomed. Mass Spectrom.* **8**, 231 (1981).

of the NCO$^-$ ion (m/z 42), which usually has a low relative intensity. The base peak in the NICI is dependent on the nature of the R and R' substituents. Thus all 5-ethylbarbiturates (R = Et) yield the base peak [M − R']$^-$ 45, while 5-allylbarbiturates (R = allyl) give [M − R]$^-$ 43 as the base peak.

C. STRUCTURE AND CONFORMATION

1. *Structure and Conformation in the Solid State*

a. *Hydrogen Bonding and Polymorphism.* In the crystalline state the trioxo structure is typical for barbituric acid and its derivatives (cf. Section II,A). Crystallographic data show that barbiturates generally have low-symmetric crystal systems, i.e., monoclinic or sometimes triclinic (Table V).

In the isolated molecule of barbituric acid and its 5-mono-, 5,5-di-, and 1,3,5,5-tetrasubstituted derivatives, all with the same 1,3-substituents, a flat or almost flat pyrimidine ring is symmetric about the C_2 line (C-5 ··· C-2)[149,150] (Fig. 4). CNDO/2 and PCILO calculations show that for the isolated molecule of 5,5-diethylbarbituric acid a distribution of atomic charges is symmetric about that C_2 line.[60,151]

TABLE V
CRYSTAL STRUCTURES OF SELECTED BARBITURATES

Substituents (trivial name)	Crystal system[a]	Z[b]	H-Bond mode[c]	References
5-Ethyl	M	4	a	152
5,5-Diethyl (Barbital I)	TG	18	a	16
5,5-Diethyl (Barbital II)	M	4	b	16
5,5-Diethyl-1-methyl (Metharbital)	M	4	c	154
5-Ethyl-5-(3',3'-dimethylbutyl) (γ-Methylamobarbital)	M	4	e	155
trans-5-But-1'-enyl-5-ethyl	T	2	d	156
5,5-Diallyl (Dial)	M	8	d	157
5,5-Diallyl-1-cyclohexyl	M	4	c	158

[a] M, monoclinic; T, triclinic; TG, trigonal.
[b] Number of molecules per cell.
[c] H-Bond modes shown in Fig. 5.

[149] B. M. Craven, C. Cusatis, G. L. Gartland, and E. A. Vizzini, *J. Mol. Struct.* **16**, 331 (1973).
[150] L. Dupont, D. Dideberg, and D. Pyzalska, *Acta Crystallogr., Sect. B* **B30**, 2447 (1974).
[151] D. Voet, *J. Am. Chem. Soc.* **94**, 8213 (1972).

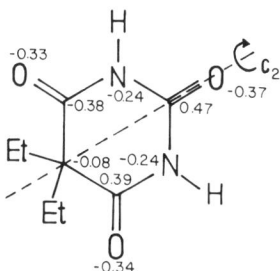

Fig. 4. Distribution of atomic charges in the 5,5-diethylbarbituric acid molecule by CNDO/2 method.[151]

In the crystalline state, barbiturates show several modes of intermolecular hydrogen bonding, which differ by the number of hydrogen bonds formed between the NH and CO groups (Fig. 5) (Table V). For various hydrogen-bonding modes (Fig. 5) small differences in bond lengths, bond angles, and atomic charges are observed.[16,17,149,152-159] These hydrogen bonds are responsible for layer packing in the crystal lattice. Weak intermolecular van der Waals interactions of carbonyl groups are also responsible for the three-dimensional packing of the molecules in the crystal structure.[17,152,154,160] Three modes of these van der Waals interactions are shown in Fig. 6, and it is noteworthy that the distance between the C-2 carbonyl groups, for example, in 5,5-diethyl-1-methylbarbituric acid (Fig. 6, interaction type A) is 3.075 Å, which is close to the sum of the calculated van der Waals radii of 3.1 Å.[154]

Both hydrogen bonding and intermolecular van der Waals interactions are responsible for the polymorphism of barbiturates. The stability of different polymorphic forms is responsible for their biological availability.[16,17,88,149,161-167] Some barbiturates have many polymorphic forms;

[152] B. M. Gatehouse and B. M. Craven, *Acta Crystallogr., Sect. B* **B27**, 1337 (1971).
[153] B. M. Craven and E. A. Vizzini, *Acta Crystallogr., Sect. B* **B27**, 1917 (1971).
[154] H. Wunderlich, *Acta Crystallogr., Sect. B* **B29**, 168 (1973).
[155] G. L. Gartland and B. M. Craven, *Acta Crystallogr., Sect. B* **B27**, 1909 (1971).
[156] G. P. Jones and P. R. Andrews, *J. Cryst. Mol. Struct.* **11**, 125 (1981).
[157] C. Escobar, *Acta Crystallogr., Sect. B* **B31**, 1059 (1975).
[158] D. Dideberg, L. Dupont, and D. Pyzalska, *Acta Crystallogr., Sect. B* **B31**, 685 (1975).
[159] J. R. Ruble, A. C. Wang, and B. M. Craven, *J. Mol. Struct.* **51**, 229 (1979).
[160] R. Anulewicz, *Pol. J. Chem.* **55**, 187 (1981).
[161] B. M. Craven and E. A. Vizzini, *Acta Crystallogr., Sect. B* **B25**, 1993 (1969).
[162] P. P. Williams, *Acta Crystallogr., Sect. B* **B29**, 1572 (1973).
[163] H. M. El-Banna, A. R. Ebian, and A. A. Ismail, *Pharmazie* **30**, 455 (1975).
[164] R. Kaliszan and J. Halkiewicz, *Pol. J. Pharmacol. Pharm.* **27**, 579 (1975).
[165] I. Grabowska and R. Kaliszan, *Pol. J. Pharmacol. Pharm.* **28**, 529 (1976).
[166] J. Caillet and P. Claverie, *Acta Crystallogr., Sect. B* **B36**, 2642 (1980).
[167] B. M. Craven, R. C. Fox, and H.-P. Weber, *Acta Crystallogr., Sect. B* **B38**, 1942 (1982).

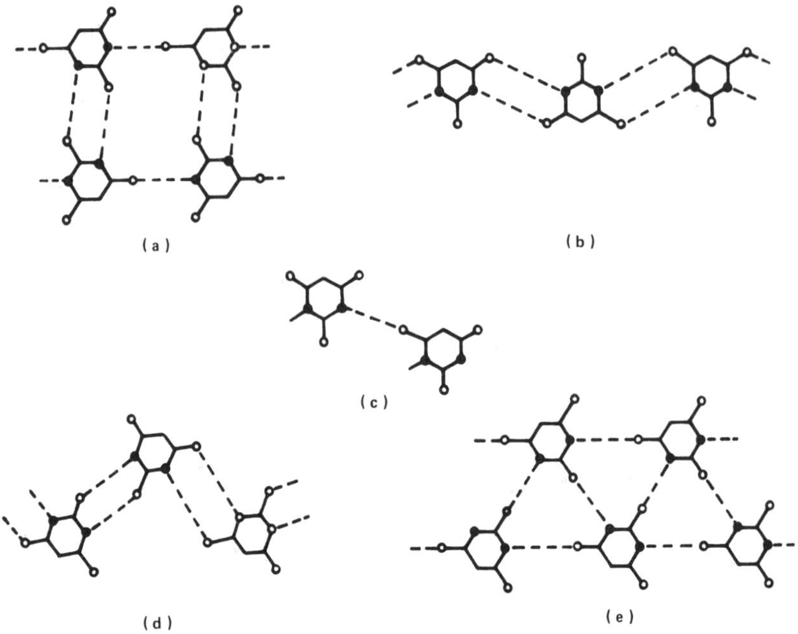

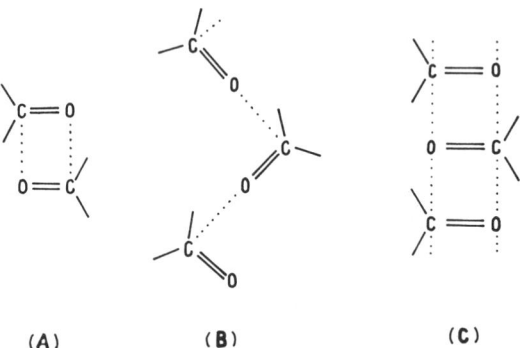

Fig. 5. Hydrogen-bonding modes in barbiturates.

Fig. 6. Three types of van der Waals interactions in barbiturates.

for example, 6 forms for 5,5-diethylbarbituric acid and 13 forms for 5-ethyl-5-phenylbarbituric acid have been reported.[16,162,163] The most detailed studies on the polymorphism and the relation between the stability of particular forms and their physicochemical properties were carried out for 5,5-diethylbarbituric acid.[16,88,96,164–171]

b. *Ring Conformations.* The 2,4,6-pyrimidinetrione ring of barbituric acid derivatives have planar or nearly planar conformations, but the kind and size of the substituents, as well as their positions in the ring, have significant influence on a deviation of atoms from the ring plane. Two types of conformations of the ring (flat and puckered) are observed. While the flat conformation is observed only in a few cases, i.e., 5,5-diethyl-1-methyl-,[154] 5,5-diallyl-1,3-dicyclohexyl-,[150] 5-(1'-cycloheptenyl)-5-ethyl-,[157] or *trans*-5-ethyl-5-(1',3'-dimethylbut-1'-enyl)barbituric acid,[172] nonplanar conformations are typical for other barbiturates.

According to Craven *et al.*, there are two types of puckered conformations of the barbituric acid ring.[149] In the symmetric puckered conformation, the C-2 and C-5 atoms lie on the C_2 line (Fig. 7). This conformation is observed for a few barbiturates.[20,149,158] The asymmetric puckered conformation has the C-5 atom displaced from the best least squares plane of the other ring atoms (Fig. 7). This usually slight deviation of the C-5 atom varies from one barbiturate to another.[149,152,155,156,159,162,173–175] For example, dihedral angles between the C-4–C-5–C-6 plane and the best least squares plane for 5-(1'-cycloheptenyl)-5-ethyl-, 5,5-diallyl-, and 5-cyclohexenyl-1,5-dimethylbarbituric acid are 3, 10.4, and 29°, respectively.[157] Deviations from the planarity of the barbituric acid ring were also predicted by MINDO/3 calculations.[175]

c. *Orientation of Substituents.* X-Ray crystallographic study shows that 5,5-diethylbarbituric acid has ethyl groups directed above the ring

(a)   (b)

FIG. 7. (a) The asymmetric puckered, and (b) the symmetric puckered conformations of the barbituric acid ring.

---

[168] K. Hollenbach, K. Pintye-Hódi, and G. Kedvessy, *Pharmazie* **34**, 164 (1979).
[169] K. Pintye-Hódi and K. Hollenbach, *Pharmazie* **34**, 807 (1979).
[170] K. Hollenbach, K. Pintye-Hódi, and G. Kedvessy, *Pharmazie* **35**, 32 (1980).
[171] K. Hollenbach, K. Pintye-Hódi, and G. Kedvessy, *Pharmazie* **35**, 95 (1980).
[172] P. R. Andrews and G. P. Jones, *J. Cryst. Mol. Struct.* **11**, 135 (1981).
[173] P. H. Smit and J. A. Kantars, *Acta Crystallogr.*, Sect. B **B30**, 784 (1974).
[174] G. P. Jones and P. R. Andrews, *J. Cryst. Mol. Struct.* **11**, 145 (1981).
[175] P. R. Andrews and G. P. Jones, *Eur. J. Med. Chem.—Chim. Ther.* **16**, 139 (1981).

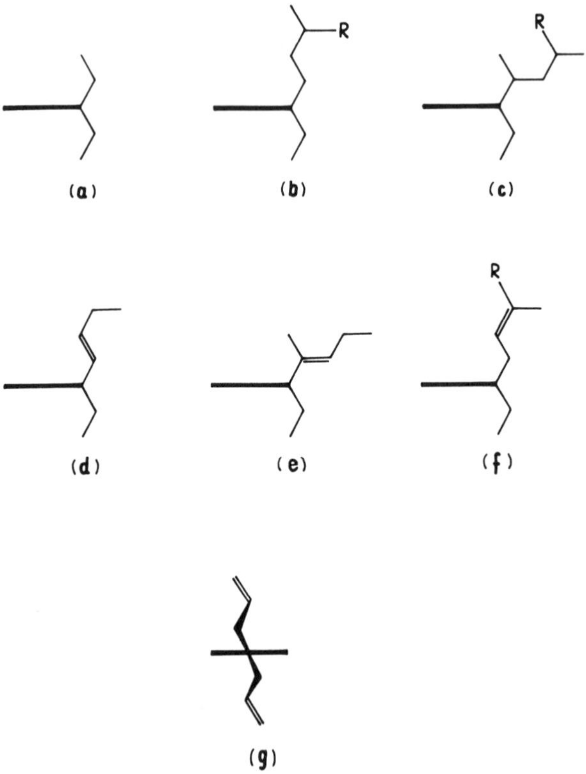

FIG. 8. Orientation of C-5 substituents in barbituric acid derivatives. The heavy line represents the plane of the barbituric acid ring.

system[16] in a plane perpendicular to the barbiturate ring (Fig. 8a), and the preferred conformation calculated by PCILO or EHT methods agrees entirely.[60,176] Both X-ray data[17,155,156,162,173,174,177] and theoretical calculations[60,175,178] indicate that the "over-ring" orientation of the 5-ethyl group, as shown in Fig. 5, is favored independently of the nature of the second substituent at this position.

For 5-alkyl substituents longer than ethyl, the extended conformation (Fig. 8b) is preferred.[60,155,175,178] However, introduction of a methyl group at the α position results in a predominance of the conformation shown in Fig. 8c, where the methyl group is situated above the barbiturate ring.[173,175,178]

[176] B. Laprade, H. Petersen, J. G. Turcotte, and A. N. Paruta, *Can. J. Pharm. Sci.* **11**, 54 (1976).
[177] P. R. Andrews and G. P. Jones, *J. Cryst. Mol. Struct.* **11**, 125 (1981).
[178] G. P. Jones and P. R. Andrews, *J. Med. Chem.* **23**, 444 (1980).

Preferred conformations of but-1'-enyl and but-2'-enyl derivatives are shown in Fig. 8d–f. The double bond of the but-1'-enyl substituent is directed above the ring and partially eclipses the C-4—C-5 ring bond, and the terminal methyl group is folded back at the end of the chain (Fig. 8d).[156,175,178] The double bond of both the cyclohexene ring in 5-(1'-cyclohexenyl)-1,5-dimethylbarbituric acid and the phenyl ring in 5-ethyl-5-phenylbarbituric acid are also partially eclipsed by the C-4—C-5 pyrimidine bond.[60,162,179,180] Similar orientations of substituents with a characteristic S-shape (Fig. 8g) were observed for 5,5-diallylbarbituric acid.[157] Introduction of a methyl group at the α position with a but-1'-enyl substituent results in a similar favored conformation as in the saturated analogues (Fig. 8c). The over-ring conformation of the but-2'-enyl substituent (Fig. 8f) is also the most stable one.[17,174,175,177,178]

In compound **47** the aromatic ring and the imide bond are coplanar but orthogonal to the ylidenebarbituric acid moiety.[181]

(47)

2. Conformation in Solution

Quantum mechanical calculations by MINDO/3 suggest that barbiturates with alkyl and/or alkenyl substituents at the 5-position show in solution the same favored conformations as in the solid state[109,182,183] (Fig. 8). This conclusion is also confirmed by the results of conformational analyses using ^{1}H- and ^{13}C-NMR spectroscopy.[109,183] Moreover, Andrews et al. conclude that these preferred conformations in solution are independent of the ionization state of the barbiturate ring.[109]

Haslinger et al. give an interesting example of weak intramolecular interaction occurring in barbiturates as evidenced by UV–VIS and ^{1}H-NMR

---

[179] P. R. Andrews and J. A. Defina, *Int. J. Quantum Chem., Quantum Biol. Symp.* **7**, 297 (1980).
[180] H.-D. Höltje, *Arch. Pharm.* (*Weinheim, Ger.*) **310**, 650 (1977).
[181] M. C. Apreda, F. H. Cano, C. Foces-Foces, and S. Garcia-Blanco, *Acta Crystallogr., Sect. B* **B37**, 1935 (1981).
[182] G. D. Daves, R. B. Belshee, W. R. Anderson, and H. Downes, *Mol. Pharmacol.* **11**, 470 (1975).
[183] P. R. Andrews and G. P. Jones, *Int. J. Quantum Chem., Quantum Biol. Symp.* **6**, 439 (1979).

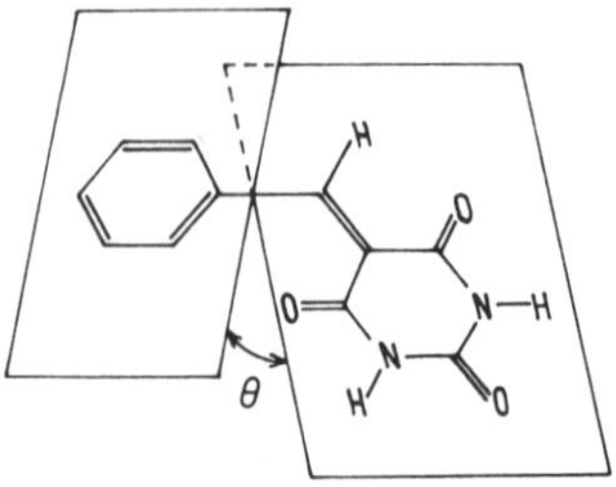

FIG. 9. Conformation of 5-arylidenebarbituric acid molecule; θ, dihedral angle between the phenyl and pyrimidine rings.

spectroscopy. They show that the phenyl ring in 5-(3′-phenylpropyl)barbituric acid (**48**) lies above the barbituric acid ring and the strength of the interaction between these rings is significantly dependent on the nature of the substituent R.[184]

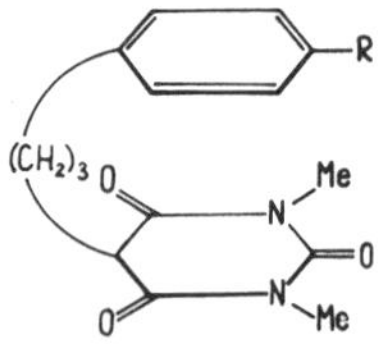

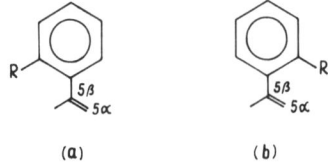

(**48**)

It was also found by ^{1}H-NMR spectroscopy that the phenyl ring in arylidenebarbiturates (**5**) is twisted in relation to the plane of the ylidenebarbiturate.[114,115] A detailed analysis of chemical shifts of particular aromatic protons and their coupling constants with the vinyl proton leads to the conclusion that the value of the dihedral angle θ (Fig. 9) depends on the nature and position of substitution in the aromatic ring. Ortho substitution in the phenyl ring, where steric effects predominate, results in hindrance to rotation. Thus of the two possible conformations conformer A (Fig. 10) is favored. Moreover, for o-nitro substituted derivative **5** the dihedral angle is 43°.[115]

FIG. 10. Two conformations of 5-arylidenebarbiturates o-substituted in the phenyl ring (only arylidene moiety shown).

[184] E. Haslinger, H. Kalchhauser, and P. Wolschann, *Monatsh. Chem.* **113**, 633 (1982).

Rotational barriers about the $=$C—N and —C$=$N bonds in **49** and **50** were also investigated; the enamine bonds have $\Delta G^{\ddagger}$ values higher by 15–20 kJ/mol than the hydrazone bonds.[185]

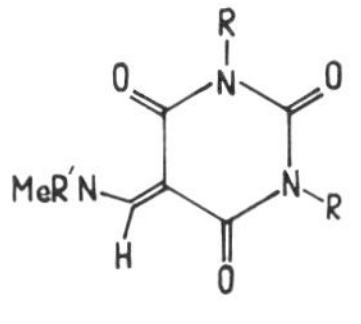

(49) 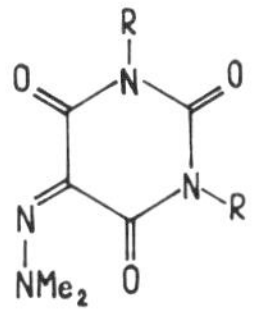   (50)

D. OPTICAL ISOMERS

Two types of optically active barbiturates attract the attention of chemists and pharmacologists. In the first, the center of chirality is outside the ring in one of the substituents at C-5 (e.g., **51**), in the second, the chirality is associated with the dissymmetry of the ring (e.g., **52**). Separation of racemic mixtures of

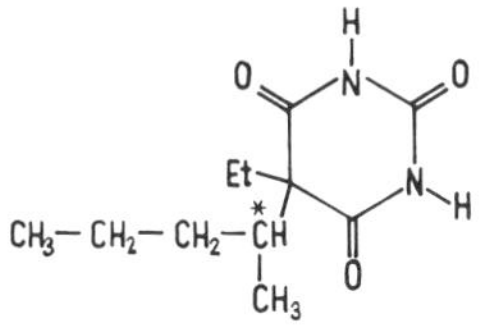

(51) 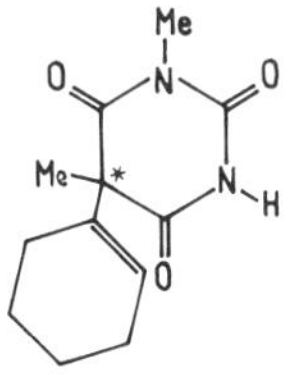   (52)

compounds of both types were reported by Knabe et al.[186-189] The enantiomers were obtained by fractional crystallization of salts prepared from barbituric acids and N-methylquininium hydroxide. This procedure has also been used by others[42,73,190-195] but sometimes was considered unattractive because of poor yields and low optical purity.[73,195]

Bobrański et al. reported the resolution of optical isomers of 5-allyl-5-(2'-hydroxypropyl)barbituric acid by the fractional crystallization of their

---

[185] U. Kölle, B. Kolb, and A. Mannschrenk, *Chem. Ber.* **113**, 2545 (1980).
[186] J. Knabe and R. Kräuter, *Arch. Pharm. (Weinheim, Ger.)* **298**, 1 (1965).
[187] J. Knabe, R. Kräuter, and K. Philipson, *Tetrahedron Lett.*, 571 (1965).
[188] J. Knabe and K. Philipson, *Arch. Pharm. (Weinheim, Ger.)* **299**, 231 (1965).
[189] J. Knabe, D. Strauss, and C. Urbahn, *Pharmazie* **23**, 522 (1968).
[190] R. S. Perry, H. Downes, and R. Karler, *Fed. Proc., Fed. Am. Soc. Exp. Biol.* **28**, 776 (1969).
[191] H. Downes, R. S. Perry, R. E. Ostlund, and R. Karler, *J. Pharmacol. Exp. Ther.* **175**, 692 (1970).
[192] E. Gordis, *Biochem. Pharmacol.* **20**, 246 (1971).
[193] K. Miyano and S. Toki, *Drug Metab. Dispos.* **8**, 111 (1980).
[194] K. Miyano, T. Ota, and S. Toki, *Drug Metab. Dispos.* **9**, 60 (1981).
[195] P. R. Andrews, G. P. Jones, and D. B. Poulton, *Eur. J. Pharmacol.* **79**, 61 (1982).

hydrophthalate ester salts with quinine and subsequent acid hydrolysis of diastereomeric salts and enantiomeric esters.[196] Camphorosulfonic acid was used for the formation of diastereomeric salts of chiral 1-methyl-, 1-ethyl-, and 1-cyclohexylbarbiturates with piperidine, pyrrolidine, and dimethylamine as basic substituents bound through their N atoms with the C-5 atom of the barbituric ring. Separation and hydrolysis of these salts yielded appropriate enantiomers with optical purity better than 95%.[197,198]

Blaschke described satisfactory separation of enantiomers of **52** and several 1-methyl-5,-5-disubstituted barbiturates by column chromatography on microcrystalline cellulose triacetate, but for some pairs of C-5 substituents (phenyl, propyl and 1-cyclopentenyl, ethyl) only partial separation was achieved, while for ethyl, propyl and 2′-pentyl, vinyl pairs no separation was observed.[199]

Synthetic approaches to optically pure enantiomers of barbiturates followed general methods of preparation of these compounds, but pure enantiomers of known configuration have been used as the substrates. Thus, R(−) and S(+) enantiomers of **51** and closely related compounds **(54)** were prepared from the enantiomers of 3-methylhexanoic acid **(53)** by the reaction sequence shown in Scheme 6.[73,200]

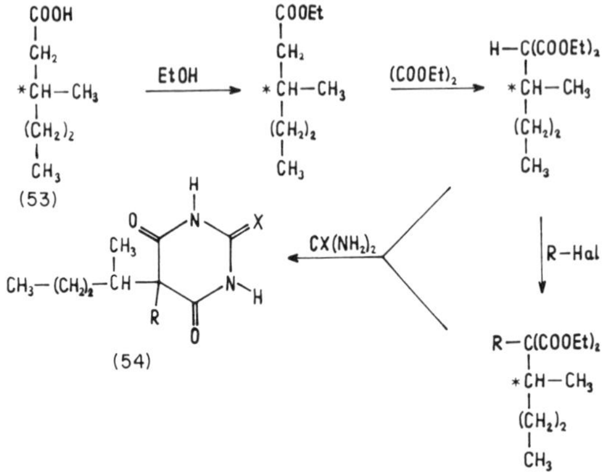

(54a) R = Et, X = O
(54b) R = Et, X = S
(54c) R = H, X = O
(54d) R = allyl, X = O
(54e) R = allyl, X = S

SCHEME 6

Sec. II.D] BARBITURIC ACID 261

A similar procedure was used for preparation of enantiomers of 5-allyl-5-(1'-methyl-2'-pentynyl)-2-thiobarbituric acid[201] and 5-(1',3'-dimethylbutyl)-5-ethylbarbituric acid.[202]
Knabe et al.[203-208] prepared **52** and a series of 5,5-disubstituted analogs (**59**) starting from the enantiomers of substituted cyanoacetic acids (**55**) previously separated via their salts with (1R,2R)-2-amino-1-phenylpropane-1,3-diol (Scheme 7). An interesting feature of this procedure is that starting from the same enantiomer of **55** both (+) and (−) forms of **59** can be obtained

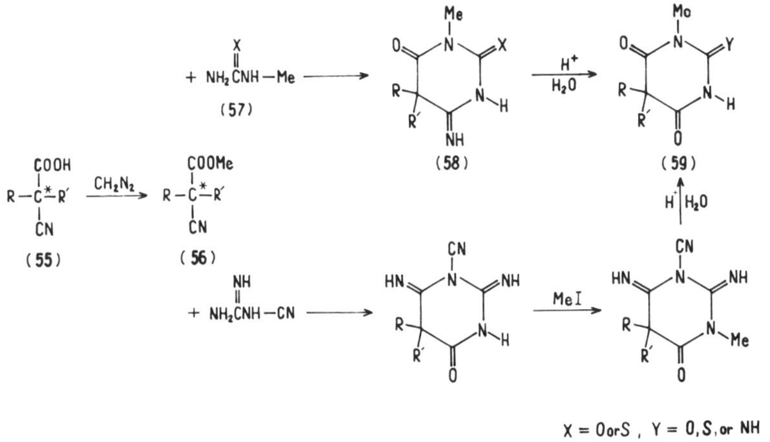

$X = O$ or S , $Y = O, S,$ or NH

SCHEME 7

---

[196] B. Bobrański, M. Wilimowski, J. Barczyńska, R. Seniuta, B. Sędzimirska, and M. Witkowska, Arch. Immunol. Ther. Exp. **21**, 299 (1973).
[197] J. Knabe and J. Reinhardt, Arch. Pharm. (Weinheim, Ger.) **315**, 706 (1982).
[198] J. Knabe and J. Reinhardt, Arch. Pharm. (Weinheim, Ger.) **315**, 772 (1982).
[199] G. Blaschke, Angew. Chem., Int. Ed. Engl. **19**, 13 (1980).
[200] C. E. Cook and C. R. Talent, J. Heterocycl. Chem. **6**, 203 (1969).
[201] F. I. Carroll, A. Philip, D. M. Naylor, H. D. Christensen, and W. C. Goad, J. Med. Chem. **24**, 1241 (1981).
[202] K. C. Rice, J. Org. Chem. **47**, 3617 (1982).
[203] J. Knabe and J. Strauss, Angew. Chem., Int. Ed. Engl. **7**, 463 (1972).
[204] J. Knabe and H. Junginger, Pharmazie **27**, 443 (1972).
[205] J. Knabe and N. Franz, Arch. Pharm. (Weinheim, Ger.) **308**, 313 (1975).
[206] J. Knabe and N. Franz, Arch. Pharm. (Weinheim, Ger.) **309**, 173 (1976).
[207] J. Knabe, H. P. Büch, V. Gradmann, and I. Wolff, Arch. Pharm. (Weinheim, Ger.) **310**, 421 (1977).
[208] J. Knabe and W. Fürst, Arch. Pharm. (Weinheim, Ger.) **312**, 86 (1979).

separately, depending on the alternative routes of synthesis. The scope and details of this method are discussed in a review published in 1978.[209]

The synthesis of 1-methyl-5-phenyl-5-propyl-2-thiobarbituric acid by the condensation of appropriate **56** with **57** proved unsuccessful but has been accomplished by a direct reaction of phenylpropylcyanoacetic acid with **57** in the presence of dicyclohexylcarbodiimide and hydrolysis of the resultant 4-imino-2-thiobarbiturates (**58**).[210]

Several barbiturates (e.g. **60–62**) having two chiral centers exist in both enantiomeric and diastereomeric forms. The stereoisomers of **60** were obtained by Doran[211] and exhibited different pharmacological activity.[212] Four possible optical isomers of **61** were obtained by Miyano and Toki by enzymatic and chemical methods.[193] Separations of diastereomeric mixtures were achieved by TLC and column chromatography on silica gel. The use of an adsorption of glucuronides of **61** on XAD-2 resin was advantageous for preliminary isolation of these compounds from urine.[194]

Carroll and Moreland[213] synthesized all stereoisomers of **62** (R = Et or allyl) and their 2-thio analogs from optical isomers of 3,5-dimethylvalerolactone by the method analogous to that in Scheme 7 and assigned absolute configurations. This permitted comparisons with appropriate metabolites isolated from *in vivo* studies. Two diastereoisomers of **62**(R = Et) [3'(R), 1'(R) and 3'(R), 1'(S)] have been found in the urine of dogs after administration of the R(+) enantiomer of **51**.[214]

Optical purity of barbituric acid enantiomers was determined by the isotope dilution method using 2-^{14}C-labeled compounds [215] or by an NMR

---

[209] J. Knabe, W. Rummel, H. P. Büch, and N. Franz, *Arzneim. Forsch.* **28**, 1048 (1978).
[210] J. Knabe and L. Schamber, *Arch. Pharm.* (*Weinheim, Ger.*) **315**, 878 (1982).
[211] W. J. Doran, *J. Org. Chem.* **25**, 1737 (1960).
[212] W. R. Gibson, W. J. Doran, W. C. Wood, and E. C. Swanson, *J. Pharmacol. Exp. Ther.* **125**, 23 (1959).
[213] F. I. Carroll and G. N. Moreland, *J. Med. Chem.* **18**, 37 (1975).
[214] K. H. Palmer, M. S. Fowler, M. E. Wall, L. S. Rhodes, W. J. Waddell, and B. Baggett, *J. Pharmacol. Exp. Ther.* **170**, 355 (1969).
[215] J. Knabe and V. Gradmann, *Arch. Pharm.* (*Weinheim, Ger.*) **306**, 306 (1973).

method using lanthanide shift reagents.[198,201,216] The optical rotations of pure optical isomers allowed estimations of optical purity of partly resolved enantiomers reported earlier.[202] The assignments of absolute configurations to optical isomers of barbiturates were made on the basis of correlations with compounds of known configuration after appropriate chemical conversions, ORD, CD and NMR spectra, and X-ray analysis.[193,198,209,213,217]

Many workers report different stereospecific pharmacological effectiveness, metabolism, or pharmacokinetic parameters for the stereoisomers of barbiturates investigated both *in vitro* and *in vivo*.[209,218–228] The results of these studies significantly contribute to a better understanding of the mechanisms of action of barbiturate drugs at the molecular level and make questionable Doran's statement that "whereas structural isomerism has a significant bearing on pharmacological activity in the barbituric acid series, stereoisomerism is apparently of lesser importance."[3]

III. Reactivity of Barbiturates

A. Reactions at the C-5 Atom

Barbituric acid has two active hydrogen atoms at the 5-position which can be replaced by various substituents to make the molecule biologically active.[229–233] The introduction of simple alkyl substituents into this position

---

[216] J. Knabe and V. Gradmann, *Arch. Pharm. (Weinheim, Ger.)* **310**, 468 (1977).
[217] J. Knabe and W. Wunn, *Arch. Pharm. (Weinheim, Ger.)* **312**, 973 (1979).
[218] I. K. Ho and R. A. Harris, *Annu. Rev. Pharmacol. Toxicol.* **21**, 83 (1981).
[219] G. Wahlström, *Eur. J. Pharmacol.* **59**, 219 (1979).
[220] D. A. Mathers and J. L. Barker, *Science* **209**, 507 (1980).
[221] K. Miyano and S. Toki, *Drug Metab. Dispos.* **8**, 104 (1980).
[222] F. Leeb-Lundberg, A. Snowman, and R. W. Olsen, *Proc. Natl. Acad. Sci. U.S.A.* **77**, 7468 (1980).
[223] M. K. Ticku, *Biochem. Pharmacol.* **30**, 1573 (1981).
[224] J. R. Holtman and J. A. Richter, *Biochem. Pharmacol.* **30**, 2619 (1981).
[225] J. Knabe, H. P. Büch, and J. Reinhardt, *Arch. Pharm. (Weinheim, Ger.)* **315**, 832 (1982).
[226] J. Baldauf, H. Wilbert, and H. P. Büch, *Arzneim.-Forsch.* **32**, 1281 (1982).
[227] P. Skolnick, K. C. Rice, J. L. Barker, and M. S. Paul, *Brain Res.* **233**, 143 (1982).
[228] M. Van der Graaff, N. P. Vermeulen, R. P. Joeres, and D. D. Breimer, *Drug Metab. Disp.* **11**, 489 (1983).
[229] T. Ukita, Y. Kato, M. Hori, and H. Nishizawa, *Chem. Pharm. Bull.* **8**, 1021 (1960).
[230] I. Weinryb, I. M. Michel, J. F. Alicino, and S. M. Hess, *Arch. Biochem. Biophys.* **146**, 591 (1971).
[231] A. J. Stuper and P. C. Jurs, *J. Pharm. Sci.* **67**, 745 (1978).
[232] B. Testa, *Pharm. Acta Helv.* **53**, 143 (1978).
[233] R. D. Budd, F. C. Yang, and K. O. Utley, *Clin. Toxicol.* **18**, 317 (1981).

is easy and is similar to that in malonic esters.[4] Complex substituents can be introduced, using indole derivatives. Rao and Chalmers reported that the reaction of barbituric acid with 3-isopropylaminoethyl-2-methylindole yields 5-(2'-isopropylamino)ethylbarbituric acid in the presence of piperidine[234] [Eq. (1)]. Suvorov et al.[235] described the C-5 substitution of barbituric and

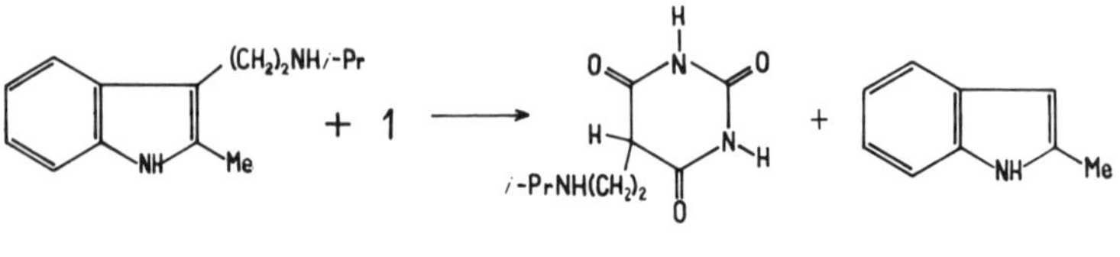

2-thiobarbituric acids (**64**; R' = Bu, Ph; X = O, S) using **63** (R = H, OCH_3) [Eq. (2)].

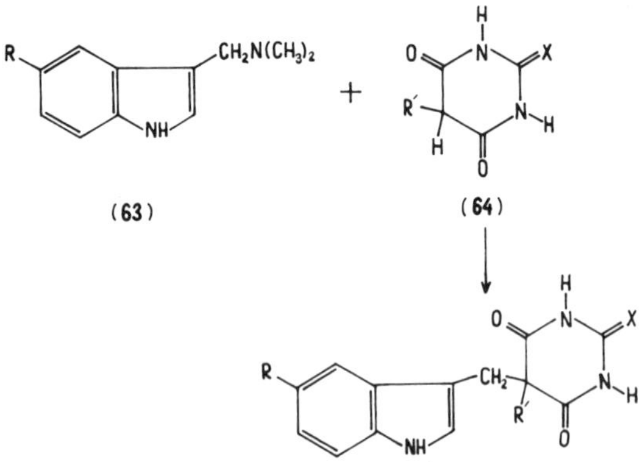

The formation of some 5-alkyl or 5-arylmethylbarbituric acids has been studied by Sekiya et al.[236-238] The barbituric acid derivatives **65** (*o*-, *m*-, *p*-substituted phenyl, benzyl, styrenyl; R^1, R^2 = H and/or methyl) containing a methylidene bond at the 5-position can be reduced by triethylammonium formate (TEAF) in high yield [Eq. (3)]. Similarly, 5-arylaminomethylene- and 5-alkylaminomethylene-substituted barbituric acids can be converted to 5-methylbarbituric acids.[238]

---

[234] R. P. Rao and A. H. Chalmers, *Indian J. Chem.* **6**, 336 (1968).
[235] N. N. Suvorov, V. S. Velezeva, V. V. Vampilova, and J. N. Gordieev, *Khim. Geterotsikl. Soedin.*, 515 (1974).
[236] M. Sekiya and C. Yanaihara, *Chem. Pharm. Bull.* **17**, 747 (1969).
[237] M. Sekiya, C. Yanaihara, and J. Suzuki, *Chem. Pharm. Bull.* **17**, 752 (1969).
[238] M. Sekiya and C. Yanaihara, *Chem. Pharm. Bull.* **17**, 810 (1969).

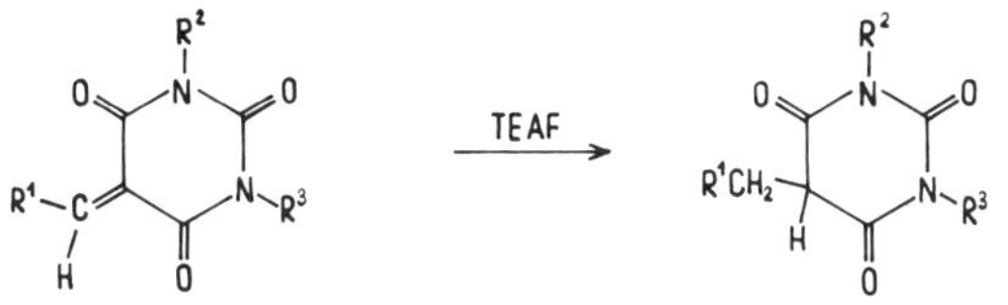

Sec. III.A]   BARBITURIC ACID   265

(3)

(65)

Ethier and Neville[239] presented the facile oxidative methylation of 5-vanillylidene and 5-benzylidenebarbituric acids in DMF with methyl iodide in the presence of Ag_2O leading to 1,3,5,5-tetramethylbarbituric acid (66). According to the authors, the oxidative cleavage with methylation followed the interaction of silver ion with the olefinic double bond.

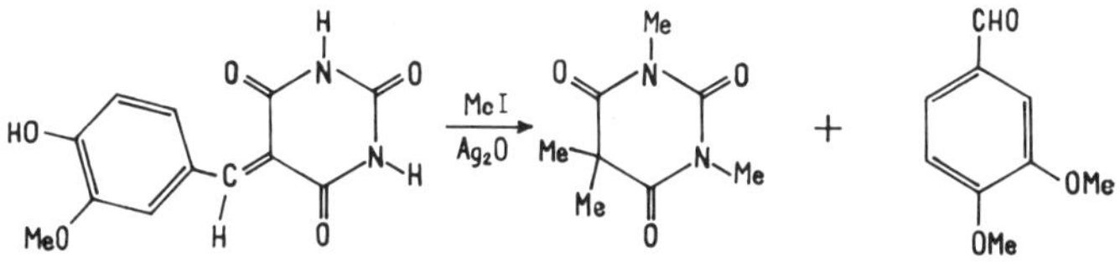

(66)

Barbiturates smoothly undergo the Mannich reaction. Although the aminomethyl group is usually introduced in the 1- and/or 3-position of the barbituric ring (see Section III,B), there are known cases of this reaction at the 5-position. Sladowska obtained the series of 1,3-dicyclohexyl-5-alkyl-5-aminomethylbarbituric acids (67; R = Et, allyl; R' = 1-piperidinyl, 1-pyrrolidinyl, 1-morpholinyl) from 1,3-dicyclohexyl-5-alkylbarbituric acids.[240]

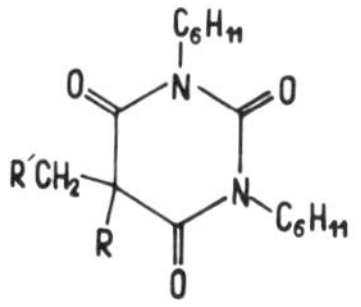

(67)

Unsubstituted, 1-mono- or 1,3-dialkyl- and/or phenyl-substituted barbituric acids (68) can be converted to 5-formimidoyl derivatives by a reaction with s-triazine[241] [Eq. (4)].

The reaction of orthoformates with CH_2-acidic molecules (diethylmalonate, malononitrile, etc.) is used to introduce the formyl (or its tautomer

[239] I. C. Ethier and G. A. Neville, *Tetrahedron Lett.* 52, 5297 (1972).
[240] H. Sladowska, *Farmaco, Ed. Sci.* 32, 866 (1977).
[241] A. Kreutzberger, *Arzneim.-Forsch. II* 28, 1684 (1978).

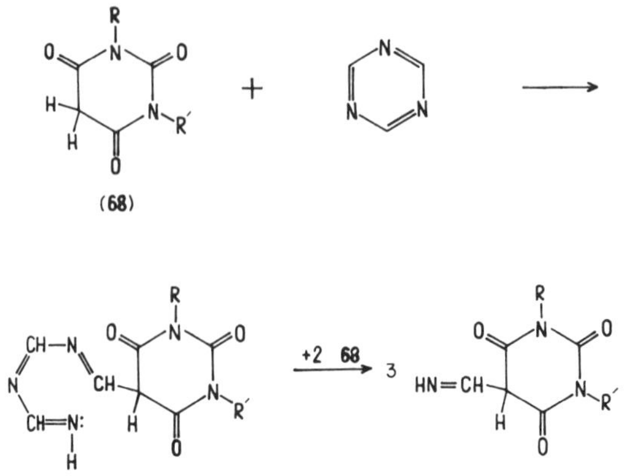

hydroxymethylene group) into the latter compounds. Wolfbeis and Junek[242] reported a simple two-step synthesis of hydroxymethylene 1,3-dicarbonyls, including barbituric acid derivatives, using orthoesters shown in Eq. (5).

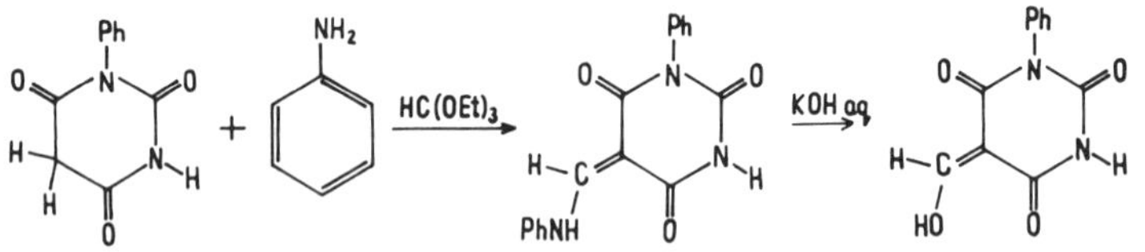

Barbituric acids condense smoothly with aldehydes under moderate conditions, yielding 5-ylidene derivatives.[236,243-245] The presence of low electron density at the C-5 position in the exocyclic C=C bond, owing to conjugation with the carbonyl groups in the 4- and 6-positions, causes the nucleophilic addition reaction in these barbiturates.

5-Arylidenebarbituric and thiobarbituric acids react with compounds containing an active methylene group under Michael reaction conditions.[246,247] Compounds **69** (R = H, p-NO$_2$, p-Cl; X = O, S) have been obtained in the reaction of 5-arylidenebarbituric or thiobarbituric acids with cyclohexanone, benzyl phenyl ketone, camphor, ethyl ester of phenylacetic acid, or nitromethane. The reaction with ethyl cyanoacetate or ethyl

---

[242] O. S. Wolfbeis and H. Junek, *Z. Naturforsch. B: Anorg. Chem., Org. Chem*, **34B**, 283 (1979).
[243] W. M. Vviedenskij, *Khim. Geterotsikl. Soedin.*, 1092 (1969).
[244] G. Haas, J. L. Stanton, A. Sprecher, and P. Wenk, *J. Heterocycl. Chem.* **18**, 607 (1980).
[245] U. Elben, H.-B. Fuchs, K. Frensch, and F. Vögtle, *Liebigs Ann. Chem.* 1102 (1979).
[246] M. El-Hashash, M. Mahmoud, and H. El-Fiky, *Rev. Roum. Chim.* **24**, 1191 (1979).
[247] A. F. Fahmy, M. M. Mohamed, A. A. Afify, M. Y. El-Kady, and M. A. El-Hashash, *Rev. Roum. Chim.* **25**, 125 (1980).

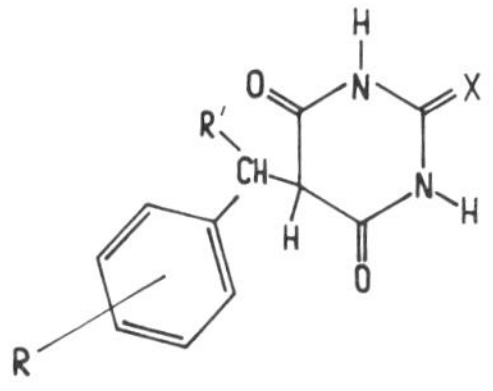

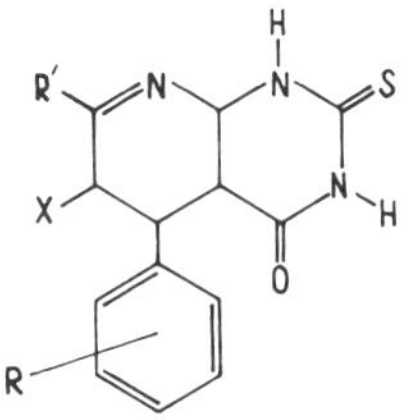

acetoacetate yielded **70** (R = H, p-NO$_2$, p-Cl; X = NH$_2$CO, CH$_3$CO; R' = NH$_2$, OH). N,N-Disubstituted barbiturates (**71**; R^1 = R^2 = Me; X = O, S) react with alkyl cyanoacetates or malononitrile in DMF in the presence of KOH to yield crystalline salts (**72**), which undergo cyclization to 1,2,3,4-tetrahydro-7H-pyrano[2,3d]pyrimidines (**73**), while N,N-unsubstituted anilinomethylenebarbituric acids (**71**, R^1 = R^2 = H) under the same conditions yield pyrimidine derivatives (**74**, R^1 = R^2 = H).[248] 5-Ylidenebarbiturates

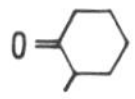

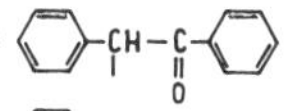

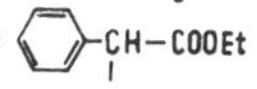

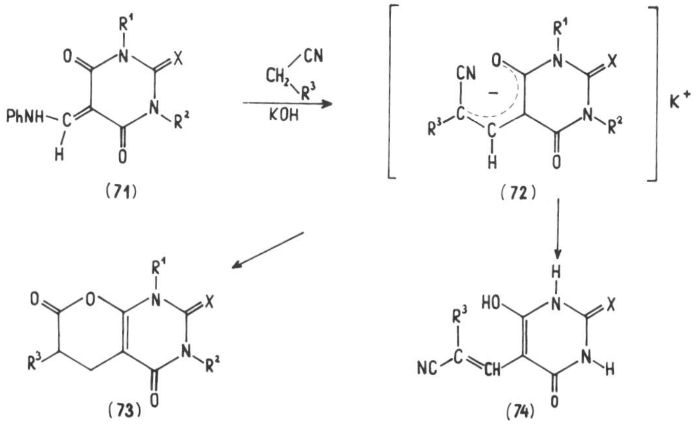

have also been postulated as intermediates formed during the condensation of barbituric acids (**75**; R^1 = H, Me, Ph; R^2 = H, Me, Ph, p-NO$_2$-C$_6$H$_4$) with aldehydes (**76**; R^3 = H, Me, Ph, m-, p-NO$_2$-C$_6$H$_4$, o-, m-, p-Cl-C$_6$H$_4$, p-MeCOO-C$_6$H$_4$, m-Me-C$_6$H$_4$, o-HO-C$_6$H$_4$) and phenylacetylene.[249] The

---

[248] H. Wipfler, E. Ziegler, and O. S. Wolfbeis, *Z. Naturforsch., B: Anorg. Chem., Org. Chem*, **33B**, 1016 (1978).
[249] K. E. Schulte, V. Weissenborn, and G. L. Tittel, *Chem. Ber.* **103**, 1250 (1970).

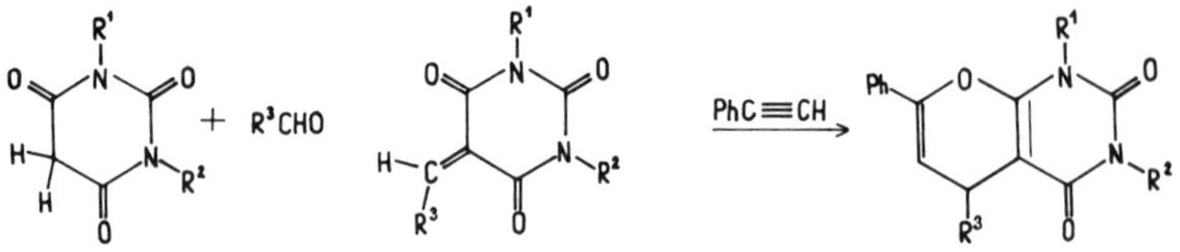

(75) (76) (77) (78)

condensed pyran systems **78** were formed by the 1,4-cycloaddition of phenylacetylene to the 5-ylidene compounds **77**. The stabilized sulfonium ylide **79** was obtained in the reaction of barbituric acid with DMSO[250,251] and was the intermediate in the formation of **80** (R = R' = H, Me).[252]

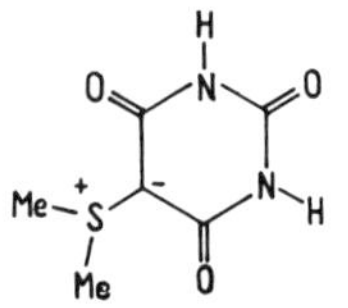

(79) 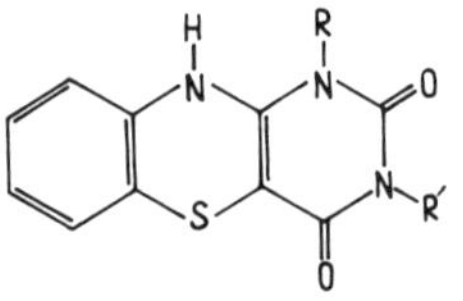 (80)

There are several examples of the addition of barbituric acid to other compounds.[253-255] Barbituric acid and its 2-thio analogue undergo nucleophilic addition to purine and its 2- and/or 8-substituted derivatives (**81**; R = H, NH$_2$; R' = H, SO$_2$Me, CF$_3$). The reaction, carried out in water in the presence of K$_2$CO$_3$, yields adducts **82** (R = H, NH$_2$; R' = H, SO$_2$Me,

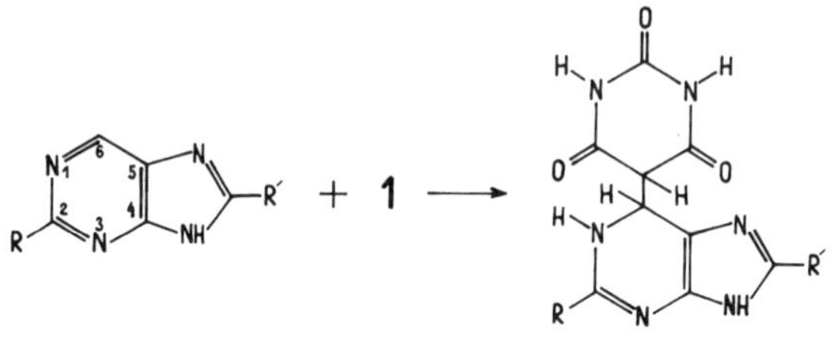

(81) (82)

CF$_3$; X = O, S).[253] As the products of addition of 5-alkyl barbituric acids (**83**; R = Me, Et, Pr, i-Pr, Bu, t-Bu) to α,β-unsaturated ketones (**84**; R' = p-OMe, p-Br, p-Cl, p-OH, p-NO$_2$), various 5,5-disubstituted derivatives (**85**) were formed. [254] The addition of barbituric or 1-substituted barbituric acids (**86**;

[250] A. F. Cook and J. G. Moffatt, *J. Am. Chem. Soc.* **90**, 740 (1968).
[251] D. Martin and H. J. Niclas, *Chem. Ber.* **102**, 31 (1969).
[252] H. Fenner, H. H. Rossler, and R. Grauert, *Arch. Pharm. (Weinheim, Ger.)* **314**, 1015 (1981).
[253] W. Pendergost, *J.C.S. Perkin I*, 2759 (1973).
[254] L. P. Zalukayev and V. L. Trostyanetskaya, *Khim. Geterotsikl. Soedin.*, 836 (1971).
[255] D. Prelicz, *Diss. Pharm. Pharmacol.* **18**, 31 (1966).

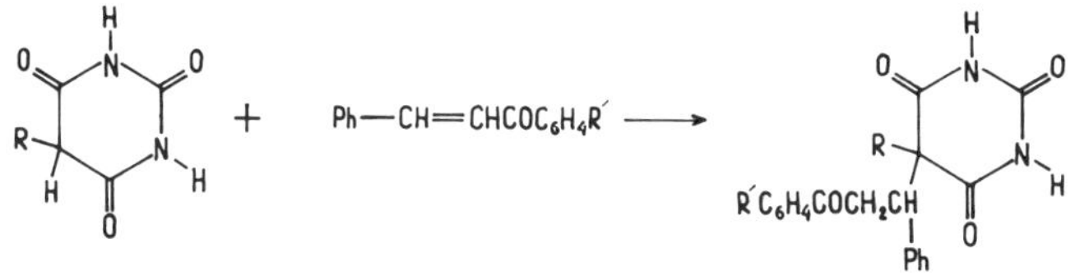

Sec. III.B]　　　　　　　BARBITURIC ACID　　　　　　　269

(83)　　　　(84)　　　　(85)

R = Me, Ph) to phenyl isocyanate leads to 5-phenylcarbamoyl barbituric acids (87).[229] Similarly, the 2-thio compounds are formed in the reaction with

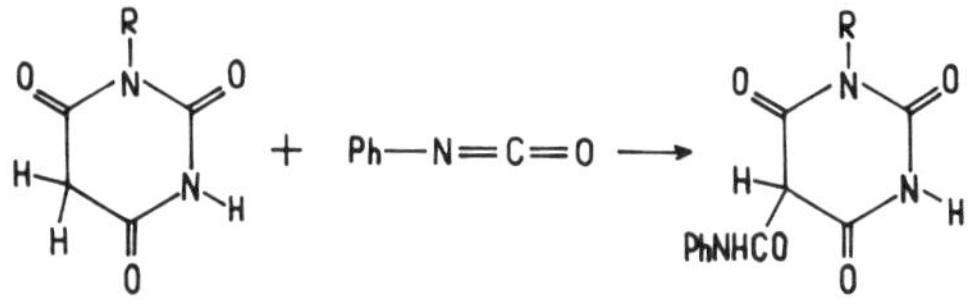

(86)　　　　　　　　(87)

isothiocyanates.[255] These reactions occur in pyridine or in ethanol–water solutions in the presence of pyridine or N,N-dimethylaniline. Recently the addition of barbituric acid to arylidene malononitrile has been described.[256] As a result of the reaction of barbituric acid with 88 (R = p-F-C_6H_4, p-Br-C_6H_4), condensed pyrano[2,3-d]pyrimidine systems (89) were obtained.

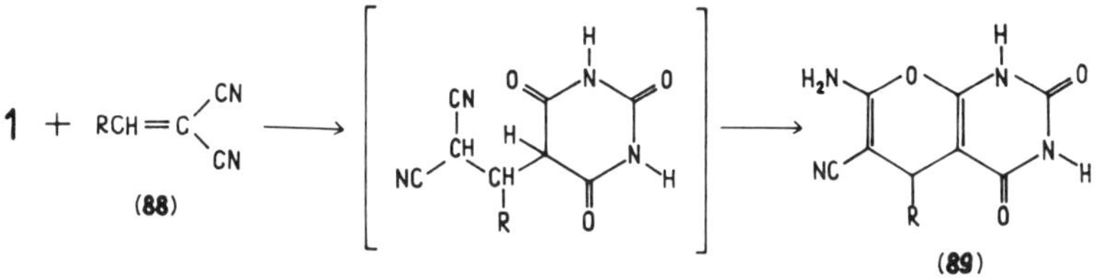

(88)　　　　　　　　　　　　　　(89)

B. Substitution at Nitrogen

The imide hydrogen atoms in barbituric acids take part in tautomerism (Section II,A) so that the alkylation of these compounds leads to a mixture of N- and O-alkyl derivatives. Levina thoroughly discussed the N-alkylation.[4]

The direct alkylation of barbituric acid preferentially occurs at the C-5 atom rather than at the 1- or 3- position. However, 1-methyl-6-methoxyuracil was obtained as the major product of the action of diazomethane on barbituric acid.[257,258] Alkylation of 1- or 5-monosubstituted barbiturates

---

[256] Yu. A. Sharanin and G. V. Klokol, *Khim. Geterotsikl. Soedin.*, 277 (1983).
[257] K. Ito, K. Nishiie, and M. Sekiya, *Yakugaku Zasshi* **90**, 188 (1970).
[258] G. A. Neville and H. W. Avdovich, *Can. J. Chem.* **50**, 880 (1972).

was found to yield the products of N- and/or O-alkylation.[257-260] On the other hand, N-methylation predominates in 5,5-disubstituted and 1,5,5-trisubstituted barbiturates.[107,138,261-264] During the methylation of 5,5-disubstituted barbituric acids (90; R = Et, Ph; R' = Et; R = R' = allyl) with diazoalkanes, N,N'-dialkyl derivatives were obtained as the major products. Small amounts of isomeric N,O-dialkyl derivatives (91 and 92) were also

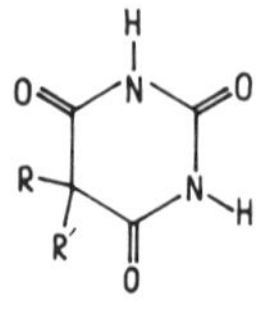

(90) 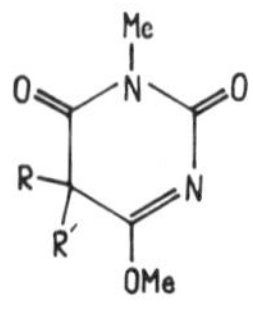   (91) 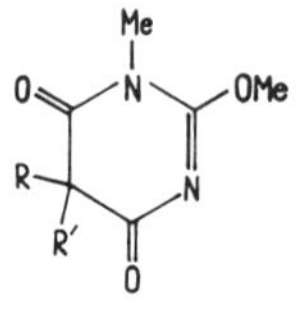   (92)

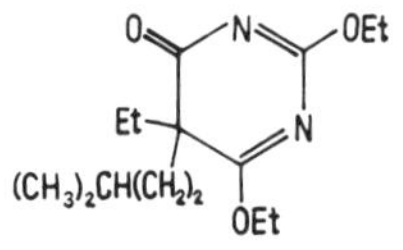

(93) 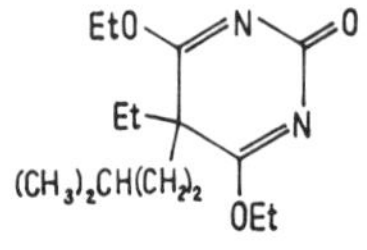   (94)

produced in these reactions. Moreover, ethylation of 90 (R = Et, R' = 3'-methylbutyl) gave additionally two O,O-dialkyl isomers (93 and 94).[262] Both mono- and disubstituted p-NO_2-benzyl derivatives have the structures of N-derivatives (95).[265] 5,5-Disubstituted thiobarbituric acid (96) can be

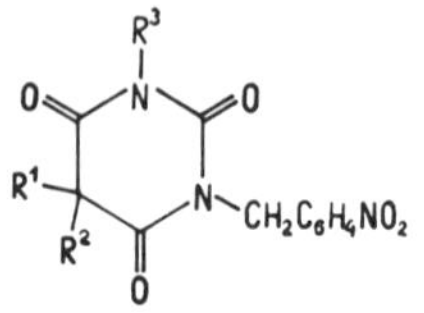

(95)

(95a) R^1 = Et, R^2 = Ph, R^3 = Me
(95b) R^1 = Me, R^2 = 1-cyclohexenyl-1'-, R^3 = Me
(95c) R^1 = Et, R^2 = Et, R^3 = H
(95d) R^1 = Et, R^2 = Et, R^3 = Me
(95e) R^1 = Et, R^2 = Et, R^3 = p-NO_2-benzyl

---

[259] R. M. Thompson, *J. Pharm. Sci.* **65**, 288 (1976).
[260] M. T. Bush and E. Sanders-Bush, *Anal. Biochem.* **106**, 351 (1980).
[261] S. Dilli and D. N. Pillai, *Aust. J. Chem.* **28**, 2265 (1975).
[262] D. J. Harvey, J. Nowlin, P. Hickert, C. Butter, O. Gansow, and M. G. Horning, *Biomed. Mass Spectrom.* **1**, 340 (1974).
[263] J. A. Vida, *Tetrahedron Lett.* **37**, 3921 (1972).
[264] J. F. Menez, F. Berthou, D. Picart, L. Bardou, and H. H. Floch, *J. Chromatogr.* **129**, 155 (1976); R. D. Budd. *Clin. Toxicol,* **16**, 189 (1980).
[265] J. Bojarski, W. Kahl, and M. Melzacka, *Rocz. Chem.* **41**, 2089 (1967); J. Bojarski, W. Kahl, and M. Melzacka, *Rocz. Chem.* **42**, 41 (1968).

Sec. III.B]   BARBITURIC ACID   271

converted to *N,S*-bis(methoxymethyl) derivatives [**97**, Eq. (6)] upon reaction with chloromethyl methyl ether. Further oxidation of **97** gave **98** in excellent yield.[266]

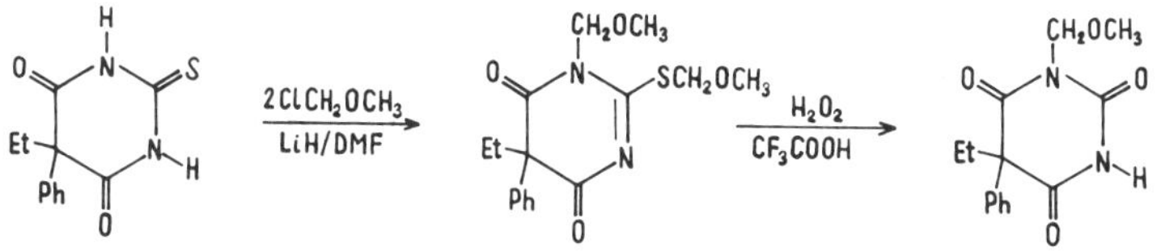

(6)

(96)   (97)   (98)

The hydrogen atom in the NH group can also be substituted by the acyl residue.[229,267–272] Reaction of benzoyl chloride with 5,5-diethyl- and 5-ethyl-5-phenylbarbituric acids proceeds in two steps. In the first step, benzoyl chloride reacts with N-nonsubstituted barbiturates, yielding N-monosubstituted derivatives (**99**; R^1 = Et, R^2 = Et, Ph; R^3 = PhCO), which then undergo the self-acylation reaction [Eq. (7)].

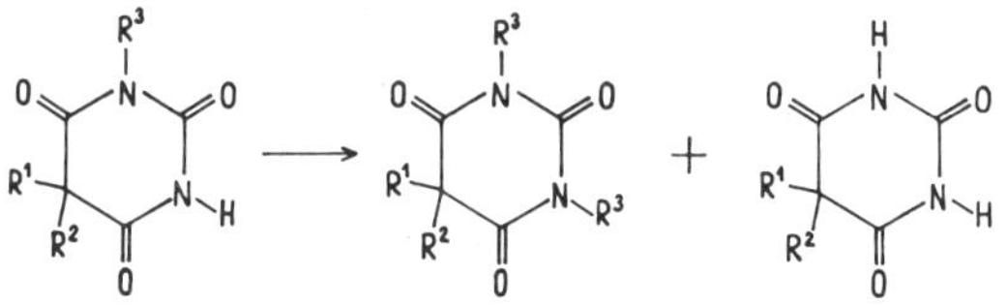

(7)

(99)

The *N*-carboxymethyl derivatives of barbituric acid can be obtained by condensation of barbiturate with appropriate chloroesters.[273] Similarly, N-substituted barbiturates containing a basic group at the nitrogen atom were formed. Treatment of 5,5-dialkylbarbituric acid with benzylaminoethyl chloride in the presence of sodium ethanolate led to *N*-(2′-benzylaminoethyl)-5,5-diallylbarbituric acid.[274]

The direct introduction of a hydroxyl group onto 1- or 3-positions in the barbituric ring system is difficult, and attempts to accomplish it by direct oxidation with hydrogen peroxide, peracetic, trifluoroperacetic, and

---

[266] J. A. Vida, *Synth. Commun.* **3**, 105 (1973).
[267] L. P. Kuliov, G. M. Stepnova, V. G. Stolarshchuk, and O. N. Neczajeva, *Zh. Obshch. Khim.* **30**, 1385 (1960).
[268] L. P. Kuliov and A. L. Shesterova, *Zh. Obshch. Khim.* **31**, 1378 (1961).
[269] J. Bojarski and W. Kahl, *Rocz. Chem.* **41**, 311 (1967).
[270] J. Bojarski, W. Kahl, M. Melzacka, and W. Pasek, *Rocz. Chem.* **42**, 337 (1968).
[271] B. Heublein, G. Heublein, B. Bockel, and H. Schutz, *Z. Chem.* **21**, 443 (1981).
[272] H. Sladowska, *Farmaco, Ed. Sci.* **35**, 60 (1980).
[273] H. Sladowska, *Farmaco, Ed. Sci.* **31**, 714 (1976).
[274] S. Bogdanova, *Rocz. Chem.* **47**, 115 (1973).

m-chloroperacetic acid were unsuccessful.[275] Nevertheless, 5,5-disubstituted-*N*-hydroxy- (**100**) and *N*,*N'*-dihydroxybarbituric acids (**101**; R = Et, Ph) were obtained indirectly by the oxidation of the appropriate amino derivatives, followed by acid hydrolysis.[275,276]

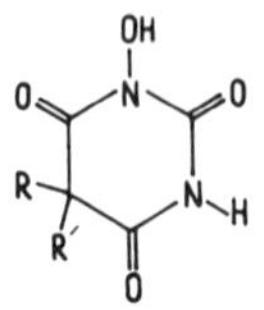

(**100**) 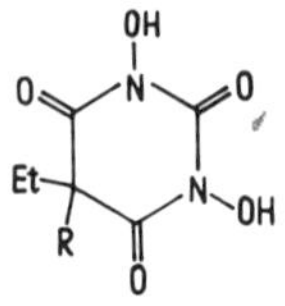   (**101**)

	R^1	R^2
(**100a**)	Me	*n*-Pr
(**100b**)	Et	Et
(**100c**)	Et	*n*-Pr
(**100d**)	Et	Ph
(**100e**)	*m*-Pr	*n*-Pr

The NH groups of barbituric acids easily take part in Mannich reactions.[277,278] 5-Ethylbarbituric acid yielded *N*,*C*-bis(piperidinemethyl) derivative **102** during aminomethylation with formaldehyde and piperidine.[279]

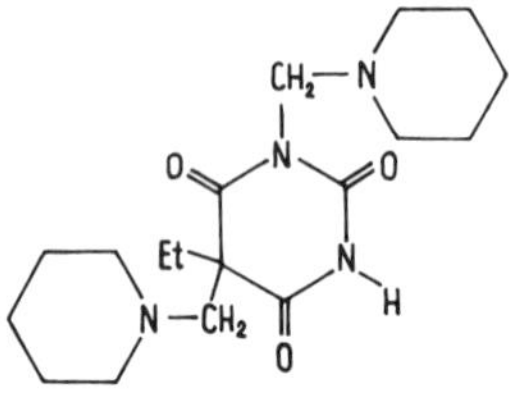

(**102**)

The second NH group in **103** (R = Me, Et; R' = Ph, cyclohexenyl) did not undergo further aminomethylation. The reaction led only to *N*-aminomethyl product **104** (R = Me, Et; R' = Ph, cyclohexenyl). This was explained by the formation of the lactim form, which renders the other hydrogen inaccessible to a second condensation reaction owing to the stabilization of the lactim moiety by an intermolecular hydrogen bond.[25]

---

[275] W. Cowden and N. Jacobsen, *Aust. J. Chem.* **31**, 2517 (1978).
[276] W. Cowden and N. Jacobsen, *Aust. J. Chem.* **35**, 795 (1982).
[277] L. Rylski, L. Senczuk, K. Falandysz, L. Konopka, and D. Zimna, *Acta Pol. Pharm.* **24**, 369 (1967).
[278] J. A. Vida and M. L. Hooker, *J. Med. Chem.* **16**, 602 (1973).
[279] H. J. Roth and R. Brandes, *Arch. Pharm.* (*Weinheim, Ger.*) **299**, 612 (1966).

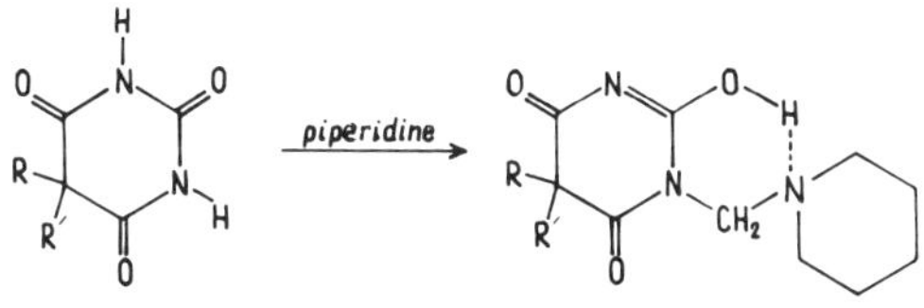

Sec. III.C]   BARBITURIC ACID   273

(103)   (104)

Danielsson and Dolby[280] showed that aminomethylation in the 1- and 3-positions was possible in both 5,5-diethyl- and 5-ethyl-5-phenylbarbituric acids when formaldehyde and morpholine were used as the reagents. Furthermore, Werner and Fritzsche[281] found that 1-aminomethylated 5,5-disubstituted barbituric acids 105 [R = piperidinyl bis(2-chloroethyl)amine] have nonchelated NH groups and that a second aminomethylation is possible at the N-3 atom.

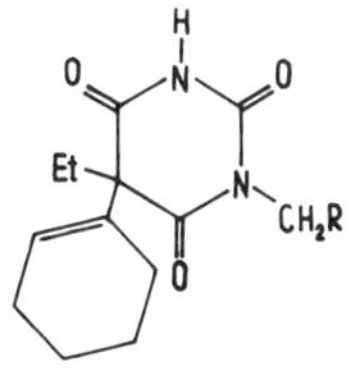

(105)

C. REACTIONS OF CARBONYL GROUPS

The carbonyl group at the 2-position shows a different character from the other carbonyl groups in the barbituric acid ring. The condensation of barbituric acid with hydrazine, phenyl-, 1,1-diphenyl-, and 2,4-dinitrophenylhydrazine led only to derivatives 106 or 107 (R = R' = Ph; R = H, R' = Ph, 2,4-di-NO_2-C_6H_4).[282]

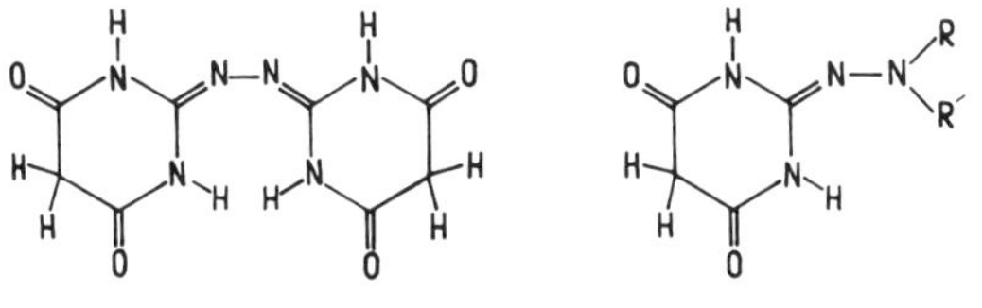

(106)   (107)

The exclusive formation of O-alkyl derivatives of barbituric acids was seldom observed, because in the course of alkylation a mixture of products is obtained (Section III,A). Nevertheless, only O-alkylation was observed in

[280] B. Danielsson and J. Dolby, *Acta. Pharm. Suec.* **1**, 91 (1964).
[281] W. Werner and H. Fritzsche, *Arch. Pharm. (Weinheim, Ger.)* **302**, 188 (1969).
[282] V. M. Vvedenskii and A. J. Zhvalevskaya, *Khim. Geterotsikl. Soedin.*, 95 (1970).

the intramolecular cyclization of 5-halogenoalkyl or N-halogenoalkyl barbituric and thiobarbituric acids[283–286] (Section III,F). Moreover, products **108** (R = H, Ph) of O-alkylation have been obtained during the action of methanol and HCl on barbituric and N-phenylbarbituric acids.[287]

(108)

During treatment with LiAlH$_4$, 5-substituted barbituric acids formed 2-oxohexahydropyrimidine derivatives.[288] N-Methylated barbiturates (**109**; R = Ph, 1-cyclohexenyl; R' = Me, Et) were reduced with LiAlH$_4$, giving **110**, while such reduction in the presence of AlCl$_3$ led to **111**.[289] It was sug-

(110)   (109)   (111)

gested that barbiturates would not be reduced with borohydrides unless they had a phenyl ring as one of the 5-substituents.[290,291] Sodium borohydride was also considered to be too mild a reducing agent for barbiturates. Nevertheless, 5-ethyl-5-phenylbarbituric acid was found to react with sodium borohydride in aqueous and organic media.[292,293] Moreover, 1,3-dimethyl-5,5-dibenzylbarbituric acid was reduced when the reaction was carried out in methanol at room temperature for 1 hr.[294] Reduction of barbiturates with sodium borohydride in various solvents was investigated in detail by

---

[283] E. E. Smissman and R. A. Robinson, *J. Org. Chem.* **35**, 3532 (1970).
[284] E. E. Smissman, R. A. Robinson, and A. J. Boyer-Matuszak, *J. Org. Chem.* **35**, 3822 (1970).
[285] E. E. Smissman and J. W. Ayres, *J. Org. Chem.* **36**, 2407 (1971).
[286] E. E. Smissman, J. W. Ayres, P. J. Wirth, and D. R. Abernethy, *J. Org. Chem.* **37**, 3486 (1972).
[287] K. Stankiewicz and B. Bobranski, *Farmaco, Ed. Sci.* **33**, 740 (1978).
[288] F. J. Marshall, *J. Am. Chem. Soc.* **78**, 3696 (1956).
[289] J. Knabe, W. Geismar, and C. Urbahn, *Arch. Pharm. (Weinheim, Ger.)* **302**, 468 (1969).
[290] H. C. Brown and B. C. Subba Rao, *J. Am. Chem. Soc.* **81**, 6433 (1959).
[291] E. E. Smissman, A. J. Matuszak, and C. N. Corder, *J. Pharm. Sci.* **53**, 1541 (1964).
[292] Y. Kondo and B. Witkop, *J. Org. Chem.* **33**, 206 (1968).
[293] L. Chafetz, T. H. Chen, and R. C. Greenough, *J. Pharm. Sci.* **62**, 512 (1973).
[294] K. H. Dudley, J. J. Davis, D. H. Kim, and F. T. Ross, *J. Org. Chem.* **35**, 147 (1979).

Sec. III.C] BARBITURIC ACID 275

Rautio.[295-297] Four main products of reduction were isolated for each barbiturate (112; R^1 = Me, Et; R^2 = Ph; $R^1 = R^2$ = allyl; R^3 = H, Me), namely, di- and tetrahydrobarbiturates 113 and 114 when one or two carbonyl groups were reduced to a secondary hydroxyl group, primary alcohols 115, which were formed via the reductive cleavage of the barbituric acid ring, and urea derivatives 116, which were formed simultaneously with the primary alcohols from the rest of the ring[298] (Scheme 8). Additionally, the reduction of 1-methyl-5,5-diallylbarbituric acid led to formation of small amounts of 117.[299]

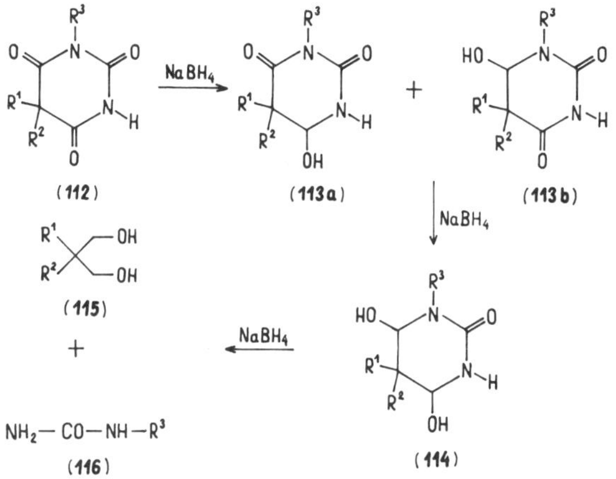

(114a) HO *trans*
(114b) HO *cis*

SCHEME 8.

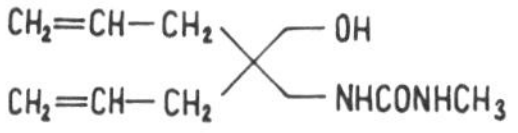

Studies on 112 (R^1 = allyl, R^2 = 2′-hydroxypropyl, R^3 = H) indicated the formation of 114a on reduction with sodium borohydride in absolute ethanol for 90 min.[300] Under the same conditions, different barbituric acids were reduced, yielding *trans*- and *cis*-4,6-diols.[298,301]

[295] M. Rautio, *Farm. Aikak.* **83**, 123 (1974).
[296] M. Rautio, *Farm. Aikak.* **84**, 143 (1975).
[297] M. Rautio, *Farm. Aikak.* **84**, 155 (1975).
[298] M. Rautio, *Ann. Acad. Sci. Fenn., Ser. A2* **178**, 1 (1976).
[299] M. Rautio, A. Hesso, and E. Rahkamaa, *Arch. Pharm. (Weinheim, Ger.)* **314**, 622 (1981).
[300] M. Rautio and K. Vuori, *Acta Chem. Scand., Ser. B* **B34**, 770 (1980).
[301] M. Rautio, *Acta Chem. Scand., Ser. B* **B33**, 770 (1979).

The treatment of 5-acylbarbituric acids (**118**; R = Pr, *i*-Pr, pentyl, pentadecyl; R' = H, Me) with sodium cyanoborohydride in acetic acid resulted in the reduction of the exocyclic carbonyl group only and the formation of the corresponding alkyl barbiturates.[302]

(**118**)

D. STABILITY OF THE PYRIMIDINE RING

Hydrolytic degradation of medicinal barbiturates and their salts is an important but undesirable process from the pharmaceutical point of view. A brief account has been given by Doran,[3] and Bojarski reviewed the literature up to 1967.[303]

The products of hydrolytic ring opening and possible pathways of their formation via hydrolytic and decarboxylation processes, summarized in Scheme 9, are dependent on the reaction conditions, i.e., temperature, pH, pressure, and the type of barbituric acid derivative. The sequence **119** → **120** → **121** involves 1–6 ring opening and is the most common degradation pathway for aqueous solutions of 5,5-di- and 1,5,5-trisubstituted barbiturates with hydrocarbon substituents.[3,304–311]

The diamides of malonic acid (**124**) confirm the 1–2 ring opening and were also detected among the degradation products.[305,308,312–315] Compounds

---

[302] C. F. Nutaitis, R. A. Schultz, J. Obaza, and F. X. Smith, *J. Org. Chem.* **45**, 4606 (1980).
[303] J. Bojarski, *Wiad. Chem.* **23**, 399 (1969).
[304] H. Aspelund and B. Eklund, *Acta Acad. Abo., Ser. B.* **21**, No. 3 (1957).
[305] F. Fretwurst, *Arzneim.-Forsch.* **8**, 44 (1958).
[306] A. J. Kapadia, J. E. Goyan, and J. Autian, *J. Am. Pharm. Assoc.* **48**, 407 (1958).
[307] J. Knabe and H. Junginger, *Dtsch. Apoth.-Ztg.* **111**, 1415 (1971).
[308] H. V. Maulding, J. Nazareno, J. Polesuk, and A. Michaelis, *J. Pharm. Sci.* **61**, 1389 (1972).
[309] M. Rautio, *Farm. Aikak.* **83**, 131 (1974).
[310] A. B. Hansen and S. S. Larsen, *Arch. Pharm. Chem., Sci. Ed.* **3**, 7 (1975).
[311] T. Hermann, *Pharmazie* **31**, 368 (1976).
[312] J. Bojarski, W. Kahl, and M. Melzacka, *Rocz. Chem.* **39**, 875 (1965).
[313] H. Aspelund, *Acta Acad. Abo., Ser. B* **29**, No. 8 (1969).
[314] J. Bojarski, *Rocz. Chem.* **47**, 1417 (1973).
[315] J. Leslie, *J. Pharm. Sci.* **68**, 639 (1979).

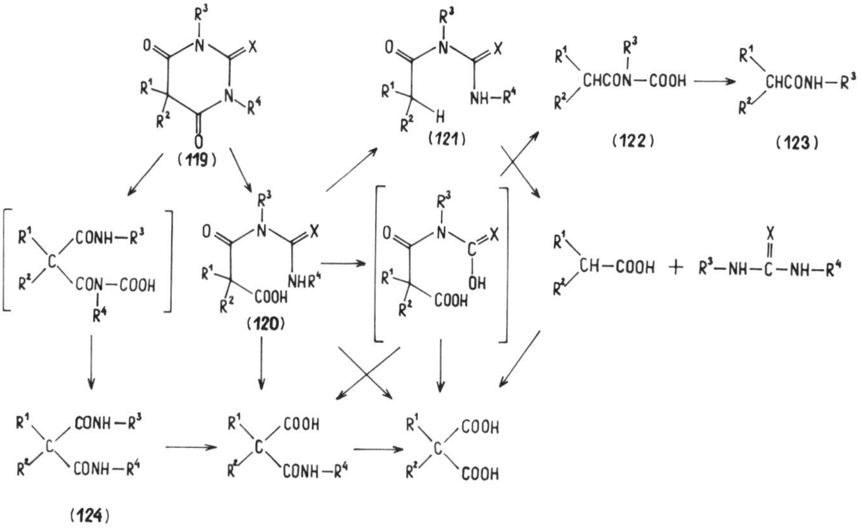

Sec. III.D] BARBITURIC ACID 277

SCHEME 9

123 were isolated when the hydrolytic degradation of barbiturates was carried out under elevated pressure.[316] N-Methyl-N-(2-phenylbutyryl)carbamic acid (125), corresponding with 122 (R^1 = Et, R^2 = Ph, R^3 = Me), was isolated among the products of alkaline hydrolysis of N-acetyl, N-benzoyl, and N-p-nitrobenzoyl derivatives of 5-ethyl-1-methyl-5-phenylbarbituric acid, and its structure was confirmed by decarboxylation to 127 and by independent synthesis of methyl ester 126 shown in Scheme 10.[317] The unusual stability of 125 was due to inter- and intramolecular hydrogen bonding as evidenced by IR studies.[318]

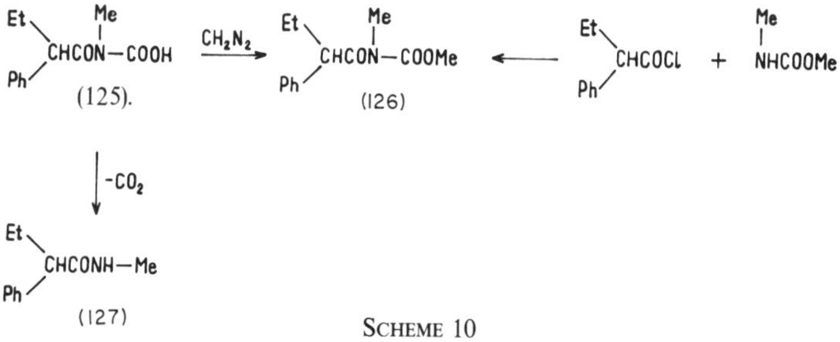

SCHEME 10

[316] M. Freifelder, A. O. Geiszler, and G. R. Stone, J. Org. Chem. 26, 203 (1961).
[317] J. Bojarski, W. Kahl, and M. Melzacka, Rocz. Chem. 40, 1465 (1966).
[318] W. Waclawek and J. Bojarski, Rocz. Chem. 43, 407 (1969).

Kinetic studies showed that undissociated and monoanionic barbiturate acid hydrolyzed with different pseudo-first-order rates at constant pH.[308,311,315,319–337] The log k–pH profiles for hydrolysis of barbiturates in neutral and alkaline media indicate that degradation takes place by hydroxide ion attack on the undissociated and monoanion species or by the kinetically equivalent attack of water on the mono- and dianion forms.[319,326,329–331,338] Several authors pointed out that this relatively simple kinetic model is complicated by some reversible processes. Eriksson presented kinetic evidence for the reversible reaction between the undissociated form of barbiturates and its tetrahedral hydroxide ion complex.[339–341] The existence of mono-, di-, and trianionic tetrahedral addition intermediates formed from the monoanionic form of barbituric acid in alkaline medium has been postulated by Khan and Khan.[336] Garrett et al. demonstrated a reversible reaction between the product of hydrolysis **120** and the parent barbiturate **119** in the case of barbital.[329] This finding was confirmed[342] and also demonstrated for other barbiturates.[308,315,330] Reversible reactions have been shown to exist in the hydrolysis of tetrasubstituted barbiturates.[314,343] Leslie reported a biexponential degradation rate of 1,3-dimethylphenobarbital in 0.02–0.32 M KOH, but at lower and higher OH$^-$ concentrations it approached first-order kinetics.[315]

[319] J. Hasegawa, K. Ikeda, and T. Matsuzawa, *Chem. Pharm. Bull.* **6**, 36 (1958).
[320] K. Ikeda, *Chem. Pharm. Bull.* **8**, 504 (1960).
[321] J. E. Goyan, Z. I. Shaikh, and J. Autian, *J. Am. Pharm. Assoc.* **49**, 627 (1960).
[322] F. Tishler, J. E. Sinsheimer, and J. E. Goyan, *J. Pharm. Sci.* **51**, 214 (1962).
[323] E. Zajta, *Acta Pharm. Hung.* **32**, 129 (1962).
[324] S. O. Eriksson and A. Holmgren, *Acta Pharm. Suec.* **2**, 293 (1965).
[325] S. O. Eriksson, *Acta Pharm. Suec.* **2**, 305 (1965).
[326] E. Goto, T. Furukawa, and S. Goto, *Arch. Pract. Pharm.* **27**, 311 (1967).
[327] S. S. Larsen and V. G. Jensen, *Dan. Tidsskr. Farm.* **43**, 47 (1969).
[328] H. V. Maulding and M. A. Zoglio, *J. Pharm. Sci.* **60**, 40 (1971).
[329] E. R. Garrett, J. T. Bojarski, and G. J. Yakatan, *J. Pharm. Sci.* **60**, 1145 (1971).
[330] J. Bojarski, *Rocz. Chem.* **48**, 619 (1974).
[331] T. Hermann, *Pharmazie* **29**, 453 (1974).
[332] H. V. Maulding, J. Polesuk, and D. Rosenbaum, *J. Pharm. Sci.* **64**, 272 (1975).
[333] A. B. Hansen and S. S. Larsen, *Arch. Pharm. Chem., Sci. Ed.* **3**, 7 (1975).
[334] T. Hermann, *Pharmazie* **31**, 618 (1976).
[335] T. Hermann and M. Piechowiak-Gwiazdowska, *Pol. J. Pharmacol. Pharm.* **28**, 205 (1976).
[336] M. N. Khan and A. A. Khan, *J.C.S. Perkin I*, 1009 (1976).
[337] S. Asada, M. Yamamoto, and J. Nishijo, *Bull. Chem. Soc. Jpn.* **53**, 3017 (1980).
[338] S. O. Eriksson, *Sci. Pharm.* **2**, 255 (1967).
[339] S. O. Eriksson and C. G. Regardh, *Acta Pharm. Suec.* **5**, 457 (1968).
[340] S. O. Eriksson, *Acta Pharm. Suec.* **6**, 139 (1969).
[341] S. O. Eriksson, *Acta Pharm. Suec.* **6**, 321 (1969).
[342] L. A. Gardner and J. E. Goyan, *J. Pharm. Sci.* **62**, 1026 (1973).
[343] J. Bojarski, J. Mokrosz, and M. Klimczak, *Pol. J. Chem.* **52**, 1457 (1978).

The hypothesis that the 1–2 hydrolytic opening of the barbiturate ring is associated with the undissociated form was not fully supported by experimental evidence.[319,322] The 1–6 ring opening is favored in all barbiturates and the 1–2 opening is of minor importance, except when some steric factors, such as a bulky C-5 substituent or the presence of some N-substituents, inhibit the 1–6 alternative.[329] The results of other studies support this view,[305,308,312,313,344] but its definite and detailed experimental proof is still lacking.

The effect of substituents on the stability of barbiturates has been widely investigated, and many qualitative and quantitative data allow some general conclusions. The steric influence of C-5 substituents exerts a dominant effect on the rate of hydrolysis. Elongation of the carbon side chain, especially a substitution of a methyl by an ethyl group, decreases the rate of hydrolysis.[307,324,325,329,345] Similar effects are produced by branching of the chain, especially at the α position,[324,325,329,346] introduction of a bromine atom into the side chain,[324,325] and enlargement of the substituent ring size.[307,329] Larger C-5 substituents shield the ring against nucleophilic attack. Thus for barbiturates where such steric shielding is impossible (5,5-dimethyl- and spirocycloalkyl-1′,5-barbituric acids), rate constants for hydrolysis are markedly higher and activation energies are lower than those for 5,5-dialkylbarbiturates.[46] Electronic effects are responsible for the enhancement of the hydrolysis rate when unsaturated or aromatic substituents are attached to the C-5 atom.[305,307,324–326,329,345]

N-Methyl-5,5-disubstituted barbiturates are less stable in alkaline medium than their nonmethylated analogs,[344,347,348] but this may not be true when the ionization of the molecule is taken into account and similar species are compared. Although the rate constants for the hydrolysis of undissociated and monoanionic forms of 5,5-diallylbarbituric acid are almost twice as high as those for 1-methyl-5,5-diallylbarbituric acid hydrolyzed under the same conditions,[341] in the case of 5,5-diethylbarbituric acid the corresponding rate constants were higher for the N-methyl analog.[329] It was claimed that the N-phenyl substituent decreases the stability of the barbituric acid ring more than the N-methyl group,[344,349] while a similar effect of the N-benzoyl group is stronger than that of the N-phenyl substituent.[350] The higher the electronegativity of the N-acyl substituent, the lower was the stability of the

[344] H. Aspelund, *Acta Acad. Abo., Ser. B* **29**, No. 7 (1969).
[345] C. Stainier, J. Bosly, F. Dutrieux, and R. Stainier, *Pharm. Acta Helv.* **38**, 587 (1963).
[346] J. M. A. Sitsen and J. A. Fresen, *Pharm. Weekbl.* **109**, 61 (1974).
[347] H. Ruhkopf, *Ber. Dtsch. Chem. Ges.* **73**, 938 (1940).
[348] H. Aspelund and K. Mäkelä, *Acta Acad. Abo. Ser. B* **23**, 1 (1963).
[349] B. Bobranski and R. Wojtowski, *Rocz. Chem.* **40**, 1707 (1966).
[350] H. Sladowska, *Rocz. Chem.* **46**, 857 (1972).

appropriate N-acyl derivative of 5-ethyl-1-methyl-5-phenylbarbituric acid.[313] The N-substituents significantly influence the ability of the barbituric acid ring with a 5-β-hydroxypropyl substituent to isomerize in alkaline medium to the corresponding allophanyl-γ-valerolactone as the result of the C—N bond fission.[351] For 1,5,5-trisubstituted derivatives (**128**) there are two possibilities of ring opening, 1–6 or 3–4, leading to the isomeric derivatives of acetylurea **129** or **130**. But ring opening yields only **130** for N-alkyl,[305,325] aryl,[350] and acyl[352,353] substituents.

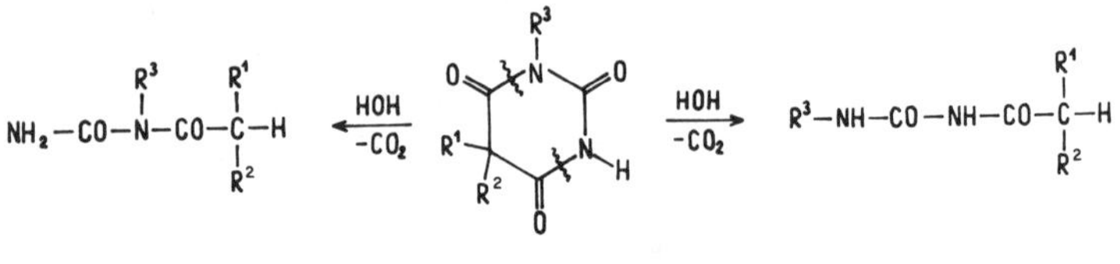

$$NH_2-CO-\underset{R^2}{\underset{|}{N}}-CO-\underset{R^2}{\underset{|}{C}}-H \quad \xleftarrow{\underset{-CO_2}{HOH}} \quad (128) \quad \xrightarrow{\underset{-CO_2}{HOH}} \quad R^3-NH-CO-NH-CO-\underset{R^2}{\underset{|}{C}}-H$$

(**129**)   (**128**)   (**130**)

R^3 = R, Ar, ArCO

For 1,3,5,5-tetrasubstituted barbiturates the 1–6 versus 1–2 or 3–4 ring opening depends on the nature of the substituents and on hydrolysis conditions.[312,313,354] 2-Thiobarbiturates are less stable than their 2-oxo analogs in an alkaline medium.[304,329,355,356] The barbituric acid ring is quite stable in an acidic medium (undissociated molecule)[4] and the only barbiturates whose ring opens under these conditions are 5-allyl-5-(2'-hydroxypropyl)barbituric acid (proxibarbal)[357–359] and 1-benzoyl-5-ethyl-5-phenylbarbituric acid (benzonal).[353]

Sayer and DePecol reported that **131** undergoes hydrolysis in aqueous acid and base solutions, yielding **132**, **133**, and **134**, respectively.[360] Ring con-

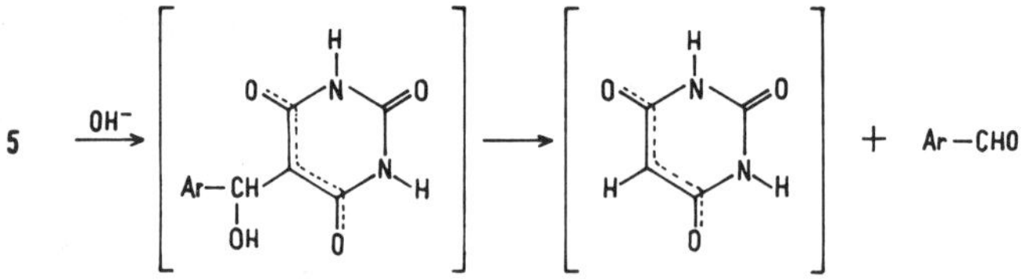

$$\mathbf{5} \quad \xrightarrow{OH^-} \quad \left[\text{Ar-CH(OH)-ring} \right] \quad \longrightarrow \quad \left[\text{ring} \right] \quad + \quad Ar-CHO$$

SCHEME 11

[351] B. Bobranski, *Farmaco, Ed. Sci.* **27**, 429 (1972).
[352] M. Melzacka and W. Kahl, *Rocz. Chem.* **45**, 1091 (1971).
[353] M. Paluchowska, L. Ekiert, K. Jochym, and J. Bojarski, *Pol. J. Chem.* **57**, 799 (1983).
[354] M. Melzacka and W. Kahl, *Rocz. Chem.* **43**, 1199 (1969).
[355] F. I. Carroll and A. Philip, *J. Med. Chem.* **14**, 394 (1971).
[356] J. Bojarski, *Rocz. Chem.* **48**, 619 (1974).
[357] R. Wojtowski, *Arch. Immunol. Ther. Exp.* **13**, 127 (1965).
[358] T. Hermann and J. Kulbiej, *Acta Pol. Pharm.* **36**, 51 (1979).
[359] T. Hermann, J. Mokrosz, and J. Bojarski, *Pol. J. Pharmacol. Pharm.* **32**, 807 (1980).
[360] J. M. Sayer and M. DePecol, *J. Am. Chem. Soc.* **99**, 2665 (1977).

tractions of substituted 5-bromobarbiturates to 4-oxazolidinone and hydantoin systems under alkaline conditions were also described, but the mechanism of these reactions is not fully elucidated.[361]

Hydrolysis of 5-arylidenebarbiturates (5) proceeds according to Scheme 11.[24] Kinetic studies revealed the isoentropic character of this reaction.[362,363]

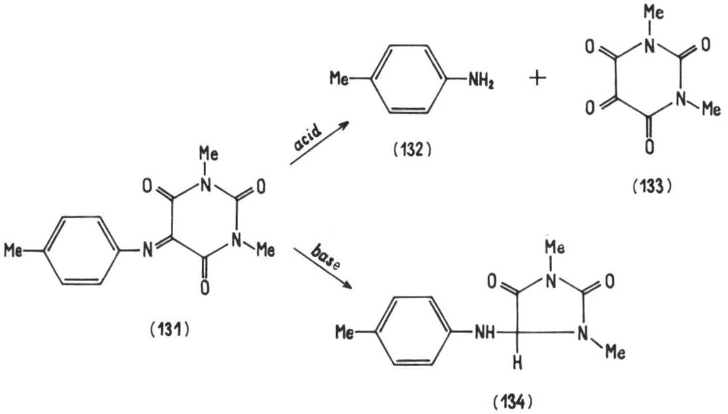

Hydrolysis was the accompanying reaction to the reduction of 5,5-diallylbarbituric acid with $NaBH_4$[309,364] and to the N-methylation of different barbiturates.[365-367] The effects of different protecting agents and solubilizers on the stability of medicinal barbiturates also deserve close attention because of their pharmaceutical importance.[335,368-375] The stability of barbiturates in frozen systems has also been investigated.[327,376,377]

---

[361] H. Aspelund, *Acta Acad. Abo. Ser. B* **27**, No. 9 (1967).
[362] B. A. Ivin, A. I. Dyachkov, I. M. Vishnyakov, and E. G. Sochilin, *Zh. Org. Khim.* **11**, 1550 (1975).
[363] A. I. Dyachkov, B. A. Ivin, I. M. Vishnyakov, and E. G. Sochilin, *Zh. Org. Khim.* **13**, 1758 (1977).
[364] M. Rautio and J. Halmekoski, *Farm. Aikak.* **83**, 83 (1974).
[365] J. Bojarski, W. Kahl, and M. Melzacka, *Rocz. Chem.* **42**, 41 (1968).
[366] R. M. Thompson and D. M. Desiderio, *Org. Mass Spectrom.* **7**, 989 (1973).
[367] R. D. Budd, *J. Chromatogr.* **237**, 155 (1982).
[368] B. M. Colombo and P. Causa, *Farmaco, Ed. Pract.* **25**, 95 (1970).
[369] B. M. Colombo, P. Primavera, and D. Lojodice, *Farmaco, Ed. Pract.* **25**, 241 (1970).
[370] S. A. Khalil, M. A. Moustafa, V. F. Naggar, and M. M. Motawi, *Can. J. Pharm. Sci.* **7**, 109 (1972).
[371] E. Pawelczyk and B. Knitter, *Acta Pol. Pharm.* **28**, 297 (1974).
[372] I. Kojo, S. Awazu, and M. Hanano, *Chem. Pharm. Bull.* **22**, 864 (1974).
[373] E. Izgü and I. Ababeyoglu, *J. Fac. Pharm. Ankara Univ.* **6**, 30 (1976).
[374] R. C. Kelly, J. C. Valentour, and I. Sunshine, *J. Chromatogr.* **138**, 413 (1977).
[375] S. Borodkin, L. Macy, G. Thompson, and R. Schmits, *J. Pharm. Sci.* **66**, 693 (1977).
[376] S. S. Larsen and V. G. Jensen, *Dan. Tidsskr. Farm.* **44**, 21 (1970).
[377] A. B. Hansen and S. S. Larsen, *Arch. Pharm. Chem., Sci. Ed.* **4**, 21 (1976).

E. Photochemical Reactions

Only a few workers have reported photochemical reactions of barbiturates. Otsuji et al. found that hydrolysis of 5,5-diethylbarbituric acid, in alkaline solution, is accelerated by UV (254 nm) light.[378] Investigations of 5,5-dialkylbarbituric acid photolysis, carried out by Bojarski and co-workers,[379,380] questioned a previous photohydrolytic mechanism[378] and implicated the photochemical ring-opening reaction within the malonyl moiety with the isocyanate derivative **135** as an intermediate[379] (Scheme 12).

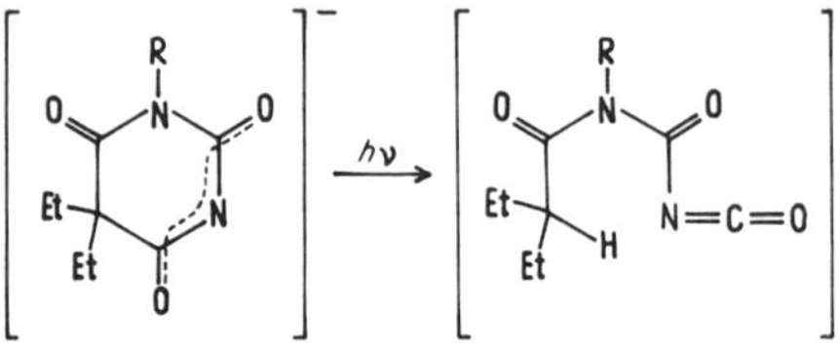

(**135**)

SCHEME 12. X = NH_2, OEt, NMe_2, NCH_2CH_3, OH (with subsequent decarboxylation)

Moreover, the isocyanate photoproduct was found to undergo hydrolysis at both the isocyanate and adjacent carbonyl groups, producing urea and amide derivatives, respectively.[381] The exact position of ring opening at the C-4—C-5 bond was evidenced by the formation of **137** during photolysis of **136** (R = Et, Ph) but the formation of hydantoin derivative **138** by an elimination

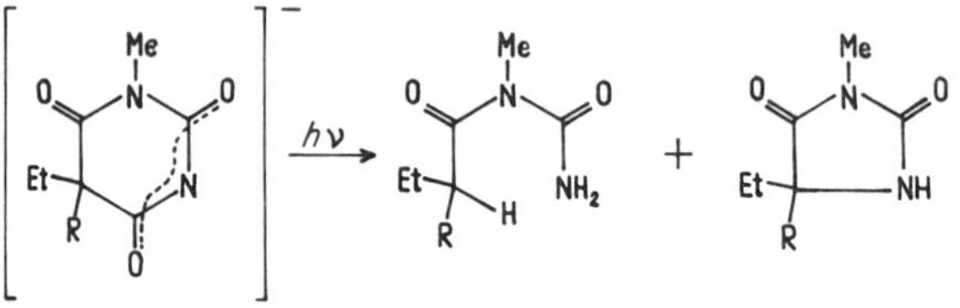

(**136**)   (**137**)   (**138**)

of the CO molecule from the barbituric acid ring was also observed.[379,382] Similar photochemical contractions of the barbituric acid ring were described for 5-methylbarbituric acid[383] and 5,5-dihydroxybarbituric acid.[384] Thus the photochemical ring opening of monoanions of barbiturates occurs at the

[378] Y. Otsuji, T. Kuroda, and E. Imoto, *Bull. Chem. Soc. Jpn.* **41**, 2713 (1968).
[379] H. Bartoń, J. Bojarski and J. Mokrosz, *Tetrahedron Lett.* **23**, 2133 (1982).
[380] H. Bartoń, J. Mokrosz, J. Bojarski, and M. Klimczak, *Pharmazie* **35**, 155 (1980).
[381] H. Bartoń, PhD. Thesis, Jagiellonian University, Kraków (1982).
[382] H. Bartoń, A. Żurowska, J. Bojarski, and W. Wełna, *Pharmazie* **38**, 268 (1983).
[383] E. Fahr, P. Fecher, G. Roth, and P. Wuestenfeld, *Angew. Chem.* **92**, 858 (1980).
[384] Y. Otsuji, S. Wake, and E. Imoto, *Tetrahedron* **26**, 4139 (1970).

ionized side of the pyrimidine ring, that is, at the side other than that found in the hydrolytic pathway (see Section III,D).

Kinetic investigations showed that the photochemical ring-opening reaction depends on the nature of the C-5 substituents, i.e., the quantum yield is the highest for short alkyl substituents and decreases for long alkyl, unsaturated, and aromatic ones. Moreover, the quantum yield of photolysis is higher for the monoanion forms than for the undissociated molecules.[385,386]

During the photolysis of barbiturates with α-branched substituents at C-5, the ring-opening reaction is accompanied by dealkylation of these substituents.[385–387] The elimination of the allyl group has also been observed in the case of photolysis of 5-allyl-5-(1'-methylbutyl)barbituric acid.[387] On the other hand, comparative studies on hydrolysis and photolysis of 5-allyl-5-(2'-hydroxypropyl)barbituric acid showed only an increase of the rate of hydrolysis by UV light with no differences in the isolated products.[359,388]

Photooxidation at the 6'-position of the cyclohexene ring was observed for 5-(1'-cyclohexen-1'-yl)-5-ethylbarbituric acid.[389,390] The influence of X-ray,[391–394] gamma ray,[395,396] and electron[397] irradiations on barbituric acid derivatives was also investigated.

F. OTHER REACTIONS

During cyclization of 5,5-di- and/or 1,5,5-trisubstituted barbiturates, formation of bicyclo compounds was observed as a result of intramolecular O-alkylation. Thus **139** (R = allyl, R' = Ph, i-Pr; R = Ph, R' = 2'-bromopropyl) formed furanopyrimidine **140**, and **139** (R = Ph, R' = 3'-bromopropyl) formed pyranopyrimidine **141**.[283,284] Similarly, cyclization of some N-(3'-halopropyl)barbiturates led to products of intramolecular O-alkylation at the 2- or 6-carbonyl group[285,286] (Eq. 8).

[385] J. Mokrosz, M. Klimczak, H. Bartoń, and J. Bojarski, *Pharmazie* **35**, 205 (1980).
[386] J. Mokrosz and J. Bojarski, *Pharmazie* **37**, 768 (1982).
[387] H. Bartoń and J. Bojarski, *Pharmazie* **38**, 630 (1983).
[388] J. Mokrosz, A. Żurowska, and J. Bojarski, *Pharmazie* **37**, 832 (1982).
[389] G. Willems, D. De Backer, R. Bouche, and C. De Ranter, *Bull. Soc. Chim. Belg.* **82**, 803 (1973).
[390] R. Bouche, M. Draguet-Brughmans, J. P. Flandre, C. Moreaux, and M. van Meerssche, *J. Pharm. Sci.* **67**, 1019 (1978).
[391] R. A. Haak and B. Benson, *J. Chem. Phys.* **55**, 3693 (1971).
[392] P. Guetierrez and B. Benson, *J. Chem. Phys.* **60**, 640 (1974).
[393] P. Guetierrez and B. Benson, *Radiat. Res.* **58**, 141 (1974).
[394] J. Hüttermann and G. Schmidt, *J. Magn. Reson.* **21**, 221 (1976).
[395] B. Benson and L. Erich, *Radiat. Res.* **86**, 411 (1981).
[396] C. Houee-Levin, M. Gardes-Albert, C. Ferradini, and J. Pucheault, *Radiat. Res.* **88**, 20 (1981).
[397] R. Boyum, E. Sagstuen, and T. Henriksen, *Radiat. Res.* **80**, 233 (1979).

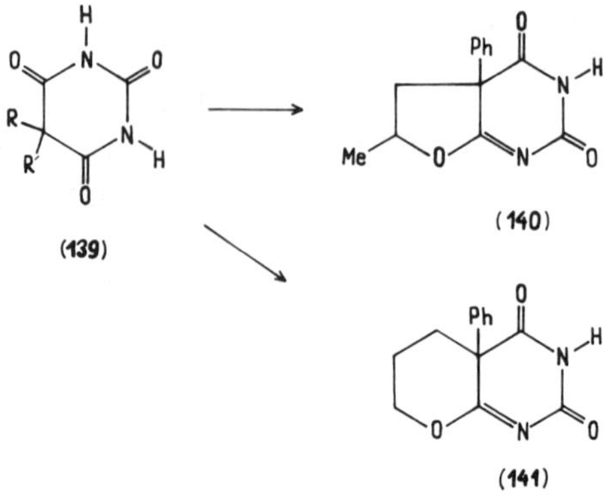

(139) → (140)

(141)

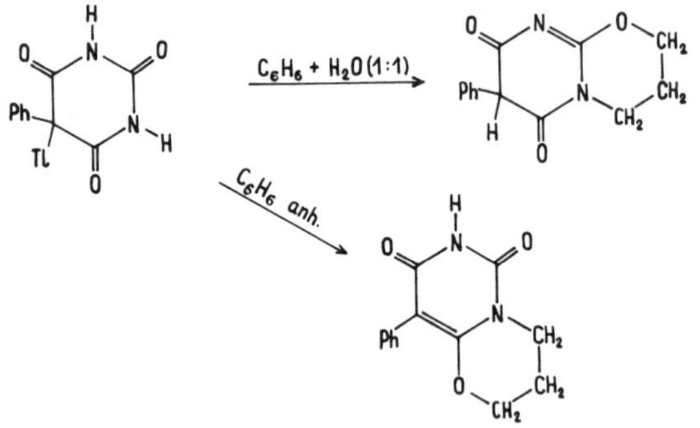

(8)

5,5-diallylbarbituric acid undergoes cyclization in the presence of Pd(II), and spiro compound **142** is formed.[398]

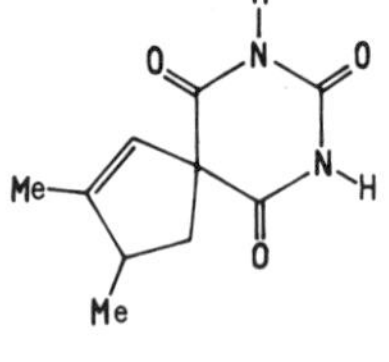

(142)

Although the reaction of barbiturates with saturated aldehydes normally yields the alkylidene compounds (Section III,A), condensation of **143** (R = R' = H, Me) with both enantiomers of **144** produced optically pure tricyclic dihydropyrans[399] (Eq. 9).

[398] R. Grigg, T. R. B. Mitchell, and A. Ramasubbu, *J.C.S. Chem. Commun.*, 666 (1979).
[399] L. F. Tietze and G. V. Kiedrowski, *Tetrahedron Lett.* **22**, 219 (1981).

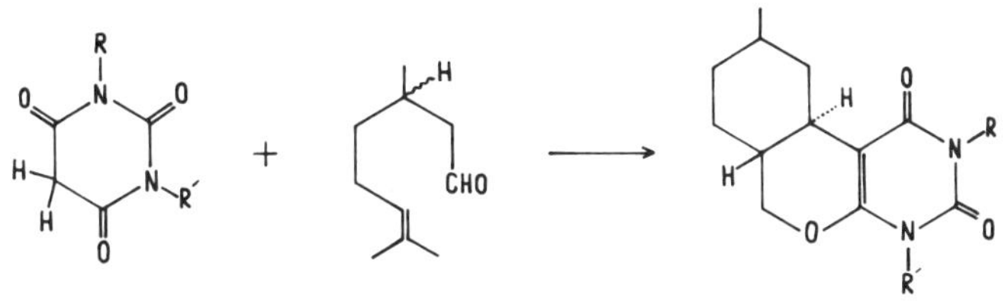

Sec. III.F] BARBITURIC ACID 285

[Scheme showing reaction of (143) + (144) → product] (9)

The treatment of 5-phenyl-5-(3'-bromopropyl)barbituric acid (**145**) with ammonium hydroxide yields 3-phenyl-3-allophanyl-2-piperidon. The reaction begins with the ammonolysis of **145**; and then the resultant primary amino function attaches to the C-6 or C-4 carbonyl group of the ring with ring opening to yield **146**, which then hydrolyzes to **147**.[400] 5-Phenyl- and 5-allyl-5-(2'-hydroxypropyl)barbituric acids undergo similar intramolecular isomerization and α-phenyl-α-allophanyl-γ-valerolactones are formed, respectively.[359,401–403]

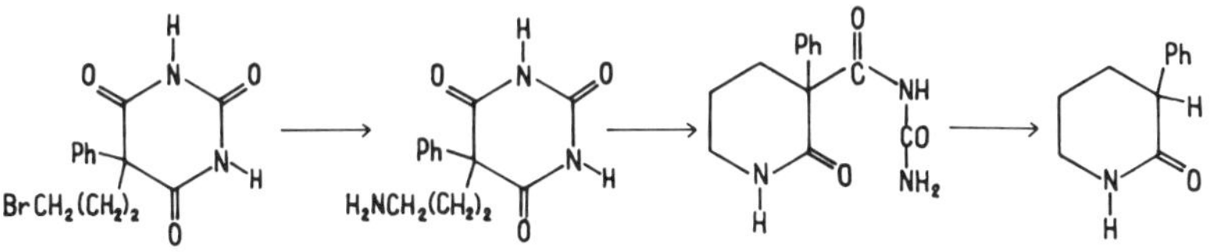

[Scheme: (145) → (146) → (147)]

N-Hydroxyacyl barbiturates isomerize to compounds of hemiketal structure. 1-Benzyl-3-α-hydroxypropionyl-5,5-diethylbarbituric acid isomerizes spontaneously to hemiketal **148**, in anhydrous THF during PdO$_2$ catalytic hydrogenation under normal pressure.[404] On the other hand, the

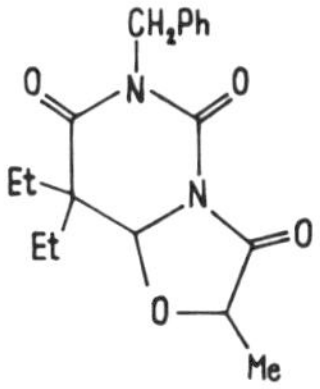

[Structure (148)]

---

[400] E. E. Smissman and P. J. Wirth, *J. Org. Chem.* **40**, 1576 (1975).
[401] B. Bobrański, D. Prelicz, L. Syper, and R. Wojtowski, *Rocz. Chem.* **37**, 795 (1963).
[402] H. Wittekind and B. Testa, *J. Chromatogr.* **179**, 370 (1979).
[403] O. Lafont, B. Lambrey, C. Jacquot, J.-R. Rapin, P. d'Athis, and D. de Lauture, *Eur. J. Med. Chem.* **14**, 143 (1979).
[404] B. Bobrański and H. Sladowska, *Rocz. Chem.* **46**, 451 (1972).

hydrogenation of 1,3-bis(α-benzyloxypropionyl)-5,5-diethyl-2-desoxybarbituric acid under the same conditions gave two isomers, which had the structures of 12-membered cyclodepsipeptides **149a** and **149b**.[405]

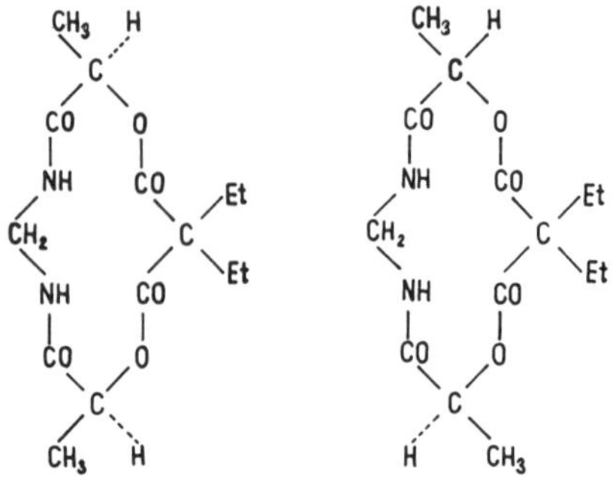

(149a) (*cis*)   (149b) (*trans*)

The isomerization with cleavage of the barbituric acid ring depends on substituents at the nitrogen atoms.[272,406,407] Introduction of a benzoyl group into the 3-position of 1-*o*-hydroxyphenyl-5,5-dipropylbarbituric acid enables this compound to isomerize to the correspondent depsipeptide (**150**), whereas 1-*o*-hydroxyphenyl-3-phenyl-5,5-dipropylbarbituric acid does not isomerize.[350,408]

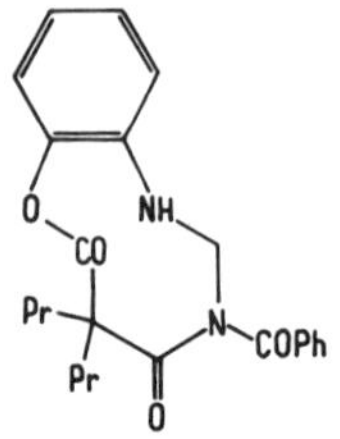

(**150**)

The ring contraction of barbiturates is observed in the course of hydrolytic degradation (Section III,D). N-Substituted barbiturates **151** ($R^1 = R^2 =$ alkyl, $R^3 = H$) yield the 5-substituted oxazolidine-2,4-dione **152**, and N,N-unsymmetrically substituted barbiturates yield a mixture of both **152** and **153**. 5-Bromobarbituric acid and its derivatives give a mixture of iminooxazolidin-4-one and hydantoin.[409]

1-Aminobarbituric acid (**154**; R = H, Me, Et, Pr, Bu) undergoes an

---

[405] B. Bobrański and H. Sladowska, *Rocz. Chem.* **46**, 459 (1972).
[406] B. Bobrański and H. Matczak, *Rocz. Chem.* **48**, 2137 (1974).
[407] H. Sladowska, *Farmaco, Ed. Sci.* **30**, 159 (1975).
[408] H. Sladowska, *Rocz. Chem.* **46**, 615 (1972).
[409] H. C. van der Plas, "Ring Transformation of Heterocycles," Vol. 2, p. 125. Academic Press, New York, 1973.

(151)   (152)   (153)

isomerization in acidic medium to 2-(5-oxo-4,5-dihydro-1H-1,2,3-triazol-3-yl) aliphatic acids (156). Reaction was found to be initiated by a simple, acid-catalyzed hydrolysis of the C-6—N-1 amide bond; then intermediate 155 cyclized to the triazole ring system.[410]

(155)   (154)   (156)

Barbiturates also form crystalline intermolecular complexes. Complexes of 5,5-diethylbarbituric acid with acetamide,[411] urea,[412] imidazole,[413] 2,4-diamino-5-(3,4,5-trimetoxybenzyl)pyrimidine,[414] and N-methyl-2-pyridone[415] have a 1:1 composition. 5-Ethyl-5-(3'-methylbutyl)barbituric acid with salicylamide forms the 1:1 complex as well.[416] The former gives a crystalline 1:2 complex with 2-aminopyridine.[417] 1-Methyl-5,5-diethylbarbituric acid also forms 1:1 complexes with dimethylacetamide and 9-ethyladenine.[418] The structures of the inclusion complexes of barbiturates with cycloheptaamylose[419] and with β-cyclodextrin[79] were investigated in detail. Moreover, there are several examples of complexation of barbiturates with Cu(II),[420,421] Co(II),[422] Ag(I),[423] lanthanides,[424] and pentachloronit-

[410] N. W. Jacobsen, B. L. McCarthy, and S. Smith, Aust. J. Chem. 32, 161 (1979).
[411] I.-N. Hsu and B. M. Craven, Acta Crystallogr., Sect. B B30, 974 (1974).
[412] G. L. Gartland and B. M. Craven, Acta Crystallogr., Sect. B B30, 980 (1974).
[413] I.-N. Hsu and B. M. Craven, Acta Crystallogr., Sect. B B30, 988 (1974).
[414] N. Shimizu, S. Nishigaki, Y. Nakai, and K. Osaki, Acta Crystallogr., Sect. B B38, 2309 (1982).
[415] I.-N. Hsu and B. M. Craven, Acta Crystallogr., Sect. B B30, 998 (1974).
[416] I.-N. Hsu and B. M. Craven, Acta Crystallogr., Sect. B B30, 843 (1974).
[417] I.-N. Hsu and B. M. Craven, Acta Crystallogr., Sect. B B30, 994 (1974).
[418] H. Loth and D. Beer, Arch. Pharm. (Weinheim, Ger.) 304, 65 (1971).
[419] A. L. Thakkar and P. V. Demarco, J. Pharm. Sci. 60, 652 (1971).
[420] J. Morvay, J. Csaszar, and V. Nikolasev, Acta Pharm. Hung. 39, 193 (1969).
[421] J. Morvay, J. Szabo, and G. Kozepesy, Acta Pharm. Hung. 39, 208 (1969).
[422] J. Morvay, J. Csaszar, and V. Nikolasev, Acta Pharm. Hung. 39, 202 (1969).
[423] A. Bult and H. B. Klasen, Pharm. Weekbl. 110, 533 (1975).
[424] M. D. Pundlik and R. W. Ramachandra, Khim. Geterotsikl. Soedin., 837 (1976).

rosylruthenate.[425] Investigations of the interaction between barbiturates and polyethylene glycol 4000 show that the hydrogen bond is responsible for the formation of these complexes.[426]

IV. Analytical Methods for Barbiturates

Analysis of barbiturates is closely associated with their use, misuse, and abuse as therapeutic agents and is of deep interest to clinical and forensic analysts and toxicologists. The results may be of value for monitoring the therapy or treatment of patients in cases of overdosing and accidental or criminal poisoning with barbiturate drugs. A review of the detection and estimation of barbituric acid derivatives, published in 1962, contains 500 references,[427] and since that time hundreds of papers have been published on these subjects. Several reviews and monographs present information about different analytical methods, their applications, and limitations for barbiturates.[428–437] Classical color reactions for the identification of barbiturates with cobalt(II) salts (Parri reaction), copper(II) salts + pyridine (Zwicker reaction), and Hg(II) salts + dithisone are still used and being improved.[438–443] Auterhoff and Streck elucidated the chemical nature of the

---

[425] S. Sueur, C. Bremard, and G. Nowogrocki, *Bull. Soc. Chim. Fr.*, 1051 (1976).
[426] B. L. Chang, N. O. Nuessle, and W. G. Haney, *J. Pharm. Sci.* **64**, 1787 (1975).
[427] G. Schmidt, in "Methods of Forensic Science" (F. Lundquist, ed.), Vol. 1, p. 373, Wiley (Interscience), New York, 1962.
[428] A. S. Curry, in "Toxicology" (C. P. Stewart and A. Stolman, eds.), Vol. 2, p. 153 Academic Press, New York, 1961.
[429] F. Amelink, "Rapid Microchemical Identification Methods in Pharmacy and Toxicology." Netherlands Univ. Press, Amsterdam, 1962.
[430] D. M. Baer, *Am. J. Clin. Pathol.* **44**, 114 (1965).
[431] E. G. C. Clarke ed., "Isolation and Identification of Drugs." Pharm. Press, London, 1975.
[432] C. C. Fulton, "Modern Microcrystal Tests for Drugs," p. 134. Wiley (Interscience), New York, 1969.
[433] K. Macek, ed., "Pharmaceutical Application of Thin Layer and Paper Chromatography," p. 213. Elsevier, Amsterdam, 1972.
[434] N. C. Jain and R. H. Cravey, *J. Chromatogr. Sci.* **12**, 228 (1974).
[435] J. Sunshine, ed., "Methodology for Analytical Toxicology," p. 34. CRC Press, Cleveland, Ohio, 1975.
[436] R. K. Müller, "Die toxikologisch-chemische Analyse," p. 239. Steinkopff, Dresden, 1976.
[437] W. Sadee and G. C. M. Beelen, "Drug Level Monitoring," p. 148. Wiley, New York, 1980.
[438] D. Bruijn, *Clin. Chim. Acta* **53**, 385 (1974).
[439] M. J. de Faubert-Maunder, *Analyst* **100**, 878 (1975).
[440] F. Pehr, *Clin. Chem. (Winston-Salem, N.C.)* **21**, 1609 (1975).
[441] A. Bult, *Pharm. Weekbl.* **111**, 157 (1976).
[442] F. Pehr, *Clin. Chem. (Winston-Salem, N.C.)* **25**, 1339 (1979).
[443] A. Osman and R. Abu-Eittah, *J. Pharm. Sci.* **69**, 1164 (1980).

color reaction of 5-phenyl-substituted barbiturates in concentrated sulfuric acid with sodium nitrite, formaldehyde, and piperonal. A π complex between the phenyl ring and the nitrosyl cation and the formation of carbocations and quinone structures without methylene or oxymethylene groups were found responsible, respectively, for these reactions.[444]

Spectroscopic methods (cf. Section II,B) have been extensively used for identification and quantitation purposes. Highly sensitive luminescence methods were particularly useful,[53,56,61-67] however, in UV absorption spectroscopy some interferences have been observed.[445-448]

Various chromatographic techniques are among the most important separation, identification, and determination methods for barbiturates, and only sample papers or reviews will be mentioned. De Zeeuw provided a survey on paper and thin-layer chromatography, collecting useful data on R_f values in different chromatographic systems.[449] In recent years, TLC techniques replaced paper chromatography to a significant degree because of high sensitivity and short time of analysis. Many improvements have been proposed in chromatographic systems for barbiturates, including the use of reversed phases, visualization reagents, and densitometry of spots for quantitative results.[450-465] Gas chromatography of barbiturates as either the free acids or their derivatives obtained by direct on-column reactions was

[444] H. Auterhoff and R. Streck, *Arch. Pharm. (Weinheim, Ger.)* **309**, 988 (1976).
[445] R. N. Gupta and P. M. Keane, *Clin. Chem. (Winston-Salem, N.C.)* **19**, 1318 (1973).
[446] P. Jatlow, *Am. J. Clin. Pathol.* **59**, 167 (1973).
[447] G. B. Schumann, *Am. J. Clin. Pathol.* **66**, 823 (1976).
[448] G. C. Khan, V. D. Goldberg, R. A. Khan, and P. T. Lascelles, *Clin. Chem. (Winston-Salem, N.C.)* **23**, 1942 (1977).
[449] R. A. de Zeeuw, *Prog. Chem. Toxicol.* **4**, 59 (1969).
[450] K. K. Kaistha and J. H. Jaffe, *J. Pharm. Sci.* **61**, 679 (1972).
[451] A. B. Hansen and S. S. Larsen, *Dan. Tiddskr. Farm.* **46**, 105 (1972).
[452] W. Dünges and H. W. Peter, in "Methods of Analysis of Antiepileptic Drugs" (J. W. A. Meijer, ed.), p. 126. Excerpta Medica, Amsterdam, 1973.
[453] M. Amin, *J. Chromatogr.* **101**, 387 (1974).
[454] U. Baeyer and D. Villumsen, *J. Chromatogr.* **115**, 493 (1975).
[455] M. Braun, G. Hanel, and G. Heinisch, *Sci. Pharm.* **43**, 168 (1975).
[456] D. B. Faber, *J. Chromatogr.* **142**, 421 (1977).
[457] J. A. Vinson, J. E. Hooyman, H. Koharched, and M. M. Holmes, *J. Chromatogr.* **140**, 71 (1977).
[458] I. C. Dijkhuis, *Clin. Chem. (Winston-Salem, N.C.)* **23**, 2171 (1977).
[459] J. C. Touchstone, M. F. Schwartz, and S. S. Levin, *J. Chromatogr. Sci.* **15**, 528 (1977).
[460] R. A. Eittah, A. Osman, and A. El-Behare, *Analyst* **103**, 1083 (1978).
[461] D. C. Fenimore, C. M. Davis, and C. J. Mayer, *Clin. Chem. (Winston-Salem, N.C.)* **24**, 1386 (1978).
[462] L. Ekiert, Z. Grodzińska-Zachwieja, and J. Bojarski, *Chem. Anal.* **24**, 1045 (1979).
[463] L. Ekiert, Z. Grodzińska-Zachwieja, and J. Bojarski, *Chromatographia* **13**, 472 (1980).
[464] W. Bress, K. Zimiński, W. M. Long, T. Manning, and L. Lubash, *Clin. Toxicol.* **16**, 219 (1980).
[465] A. H. Stead, R. Gill, T. Wright, J. P. Gibbs, and A. C. Moffat, *Analyst* **107**, 1106 (1982).

excellently reviewed by Dilli and Pillai,[466] and other workers report progress.[467–471] High performance liquid chromatography (HPLC) gained wide acceptance, and numerous studies were devoted to analytical applications of this method to barbiturates.[472–490] Popova studied analytical applications of gel chromatography to barbiturates,[491–493] and reports on simultaneous use of several chromatographic techniques were also published.[494,495] Plasma chromatography was applied as a very sensitive method for qualitative identification of barbiturates at nanogram levels.[496]

Relatively new but quickly adopted and developed analytical methods for barbiturates are the enzyme and radioimmunoassay techniques.[497] The principles and procedures have been reviewed[498–502] and comparisons between these and other analytical methods are available.[233,503–512]

[466] D. N. Pillai and S. Dilli, *J. Chromatogr.* **22**, 253 (1981).
[467] B. Kinberger, A. Holmen, and P. Wahrgren, *J. Chromatogr.* **224**, 449 (1981).
[468] A. Turcant, A. Premel-Cabic, A. Cailleux, and P. Allain, *J. Chromatogr.* **229**, 222 (1982).
[469] L. Zoccobillo, G. Cartoni, and L. Lozzi, *J. Chromatogr.* **230**, 339 (1982).
[470] D. B. Black, H. Kolasinski, E. G. Lovering, and J. R. Watson, *J. Assoc. Off. Anal. Chem.* **65**, 1054 (1982).
[471] R. D. Budd and D. F. Mathis, *J. Anal. Toxicol.* **6**, 317 (1982).
[472] R. W. Roos, *J. Pharm. Sci.* **61**, 1979 (1972).
[473] W. Dünges, G. Naundorf, and N. Seiler, *J. Chromatogr. Sci.* **12**, 655 (1974).
[474] U. R. Tjenden, J. C. Kraak, and J. F. K. Huber, *Z. Anal. Chem.* **279**, 131 (1976).
[475] U. R. Tjenden, J. C. Kraak, and J. F. K. Huber, *J. Chromatogr.* **143**, 183 (1977).
[476] B. B. Wheals and I. Jane, *Analyst* **102**, 625 (1977).
[477] C. R. Clark and J. L. Chan, *Anal. Chem.* **50**, 635 (1978).
[478] M. C. Bowman and L. G. Rushing, *J. Chromatogr. Sci.* **16**, 23 (1978).
[479] A. Hulshoff, H. Rosenboom, and J. Renema, *J. Chromatogr.* **186**, 535 (1979).
[480] P. C. White, *J. Chromatogr.* **200**, 271 (1980).
[481] H. R. Schulten and D. Kuemmler, *Anal. Chim. Acta* **113**, 253 (1980).
[482] I. S. Lurie and S. M. Demchuk, *J. Liq. Chromatogr.* **4**, 337 (1981).
[483] I. S. Lurie and S. M. Demchuk, *J. Liq. Chromatogr.* **4**, 357 (1981).
[484] R. Gill, A. A. T. Lopes, and A. C. Moffat, *J. Chromatogr.* **226**, 117 (1981).
[485] P. Jandera, J. Churacek, and D. Szabo, *Chromatographia* **14**, 7 (1981).
[486] W. F. Schmidt and L. J. Pennington, *J. Pharm. Sci.* **71**, 954 (1982).
[487] R. N. Gupta, P. T. Smith, and F. Eng, *Clin. Chem. (Winston-Salem, N.C.)* **28**, 1172 (1982).
[488] J. C. Kraak and J. P. Crombeen, *J. Liq. Chromatogr.* **5**, 273 (1982).
[489] R. H. Drost, T. A. Plamp, and R. A. A. Maes, *J. Clin. Toxicol.* **19**, 303 (1982).
[490] H. L. Levine, M. E. Cohen, P. K. Duffner, K. A. Kustas, and D. D. Shen, *J. Pharm. Sci.* **71**, 1281 (1982).
[491] V. I. Popova, V. P. Kramarenko, and A. M. Medved, *Farm. Zh. (Kiev)*, 80 (1976).
[492] V. I. Popova, *Farm. Zh. (Kiev)*, **70** (1977).
[493] V. I. Popova, *Farmatsija (Moscow)* **27**, 34 (1978).
[494] H. J. Battista and G. Machata, *Chromatographia* **1**, 104 (1968).
[495] R. Gill, A. H. Stead, and A. C. Moffat, *J. Chromatogr.* **204**, 275 (1981).
[496] D. S. Ithakissios, *J. Chromatogr. Sci.* **18**, 88 (1980).
[497] S. Spector and E. J. Flynn, *Science* **174**, 1036 (1971).
[498] J. Landon and A. C. Moffat, *Analyst* **101**, 225 (1976).

One of the problems of barbiturate analysis is the isolation of the compound from biological material.[427,428,437,449,513] Extraction is the most accepted method, but some assays take advantage of omitting the isolation and purification step. Special attention was devoted to separation and assay of particular barbiturate drugs in combination with other compounds. For example, many methods were devised for the determination of phenobarbital and hydantoin used together in anticonvulsive therapy.[514] Several procedures were proposed for the determination of barbiturates along with their metabolites in body fluids.[437]

The literature about differential thermal analysis methods for barbiturates was collected by Wollman and Braun.[515] Polarographic assays have been reported,[516,517] and other electrochemical methods were also used.[518]

This short survey is intended only to outline major developments in the analysis of barbiturates. Methods are directed toward better sensitivity, specificity, rapidity, and discriminating power to meet many demands, but mostly those of trace analysis in blood, urine, and other body fluids and tissues, often in postmortem material, and in combinations with other drugs

---

[499] R. Cleeland, J. Christenson, M. Usategui-Gomez, J. Heveran, R. Davis, and E. Grunberg, *Clin. Chem.* (*Winston-Salem, N.C.*) **22**, 712 (1976).

[500] V. P. Butler, *Pharmacol. Rev.* **29**, 103 (1977).

[501] C. B. Walberg, *Clin. Chem.* (*Winston-Salem, N.C.*) **20**, 305 (1974).

[502] V. Spiehler, L. Sun, D. S. Miyada, S. G. Sarandis, and E. R. Walwick, *Clin. Chem.* (*Winston-Salem, N.C.*) **22**, 749 (1976).

[503] M. Usategui-Gomez, J. E. Heveran, R. Cleelvsad, B. McGhee, Z. Telischak, T. Awdziej, and E. Grunberg, *Clin. Chem.* (*Winston-Salem, N.C.*) **21**, 1378 (1975).

[504] J. J. Turri, *Clin. Chem.* (*Winston-Salem, N.C.*) **23**, 1510 (1977).

[505] P. Krugers-Dagneaux, J. T. Kleineihorst, and F. M. F. Olthuis, *Pharm. Weekbl.* **112**, 1169 (1977).

[506] C. Delmonico, C. Bory, P. Baltassat, and C. Lahet, *Clin. Chim. Acta* **86**, 1 (1978).

[507] B. Quednow and M. Walther, *Zentralbl. Pharm., Pharmakother. Laboratoriumsdiagn.* **118**, 322 (1979).

[508] K. Aoki, V. Terasawa, and Y. Kurokiwa, *Chem. Pharm. Bull.* **28**, 3291 (1980).

[509] T. Nishikawa, M. Saito, and H. Kubo, *J. Pharmacobio-Dyn.* **3**, 21 (1980).

[510] I. E. Kovalev, O. Y. Polevaja, H. P. Danilova, W. I. Ivanov, W. W. Nazarov, and T. I. Uljankina, *Khim. Farm. Zh.* **15**, 14 (1981).

[511] P. Stolba and M. Lutovska, *Cesk. Farm.* **30**, 111 (1981).

[512] P. A. Mason, B. Law, K. Pocock, and A. C. Moffat, *Analyst* **107**, 629 (1982).

[513] J. Balkon, D. Prendes, and J. Viola, *J. Anal. Toxicol.* **6**, 228 (1982).

[514] P. Aymard, R. G. Boulu, and F. Sarhan, in "Drug Measurement and Drug Effects in Laboratory Health Science" (G. Siest and D. S. Young, eds.), p. 20. Karger, Basel, 1980.

[515] H. Wollmann and V. Braun, *Pharmazie* **38**, 5 (1983).

[516] M. A. Brooks, J. A. F. de Silva, and M. R. Hackman, *Anal. Chim. Acta* **64**, 165 (1973).

[517] M. Romer, L. G. Donaruma, and P. Zuman, *Anal. Chim. Acta* **88**, 261 (1977).

[518] G. D Carmack and H. Freisner, *Anal. Chem.* **49**, 1577 (1977).

In 1962 Schmidt expressed the opinion that "because of an overflooding of the market with synthetic drugs of increased structural complexity, the barbiturates, therapeutically approved for decades, are in danger of being analytically neglected."[427] Now we can conclude that his fears were unjustified with respect to the past and most probably will remain so.

V. Correlation Analysis in Barbituric Acid Chemistry

Correlation analysis may reveal and discriminate important factors underlying physicochemical properties, chemical reactivity, and/or biological activity.

Although observations of qualitative substituent effects on the hypnotic activity of these compounds date back to the work of Nielsen in 1926,[519] quantitative structure–activity relationship (QSAR) studies were initiated by Hansch during the 1960s.[520–522] Various biological activities of barbiturates strongly depend on their lipophilic character expressed by the octanol–water partition coefficients, while steric and electronic effects of substituents play a minor role. The dependence has parabolic character, and a similar type of relationship was also evidenced for some pharmacokinetic parameters of barbiturates.[523,524] Partition coefficients were sometimes substituted by chromatographic (retention index, capacity factors) or topological (molecular connectivity indices) parameters.[525–527] Some sets of biological activity data showed linear correlations with such parameters as partition coefficients,[521,522,528–534] chromatographic R_M values,[535–537] and molecular

---

[519] C. Nielsen, J. A. Higgins, and H. S. Spruth, *J. Pharmacol. Exp. Ther.* **26**, 371 (1926).
[520] C. Hansch, A. R. Steward, and J. Iwasa, *Mol. Pharmacol.* **1**, 87 (1965).
[521] C. Hansch and S. M. Anderson, *J. Med. Chem.* **10**, 745 (1967).
[522] C. Hansch, A. R. Steward, S. M. Anderson, and D. Bentley, *J. Med. Chem.* **11**, 1 (1968).
[523] T. D. Yih and J. M. van Rossum, *Biochem. Pharmacol.* **26**, 2117 (1977).
[524] T. D. Yih and J. M. van Rossum, *J. Pharmacol. Exp. Ther.* **203**, 184 (1977).
[525] J. K. Baker, D. O. Rauls, and R. F. Borne, *J. Med. Chem.* **22**, 1301 (1979).
[526] M. J. M. Wells, C. R. Clark, and R. M. Patterson, *J. Chromatogr. Sci.* **19**, 573 (1981).
[527] W. J. Murray, L. B. Kier, and L. H. Hall, *J. Med. Chem.* **19**, 573 (1973).
[528] K. Kakemi, T. Arita, R. Hori, and R. Konishi, *Chem. Pharm. Bull.* **15**, 1534 (1967).
[529] F. Helmer, K. Kiehs, and C. Hansch, *Biochemistry* **7**, 2858 (1968).
[530] H. E. Spiegel and W. W. Wainio, *J. Pharmacol. Exp. Ther.* **165**, 23 (1969).
[531] C. Hansch, *Drug Metab. Rev.* **1**, 1 (1972).
[532] S. A. Khalil, M. A. Moustafa, and O. Y. Abdallah, *Can. J. Pharm. Sci.* **11**, 121 (1976).
[533] N. Kaneniwa, M. Hiura, and S. Nagakawa, *Chem. Pharm. Bull.* **27**, 1501 (1979).
[534] A. C. Moffat and A. T. Sullivan, *J. Forensic Sci.* **21**, 239 (1981).
[535] M. C. Bonjean, J. Alary, and C. Luu Duc, *Chim. Ther.* **8**, 93 (1973).

connectivity index.[538] Henry et al. used the logarithms of the retention volume from HPLC of barbiturates to correlate their biological activity, and both parabolic and linear dependences were found.[539]

Molecular volumes[540] and fragmental constants[541] of 5,5-substituents were used by Testa for elegant multiparameter correlations of the affinity for cytochrome P-450 and the hepatic clearance of 32 barbiturates.[232] The results showed that the factors controlling these processes were lipophilic and steric in nature; only the first effect was anticipated from the same metabolic data analyzed by the single parameter correlation.[523]

Tong and Lien reported significant correlation between the inhibition of beef-heart submitochondrial NADH oxidation by 14 barbiturates and their corn oil–water partition coefficient. However, when the data for stimulation of rat liver microsomal NADPH oxidation by the same barbiturates were correlated, a good correlation was observed only after exclusion of 5-allylbarbiturates, which indicates different activity of the allyl group in this biotransformation process.[542] The relationship between molecular structure and duration of depressant effect for 160 5,5-disubstituted barbiturates was investigated by the so-called "Pattern Recognition Technique." The results were consistent with those obtained via the Hansch approach.[543]

Although QSAR studies offer a mathematical approach to explain biological activity, semiquantitative correlations of biological effects of barbiturates with their structures are still in use.[544–547]

Since the lipophilic character was found important for physiological dispositions of barbiturates, several workers investigated the relationship between partition coefficient and structural parameters (Taft's polar and steric substituents constants, number of C-atoms) of these compounds.[548,549] Partitioning of barbiturates was also proved to be a significant factor for their

---

[536] J. M. Plá-Delfina, J. Moreno, and A. del Pozo, *J. Pharmacokinet. Biopharm.* **1**, 243 (1973).
[537] J. M. Plá-Delfina, J. Moreno, J. Duran, and A. del Pozo, *J. Pharmacokinet. Biopharm.* **3**, 115 (1975).
[538] M. C. Bonjean and C. Luu Duc, *Eur. J. Med. Chem.* **13**, 73 (1978).
[539] D. Henry, J. H. Block, J. L. Anderson, and G. R. Carlson, *J. Med. Chem.* **19**, 619 (1976).
[540] A. Bondi, *J. Phys. Chem.* **68**, 441 (1964).
[541] G. G. Nys and R. F. Rekker, *Chim. Ther.* **8**, 521 (1973).
[542] G. L. Tong and E. Lien, *J. Pharm. Sci.* **65**, 1651 (1976).
[543] A. J. Stuper and P. C. Jurs, *J. Pharm. Sci.* **67**, 745 (1978).
[544] J. C. Topham, *Biochem. Pharmacol.* **19**, 1695 (1970).
[545] T. L. Breon and A. N. Paruta, *J. Pharm. Sci.* **59**, 1306 (1970).
[546] C. Ioannides and D. V. Porke, *J. Pharm. Pharmacol.* **27**, 739 (1975).
[547] S. Ozeki and K. Tejima, *Chem. Pharm. Bull.* **25**, 1952 (1977).
[548] D. J. Lamb and L. E. Harris, *J. Am. Pharm. Assoc.* **49**, 583 (1960).
[549] K. Kakemi, T. Arita, R. Hori, and R. Konishi, *Chem. Pharm. Bull.* **15**, 1705 (1967).

transfer phenomena,[550,551] solubility behavior,[552] HPLC and GLC retention data,[553-555] and activity coefficients.[556]

Both partition and adsorption phenomena of barbiturates are involved in the chromatograhic behavior of these compounds, and numerous workers report significant, linear correlations of chromatographic retention data with structural parameters such as the number of carbon atoms in the C-5 substituents,[463,557] connectivity indices,[526,538,558-562] and molecular volumes.[463] Correlations between retention indices from different chromatographic techniques have also been observed, but the results were better within the groups of closely structurally related compounds (e.g., 5,5-dialkyl- or 5-alkyl-5-allylbarbiturates).[495] The relationship between parameters of adsorption estimated from TLC of barbiturates and a number of different alkyl and alkenyl groups in the C-5 substituents was reported.[563] Nogami et al. investigated absorption of barbiturates by carbon black and related the results to biopharmaceutical data.[564,565]

The correlations between ^{13}C-NMR chemical shifts of the ring skeletal carbon and the pK_a values of 5,5-di- and 1-methyl-5,5-disubstituted barbiturates were examined by Asada and Nishijo, and a significant relationship has been found only for the C-4 and/or C-6 atoms.[125]

The logarithms of the second-order rate constants for hydrolysis of 1-methyl-5,5-disubstituted barbiturates correlated well with Newman's "six numbers" for both C-5 substituents.[329] Linear relationships between enthalpy and entropy of activation for hydrolysis of barbiturates have been described,[325,337,566,567] and multiparameter correlations of hydrolysis rate

---

[550] P. R. Byron, R. T. Guest, and R. E. Notari, *J. Pharm. Sci.* **70**, 1265 (1981).
[551] P. G. Ruifrok, *Naunyn-Schmiedeberg's Arch. Pharmacol.* **319**, 185 (1982).
[552] C. Treiner, C. Vaution, and G. N. Cave, *J. Pharm. Pharmacol.* **34**, 539 (1982).
[553] S. Toon and M. Rowland, *J. Pharm. Pharmacol.* **32**, Suppl., 8P (1980).
[554] B. Rittich and H. Dubsky, *J. Chromatogr.* **209**, 7 (1981).
[555] K. Valko and A. Lopata, *J. Chromatogr.* **252**, 77 (1982).
[556] U. Avico, E. Ciranni-Signoretti, and P. Zuccaro, *Farmaco, Sci. Ed.* **35**, 590 (1980).
[557] R. D. Budd, *Clin. Toxicol.* **17**, 375 (1980).
[558] J. S. Millership and A. D. Woolfson, *J. Pharm. Pharmacol.* **30**, 483 (1978).
[559] J. S. Millership and A. D. Woolfson, *J. Pharm. Pharmacol.* **32**, 610 (1980).
[560] J. Bojarski and L. Ekiert, *Chromatographia* **15**, 172 (1982).
[561] A. H. Stead, R. Gill, A. T. Evans, and A. C. Moffat, *J. Chromatogr.* **234**, 277 (1982).
[562] J. Bojarski and L. Ekiert, *J. Liq. Chromatogr.* **6**, 73 (1983).
[563] M. Sarsunowa and Z. Perina, *Pharmazie* **32**, 476 (1977).
[564] H. Nogami, T. Nagai, and H. Uchida, *Chem. Pharm. Bull.* **17**, 168 (1969).
[565] H. Nogami, T. Nagai, and H. Uchida, *Chem. Pharm. Bull.* **17**, 176 (1969).
[566] S. Asada, M. Yamamoto, and Y. Hamada, *Chem. Pharm. Bull.* **27**, 1663 (1979).
[567] A. I. Dyachkov, B. A. Ivin, I. M. Vishnyakov, and E. G. Sochilin, *Zh. Org. Khim.* **11**, 2182 (1975).

constants with "six number", ^{13}C-NMR chemical shifts of C-5, dissociation constants, and molecular connectivity indices have been evaluated.[337,566,568]

The Hammett-type correlation for the rate constants of 5-allyl-5-R-barbiturates has been reported by Carstensen et al. and suggested for use in stability predictions.[569] Similar correlations were also found for the hydrolysis of 5-arylidenebarbituric acids.[363,567] Linear free energy relationships have also been reported for dissociation constants,[45,51] polarographic half-wave potentials,[570] fluorescence[70] and luminescence phenomena,[71] and ^{13}C-NMR chemical shifts[129] for different classes of barbituric acid derivatives. Application of the dual substituent parameters method in LFER analysis of barbiturates, using Taft's polar and steric constants for various chemical and physicochemical properties, was also evaluated.[571]

The results of correlation analysis for barbituric acid derivatives should be interpreted carefully, bearing in mind that the majority of results deal only with compounds (sometimes a few) used in therapy, which restrict the full representation of the class. Although highly significant correlations are observed for closely related compounds, the generalization of such correlations often lowers their significance. Some problems can also be encountered with appropriate determination and/or selection of experimental data and parameters used in correlation analysis, but one can assume that this method will be further developed and will expand the range of its applicability, as indicated in more recent papers.[572,573]

VI. Closing Remarks

Although the chemistry of barbiturates has been explored for a very long time, these compounds still attract the attention of chemists, as exemplified by a recently published synthesis for **157** as a potential cerebral perfusion agent for tomography[574] or by studies of the chemical and metabolic degradation of **158**.[575] However, it is the medicinal application of barbiturate drugs and

---

[568] H. van der Veen, Ph.D. Thesis, State University of Leiden (1977).
[569] J. T. Carstensen, E. G. Serenson, and J. J. Vance, *J. Pharm. Sci.* **53**, 1547 (1964).
[570] W. Welna and J. Bojarski, *Pol. J. Chem.* **52**, 987 (1978).
[571] J. L. Mokrosz, *Arch. Pharm. (Weinheim, Ger.)* **317**, 718 (1984).
[572] S. Hada, S. Neya, and N. Funasaki, *J. Med. Chem.* **26**, 686 (1983).
[573] S. C. Basak, C. Raychaudchury, A. B. Roy, and J. J. Gosh, *Arzneim.-Forsch.* **33**, 352 (1983).
[574] P. C. Srivastava, A. P. Callahan, E. B. Cunningham, and F. F. Knapp, *J. Med. Chem.* **26**, 742 (1983).
[575] O. Lafont, C. Cave, B. Lambrey, J. P. Briffaux, C. Jacquod, and M. Miocque, *Eur. J. Med. Chem.* **18**, 163 (1983).

(157)   (158)

associated chemical problems that are the main area of current research. For instance, pharmaceutical properties (stability) of solid forms of amobarbital,[576] aminopyrine–barbital complex,[577] and thiopental-γ-cyclodextrin complex[578] were studied recently in Japan. Studies on the adsorption of phenobarbital on activated charcoal[579] and on the interaction between barbiturates and ethanol[580] are interesting from a toxicological point of view.

Structural features of the barbiturate molecule which are required for its hypnotic, convulsant, and/or anticonvulsant properties have been summarized by Vida and Gerry,[581] and numerous studies were devoted to physiological effects and pharmacological mechanisms of action of barbiturates in relation to chemical and stereochemical factors. Some recent articles and reviews are suggested on this fascinating and still incompletely elucidated subject.[582] Clearly, the history of barbiturates[583] has many paragraphs yet to be written.

---

[576] A. Ikekawa and S. Hayakawa, *Bull. Chem. Soc. Jpn.* **55**, 3123 (1982).
[577] Y. Kawashima, S. Y. Lin, M. Ueda, and H. Takenaka, *Drug Dev. Ind. Pharm.* **9**, 285 (1983); Y. Kawashima, S. Y. Lin, M. Ueda, H. Takenaka and Y. Ando, *J. Pharm. Sci.* **72**, 514 (1983).
[578] K. Arimori, R. Iwaoku, M. Nakano, Y. Uemura, M. Otagiri, and K. Uekama, *Yakugaku Zasshi* **103**, 553 (1983).
[579] K. A. Javaid and B. H. El-Mabrouk, *J. Pharm. Sci.* **72**, 82 (1983).
[580] A. H. Stead and A. C. Moffat, *Hum. Toxicol.* **2**, 5 (1983).
[581] J. A. Vida and E. H. Gerry, *Med. Chem.* **15**, 151 (1977).
[582] J. A. Richter and J. R. Holtman, *Prog. Neurobiol.* **18**, 275 (1982); F. Leeb-Lundberg and R. W. Olsen, *Mol. Pharmacol.* **21**, 320 (1982); P. R. Andrews and E. J. Lloyd, *Med. Res. Rev.* **2**, 355 (1982); P. R. Andrews and L. C. Mark, *Anesthesiology* **57**, 314 (1982); R. J. Macdonald and M. J. McLean, *Epilepsia* **23**, Suppl. 1, 7 (1982); G. A. R. Johnston and M. Willow, *Trends Pharmacol. Sci.* **3**, 328 (1982); F. Leeb-Lundberg and R. W. Olsen, *Mol. Pharmacol.* **23**, 315 (1983); S. H. Roth, K. S. Tan, and J. Tooth, *Proc. West. Pharmacol. Soc.* **26**, 243 (1983); S. Toon and M. Rowland, *J. Pharmacol. Exp. Ther.* **225**, 752 (1983).
[583] J. W. Dundee and P. D. A. McIlroy, *Anaesthesia* **37**, 726 (1982).

VII. Addendum

This Section summarizes recent important publications that have appeared in the literature. Carroll *et al.* investigated solution conformations of barbiturates,[584] while studies on crystal structures of a variety of barbiturate H-bonded complexes indicated that the pharmacologically active barbiturates are more effective as donors than as acceptors in H-bonding.[585] Structure–activity relationships,[586] transport phenomena,[587] TLC separation on different reversed-phase chromatoplates,[588,589] and chromatographic identification of 5,5-disubstituted barbiturates[590] may be of interest for both medicinal and analytical chemists. Photostability of barbiturates in the solid state,[591] properties of isomeric N and O isopropyl derivatives of 2 (R = ethyl, R^1 = ethyl or phenyl),[592,593] enzymatic chlorination[594] and gas phase synthesis[595] of 1, and chromatographic resolution of enantiomers of some trisubstituted barbiturates have also been studied.[596]

ACKNOWLEDGMENT

The authors are grateful to Professor Mitchell L. Borke for reading the manuscript and helpful comments.

---

[584] F. I. Carroll, A. H. Lewin, E. E. Williams, J. A. Berdasco, and C. G. Moreland, *J. Med. Chem.* **27**, 1191 (1984).
[585] B. M. Craven, R. O. Fox, and H.-P. Weber, *Acta Crystallogr., Sect. B* **38**, 1942 (1982).
[586] P. R. Andrews, L. C. Mark, D. A. Winkler, and G. P. Jones, *J. Med. Chem.* **26**, 1223 (1983).
[587] S. A. Khalil, O. Y. Abdallah, and M. M. Moustafa, *Pharm. Acta Helv.* **58**, 307 (1983).
[588] G. Grassini-Strazza and M. Crystalli, *J. Chromatogr.* **214**, 209 (1981).
[589] T. Cserháti, B. Bordás, L. Ekiert, and J. Bojarski, *J. Chromatogr.* **287**, 385 (1984).
[590] R. Dybowski and T. A. Gough, *J. Chromatogr. Sci.* **22**, 104 (1984).
[591] J. Reisch and MM. Müller, *Pharm. Acta Helv.* **59**, 56 (1984).
[592] J. F. Ménez, E. Gentric, J. Lauransan, and L. G. Bardou, *Can. J. Spectrosc.* **27**, 157 (1982).
[593] J. F. Ménez, M. Bourin, M. C. Colombel, C. Benamou, C. Larousse, and L. Bardou, *Eur. J. Med. Chem.—Chim. Ther.* **18**, 521 (1983).
[594] M. C. R. Fransen and H. C. van der Plas, *Recl. Trav. Chim. Pays-Bas* **103**, 99 (1984).
[595] D. J. Burinsky and R. Graham Cooks, *Org. Mass Spectrom.* **18**, 410 (1983).
[596] G. Blaschke and H. Markgraf. *Arch. Pharm.* **317**, 455 (1984).

Heterocyclic β-Enamino Esters, Versatile Synthons in Heterocyclic Synthesis

HEINRICH WAMHOFF

Institut für Organische Chemie und Biochemie der Universität Bonn, Bonn, Federal Republic of Germany

I. Introduction	300
A. General Remarks. Reviews	300
B. The Concept of "Heterocyclic β-Enamino Esters and Nitriles"	301
C. Heteroaromatic Enamino Esters (Nitriles)	301
II. Synthetic Approaches	301
A. General	301
B. Partially Saturated Heterocyclic β-Enamino Esters by Rearrangement Reactions	302
C. Partially Saturated Heterocyclic β-Enamino Esters by Direct Syntheses	302
D. Miscellaneous Syntheses	304
E. Heteroaromatic β-Enamino Esters	305
III. Structure and Spectral Properties	309
A. Protonation	309
B. Structures from UV, IR, and ^{1}H-NMR Data	310
C. ^{13}C-NMR Data and Electron Distribution in the Enaminocarbonyl Chromophore	311
D. The Push–Pull Stabilization	315
IV. General Considerations of Chemical Reactivity	316
V. Electrophilic Attacks on the 2-Amino Group (or the Carbon-3 Atom). Synthesis of Polynuclear Heterocyclic Systems	318
A. With Isocyanates, Isothiocyanates, and Sulfenyl Chlorides	318
B. Reactions with Imido Esters, O-Alkyllactim Ethers, Formamide, and Formaldehyde	324
C. Heterocondensed 1,3-Oxazinones from 2-Acylamino Derivatives and Dihalotriphenylphosphoranes	330
D. Heterocondensed [1,2-a]Pyrimidines	333
E. Electrophilic Addition to Carbon-3	337
F. Miscellaneous Reactions	339
VI. Nucleophilic Attacks of Amines and Diamines	343
VII. Cycloadditions and Reactions with Acetylenic Esters	347
VIII. 2-(Triphenylphosphorylidenamino) Esters. The Cycloaddition–Ring Enlargement Sequence	351
A. Formation and Properties	351
B. Polar Cycloaddition–Ring-Enlargement Reactions with Acetylenic Esters	352
C. Subsequent Reactions	354
D. Miscellaneous Reactions Employing Phosphorus Compounds	355
IX. Photochemistry	355

X. Heterocyclic β-Enamino Nitriles 357
   A. Syntheses 357
   B. Properties and Comparison with Enamino Esters 360
   C. Heterocyclization Reactions 360
   D. Dimroth Rearrangements to Thieno- and Furo[2,3-d]pyrimidophanes . . . 363
   E. Miscellaneous Reactions of Heteroaromatic β-Enamino Nitriles 365
XI. Conclusion and Outlook 367

I. Introduction

A. General Remarks. Reviews

Many condensed heterocyclic systems, especially when linked to a pyrimidine ring, play an important role in biologically active compounds, e.g., as potential medicinal agents,[1] in the plant protection area[1a], and in cancer and virus research.[1b] This article illustrates the rather versatile role of a certain class of compounds, which are ideally suited for the purpose of serving as multifunctional building units for new and promising compounds in one or two easy reaction steps.

Although in the last decade the chemistry and wide applicability of enamines in organic synthesis[2] have been intensively studied and belong to the classical arsenal of preparative organic chemistry, little attention has been paid to the special class of heterocyclic β-enamino esters and nitriles. An exception is a survey on heterocyclic enamines.[3] There also exists one comprehensive review of heterocyclic β-enamino esters[4] and another approach on annelation of a pyrimidine ring to an existing ring.[5]

---

[1] R. K. Robins, P. C. Srinivasta, G. R. Revankar, T. Novinson, and J. P. Miller, *Lect. Heterocycl. Chem.* **6**, 93 (1982); H. Rapoport, *ibid.* **4**, 47 (1978); W. J. Irwin and D. G. Wibberley, *Adv. Heterocycl. Chem.* **10**, 149, esp. 197 (1969).

[1a] K. H. Büchel, ed., "Chemistry of Pesticides." Wiley (Interscience), New York, 1983.

[1b] H. E. Skipper, R. K. Robins, J. R. Thomson, C. C. Cheng, R. W. Brockman, and F. M. Schmahl, Jr., *Cancer Res.* **17**, 579 (1957) [*CA* **52**, 4850 (1958)]; A. D. Broom, J. L. Shim, and G. L. Anderson, *J. Org. Chem.* **41**, 1095 (1976); E. Hayashi *et al.*, *Yakugaku Zasshi* **98**, 1560 (1978) [*CA* **90**, 15a111h (1979)]; J. Ashby and B. M. Elliott, *in* "Comprehensive Heterocyclic Chemistry" (A. R. Katritzky and C. W. Rees, eds.), Vol. 1, p. 111. Pergamon, Oxford, 1984; J. K. Landquist, *ibid*, p. 144.

[2] A. G. Cook, "Enamines, Synthesis, Structure, and Reactions." Dekker, New York, 1969; M. E. Kuehne, *Synthesis*, 510 (1970); S. Hünig and H. Hoch, *Fortschr. Chem. Forsch.* **14**, 235 (1970); S. F. Dyke, "Chemistry of Enamines." Cambridge Univ. Press, London and New York 1973; P. W. Hickmott, *Tetrahedron* **38**, 1975, 3363 (1982).

[3] K. Blaha and O. Červinka, *Adv. Org. Chem.* **4**, 1 (1963).

[4] H. Wamhoff, *Lect. Heterocycl. Chem.* **5**, 61 (1980).

[5] A. Albert, *Adv. Heterocycl. Chem.* **32**, 1 (1982).

B. The Concept of "Heterocyclic β-Enamino Esters and Nitriles"

If an enaminocarbonyl or enaminonitrile moiety exists as part of a partially saturated heterocyclic ring system, it is called a "heterocyclic β-enamino ester" or "nitrile," to acknowledge that these cyclic variants possess an enamino double bond located between an ester or nitrile group and an adjacent heteroatom (see Scheme 1). This concept covers a great deal of a rather heterogenous number of different heterocycles, but all have the common structural feature and are closely related to both enaminocarbonyl and enaminonitrile compounds and to heterocycles.

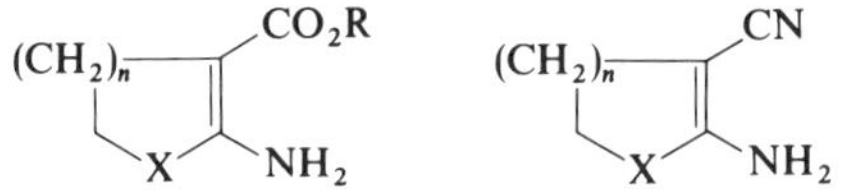

(1)   (2)

SCHEME 1

C. Heteroaromatic Enamino Esters (Nitriles)

If the heterocyclic nucleus is fully aromatic, the compounds are called "heteroaromatic enamino esters" and "nitriles" (see Scheme 2). Because of the aromatic ring there might be a modified electronic situation with respect to the enaminocarbonyl chromophore (cf. Section III).

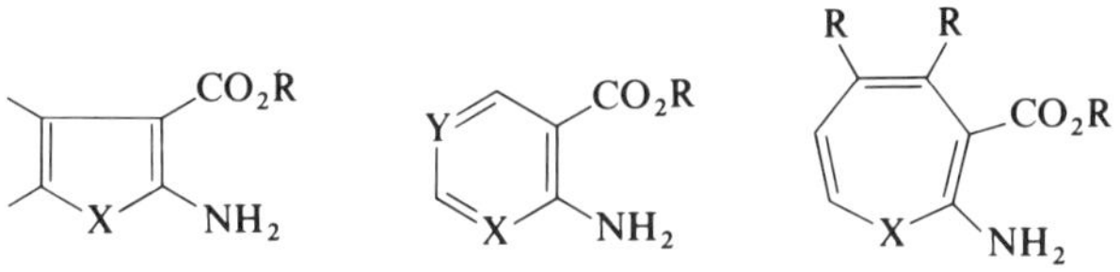

(3)   (4)   (5)

SCHEME 2

II. Synthetic Approaches

A. General

While most classical enamines and open-chain enamino esters[2] are obtained by condensation, e.g., of a secondary amine with a keto compound, heterocyclic β-enamino esters can be obtained only by special reactions as, e.g., rearrangement or direct syntheses. Some of these are depicted in this section.

B. Partially Saturated Heterocyclic β-Enamino Esters by Rearrangement Reactions

Heterocyclic β-enamino esters which are dihydrofurans,[6] dihydro-4H-pyrans,[6] and dihydro-4H-thiopyrans[7] are easily accessible in a special case of the so-called "acyl lactone rearrangement."[8] Starting from the appropriate α-cyano-γ- and δ-lactones and thiolactones (6), catalytic amounts of alkoxide ions lead to ring cleavage, rearrangement, and ring closure of the terminal X⁻ group with the cyano group (7, Scheme 3). Owing to the ring cleavage mechanism of styrene oxide—via rearrangement of the intermediary α-cyanolactone—a 4-phenyl-substituted 2-amino furan-3-carboxylic ester (9) is obtained, as shown by ^{1}H-NMR.[9,10]

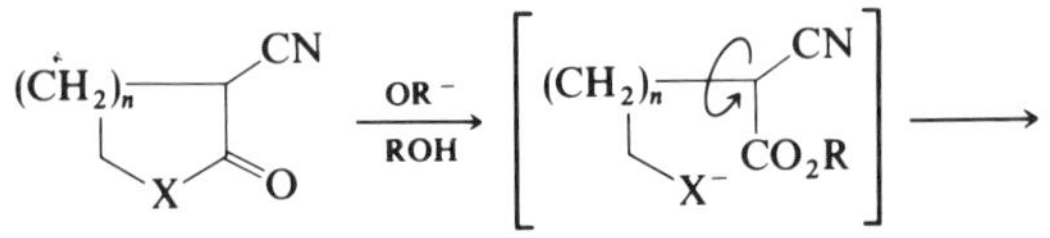

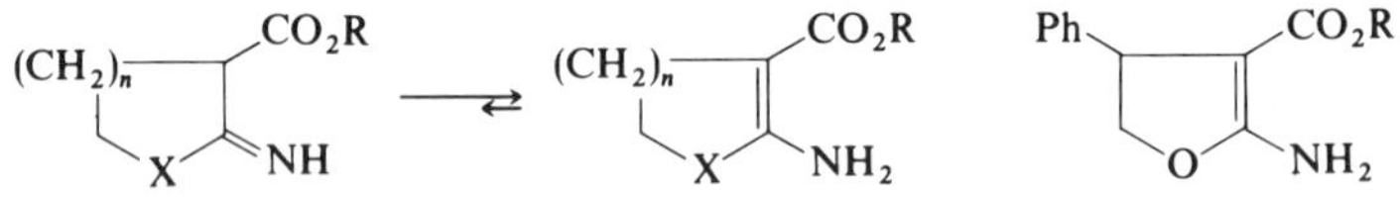

SCHEME 3

C. Partially Saturated Heterocyclic β-Enamino Esters by Direct Syntheses

While oxiranes afford with sodium ethylcyanoacetate only α-cyano-γ-butyrolactone, i.e., the insertion of the hydroxyethyl moiety into ethyl cyanoacetate,[6] thiiranes give with sodium cyanoacetate, in a one-step

---

[6] S. E. Glickman and A. C. Cope, *J. Am. Chem. Soc.* **67**, 1012 (1945); F. Korte and K. Trautner, *Chem. Ber.* **95**, 281, 295 (1962); P. L. Pacini and R. G. Ghirardelli, *J. Org. Chem.* **31**, 4133 (1966).
[7] F. Korte and H. Wamhoff, *Chem. Ber.* **97**, 1970 (1964).
[8] F. Korte and K. H. Büchel, *Angew. Chem.* **71**, 709 (1959); in "Newer Methods of Preparative Organic Chemistry" (W. Foerst, ed.), Vol. 3, p. 199. Academic Press, New York, 1964; H. Wamhoff and F. Korte, *Synthesis*, 151 (1972).
[9] E. Dradi and V. Vecchietti, *Chim. Ind.* (*Milan*) **53**, 1040 (1971) [*CA* **76**, 59336m (1972)]; for more examples, cf. Campaigne et al.[10]
[10] E. Campaigne, R. L. Ellis, M. Bradford, and J. Ho, *J. Med. Chem.* **12**, 339 (1969).

procedure, ethyl 2-amino-4, 5-dihydrothiophene-3-carboxylate (**10**) and some polymeric side products.[11] This simple procedure can be also transferred to 1-tosylaziridine, which is cleaved by sodium alkyl cyanooacetates to give 2-amino-1-tosyl-Δ^2-pyrroline-3-carboxylates (**11**)[12] (Scheme 4).

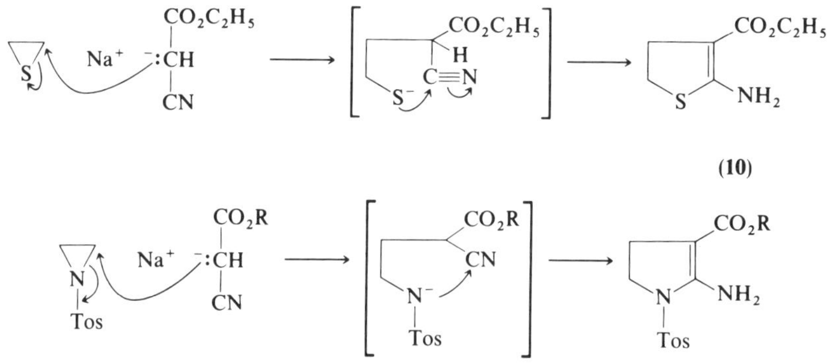

(**10**)

$R = CH_3, C_2H_5$ (**11**)

SCHEME 4

Ethyl 2-amino-1,4,5,6-tetrahydropyridine-3-carboxylate (**13**) is obtained under mild conditions by mutual treatment of 3-carbethoxy-2-piperidinone (**12**) with triethyloxonium tetrafluoroborate and then ammonia.[13] The synthesis of β-enamino esters such as 2-amino-1-aryl-3-carbethoxypyrrolin-4-one (**15**) possessing a carbonyl function in the partially saturated ring is accomplished by cyclization of ethyl γ-chloro-α-cyano-acetoacetate (**14**) with arylamines.[14] As ^{1}H-NMR spectra show, **15** exists predominantly in the β-dicarbonylenamine form (Scheme 5). Furthermore, furanones of the general

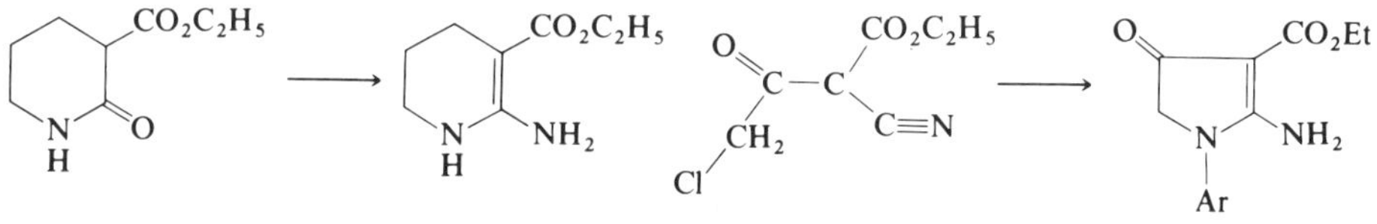

(**12**) (**13**) (**14**) (**15**)

SCHEME 5

[11] H. R. Snyder and W. Alexander, *J. Am. Chem. Soc.* **70**, 217 (1948).
[12] J. Lehmann and H. Wamhoff, *Synthesis*, 546 (1973).
[13] H. Wamhoff and L. Lichtenthäler, *Synthesis*, 426 (1975).
[14] K. J. Boosen, *Helv. Chim. Acta* **60**, 1256 (1977).

type **17** are accessible by base-catalyzed cyclocondensation of ethyl 4-chloroacetoacetate (**16**) with isocyanates[15] (Scheme 6).

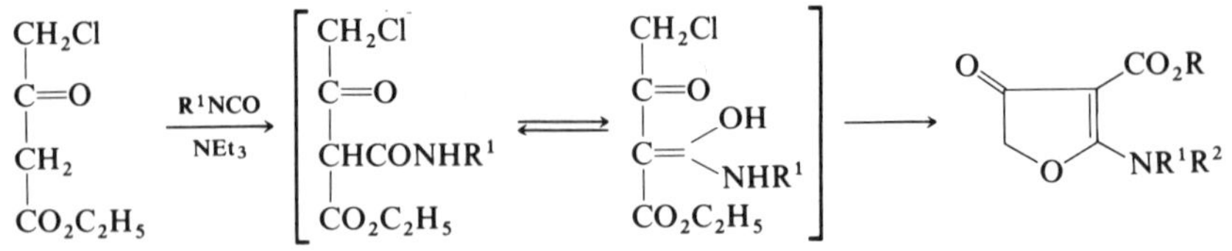

SCHEME 6

D. MISCELLANEOUS SYNTHESES

Extending the Hantzsch synthesis and in the course of producing new 4-aryl-1,4-dihydropyridines related to the powerful calcium antagonist Nifedipine (Adalat),[16] arylalkylideneacetoacetates, ketones, and malonic esters (**18** and **21**) have been treated with ketenaminals (**19**) to give 2-amino-1,4-dihydropyridine-3-carboxylic esters **20** and **22**[17] (Scheme 7). This reaction was also applied to 4-arylalkylidene-2-methyl-1,3-oxazolinone-5.[18]

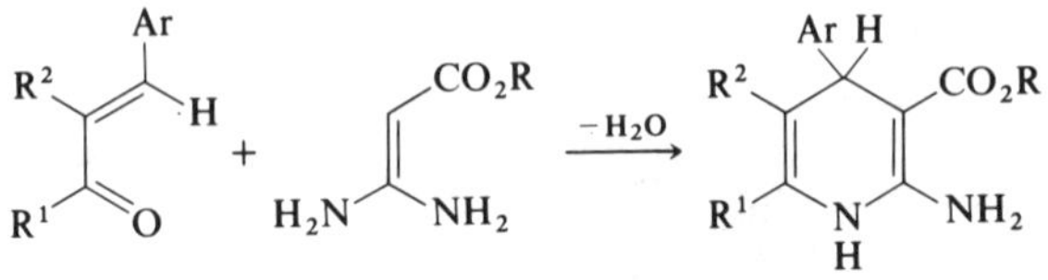

SCHEME 7

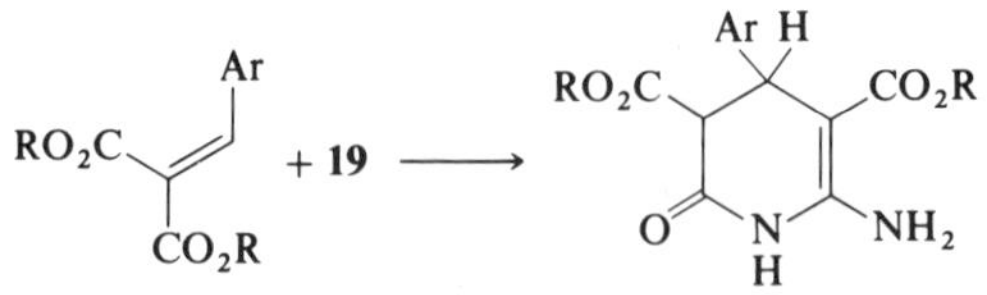

[15] L. Capuano and W. Fischer, *Chem. Ber.* **109**, 212 (1976).
[16] Cf. review: F. Bossert, H. Meyer, and E. Wehinger, *Angew. Chem.* **93**, 755 (1981); *Angew. Chem., Int. Ed. Engl.* **20**, 762 (1981).
[17] H. Meyer, F. Bossert, and H. Horstmann, *Liebigs Ann. Chem.*, 1895 (1977).
[18] H. Meyer, F. Bossert, and H. Horstmann, *Liebigs Ann. Chem.*, 1483 (1978).

Sec. II.E] HETEROCYCLIC β-ENAMINO ESTERS 305

In the presence of piperidine, compound 18, and ethyl cyanoacetate, ethyl 2-amino-4H-pyran-3-carboxylates (23) are formed.[19] In a similar way, α-benzoylcinnamonitriles (24) react with ethyl cyanoacetate to achieve ethyl 2-amino-4H-pyrane-3-carboxylates (25)[20] (Scheme 8).

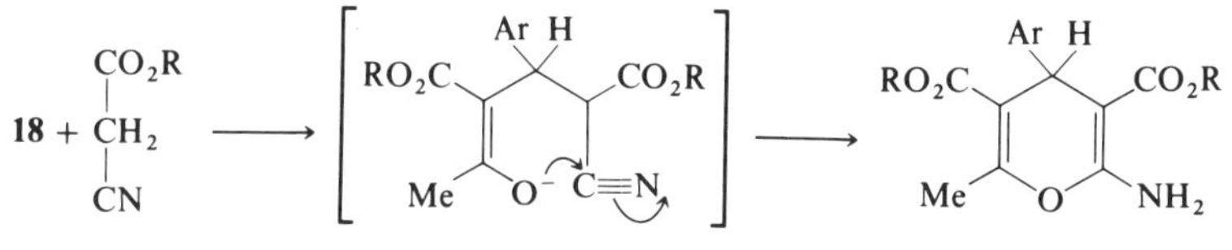

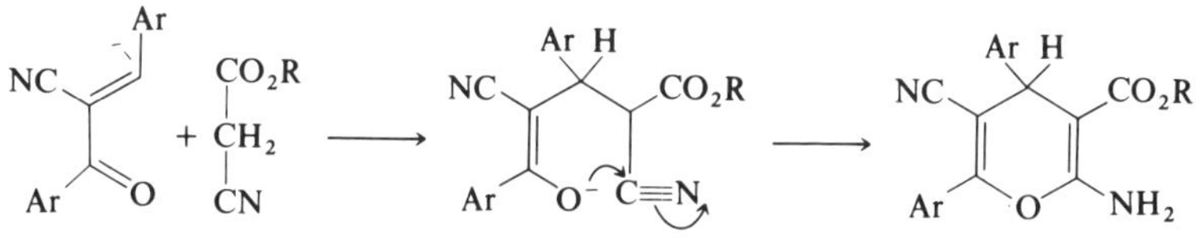

SCHEME 8

E. HETEROAROMATIC β-ENAMINO ESTERS

Numerous heteroaromatic β-enamino esters are given with appropriate references in Schemes 9 and 10.[21-52] Of the numerous preparative approaches to heteroaromatic β-enamino esters, only two very general routes are briefly discussed.

---

[19] Bayer AG, Ger. Offen. 2,235,406 (1974) (inventors: H. Meyer, F. Bossert, W. Vater, and K. Stoepel) [CA 80, 120765 (1974)].
[20] M. Quinteiro, C. Seoane, and J. L. Soto, Tetrahhedron Lett., 1835 (1977); J. Heterocycl. Chem. 15, 57 (1978); M. R. H. Elmoghayer, M. A. E. Khalifa. M. K. A. Ibraheim, and M. H. Elnagdi, Monatsh. Chem. 113, 53 (1982).
[21] K. Gewald, Chem. Ber. 98, 3571 (1965).
[22] K. Gewald and A. Martin, J. Prakt. Chem. 323, 843 (1981).
[23] K. Gewald, M. Kleinert, B. Thiele, and M. Hentschel, J. Prakt. Chem. 314, 303 (1972).
[24] H. Wamhoff and B. Wehling, Synthesis, 51 (1976).
[25] F. Langer, F. Wessely, W. Specht, and P. Klezl, Monatsh. Chem. 89, 239 (1958); J. Derkosch and I. Specht, ibid. 92, 542 (1961).
[26] K. Gewald, E. Schinke, and H. Böttcher, Chem. Ber. 99, 94 (1966); K. Gewald, ibid., 1002.
[27] C. A. Grob and O. Weissbach, Helv. Chim. Acta 44, 1748 (1961).
[28] H. Kano, Y. Makisumi, and K. Ogata, Chem. Pharm. Bull. 6, 105 (1958); [CA 53, 7140 (1959)].
[29] A. Dornow and H. Teckenburg, Chem. Ber. 93, 1103 (1960); F. Korte and K. Störiko, ibid. 94, 1956 (1961); G. Desimoni and P. Grünanger, Gazz. Chim. Ital. 97, 25 (1967).
[30] J. Goerdeler and H. Horn, Chem. Ber. 96, 1551 (1963).
[31] P. Schmidt and J. Druey, Helv. Chim. Acta 39, 986 (1956).

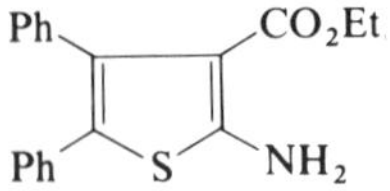

HEINRICH WAMHOFF

(26)[21]

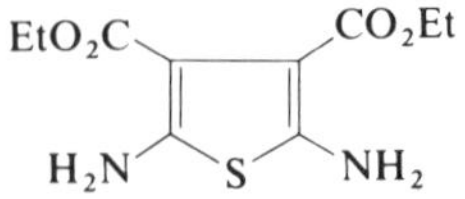

(27)[22]

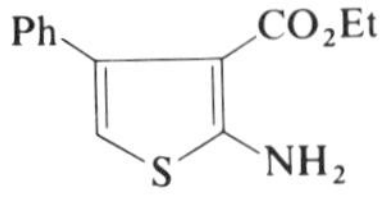

(28)[21]

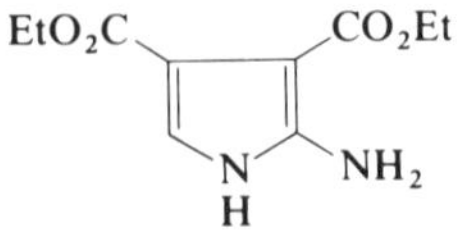

(29)[23]

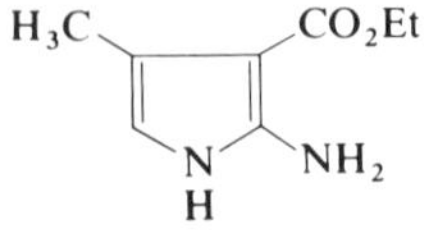

(30)[24]

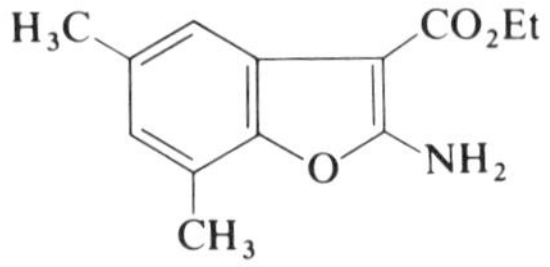

(31)[25]

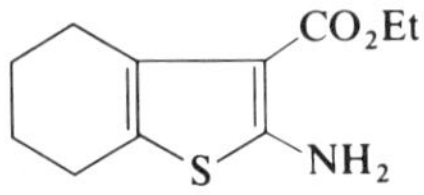

(32)[26]

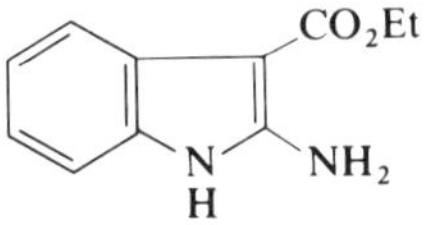

(33)[27]

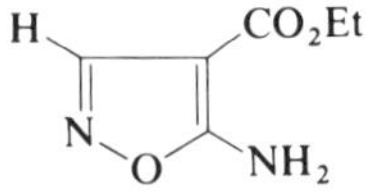

(34)[28]

SCHEME 9

---

[32] H. Dorn, G. Hilgetag, and A. Zubeck, *Chem. Ber.* **98**, (1968); C. C. Cheng, *J. Heterocycl. Chem.* **5**, 195 (1968); R. K. Robins, *J. Am. Chem. Soc.* **78**, 784 (1956).

[33] H. Wamhoff and A. Atta, unpublished.

[34] J. Goerdeler and G. Gnad, *Chem. Ber.* **99**, 1618 (1966).

[35] O. Dimroth, *Ber. Dtsch. Chem. Ges.* **35**, 4059 (1902); *Justus Liebigs Ann. Chem.* **364**, 203 (1908).

[36] H. Wamhoff, S.-Y. Yang, and J. Bohlen, unpublished.

[37] H. H. Fox, *J. Org. Chem.* **17**, 547 (1952).

[38] H. Wamhoff, unpublished results; W. F. Keir, A. H. MacLennan, and C. S. H. Wood, *J.C.S. Perkin I*, 1002 (1978).

[39] R. Grewe, *Hoppe-Seyler's Z. Physiol. Chem.* **242**, 89 (1936).

[40] J. L. Bernier, A. Lefebvre, J. P. Henichat, R. Houssin, and C. Lespagnol, *Bull. Soc. Chim. Fr.*, 616 (1976).

[41] K. Gewald, *J. Prakt. Chem.* **31**, 205 (1966).

[42] G. V. Boyd, *J. Chem. Soc., C*, 3873 (1971).

[43] K. H. Bauer, *Ber. Dtsch. Chem. Ges.* **71**, 2226 (1938).

[44] K. Gewald and J. Schael, *J. Prakt. Chem.* **315**, 39 (1973).

[45] R. Matusch and K. Hartke, *Chem. Ber.* **105**, 2594 (1972).

[46] G. B. Bennett, R. B. Mason, and M. J. Shapiro, *J. Org. Chem.* **43**, 4383 (1978).

[47] K. Gewald and H. J. Jänsch, *J. Prakt. Chem.* **318**, 313 (1976).

[48] K. Gewald and G. Heinhold, *Monatsh. Chem.* **107**, 1413 (1976).

[49] K. Gewald, P. Bellmann, and H. J. Jänsch, *Z. Chem.* **15**, 18 (1975); K. Gewald and P. Bellmann, *Liebigs Ann. Chem.*, 1534 (1979).

[50] K. Saito, S. Kambe, A. Sakurai, and H. Midorikawa, *Synthesis*, 1056 (1982).

[51] L. C. Cheney and J. R. Piening, *J. Am. Chem. Soc.* **67**, 729 (1945); O. Hromatka, D. Binder, and K. Eichinger, *Monatsh. Chem.* **104**, 1513 (1973).

[52] K. Gewald, P. Bellmann, and H. J. Jänsch, *Liebigs Ann. Chem.*, 1623 (1980).

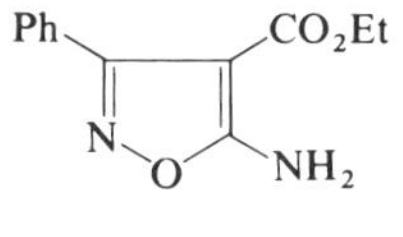

Sec. II.E] HETEROCYCLIC β-ENAMINO ESTERS 307

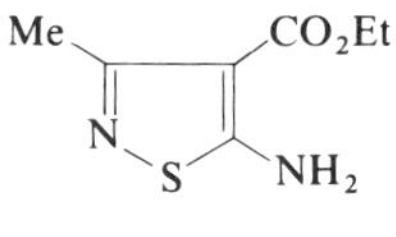

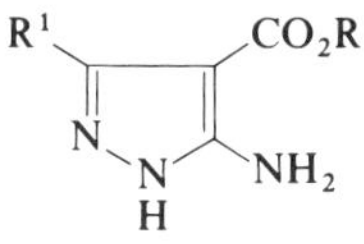

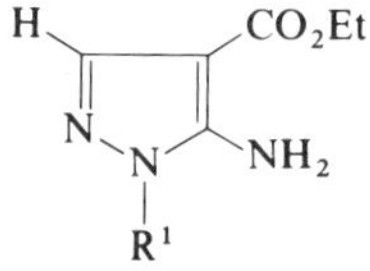

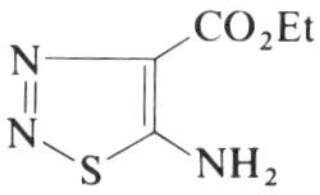

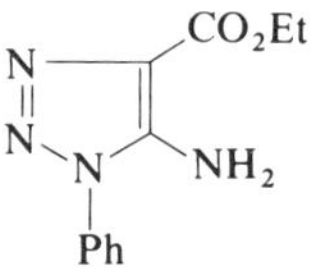

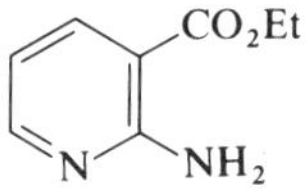

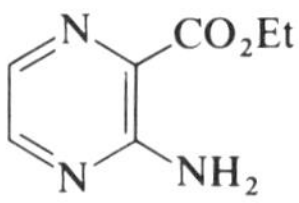

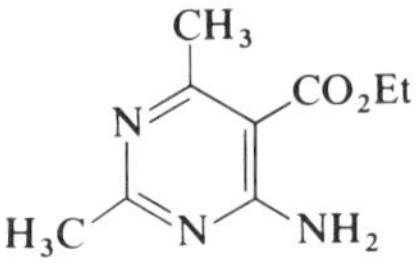

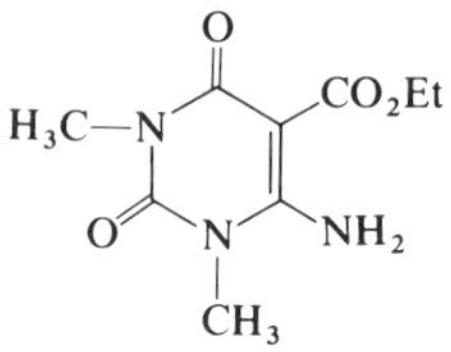

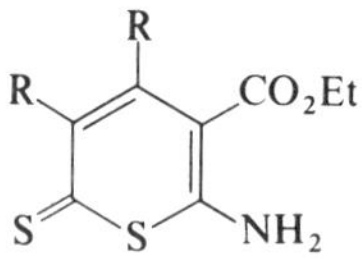

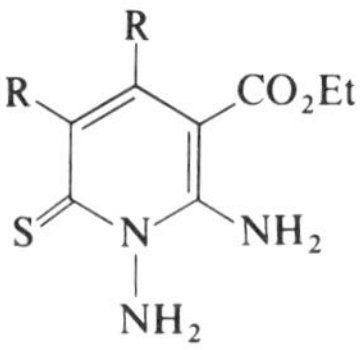

SCHEME 9 (*Continued*)

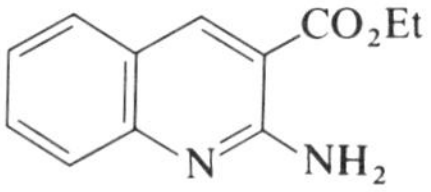

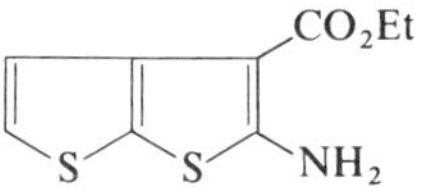

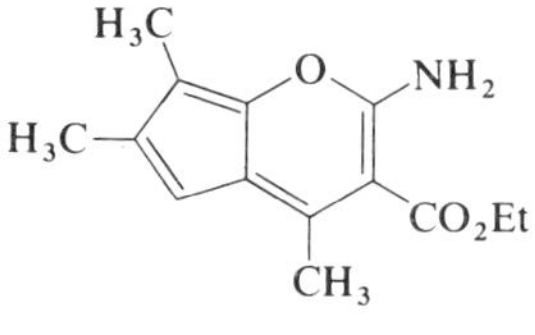

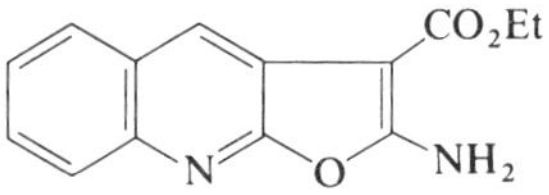

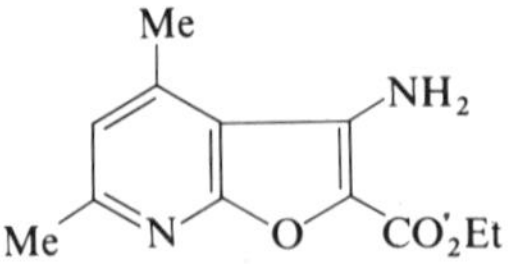

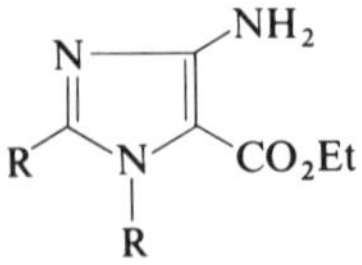

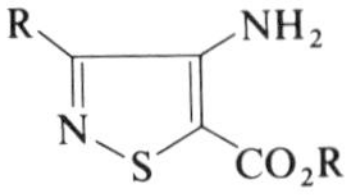

(51)[47]   (52)[48]   (53)[49]

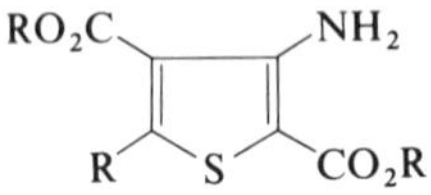

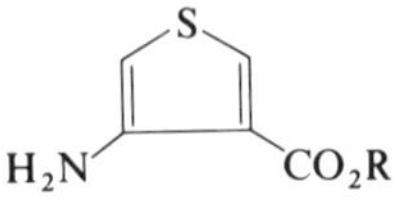

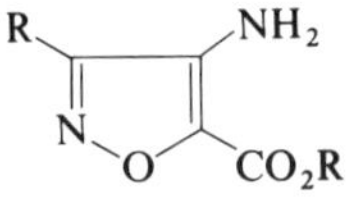

(54)[50]   (55)[51]   (56)[52]

SCHEME 10. Some "reverse" heteroaromatic enamino esters.

2-Aminofuran-, -pyrrole-, and -thiophene-3-carboxylates and -3-nitriles such as **58** are obtained in a simple, one-step reaction from α-oxo-alcohols, -amines, and -thiols (**57**) with methylene-active nitriles.[53] Thus ethyl 2-amino-4,5,6,7-tetrahydrobenzo[*b*]thiophene-3-carboxylate (**59**) results from cyclohexanone, sulfur, and ethyl cyanoacetate[26,53] (Scheme 11). The accessible

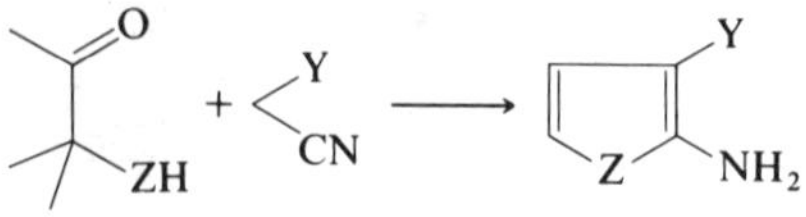

(57)   (58)

a Z = O; Y = CN   **b** Z = NH; Y = CN, CO_2R
c Z = S; Y = CN, CO_2R, $CONH_2$, COR

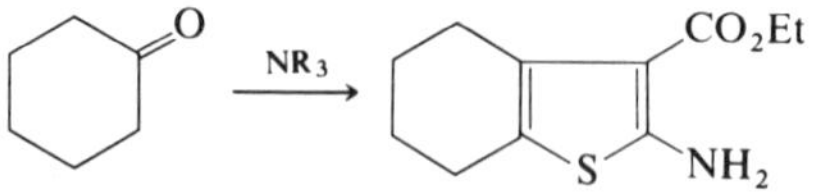

(59)

SCHEME 11

[53] K. Gewald, *Chimia* **34**, 101 (1980); *Lect. Heterocycl. Chem.* **6**, 121 (1981), and references cited therein.

alkoxymethylenecyanoacetates (**60**)[54] are smoothly cyclized with hydroxylamine, hydrazine, and their substituted analogues to afford five-membered heteroaromatic β-enamino esters (**61**)[29,55] (Scheme 12). This review mainly covers partially saturated heterocyclic β-enamino esters; but some heteroaromatic β-enamino esters are also discussed.

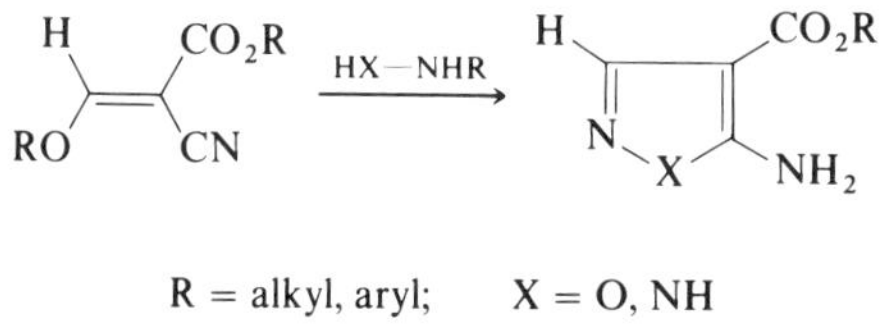

$$\underset{\text{RO}}{\overset{\text{H}}{\diagdown}}\text{C}=\underset{\text{CN}}{\overset{\text{CO}_2\text{R}}{\diagup}} \xrightarrow{\text{HX—NHR}} \underset{\text{N}_{\diagdown\text{X}}}{\overset{\text{H}}{\diagdown}}\underset{\diagdown\text{NH}_2}{\overset{\text{CO}_2\text{R}}{\diagup}}$$

R = alkyl, aryl;   X = O, NH

(**60**)   (**61**)

SCHEME 12

III. Structure and Spectral Properties

Heterocyclic β-enamino esters differ in many respects from enamines,[2] open-chain β-enamino esters,[2] and heterocyclic enamines,[3] all of which point to a specific electron distribution in their enaminocarbonyl chromophore. In the following, particular attention is paid to spectroscopic studies of these title compounds and to the consequences for their structures, protonation behavior, tautomerism, stability, and mesomeric forms.

A. PROTONATION

Consistent with previous IR investigations (in CDCl_3/TFA),[6,7] ^{1}H-NMR spectroscopy shows that protonation of heterocyclic β-enamino esters, such as **62**, occurs exclusively at C-3 ("C_β-protonation") to give oximmonium and thiaimmonium salts **63**[55a] (Scheme 13). In these ^{1}H-NMR spectra, H^X (cf. **63**) couples with the neighboring methylene protons to display a characteristic triplet signal; depending on ring size and heteroatom X; $\delta = 4.20$–4.62 ($J = 7.0$–9.0 Hz). However, furan- and thiopheneenamino esters (**64**) are protonated at C-5 to form 2,5-dihydro heterocycles (**65**).[26,56]

---

[54] R. G. Jones, *J. Am. Chem. Soc.* **73**, 3684 (1951); **74**, 4889 (1952); Roche Products Ltd., U.S. Patent 2,375,185 (1945) (inventors: F. Bergel, A. Cohen, and J. W. Haworth) [*CA* **39**, 3307 (1945)].
[55] A. Michaelis and E. Remy, *Ber. Dtsch. Chem. Ges.* **40**, 1020 (1907); C. C. Cheng and R. K. Robins, *J. Org. Chem.* **21**, 1240 (1956).
[55a] H. Wamhoff, *Tetrahedron* **26**, 3849 (1970).
[56] C. T. Wie, S. Sunder, and C. DeWitt Blanton, Jr., *Tetrahedron Lett.*, 4605 (1968).

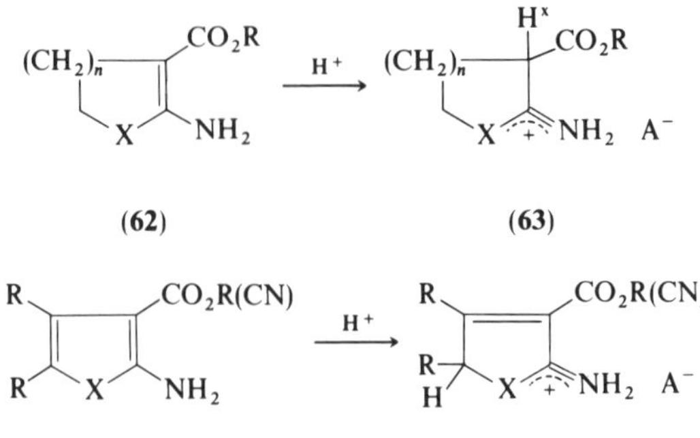

(62) (63) (64) (65)

SCHEME 13

B. STRUCTURES FROM UV, IR, AND ^{1}H-NMR DATA

The bathochromic shifted, intense UV maxima of the partially saturated heterocyclic β-enamino esters (λ_{max} 270–297.5 nm; log ε = 4.07–4.28) confirm a considerable π-electron delocalization (p–π-overlap) in the enaminocarbonyl chromophore,[57] which can be depicted by dipolar mesomeric canonical formulae **66a–66e** possessing a high transition moment[58] (Scheme 14). A strong participation of the ring heteroatom can be discerned.

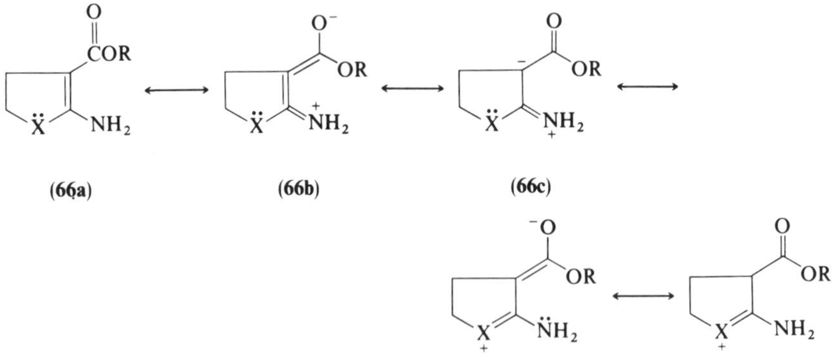

(66a) (66b) (66c) (66d) (66e)

SCHEME 14

[57] N. H. Cromwell and W. R. Watson, *J. Org. Chem.* **14**, 411 (1949).

[58] H. Wamhoff, H. W. Dürbeck, and P. Sohár, *Tetrahedron* **27**, 5873 (1971); about the conformation, cf. E. Dradi and G. Gatti, *Org. Med. Reson.* **3**, 479 (1971).

This resonance involving dipolar canonical formulas is unambiguously supported by a significant positive solvatochromism as shown by the wavelength dependence of the UV absorptions in different solvents with increasing dielectric constants (Table I).[59] In the IR spectra,[58] characteristic absorption is found in the regions 3500–3000 and 1750–1500 cm^{-1}. The position and shape of the v_{NH_2} vibrations (v_s and v_{as}) point to free rather than chelated (to the neighboring ester carbonyl) amino groups. In the region of the enaminocarbonyl bands (1700–1500 cm^{-1}), three characteristic, strong bands are found. Deuterium exchange experiments indicate that these bands belong to absorptions of group frequencies. Thus following Dabrowski[60] who investigated open-chain enamino esters, these absorptions have been named "enamino-ester bands EI, EII, and EIII" (i.e., coupled vibrations of $v_{C=O}$, $v_{C=C}$, and δ_{NH_2} bands). For detailed descriptions of the IR spectra see references 12, 13, and 58.

Because of the participation of the enamine double bond in the resonance of the aromatic ring of heteroaromatic β-enamino esters, the group frequency character of the individual bands is low. Thus the EI band is essentially $v_{C=O}$. Similarly, β-enamino nitriles do not exhibit coupled vibrations since the nitrile absorption does not participate.[24,61]

The ^{1}H-NMR spectra indicate that these potentially tautomeric compounds exist exclusively in the enamine form **67a** (Scheme 15). The amine protons are found as broad resonance signals in the region δ 5.25–6.85 and are similar to conjugated primary amines. From these values, an intramolecular hydrogen bridge (cf. **67c**) to the neighboring ester carbonyl can be excluded; all data are given in references 12, 13, and 58.

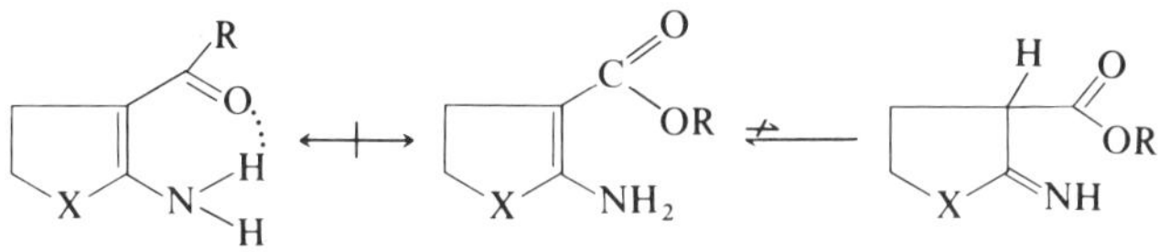

(67a)   (67c)   (67b)

SCHEME 15

C. ^{13}C-NMR Data and Electron Distribution in the Enaminocarbonyl Chromophore

The ^{13}C-NMR spectra support the results of the other spectroscopic investigations described above. In addition, this method allows direct

---

[59] H. Wamhoff and E. Gierke, unpublished results.
[60] J. Dabrowski, *Spectrochim. Acta* **19**, 475 (1963).
[61] H. Wamhoff and H. A. Thiemig, *Chem. Ber.* **118** (1985), in press.

TABLE I
POSITIVE SOLVATOCHROMISM[a]

	Isooctane (1.9)	THF (7.4)	CH_3CN (37.5)	DMF (37.6)	DMSO (48.9)	CH_3OH (33.6)	$HCONH_2$ (109.5)
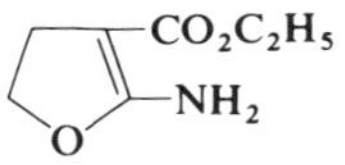 ![structure 1]	265 (3.69)	269 (4.18)	269 (4.17)	270 (4.15)	272 (4.19)	273 (4.20)	274 (4.17)
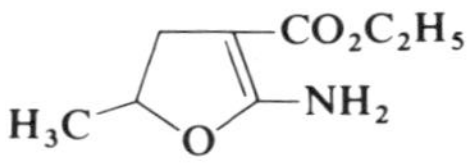 ![structure 2]	265 (3.95)	269 (4.23)	269 (4.17)	270 (4.27)	272 (4.19)	273 (4.33)	274 (4.50)
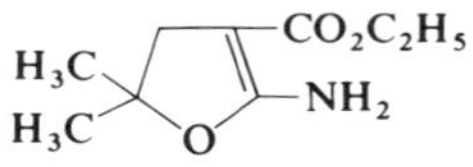 ![structure 3]	265 (3.82)	268 (4.12)	269 (4.05)	272 (4.12)	271 (4.07)	272 (4.25)	274 (4.23)
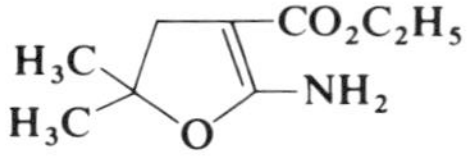 ![structure 4]	266 (4.13)	269 (4.19)	268 (4.22)	270 (4.21)	272 (4.32)	272 (4.30)	274 (4.18)

[a] For each absorbance maximum, given in nm, the logarithm of the molar absorptivity is included in parentheses.

TABLE II
^{13}C-NMR DATA OF ETHYL 2-AMINO-4,5-DIHYDROFURAN-3-CARBOXYLATES **68a–d**
(SEE SCHEME 16)^a

Compound	C-2	C-3	C-4	C-5	C-6
68a	168.11s	73.64s	27.50t	70.70t	168.11s
68b	167.00s	73.61s	40.94t	87.55t	169.06s
68c	167.60s	80.63s	37.08d	82.28d	168.26s
			(135.6)	(143.0)	
68d	166.10s	72.70s	34.10t	70.10d	167.40s

Compound	C-7	C-8	C-9	C-10
68a	58.71t	14.90q	—	—
68b	58.80t	14.90t	28.32q	28.32q
68c	57.96t	14.39q	14.39q	13.76q
	(144.0)	(126.4)	(126.4)	(126.4)
68d	58.20t	14.60q	21.50q	—

^{a} Solvent, CDCl$_3$; TMS as internal standard; the coupling constant J (in Hz) is given in parentheses.

measurements of the electron density of the individual carbon atoms, and gives the electron distribution in the enaminocarbonyl chromophore.

In Table II ^{13}C-NMR data are given for a number of 4,5-dihydrofurans (Scheme 16).[62] Schemes 17 and 18 present more ^{13}C-NMR data of additional heterocyclic β-enamino esters, which are, however, limited to unsaturated atoms C-2 and C-3.[63] All of the enamino esters **68a–d, 69–72** show the common feature of a strongly electron deficient C-2, as in an amidine-C or a ketene-S, N-(-O,N-)aminal,[65] while C-3 does not appear as an olefinic carbon

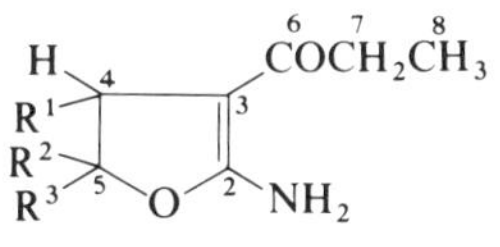

68	R^1	R^2	R^3
a	H	H	—
b	H	CH$_3$	CH$_3$
c	CH$_3$	H	CH$_3$
d	H	H	CH$_3$

SCHEME 16

[62] H. Wamhoff and G. Haffmanns, unpublished results.
[63] H. Wamhoff and L. Lichtenthäler, *Chem. Ber.* **111**, 2813 (1978); H. Wamhoff and G. Hendrikx, *ibid.* (in press).

Structure	C-2	C-3
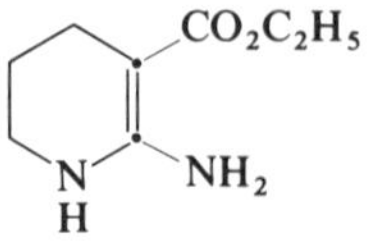 (69) tetrahydropyridine with CO$_2$C$_2$H$_5$ and NH$_2$, N-H	159.28	72.05
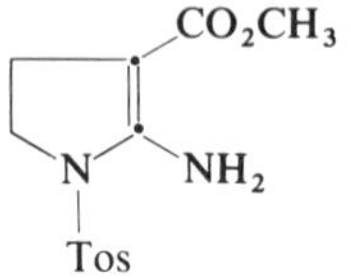 (70) pyrrolidine with CO$_2$CH$_3$ and NH$_2$, N-Tos	154.50	80.43
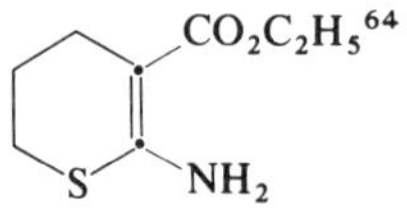 (71) thiane with CO$_2$C$_2$H$_5$[64] and NH$_2$	155.63	90.66
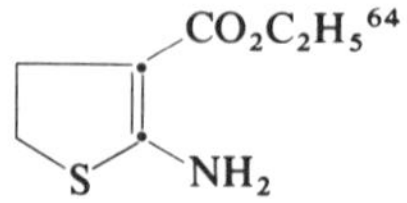 (72) thiolane with CO$_2$C$_2$H$_5$[64] and NH$_2$	163.33	92.12

SCHEME 17. ^{13}C-NMR data of C-2 and C-3 for **69–72**.

atom but instead as a highly shielded trigonal carbon atom, owing to the electron-donating effects of the heteroatoms.[66] Thus as a consequence, these ^{13}C-NMR values point to a high degree of polarization in the enamine double bond, as shown in the canonical formula **66b–e**. These ^{13}C-NMR values agree with the reported data of methyl and ethyl β-aminocrotonate.[66a] Even for anthranilic esters, characteristic polarization of the aromatic ring for C-1 and C-2 can be observed[66b] (see Scheme 18). As a consequence, the electron density

---

[64] H. Wamhoff and J. Nagelschmitz, unpublished results.

[65] Cf. R. L. Smith, D. W. Cochran, P. Gund, and E. J. Cragoe, Jr., *J. Am. Chem. Soc.* **101**, 191 (1979); C. Skötsch and E. Breitmaier, *Chem. Ber.* **113**, 795 (1980).

[66] The ^{13}C NMR values of the OH-form of 3-methyl-5-pyrazolone point to a similar electron distribution; (in DMSO): C-3 140.1, C-4 89.2, C-5 161.6 ppm; E. Breitmaier, unpublished.

[66a] Sadtler Standard Carbon-13 Spectra, No. 2331 and 2647. Sadtler Research Laboratories, Philadelphia, Pennsylvania, *In* "Bruker C-13 Data Bank," Vol. 1. Bruker-Physik, Karlsruhe.

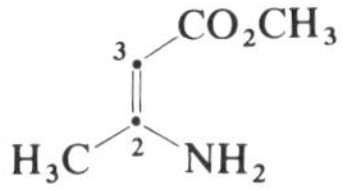

Sec. III.D] HETEROCYCLIC β-ENAMINO ESTERS 315

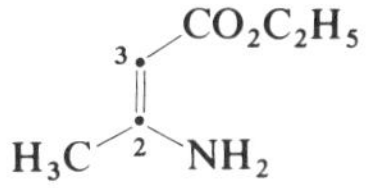

```
      CO₂CH₃                    CO₂C₂H₅
    3⁄                        3⁄
H₃C²NH₂                  H₃C²NH₂
```

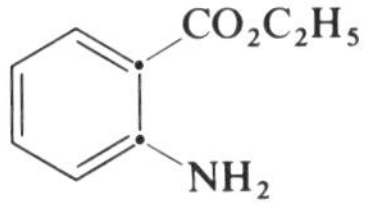

C-2: 160.00   C-2: 160.70
C-3: 83.60   C-3: 83.30

(2-aminobenzoate methyl ester)   (2-aminobenzoate ethyl ester)

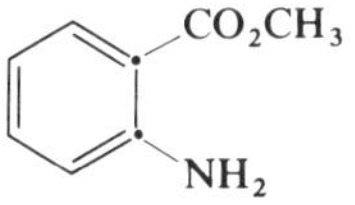

C-1: 150.60   C-1: 150.80
C-2: 110.00   C-2: 110.10

SCHEME 18. ^{13}C-NMR data of C-1, C-2, and C-3.

at the 2-NH₂ group is quite low, and therefore the nucleophilicity of the nitrogen atom is rather weak. Thus many typical reactions for enamines cannot be expected.

D. THE PUSH–PULL STABILIZATION

From the spectroscopic data of Sections II,A–C it can be concluded that partially saturated heterocyclic β-enamino esters possess a specific push–pull stabilization owing to the conjugation of electron-donating amino group and electron-withdrawing ester function.[67] Push–pull stabilization is responsible for the remarkable stability of this class of compounds; similar stabilization allows isolation of certain antiaromatic cyclobutadienes or nonaromatic o-quinodimethanes.[68] 2-Aminofurans, -pyrroles, and -thiophenes, because of their extremely large electron density, are generally unstable and sensitive

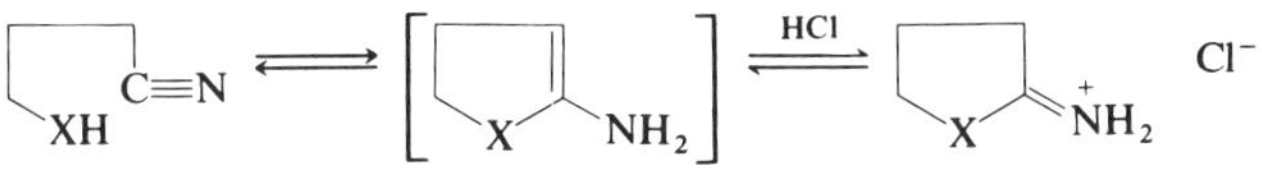

```
   C≡N    ⇌    [       ]    ⇌(HCl)    [       ]⁺ Cl⁻
 XH              X   NH₂                X   NH₂
 
 (73)           (74)                    (75)
```

SCHEME 19

[66b] Sadtler Standard Carbon-13 Spectra, No. 150. Sadtler Research Laboratories, Philadelphia, Pennsylvania, In L. F. Johnson and W. C. Jankowski, "Carbon-13 NMR Spectra," No. 296. Wiley (Interscience), New York, 1972.

[67] Push-pull alkenes: Cf. H. Kristen and K. Peseke, Wiss. Z. Wilhelm-Pieck-Univ. Rostock, Math.-Naturwiss. Reihe **25**, 1123 (1976) [CA **88**, 1055187m (1978)].

[68] Cf. R. Gompper, Bull. Soc. Chim. Belg. **92**, 781 (1983), and references cited therein.

toward many reagents;[69] and most classical synthetic approaches to them have failed. 2-Amino-4,5-dihydrofurans and -thiophenes (74) have not yet been isolated in spite of many attempts, e.g., by cyclization of 4-hydroxy- and 4-mercaptobutyronitriles (73). In all cases, only the 2-imino hydrochloride 75 was isolated[70] (Scheme 19).

However, a novel synthesis of 5-amino-2,3- or 5-imino-2,5-dihydrofurans 78B from 4-hydroxy-4,4-dialkylbut-2-yne nitriles 76A and primary amines has been reported[70a]; but in agreement with the resonance formulas shown for 76B, these compounds might be considered to be once again (internally) push–pull stabilized, like the furanones 17.[15]

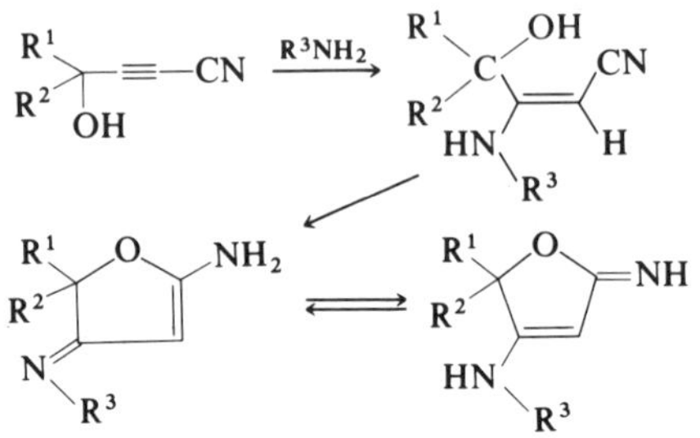

(76B)

SCHEME 19a

IV. General Considerations of Chemical Reactivity

Based on the spectroscopic information presented in Sections III,A–D, some general reaction pathways can be deduced for the partially saturated heterocyclic β-enamino esters. Electrophiles are expected to attack at the electron-rich position 3 (cf. 77). If the attacking agent possesses an additional (terminal) nucleophilic group or one is formed during the reaction, then ring closure may take place at the electron-deficient C-2 (cf. 78). The electrophile is also capable of attacking the 2-amino group (79), which exhibits weak nucleophilic behavior, but reacts with electrophiles of high electron defi-

---

[69] A. P. Dunlop and F. N. Peters, "The Furanes," p. 183ff. Van Nostrand: Reinhold, Princeton, New Jersey, 1950; F. F. Blicke, in "Heterocyclic Compounds" (R. C. Elderfield, ed.), Vol. 1, p. 228. Wiley, New York, 1950; S. Gronowitz, *Adv. Heterocycl. Chem.* **1**, 1 (1963).

[70] K. S. Topchiev and M. L. Kirmalova, *Dokl. Akad. Nauk SSSR* **63**, 281 (1948) [*CA* **43**, 2579 (1949)]; H. Nohira, Y. Nishikawa, Y. Furuya, and T. Mukaiyama, *Bull. Chem. Soc. Jpn.* **38**, 897 (1965) [*CA* **63**, 7012 (1965)]; American Cyanamid Co., U.S. Patent 3,223,585 (1965) inventor: W. Addor) [*CA* **63**, 7012 (1965)].

[70a] Z. T. Fomum, P. F. Asobo, S. R. Landor, P. D. Landor, *J. Chem. Soc., Chem. Commun.*, p. 1455 (1983).

ciency, e.g., heterocumulenes (Scheme 20). A terminal nucleophilic group present from the beginning or formed during the reaction might now attack the ester function (cf. **80**). In both cases, condensed heterocyclic systems are formed (**78, 80**).

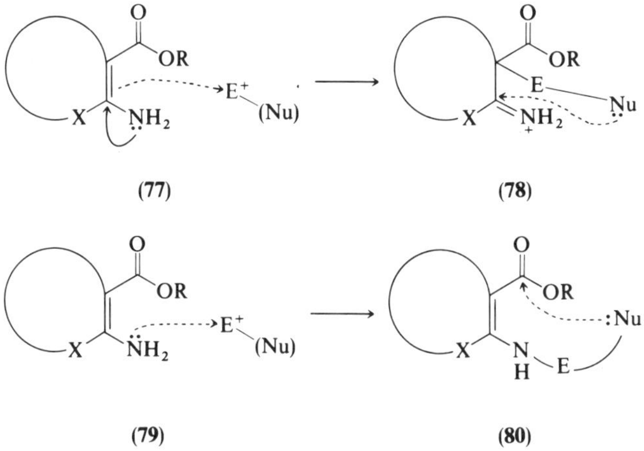

SCHEME 20

Nucleophiles can attack competitively both the 3-ester group (**81**) and the electron-poor C-2 (**82**); in the latter case, subsequent ring-cleavage reactions (**83**) or other elimination processes (**84**) might occur in order to regenerate the enaminocarbonyl moiety in a thermodynamically downhill process (Scheme 21). When the enaminocarbonyl group is part of an aromatic ring, as

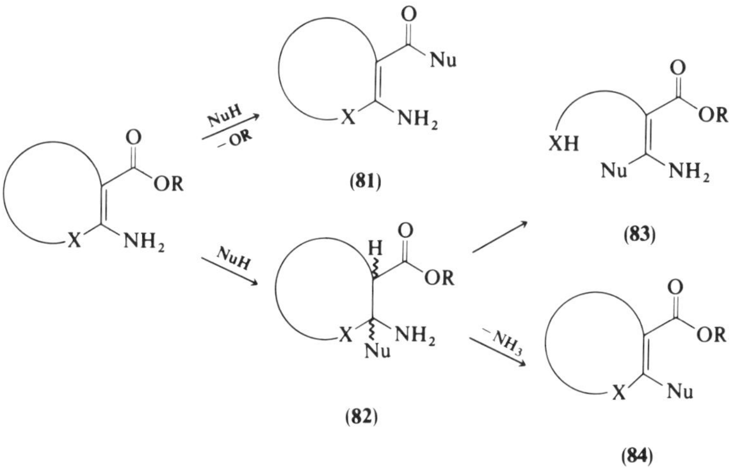

SCHEME 21

in heteroaromatic β-enamino esters **85**, reactivity comparisons should be made with anthranilic esters **86** and related compounds (Scheme 22).

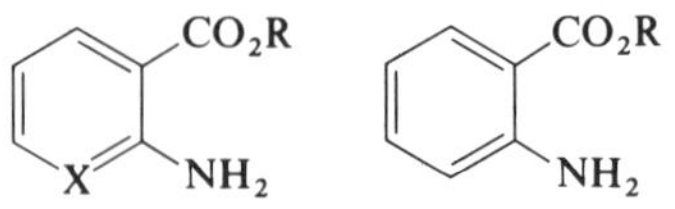

(**85**)   (**86**)

SCHEME 22

V. Electrophilic Attacks on the 2-Amino Group (or the Carbon-3 Atom). Synthesis of Polynuclear Heterocyclic Systems

A. WITH ISOCYANATES, ISOTHIOCYANATES, AND SULFENYL CHLORIDES

One of the most versatile transformations with subsequent heterocyclization is the smooth and mostly exothermic addition of electron-poor isocyanates to the 2-amino function of almost all types of heterocyclic enamino esters (cf. **87**). A mixed urea (**88**) is first formed; it can be easily isolated but in most cases is cyclized *in situ* with aqueous base to afford heterocondensed uracils (**89**) in high yields[71] (Scheme 23). This reaction

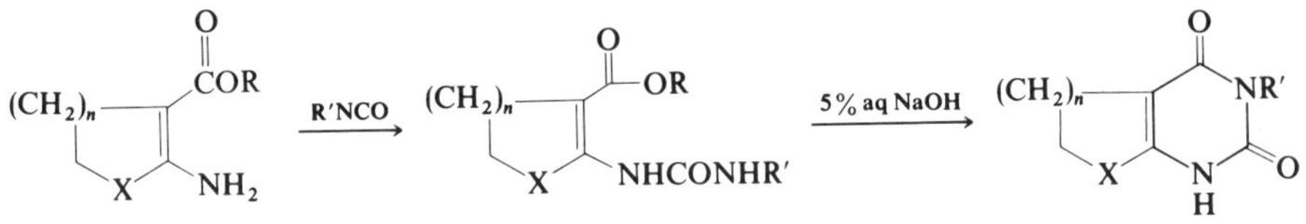

(**87**)   (**88**)   (**89**)

SCHEME 23

pathway is based on a very early communication of Behrend *et al.*[72] This method has been used to obtain highly substituted furo[2,3-*d*]pyrimidines,

---

[71] H. Wamhoff, *Chem. Ber.* **101**, 3377 (1968); Thieno[2,3-d]pyrimidines: F. Sauter and W. Deinhammer, *Monatsh. Chem.* **104**, 1593 (1973).

[72] R. Behrend, F. C. Meyer, and Y. Buckholz, *Justus Liebigs Ann. Chem.* **314**, 200 (1901).

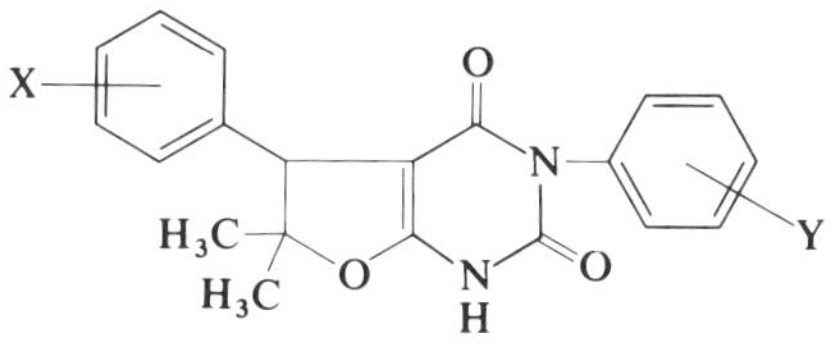

(90)

SCHEME 24

e.g., **90** (Scheme 24) as potential antimalarials.[10] Reactive aryl isocyanates[73] give urea **88** in an exothermic reaction; but for more electron-rich, less reactive alkyl isocyanates, the reaction mixture must be warmed in order to complete addition. Scheme 25 depicts the rather broad applicability (Formulas **91**–**107**).[74-80] The caffeine and isocaffeine isomers derived from **103** show diuretic and cardiac activities.[78]

Isothiocyanates normally act as weaker electrophiles and need more drastic conditions to give the thioxo derivatives;[76] also organic bases, such as pyridine, have been successfully employed[76] for the cyclization step, so that this heterocyclization procedure may be carried out as a one-pot, three-component (enamino ester, isocyanate, organic base) reaction. Higher carbamoylated products, such as **109**, are smoothly hydrolyzed in aqueous 5% KOH to the pyrido[2,3-*d*]pyrimidines **110**[76] (Scheme 26).

After chlorolysis of furo[2,3-*d*]pyrimidines (**111**) with phosphorus oxychloride, 5-(2-chloroethyl)uracils (**112**) are obtained. These can be transformed to thieno[2,3-*d*]pyrimidines (**113**) or rearranged with triethylamine to give novel 4-chlorofuro[2,3-*d*]pyrimidines (**114**), which are useful starting materials for novel tetracyclic systems (cf. **115**) via photolysis of the 4-azido derivative[81] (Scheme 27).

---

[73] Cf. S. Petersen, *in* "Methoden der Organischen Chemie" (Houben, Weyl, Müller, eds.), 4th ed., p. 192ff. Thieme, Stuttgart, 1952; R. G. Arnold, J. A. Nelson, and J. J. Verbane, *Chem. Rev.* **57**, 50 (1957).
[74] H. Wamhoff and B. Wehling, *Chem. Ber.* **109**, 2983 (1976).
[75] H. Wamhoff, unpublished results.
[76] H. Wamhoff and L. Lichtenthäler, *Chem. Ber.* **111**, 2297 (1978).
[77] J. Goerdeler and H. Horn, *Chem. Ber.* **96**, 1551 (1963).
[78] P. Schmidt, K. Eichenberger, and J. Druey, *Helv. Chim. Acta* **41**, 1052 (1958); P. Schmidt, K. Eichenberger, M. Wilhelm, and J. Druey, *ibid.* **42**, 349 (1959).
[79] H. Wamhoff, S.-Y. Yang, and J. Bohlen, unpublished.
[80] J. Lehmann and H. Wamhoff, *Chem. Ber.* **106**, 3533 (1973).
[81] H. Wamhoff and C. von Waldow, *Chem. Ber.* **107**, 2265 (1974); **110**, 1730 (1977).

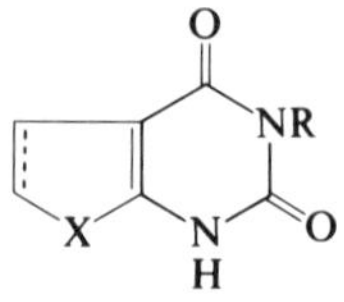

(91)[71] X = O
(92)[71] X = S
(93)[74] X = NR

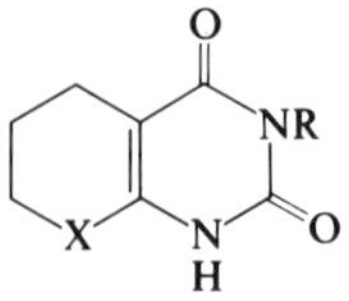

(94)[75] X = O
(95)[71] X = S
(96)[76] X = NH

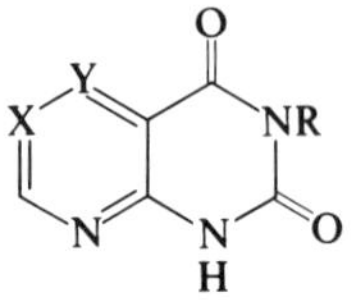

(97)[75] X = Y = CH
(98)[75] X = N
      Y = CH
(99)[75] X = CH
      Y = N

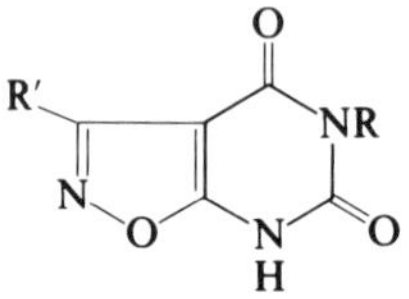

(100)[75]

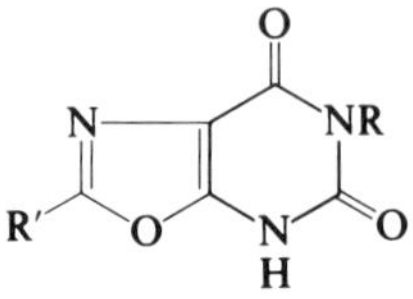

(101)[75]

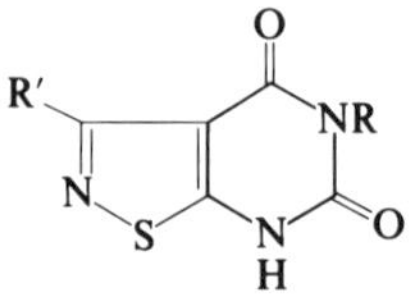

(102)[77]

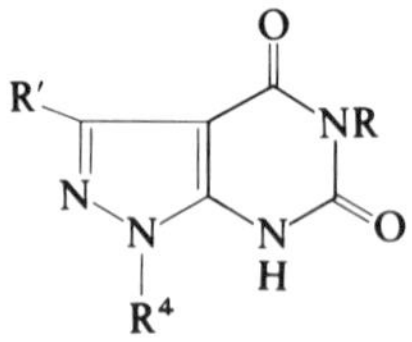

(103)[78]

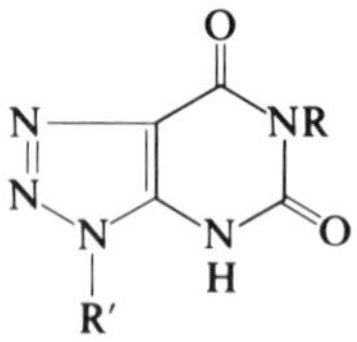

(104)[79]

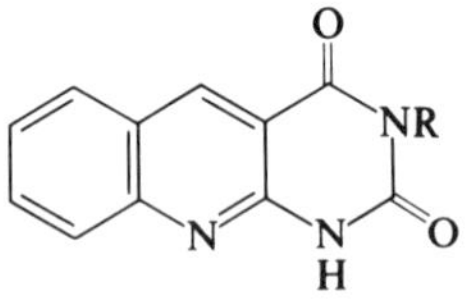

(105)[80]

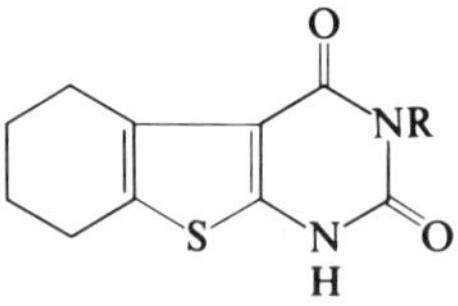

(106)[71,75]

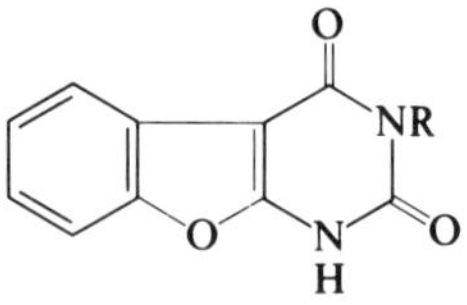

(107)[75]

SCHEME 25

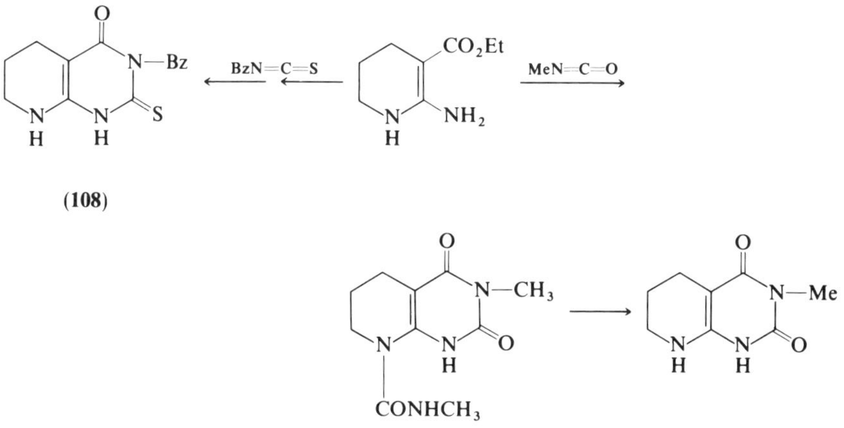

Sec. V.A]   HETEROCYCLIC β-ENAMINO ESTERS   321

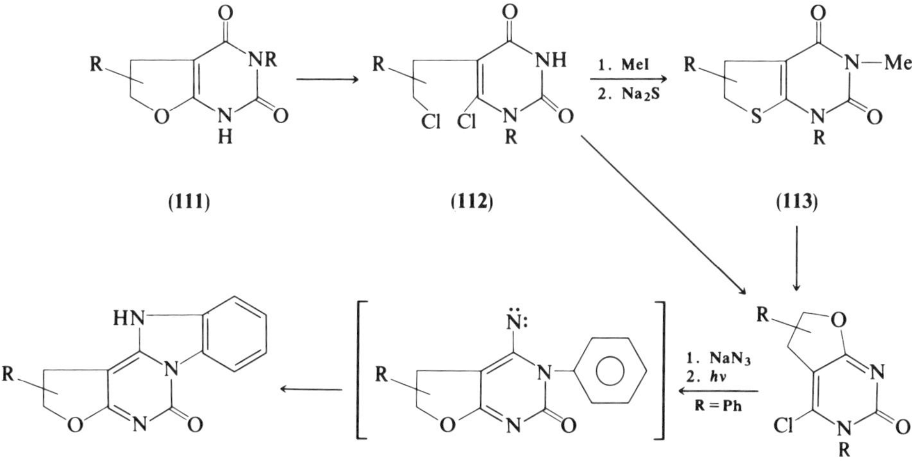

SCHEME 26

SCHEME 27

Among the many new condensed systems, many of potential biological interest, pyrrolo[2,3-*d*]pyrimidines[82] deserve special attention as aglycones of antibiotic and cytostatic principles, such as Tubericidine, Toyocamycine, and Sangivamycine.[83] Treatment of suitably substituted pyrrole- and pyrrolinenamino esters (**116** and **118**) with isocyanates represents a new and versatile approach to this interesting class of compounds (**117**, **119**)[74] (Scheme 28). Treatment of the enamino nitrile **120** with methyl isocyanate

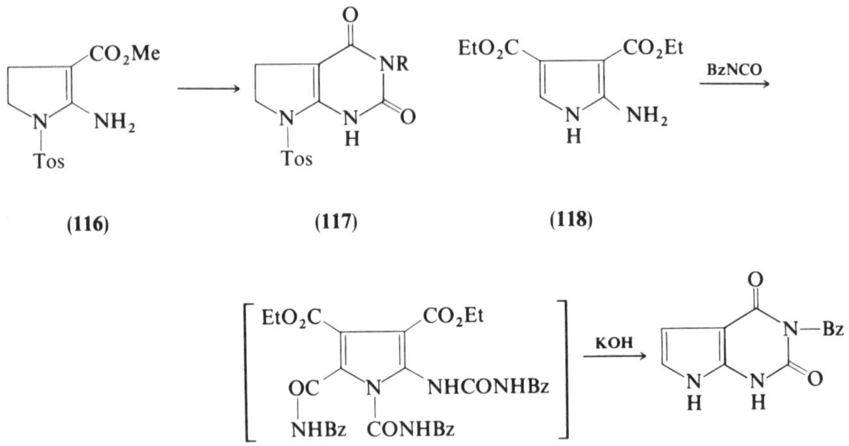

(116) **(117)** **(118)** **(119)**

SCHEME 28

under more energetic conditions leads to the triscarbamoyl derivative **121**, which is cyclized in turn in sodium methoxide to afford pyrrolo[1,2-*a*]-1,3,5-

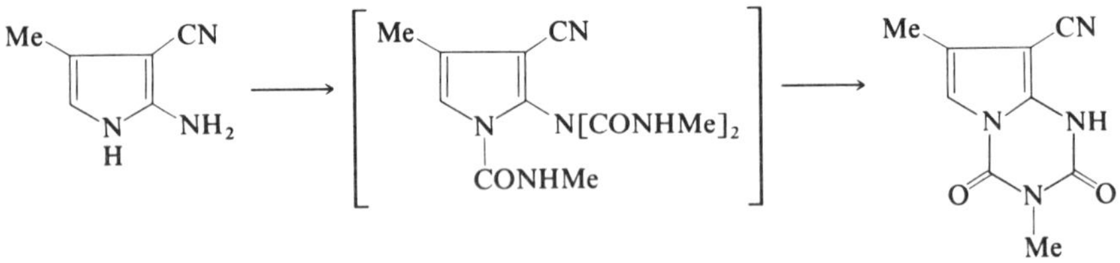

(120) **(121)** **(122)**

SCHEME 29

[82] V. Amarnath and R. Madhav, *Synthesis*, 837 (1974).

[83] G. Acs and E. Reich, *in* "Antibiotics" (P. Gottlieb and P. D. Shaw, eds.), Vol. 1, p. 494ff. Springer-Verlag, Berlin and New York, 1967.

triazine (122) in the first preparation[74] of the ring system of the antibiotic Viomycine[84] (Scheme 29). It is noteworthy that ethyl 2-aminoindole-3-carboxylate (33)[27] does not show any readiness to add isocyanates to give the benzo[b]pyrrolo[2,3-d]pyrimidines.[85]

In a related reaction, isopropylsulfamoyl chloride (123) reacts with the enamino esters 32 and 38 to give the sulfonyl analogues of ureas, e.g., 124. Basic ring closure leads then to a benzo[b]thieno[3,4-e]-2,1,3-thiadiazine 2,2-dioxide (125) and a pyrazolo[3,4-e]-2,1,3-thiadiazine 2,2-dioxide (126), respectively[86] (Scheme 30). Similarly, chlorosulfonyl isocyanate (127)[87] affords with several enamino esters (after saponification and ring closure

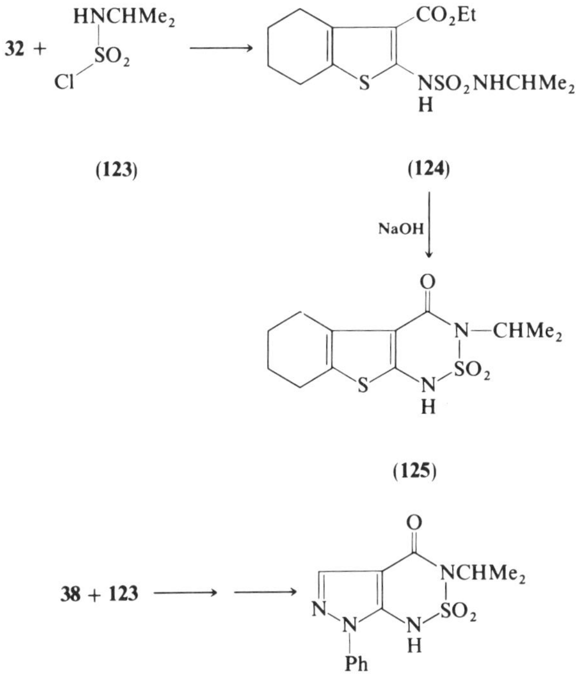

SCHEME 30

---

[84] J. R. Dyer, C. K. Kellogg, R. F. Nassar, and W. E. Streetman, *Tetrahedron Lett.*, 585 (1965).
[85] H. Wamhoff and B. Wehling, *Chem. Ber.* **108**, 2107 (1975).
[86] H. Wamhoff and M. Ertas, *Synthesis*, p. 190 (1985).
[87] R. Graf, *Chem. Ber.* **89**, 1071 (1956); *Angew. Chem.* **80**, 179 (1968); *Angew. Chem., Int. Ed. Engl.* **7**, 172 (1968); J. K. Rasmussen and A. Hassner, *Chem. Rev.* **76**, 389 (1976).

of the intermediate ureas **128**) heterocondensed and unsubstituted uracils **129–131**[86] (Scheme 31).

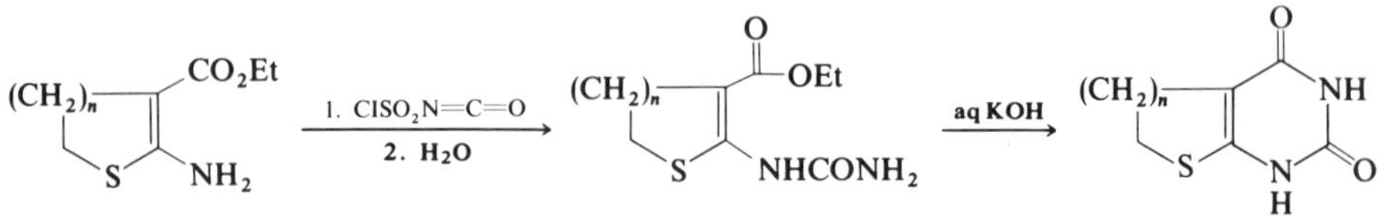

(**128**)   (**129**)

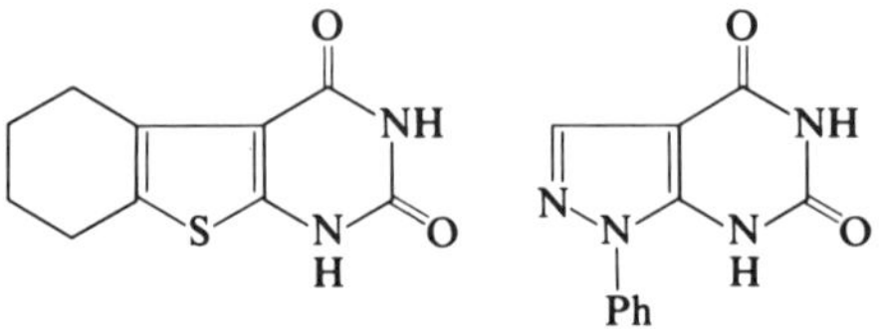

(**130**)   (**131**)

SCHEME 31

B. REACTIONS WITH IMIDO ESTERS, O-ALKYLLACTIM ETHERS, FORMAMIDE, AND FORMALDEHYDE

Imido esters, lactim ethers, and formamide are less reactive than isocyanates and isothiocyanates. Thus a higher temperature must be used. When the desired compounds are sufficiently thermostable, condensed products can be obtained via **132** and **133** in high yields (Scheme 32). Thus with imido esters, the bicyclic 4-pyrimidinones **134**, and with lactim ethers, tricyclic systems **135** are smoothly formed in high yield.[71,88] This heterocyclization method has been employed in many syntheses (see Scheme 33).[89–93]

---

[88] H. Wamhoff, *Chem. Ber.* **102**, 2739 (1969).
[89] H. Wamhoff and B. Wehling, *Chem. Ber.* **108**, 2107 (1975).
[90] H. Wamhoff, *Chem. Ber.* **105**, 743 (1972).
[91] S. Rajappa, B. G. Advani, and R. Sreenivasan, *Indian J. Chem., Sect. B* **14B**, 391 (1976) [*CA* **85**, 192651n (1976)].
[92] V. I. Shvedov, I. A. Kharizomenova, and A. N. Grinev, *Khim. Geterosikl. Soedin.*, 765 (1975) [*CA* **83**, 164119k (1975)].
[93] M. S. Manhas and S. G. Amin, *J. Heterocycl. Chem.* **13**, 903 (1976).

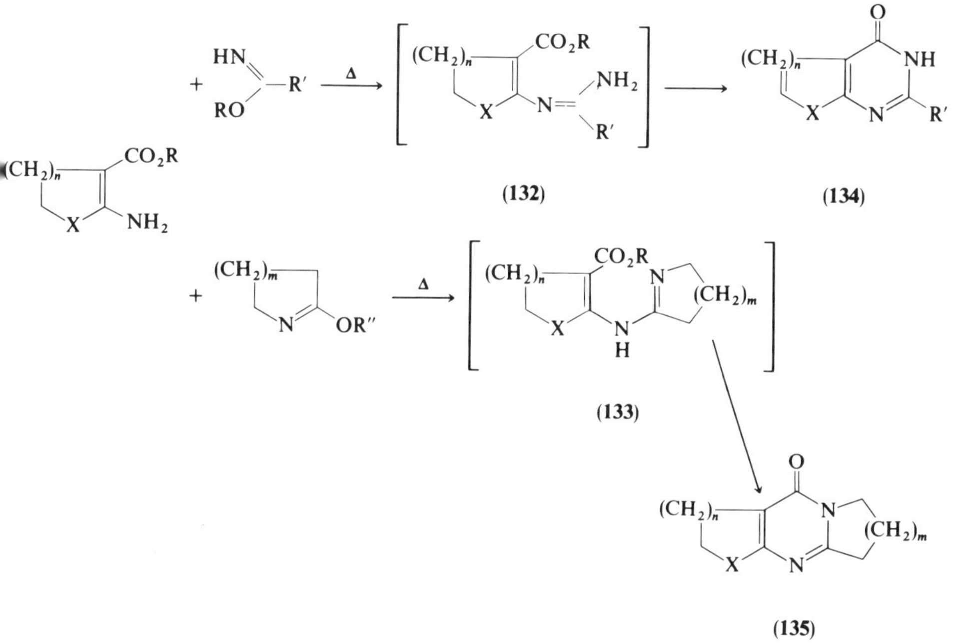

SCHEME 32

A special modification of this heterocyclization is the treatment of pyrimidinenamino esters (**151**) with lactam acetals such as **152**[94]; **153** is the primary condensation product, and basic ring closure affords the tricycle **154** (Scheme 34).

Formamide is a powerful C-N-heterocyclization reagent. 2-Formamidino derivatives are intermediates. Robins transformed 3-aminopyrazole-4-carboxamide to a 4-oxopyrazolo[3,4-d]pyrimidine (**155**)[95] by heating in formamide. In boiling formamide, **155** was obtained from ethyl 3-aminopyrazole-4-carboxylate (**38**)[31] (Scheme 35). From the enamino ester **59** and formamide a 5,6,7,8-tetrahydrobenzo[b]thieno[2,3-d]pyrimidin-4-one (**156**) was obtained,[96] as well as the aromatic thionaphtheno[2,3-d]-pyrimidine (**157**).[97] This heterocyclization method shows broad applicability

---

[94] V. G. Granik, N. B. Marchenko, and R. G. Glushkov, *Khim. Geterosikl. Soedin.*, 1549 (1978) [*CA* **90**, 121529w (1979)]; E. O. Sochneva, N. P. Solov'eva, and V. G. Granik, *ibid.*, 1671 [*CA* **90**, 121530q (1979)].

[95] R. K. Robins, *J. Am. Chem. Soc.* **78**, 784 (1956).

[96] V. I. Shvedov, V. K. Ryzhkova, and A. N. Grinev, *Khim. Geterosikl. Soedin.*, 459 (1967) [*CA* **68**, 59519h (1968)].

[97] K. Gewald and G. Neumann, *Chem. Ber.* **101**, 1933 (1968).

SCHEME 33. Heterocyclization with

Starting Material	Product	
13	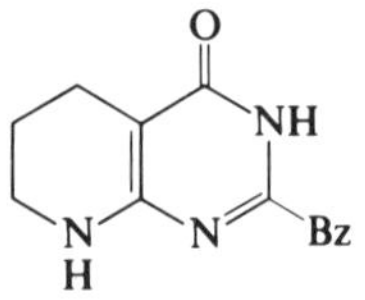 (structure)	(136)[76]
13	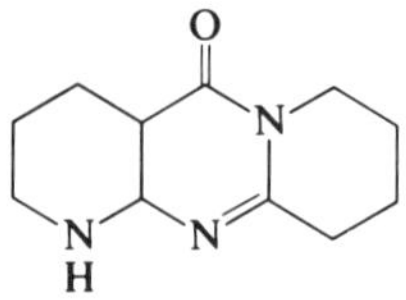 (structure)	(137)[76]
13	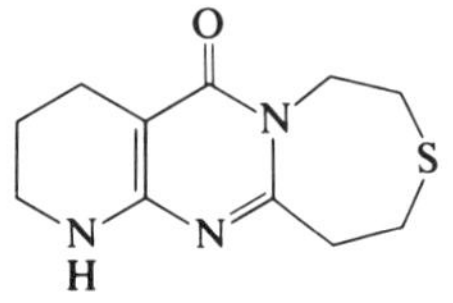 (structure)	(138)[76]
37	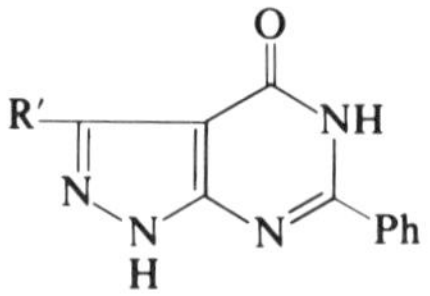 (structure)	(139)[71]
37	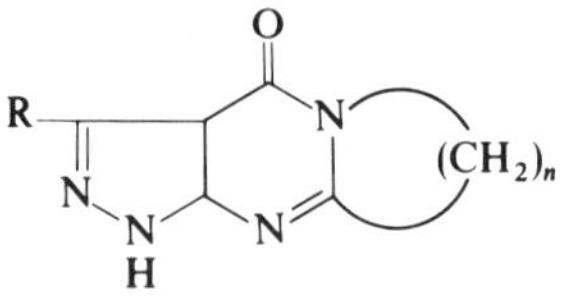 (structure)	(140)[88]
33	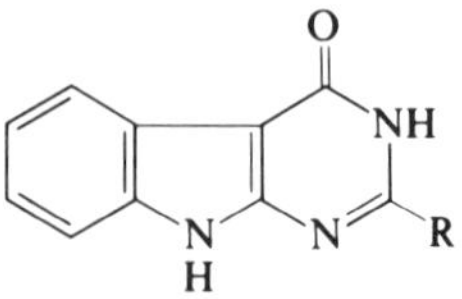 (structure)	(141)[89]
33	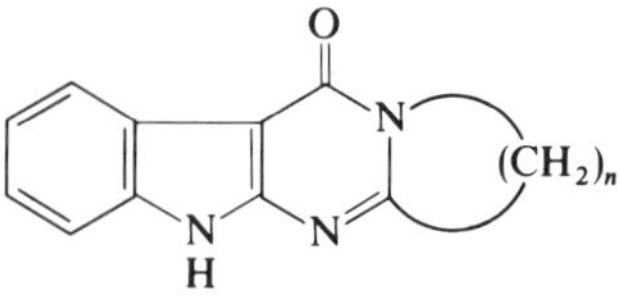 (structure)	(142)[89]

imido esters and lactam ethers

Starting Material	Product	
47	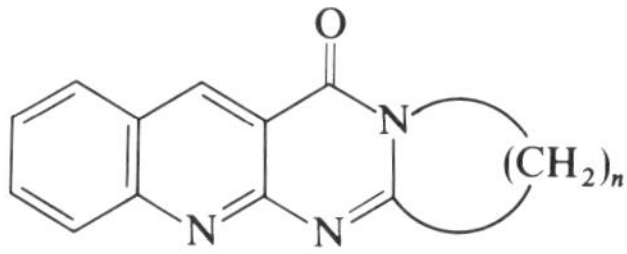 [structure]	(143)[80]
47	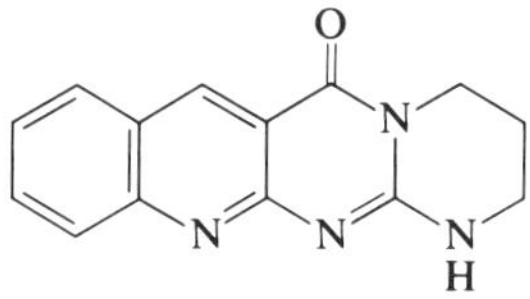 [structure]	(144)[80]
42	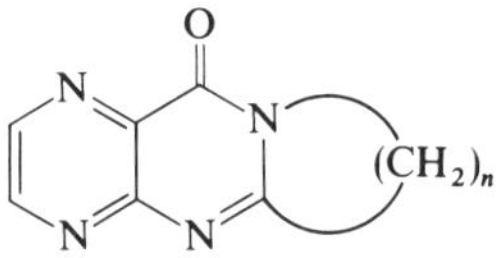 [structure]	(145)[90]
43	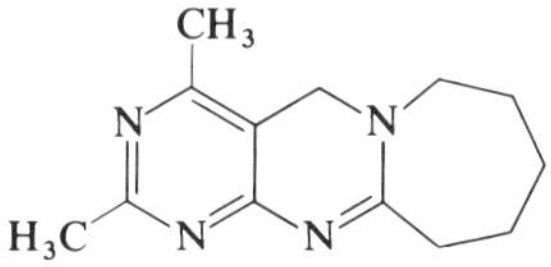 [structure]	(146)[90]
36	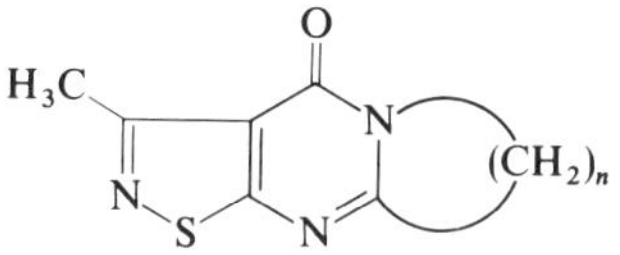 [structure]	(147)[91]
36	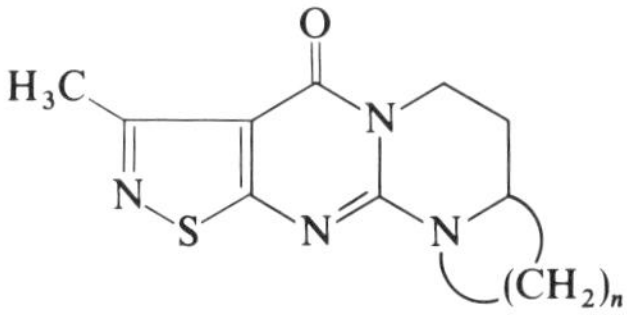 [structure]	(148)[91]
32	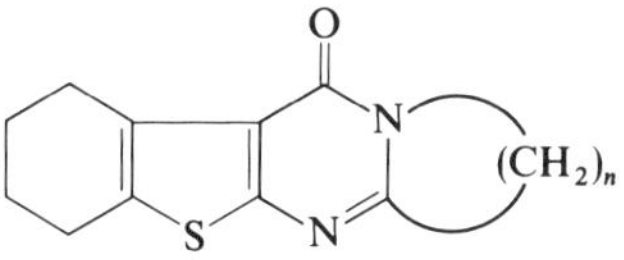 [structure]	(149)[92]
32	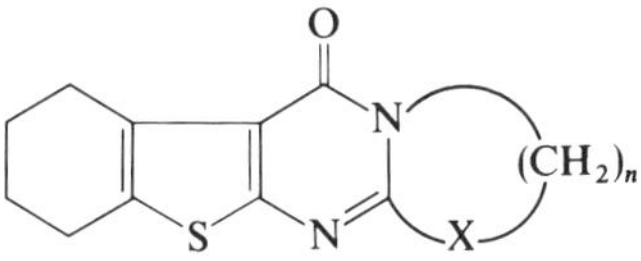 [structure]	(150)[93]

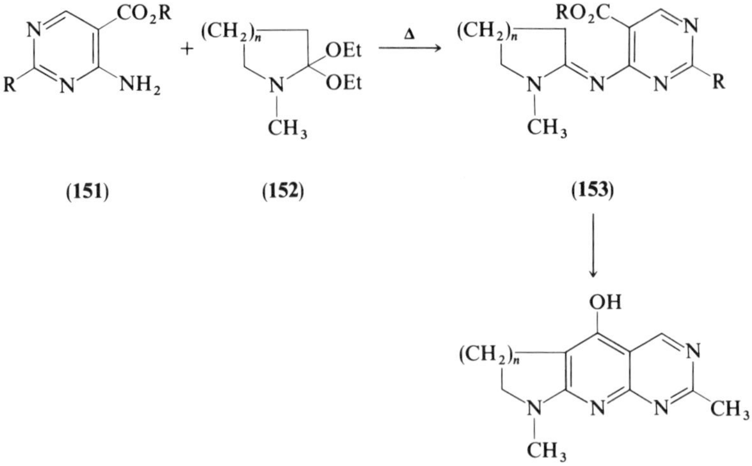

(151) (152) (153) (154)

SCHEME 34

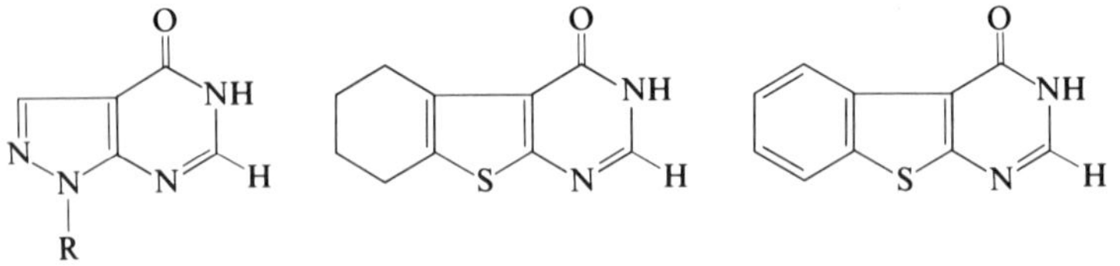

(155) (156) (157)

SCHEME 35

both for partially saturated and aromatic enamino esters, as shown in Scheme 36.[98–102] Compound **156** was independently obtained from 2-amino-4,5,6,7-tetrahydrobenzo[*b*]thiophene-3-carboxamide and formic acid.[103]

[98] W. Ried and R. Giesse, *Justus Liebigs Ann. Chem.* **713**, 149 (1968).

[99] K. Gewald and O. Calderon, *Monatsh. Chem.* **108**, 611 (1977).

[100] W. Ried and R. Giesse, *Justus Liebigs Ann. Chem.* **713**, 143 (1968); V. P. Arya and S. P. Ghate, *Indian J. Chem.* **9**, 1209 (1971) [*CA* **76**, 99604d (1972)].

[101] R. G. Glushkov, V. A. Volokova, and O. Yu. Magidson, *Khim.-Farm. Zh.* **1**, 25 (1967) [*CA* **68**, 105, 143f (1968)].

[102] K. Gewald and H. J. Jänsch, *J. Prakt. Chem.* **315**, 779 (1973).

[103] F. Sauter, *Monatsh. Chem.* **99**, 1507 (1968); **101**, 535 (1970).

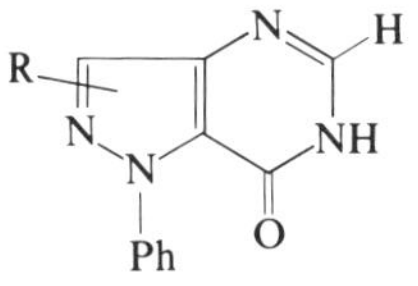

Sec. V.B]   HETEROCYCLIC β-ENAMINO ESTERS   329

(158)[98,99] 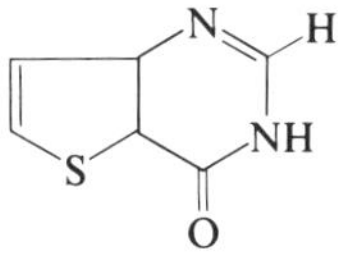   (159)[100] 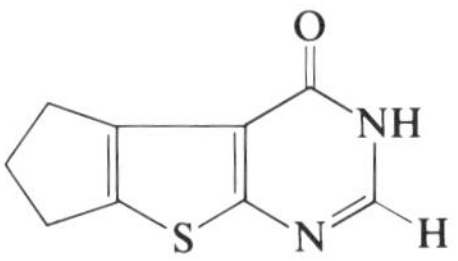   (160)[100]

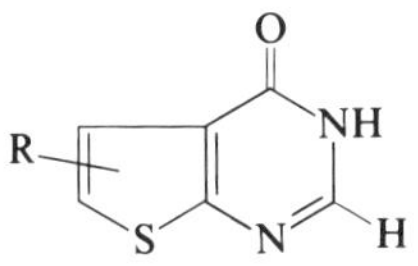

(161)[100] 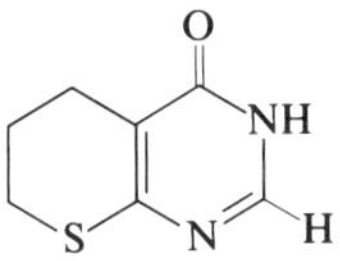   (162)[88] 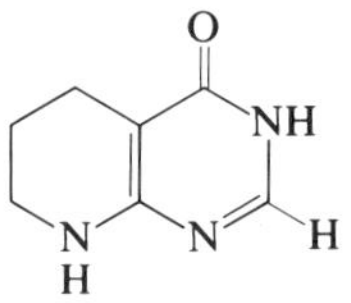   (163)[78]

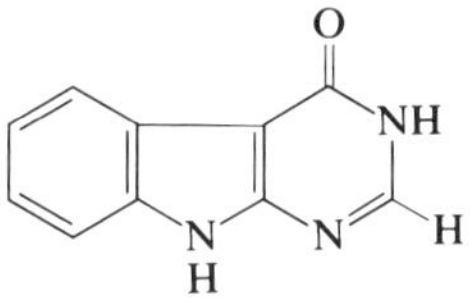

(164)[101] 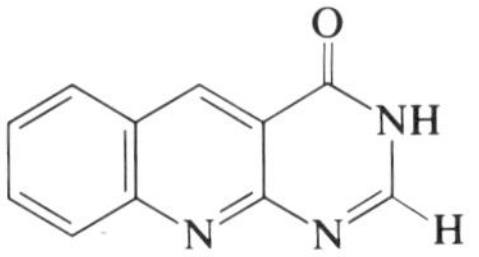   (165)[80]

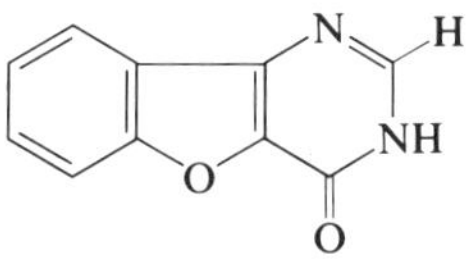

(166)[102] 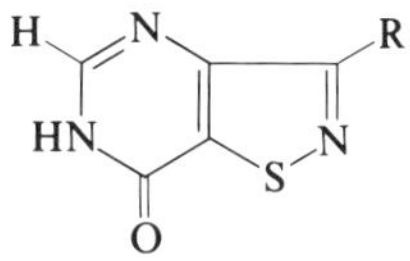   (167)[49]

SCHEME 36

Some partially saturated and heteroaromatic β-enamino esters, such as **8** (X = S, $n = 2$), **32**, and **36**, with formaldehyde and acetaldehyde give the methylenediamines **168–170** by N-methylene linkage[104] (Scheme 37). Double head-to-tail condensation of the pyrazolenamino ester **37** leads to a symmetrical tetrazocane (**171**); a 3:3-adduct (**172**) was detected[104] as a side product. From the IR and ^{1}H-NMR data, it follows that **168–170** display a chelate bridge from the 2-amino group to the adjacent 3-ester function, owing to the electron-donating influence of the N-alkyl group[104] (Scheme 38).

[104] H. Wamhoff, G. Hendrikx, and M. Ertas, *Liebigs Ann. Chem.*, 489 (1982).

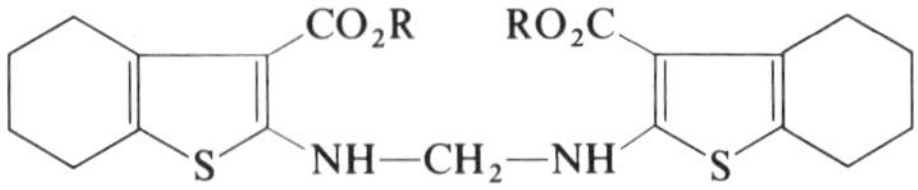

(168)

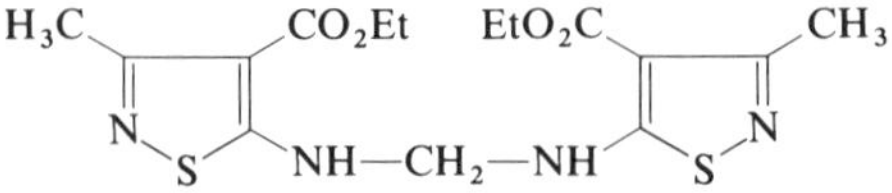

(169)

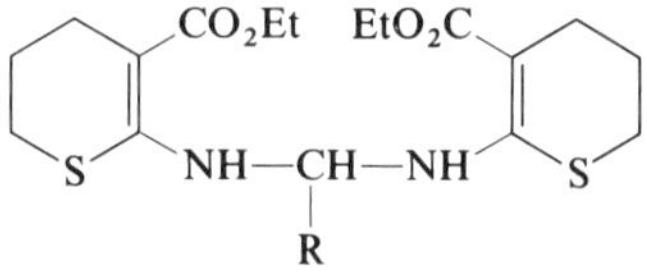

(170)

SCHEME 37

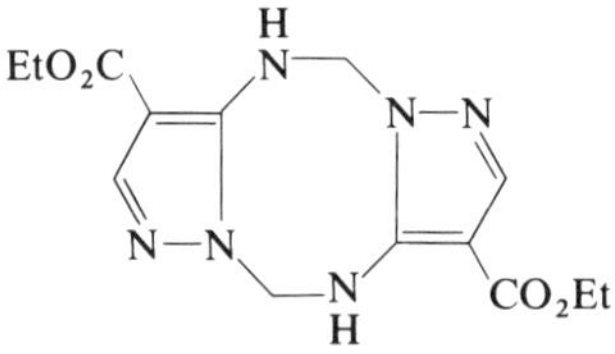

(171)

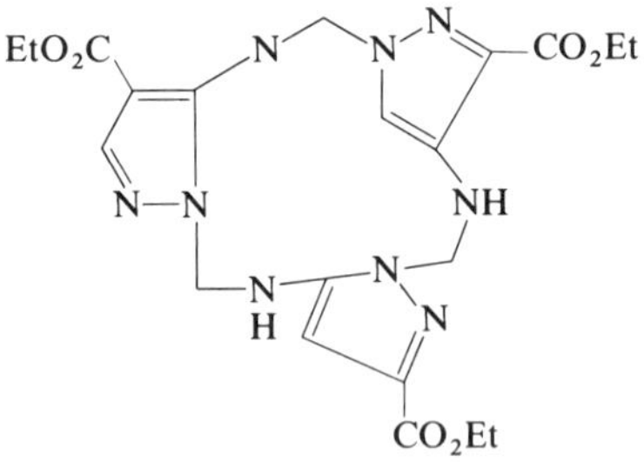

(172)

SCHEME 38

C. Heterocondensed 1,3-Oxazinones from 2-Acylamino Derivatives and Dihalotriphenylphosphoranes

Dihalogentriphenylphosphoranes (X_2PPh_3) formed *in situ* from triphenylphosphanes, hexachloroethane,[105] dibromotetrachloroethane,[106] and a base[107] are powerful reagents for heterocyclic synthesis.[108] By treatment of

[105] R. Appel, *Angew. Chem.* **87**, 863 (1975); *Angew. Chem., Int. Ed. Engl.* **14**, 801 (1975); R. Appel and M. Halstenberg, *in* "Organophosphorus Reagents in Organic Synthesis" (J.I.G. Cadogan, ed.), p. 387ff. Academic Press, London, 1979.
[106] G. Bringmann and S. Schneider, *Synthesis*, 139 (1983).
[107] Standard base: triethylamine[105]; in some cases improved yields with 1,5-diazabicyclo[4.3.0]-5-nonene (H. Wamhoff, A. Böhle, and R. Lüttgens, unpublished results).
[108] E. g., L. Farkas, J. Keuler, and H. Wamhoff, *Chem. Ber.* **113**, 2566 (1980).

the N-acylated S-heterocyclic β-enamino esters **8** (X = S, n = 1, 2) and **32** with dihalogentriphenylphosphoranes, heterocondensed 1,3-oxazinones **173** and **174** are formed smoothly.[109] This facile ring closure (instead of dehydration to ketenimines[105]) is explained either by rapid and preferred attack of an intermediate imide chloride (**175**) on the neighboring ester group or by chlorolysis of the 3-ester moiety **176**[110] and subsequent cyclization (Scheme 39). Bisacylenamino esters, such as **27** or **178**, are cyclized in a similar way to give bis-1,3-oxazinones **177** and **179**[104,109] (Scheme 40). By means of P_4S_{10}, oxazinones **173** and **174** are smoothly converted to the novel, deep-red 1,3-thiazin-6-thiones **180** and **181**, which are in turn selectively photooxidized with singlet-oxygen, via spiro-1,2,3-dioxathietanes (**182**), to afford 1,3-thiazin-6-ones **183**[111] (Scheme 41).

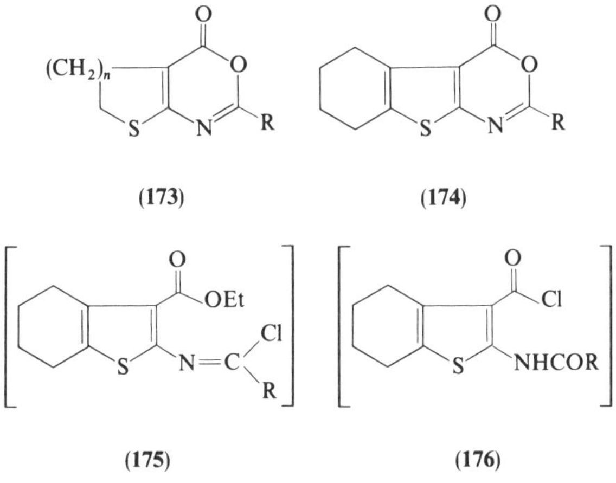

SCHEME 39

With hydrazine, **180** affords **184** and in a subsequent elimination, 4-hydrazinobenzo[b]thieno[2,3-d]pyrimidine (**185**), which can be smoothly cyclized with the orthoester R(OEt)₃ in a known procedure[112] to form a 1,2,4-triazolo[4,3-c]benzo[b]thieno[3,2-e]pyrimidine (**186**).[113] Upon heating with

---

[109] D. Achakzi, M. Ertas, R. Appel, and H. Wamhoff, *Chem. Ber.* **114**, 3188 (1981).
[110] D. J. Burton and W. M. Koppes, *J. Org. Chem.* **40**, 3026 (1975).
[111] H. Wamhoff and M. Ertas, *Angew. Chem.* **94**, 800 (1982); *Angew. Chem., Int. Ed. Engl.* **21**, 794 (1982); S. Leistner and G. Wagner, *Z. Chem.* **17**, 95 (1977).
[112] D. J. Brown and K. Shinozuka, *Aust. J. Chem.* **33**, 1147 (1980); C. J. Shoshoo, M. B. Devani, G. V. Ullas, S. Ananthan, and V. S. Bhadti, *J. Heterocycl. Chem.* **18**, 43 (1981).
[113] H. Wamhoff and M. Ertas, *Chem.-Ztg.* **107**, 344 (1983).

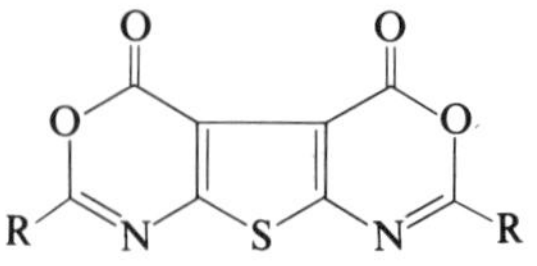

(177)

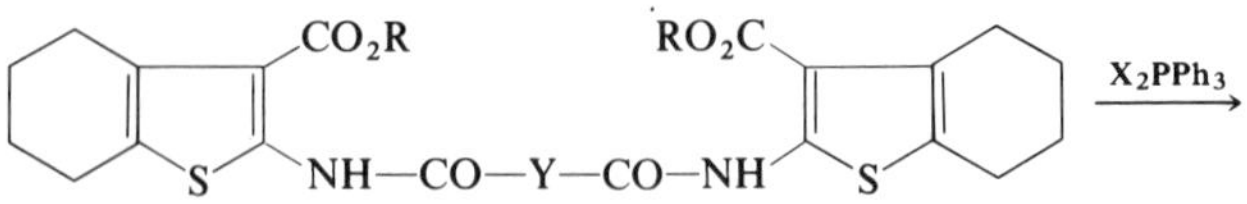

(178)

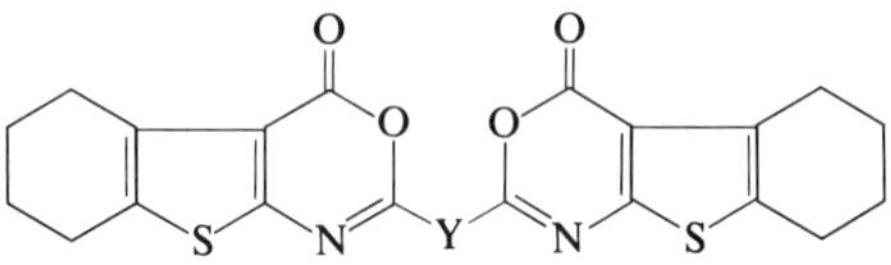

(179)

SCHEME 40

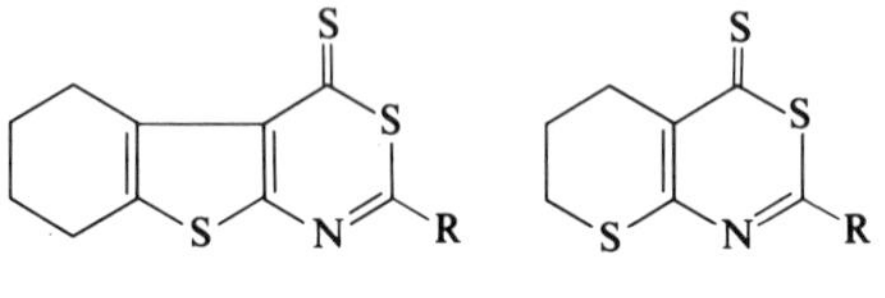

(180) (181)

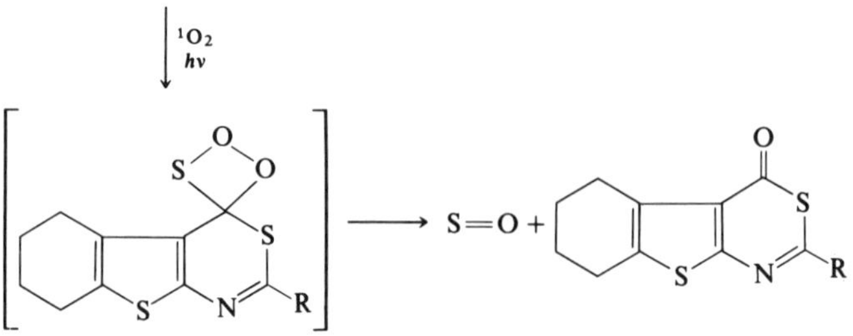

$^1O_2$
$h\nu$

(182) → S=O + (183)

SCHEME 41

N-dichloromethylenesulfonamide (**186a**), 5-aminopyrazole-4-carboxylic acid (**38**, Et = H) is converted to a pyrazolo[3,4-d][1,3]oxazine (**186b**).[113a]

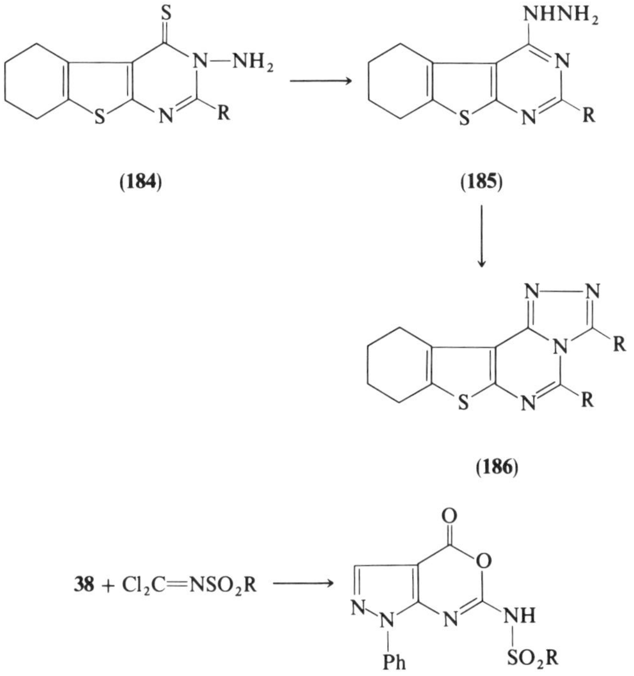

(**184**)   (**185**)

(**186**)

38 + Cl$_2$C=NSO$_2$R ⟶

(**186a**)   (**186b**)

SCHEME 42

D. HETEROCONDENSED [1,2-a]PYRIMIDINES

N-Heterocyclic β-enamino esters which are unsubstituted in the 1-position are well suited for a double condensation reaction when treated with 1,3-dicarbonyl compounds to give annelation of a [1,2-a]pyrimidine ring. Thus, the condensation of 2-aminopyrroline affords pyrrolo[1,2-a]-pyrimidines.[82,114] Shvedov et al. treated 2-amino-3-pyrrolecarbonitrile **187** with β-diketones or ethyl acetoacetate[74,115] to give **188**. No ring closure

---
[113a] F. L. Merchán, *Synthesis*, 965 (1981).
[114] A. Le Berre and C. Renault, *Bull. Soc. Chim. Fr.*, 3139 (1969).
[115] V. I. Shvedov, I. A. Kharizomenova, L. B. Altrukhova, and A. N. Grinev, *Khim. Geterosikl. Soedin.*, 428 (1970) [*CA* **73**, 25403 (1970)].

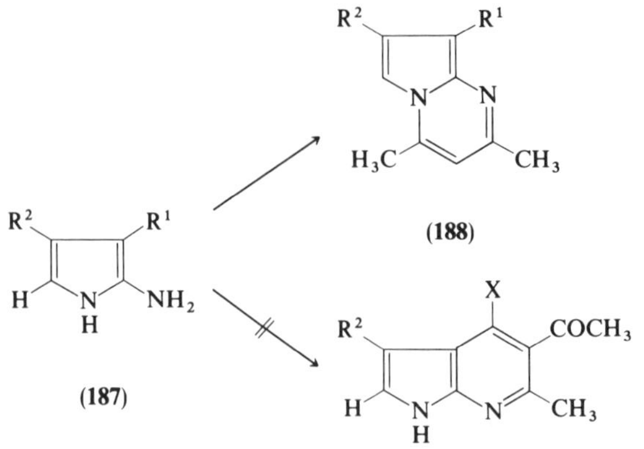

(187)

R¹ = CO₂R, CN
R² = CH₃, CO₂R

(188)

(189) X = OH, NH₂

SCHEME 43

(190)

COCH₃
|
CH₂CO₂Et
(NaOEt)

NCCH₂CO₂Et
NaOR

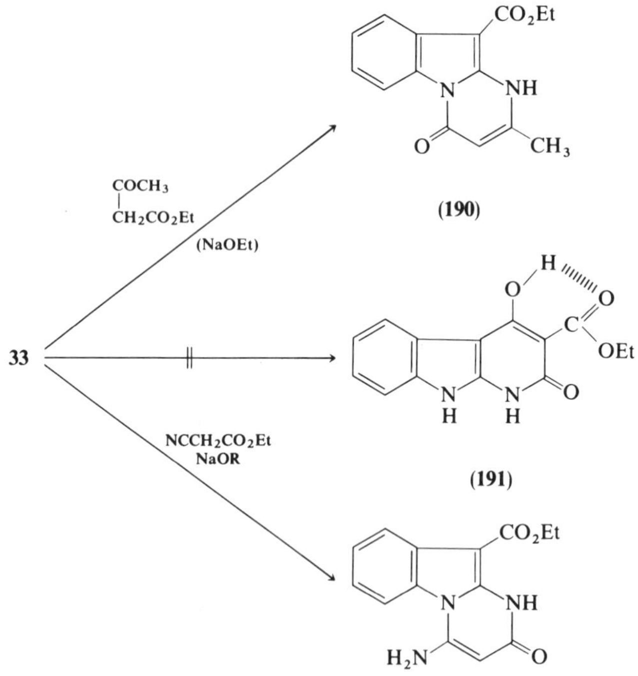

(191)

(192)

SCHEME 44

Sec. V.D] HETEROCYCLIC β-ENAMINO ESTERS 335

involving the 3-ester group to form **189** is observed[74] (Scheme 43). In a similar way, ethyl 2-aminoindole-3-carboxylate **33** affords, with several 1,3-dicarbonyl compounds, pyrimido[1,2-*a*]indoles **190**; again no pyrido[2,3-*b*]indole **191** could be detected[89] (Scheme 44). Accordingly, ethyl cyanoacetate gives the expected tricyclic amino derivative **192**, but now with a reverse regiospecificity.[89]

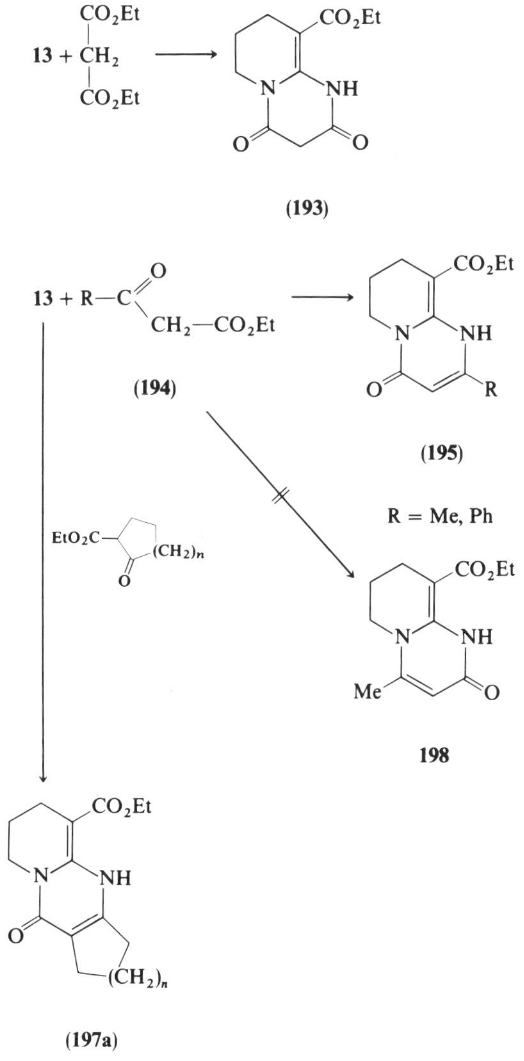

(193)

(194)

(195)

R = Me, Ph

198

(197a)

SCHEME 45

Similarly, by treatment with malonic ester, 2-amino-1,4,5,6-tetrahydropyridine-3-carboxylate (13) is smoothly and exclusively converted to the pyrido[1,2-a]pyrimidine system 193.[116] Basic catalysts convert asymmetrical 1,3-dicarbonyl compounds, such as acetoacetates 194 and cycloalkanone-2-carboxylates, to pyrido[1,2-a]pyrimidines 195 and 197a; isomers such as 198 were not found.[116] Structures 195 and 197a are supported by their UV spectra.[117,118] ^{1}H-NMR spectroscopy of 195 reveals a tautomeric equilibrium (195I ⇌ 195II)[116] (Scheme 46).

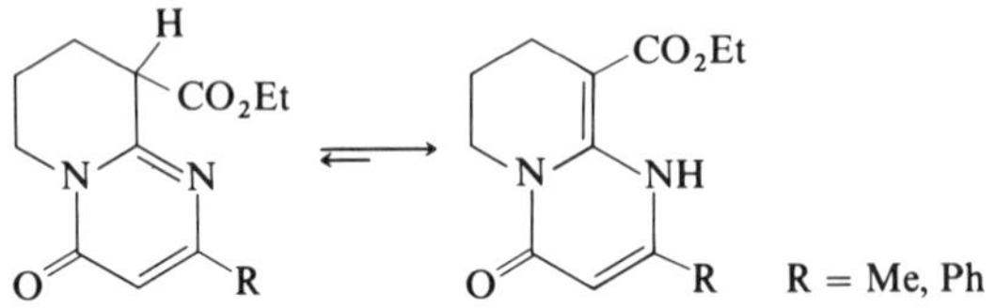

[20%]   [80%]

(195II)   (195I)

SCHEME 46

In line with the above, 13 affords with diethyl oxalate the imidazo[1,2-a]pyrimidine 196[116] (Scheme 47). However, ethyl 2-amino-3-quinoline-

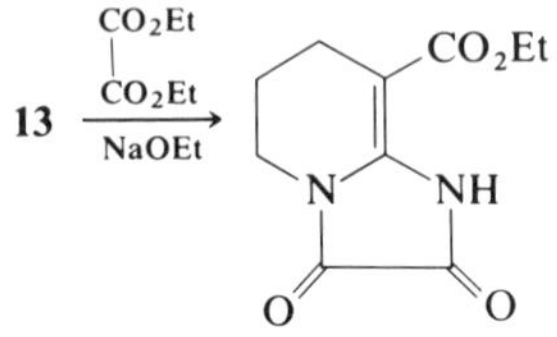

(196)

SCHEME 47

carboxylate (47) with ethyl malonate yields—similar to the Dornow and Neuse cyclization[119]—the potentially tautomeric benzo[b]-1,8-naphthyridine 197b[80] (Scheme 48). 3,5-Diamino-4-ethoxycarbonylpyrazole

---

[116] H. Wamhoff and L. Lichtenthäler, *Chem. Ber.* **111**, 2813 (1978).
[117] C. F. H. Allen, H. R. Beilfuss, D. M. Burness, G. A. Reynolds, J. F. Tinker, and J. A. Van Allan, *J. Org. Chem.* **24**, 779 (1959).
[118] H. L. Yale and E. R. Spitzmiller, *J. Heterocycl. Chem.* **13**, 797 (1976).
[119] A. Dornow and E. Neuse, *Arch. Pharm. Ber. Dtsch. Pharm. Ges.* **288**, 174 (1955).

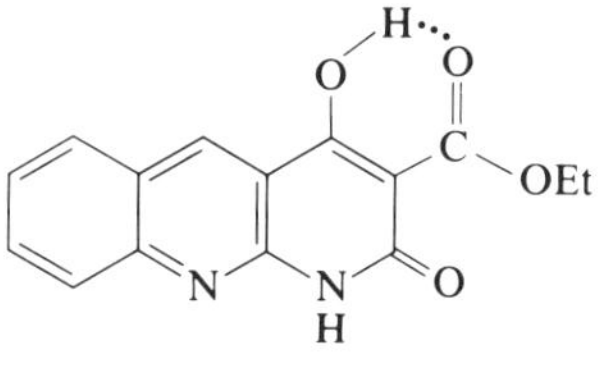

Sec. V.E] HETEROCYCLIC β-ENAMINO ESTERS 337

(197b)

SCHEME 48

(37) reacts with β-bifunctional reagents to form pyrazolo[1,5-a]pyrimidines 198 and 199 (Scheme 49).[120]

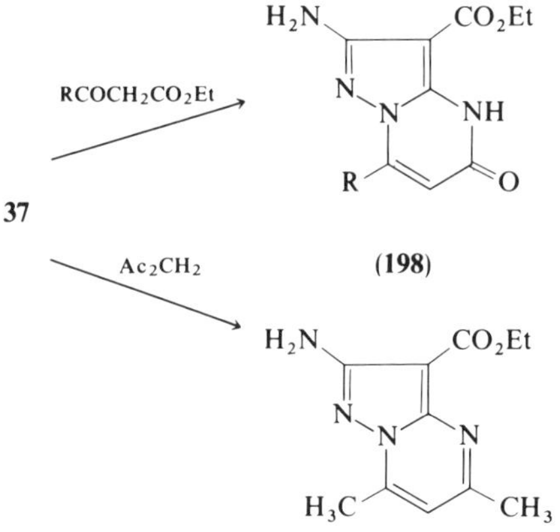

SCHEME 49

E. ELECTROPHILIC ADDITION TO CARBON-3

The highly electron-deficient 1,2,4-triazoline-3,5-diones 200[121,122] readily attack the 3-position of ethyl 2-amino-4,5-dihydrofuran-3-carboxylates (8, X = O, n = 1) at room temperature to give the imino lactone adducts

---

[120] E. M. Kandeel, V. B. Baghos, I. S. Mohareb, and M. H. Elnagdi, *Arch. Pharm. Ber. Dtsch. Pharm. Ges.* **316**, 713 (1983).
[121] H. Wamhoff and K. M. Wald, *Org. Prep. Proced. Int.* **7**, 251 (1975); cf. M. Fieser and L. F. Fieser, "Reagents for Organic Synthesis," Vol. 6, p. 75ff. Wiley (Interscience), New York 1977.
[122] H. Wamhoff and G. Kunz, *Angew. Chem.* **93**, 832 (1981); *Angew. Chem., Int. Ed. Engl.* **20**, 797 (1981), and references cited therein; H. Wamhoff and G. Kunz, unpublished results.

201[123] Saponification, methylation, and heating in DMSO/NaCl[124] afford rearrangement to the spiro lactones **202**[123] (See Scheme 50). However, direct treatment of lactone **203** with DMSO/NaCl leads under elimination of HNCO to phenylureidobutenolides **204**.[123]

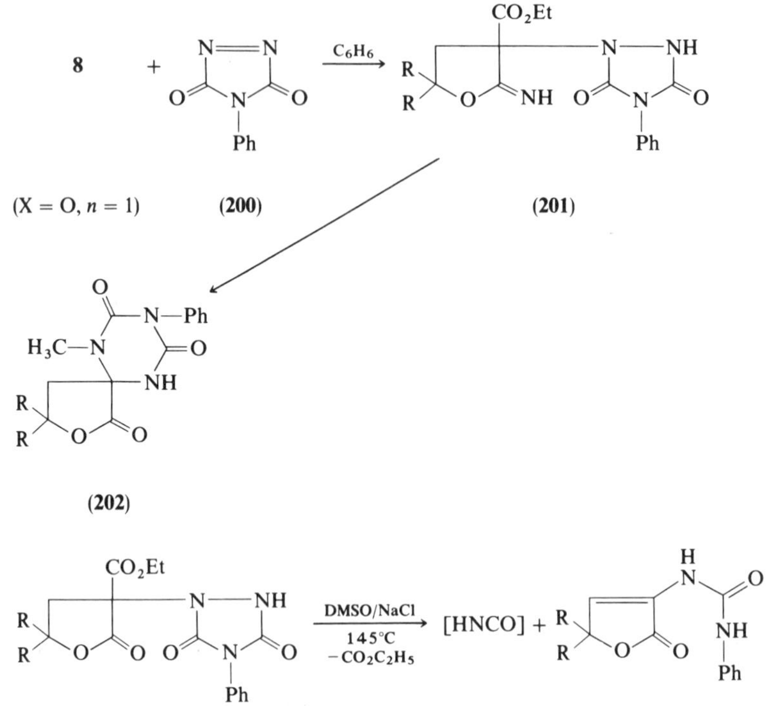

(X = O, *n* = 1) (**200**) (**201**)

(**202**)

(**203**) (**204**)

SCHEME 50

In tetrahydrofuran the dihydrofurans **8** (X = O, *n* = 1) react with maleic anhydride at room temperature to yield pyrano[4,3-*c*]pyridines **205** via a complex addition–rearrangement scheme[125] involving betaine **206** (Scheme 51). Furthermore, in the formation of the naphthyridines **207** and **208** from **13** and acetylenic esters or methyl acrylate, the first step consists of a Michael-type attack on the electron-rich C-3[116] (Scheme 52).

[123] H. Wamhoff and K. Wald, *Chem. Ber.* **110**, 1716 (1977).

[124] S. Takei and Y. Kawano, *Tetrahedron Lett.*, 4389 (1975); cf. also A. P. Krapcho, *Synthesis*, 805, 893 (1982), and references cited therein.

[125] G. Szilágyi and H. Wamhoff, *Synthesis*, 698 (1980).

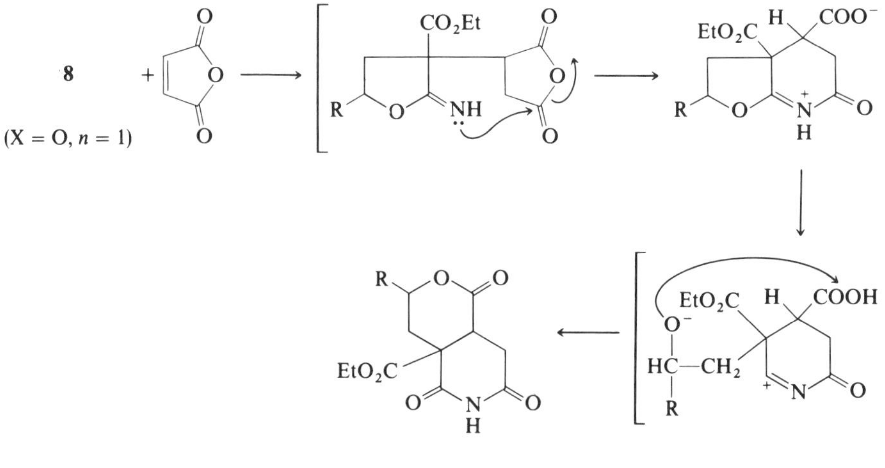

(205) (206)

SCHEME 51

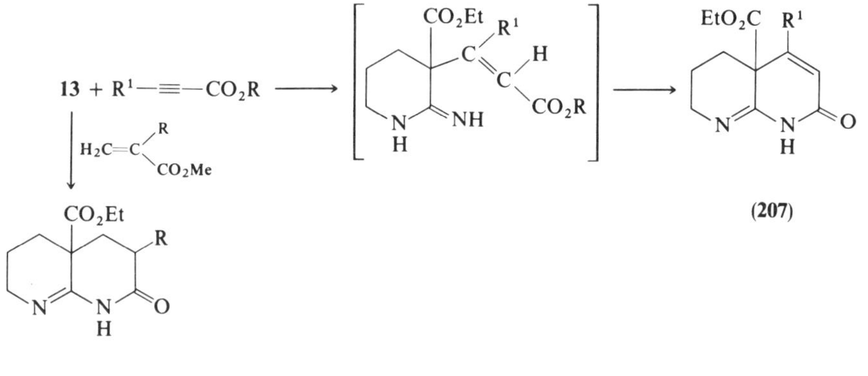

(207)

(208)

SCHEME 52

F. MISCELLANEOUS REACTIONS

Recently, some work has appeared which deals with reactions of heterocyclic β-enamino esters with bifunctional reaction partners leading in two or more consecutive steps to polycyclic heterocyclic systems. Thus, allyl isothiocyanate gives with thiophene **32** after attack of the 2-amino group in a double cyclization reaction, a benzo[*b*]thieno[2,3-*d*]thiazolo[3,2-*a*]pyrimidine (**209**)[126] (Scheme 53). 1,2-Tri- and -tetramethylenethieno[2,3-*d*]pyrimidin-4-

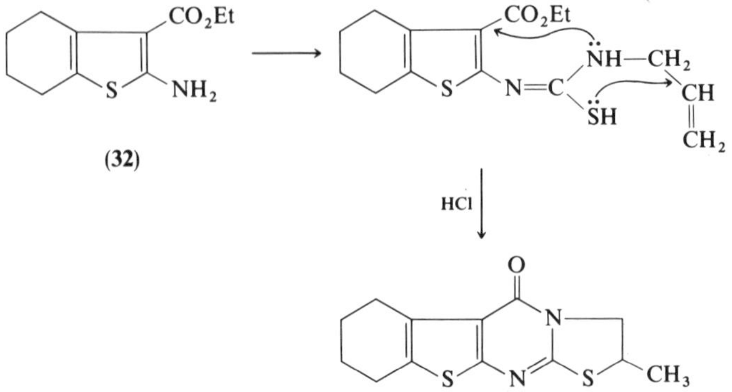

SCHEME 53

ones **211** are formed by heating of **32** with 3- and 4-chloroalkyl nitriles **210**[127] (Scheme 54).

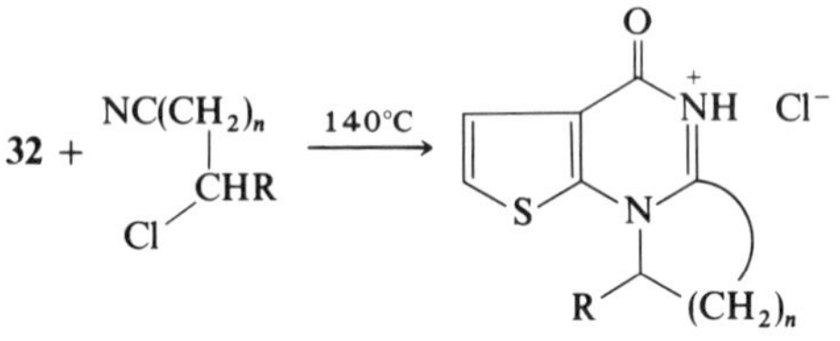

SCHEME 54

1,3-Thiazolenamino esters **212** give thiazolo[3,2-*a*]thiazolo[5,4-*d*]-pyrimidine systems such as **215**[128] via several steps involving intermediates **213** and **214** (Scheme 55).

[126] H. K. Gakhar, A. Madan, A. Khanna, and M. Kumar, *J. Indian Chem. Soc.* **55**, 705 (1978)[*CA* **90**, 6346x (1979)].

[127] V. A. Kovtunenko, L. V. Soloshonok, A. K. Tyltin, and F. S. Babichev, *Ukr. Khim. Zh.* (*Russ. Ed.*) **49**, 855 (1983) [*CA* **99**, 175 709 j(1983)].

[128] P. B. Talukdar, S. K. Sengupta, and A. K. Datta, *Indian J. Chem., Sect. B* **22B**, 243 (1983) [*CA* **99**, 122 400v (1983)].

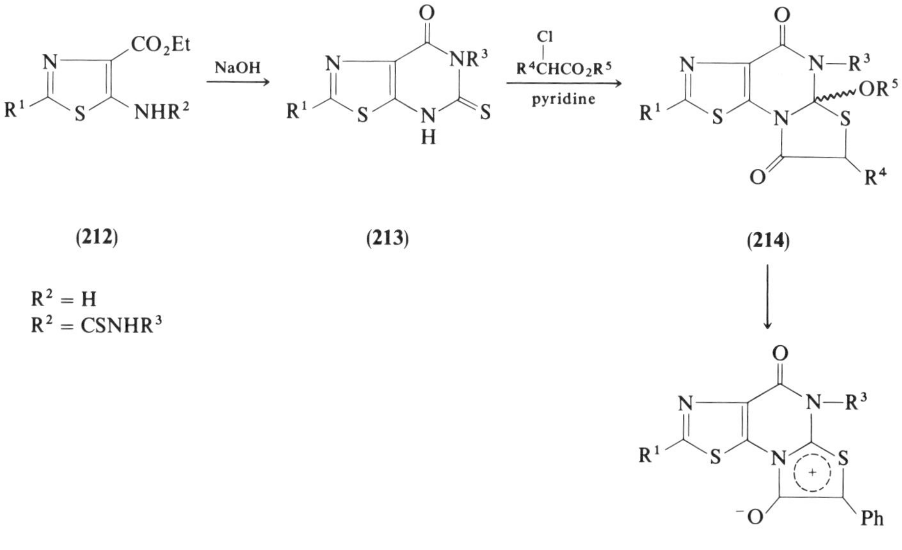

Sec. V.F] HETEROCYCLIC β-ENAMINO ESTERS 341

$R^2 = H$
$R^2 = CSNHR^3$

SCHEME 55

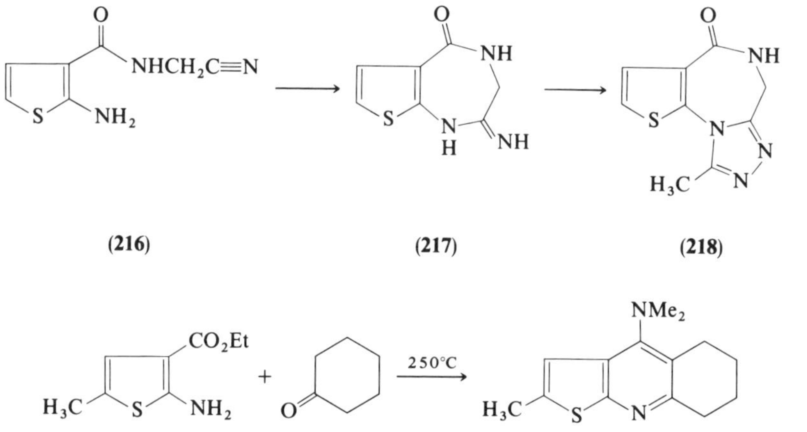

SCHEME 56

2-Amino-*N*-cyanomethyl-3-thiophenecarboxamide (**216**), easy accessible according to the Gewald method,[21] cyclizes in the presence of sodium methoxide to give 2-amino-3,4-dihydrothieno[2,3-*f*]-1,4-diazepine-5-one (**217**), which is converted with acetylhydrazine to a thieno[3,2-*f*][1,2,4]-triazolo[4,3-*a*][1,4]diazepine-6-one (**218**).[129] These compounds are of interest because of their hypnogenic and anxiolytic properties (Scheme 56). By a one-step procedure, 2-aminothiophene-3-carboxylates **219** are transformed with cyclohexanone by heating with HMPT and catalytic amounts of PPA to 4-dimethylamino-2-methyl-5,6,7,8-tetrahydrothieno[2,3-*b*]quinolines such as **220**.[130]

Tetrazolo[1,5-*c*]thieno[3,2-*c*]pyrimidines (**222**) are obtained from the thiophenenamino esters **32** and nitriles, via the thienopyrimidinones, the

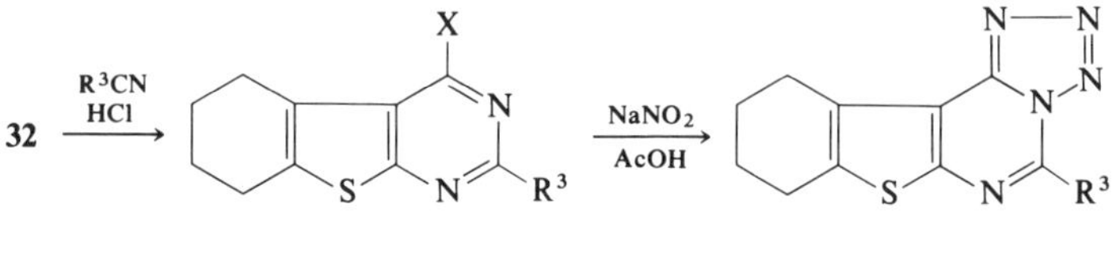

(**221**) X = OH
X = Cl
X = NHNH$_2$

(**222**)

SCHEME 57

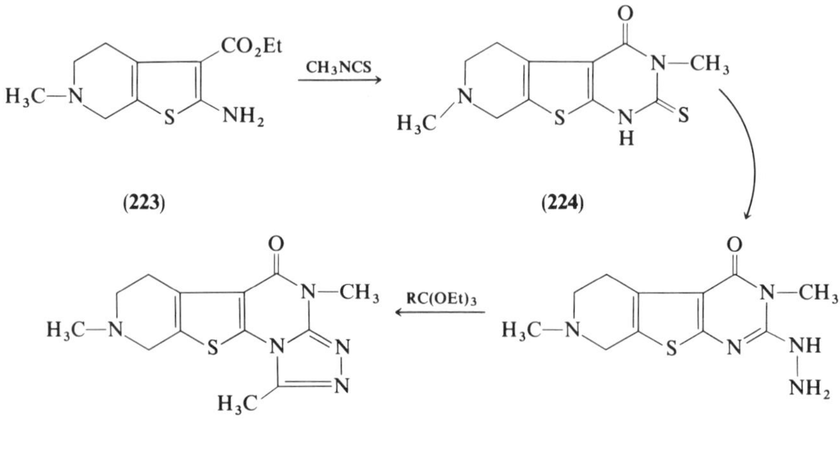

(**223**)

(**224**)

(**226**)

(**225**)

SCHEME 58

[129] K. H. Weber and H. Daniel, *Liebigs Ann. Chem.*, 328 (1979).
[130] A. Osbirk and E. B. Pedersen, *Acta Chem. Scand., Ser. B* **B33**, 313 (1979).

4-chloro derivatives, and by diazotization of the 4-hydrazino derivatives **221**[131] (Scheme 57).

In a series of consecutive steps involving **224** and **225** the new polyheterocyclic ring system **226** has been made from **223**[132] (Scheme 58). N-Heteroaromatic enamino esters, such as **227**, and ethoxymethylene cyanoacetate condense to afford the vinylenamino ester **228**; with hydrazines a 3-aminopyrimidine is annelated (**229**)[132a] (Scheme 59).

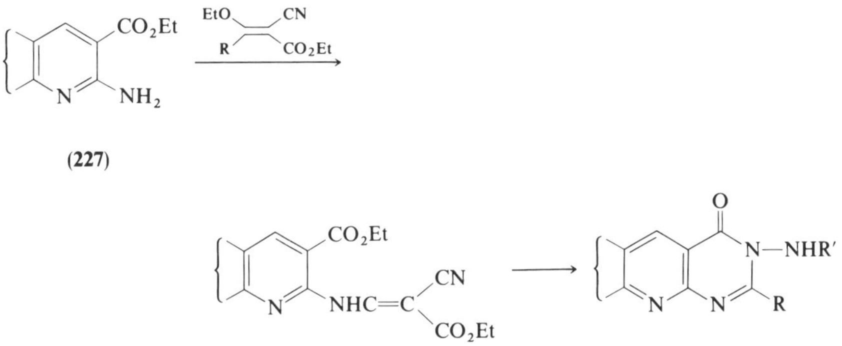

SCHEME 59

VI. Nucleophilic Attacks of Amines and Diamines

Reaction of alkanediamines **230** with ethyl 2-amino-4,5-dihydro-3-furancarboxylates (**8**: X = O, n = 1) does not give furo[2,3-e]-1,4-diazepine-5-ones **231**.[133,134] Instead, as high-resolution ^{13}C-NMR spectra show, in a sequence of addition and elimination steps (**232–233**), 3-(2-imidazolidinylidene)- and 3-(hexahydro-2-pyrimidinylidene)dihydro-2-(3H)-furanones **234** are formed[135] (Scheme 60). The constitutions of **234** were confirmed by ^{13}C-NMR gated decoupling.[135]

[131] C. J. Shishoo, M. B. Devani, M. B. Karvekar, G. V. Ullas, G. V. Ananthan, V. S. Bhadti, R. B. Patel, and T. P. Gandhi, *Indian J. Chem., Sect. B* **21B**, 666 (1982) [*CA* **97**, 198154v (1982)].
[132] I. Lalezari and M. H. Jabari Sahbari, *J. Heterocycl. Chem.* **15**, 873 (1978).
[132a] J. Y. Merour, *J. Heterocycl. Chem.* **19**, 1425 (1982).
[133] P. L. Pacini and R. G. Ghirardelli, *J. Org. Chem.* **31**, 4133 (1966).
[134] H. Wamhoff and C. Materne, *Justus Liebigs Ann. Chem.* **754**, 113 (1971); H. Wamhoff, C. Materne, and F. Knoll, *Chem. Ber.* **105**, 753 (1972); H. Wamhoff and C. Materne, *ibid.* **107**, 1784 (1974).
[135] Z. T. Huang and H. Wamhoff, *Chem. Ber.* **117**, 622 (1984).

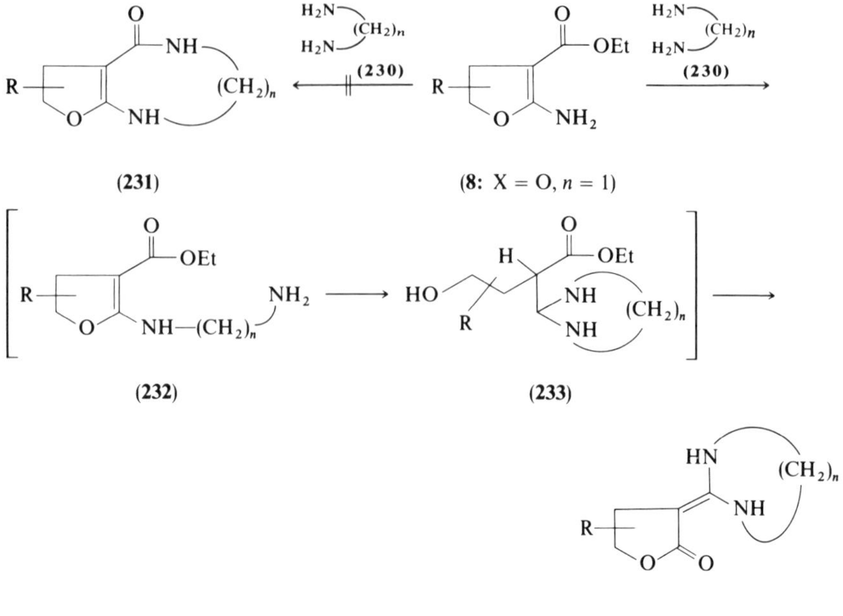

(231) (8: X = O, n = 1)

(232) (233)

(234)

SCHEME 60

Methylamine reacts with **8** (X = O; $n = 1$) to afford a mixture (40:60) of *E*,*Z*-**235**[135] (Scheme 61). Similarly, *o*-phenylenediamine, 1,8-diaminonaph-

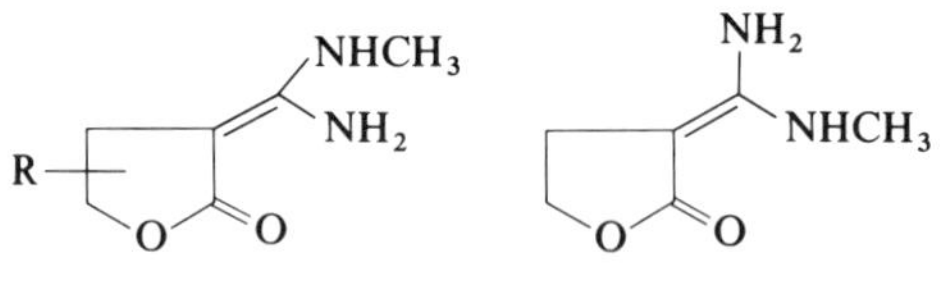

(*E*-**235**) (*Z*-**235**)

SCHEME 61

thalene, 2-aminophenol, and 2-aminoethanol lead, after ring cleavage and subsequent lactonization, to α-(2-benzimidazolyl)- (**236**), α-(2-perimidinyl)- (**237**), α-(2-benzoxazolyl)- (**238**), and α-(2-oxazolin-2-yl)-γ-lactones (**239**)[124] (Scheme 62). Thus reaction of dinucleophiles with **8** (X = O, $n = 1$) provides a simple and excellent access to α-heterocyclic substituted γ-lactones.

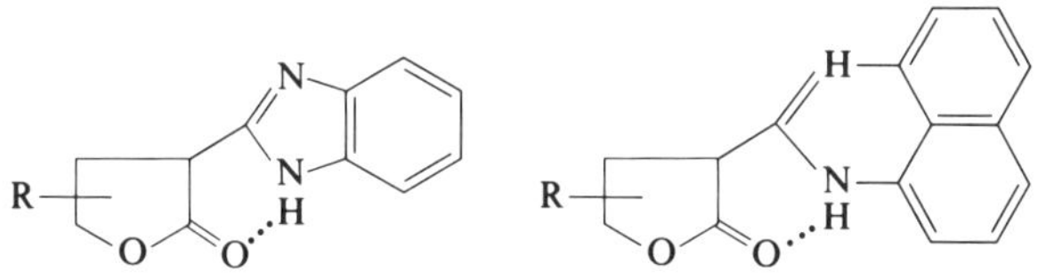

Sec. VI] HETEROCYCLIC β-ENAMINO ESTERS 345

(236) (237)

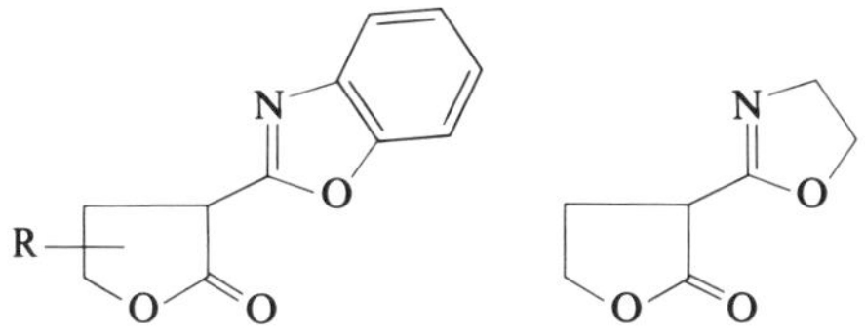

(238) (239)

SCHEME 62

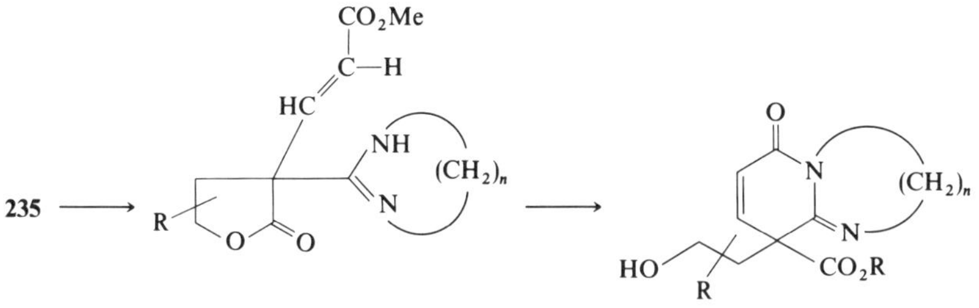

235 →

(240) (241)

235 → [...] −CH₃OH →

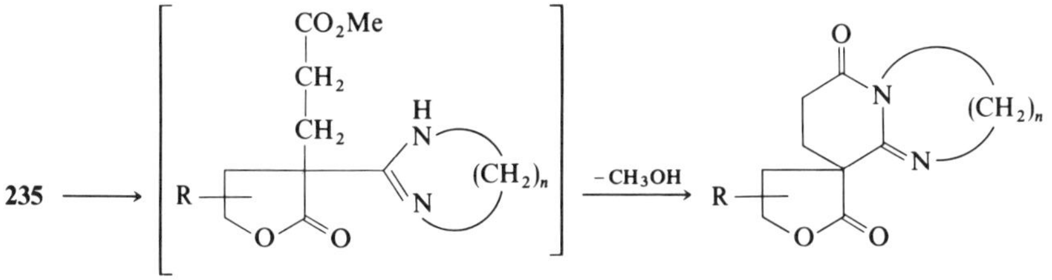

(242)

SCHEME 63

Furthermore, it has been shown that compounds of type **235** are versatile starting materials for addition and cyclization reactions employing reactive triple and double bonds. With methyl propiolate, syn addition takes place to afford the adducts **240**. In the presence of alcohols, compounds **240** are cyclized to give the imidazo[1,2-*a*]pyridines and pyrido[1,2-*a*]pyrimidines **241**[136] (Scheme 63). By treatment with methyl acrylate, the lactone derivatives **235** give the spiro compounds **242** in an addition and cyclocondensation sequence.[137] With dimethyl acetylenedicarboxylate, **235** forms several condensed and unsaturated spiro compounds **243**, similar to **242**,[137] as well as the tricyclic δ-lactone **244** (Scheme 64).

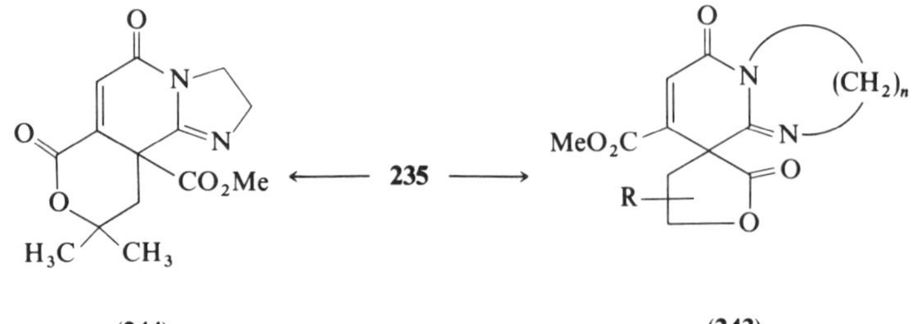

SCHEME 64

Furthermore, the dihydrofurans **8** (X = O, n = 1) condense with five-membered heterocyclic amidine derivatives via elimination of ethanol to

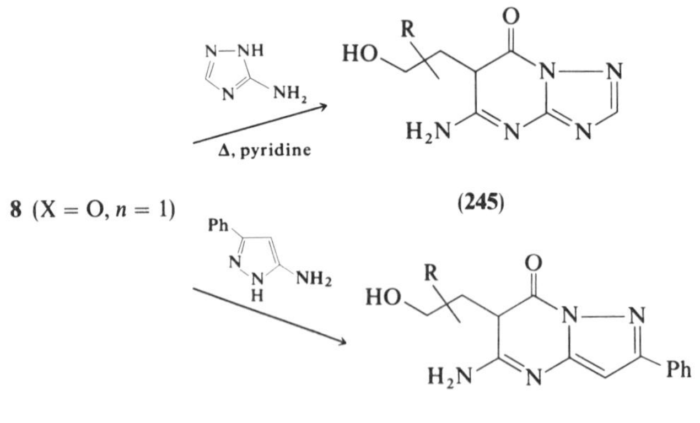

SCHEME 65

---

[136] Z. T. Huang and H. Wamhoff, *Chem. Ber.* **117**, 1856 (1984).
[137] Z. T. Huang and H. Wamhoff, *Chem. Ber.* **117**, 1926 (1984).

afford the azolopyrimidines **245** and **246**[138,139] (Scheme 65). In some cases, subsequent heterocyclization steps may be easily added. Compounds **244** and **248**, obtained with 2-amino-s-triazole and 2-aminobenzimidazole (**247**), produce, on heating in a mixture of acetic and sulfuric acids, the novel furo[2,3:5',6']-1,2,4-triazolo[1,5-a]pyrimidine (**249**) and furo[2,3:2',3']-benzoimidazo[1,2-a]pyrimidine (**250**) (Scheme 66).

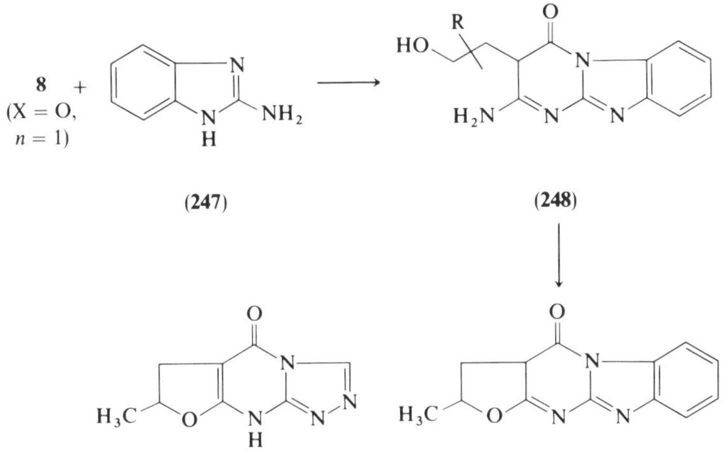

SCHEME 66

VII. Cycloadditions and Reactions with Acetylenic Esters

Although cycloadditions to enamines are well known,[2] the partially saturated heterocyclic β-enamino esters, as, e.g., **8**, display only a very weak tendency toward cycloaddition reactions, and Diels–Alder-type (4 + 2) cycloadditions have not yet been observed. From the 1,3-dipoles known today,[140] 170 hr of treatment of the enamino esters **8** (X = O, n = 1) with

---

[138] M. H. Elnagdi and H. Wamhoff, *Chem. Lett.*, 419 (1981).
[139] M. H. Elnagdi and H. Wamhoff, *J. Heterocycl. Chem.* **18**, 1287 (1981).
[140] R. Huisgen, *Angew. Chem.* **75**, 604, 742 (1963); **80** 329 (1968); *Angew. Chem., Int. Ed. Engl.* **2**, 565, 633 (1963); **7**, 321 (1968); *J. Org. Chem.* **33**, 2291 (1968); R. A. Firestone, *ibid.* **33**, 2285 (1968); C. Grundmann, *Fortschr. Chem. Forsch.* **7**, 62 (1966); A. Padwa, *Angew. Chem.* **88**, 131 (1976); *Angew. Chem., Int. Ed. Engl.* **15**, 123 (1976).

phenylazide leads—similar to the reaction of 2,3-dihydrofuran[141]—presumably to a cycloadduct **251**, which is stabilized by an eliminating ring cleavage to give 5-amino-4-(2-ethoxycarbonyloxyalkyl)-1-phenyl-1,2,3-triazole **252**[142] (Scheme 67). Benzphenylhydrazide chloride **253**, the precursor

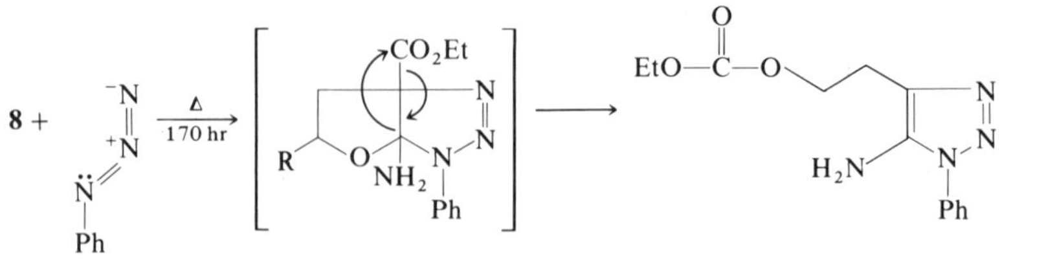

(251)   (252)

SCHEME 67

of diphenylnitrilimine,[143] reacts with **8** (X = O, n = 1) by acylation of the 2-amino group (**254**); rearrangement involving C-2 then gives α-(s-triazolyl)-γ-lactones **255**[142] (Scheme 68). Treatment of **8** (X = O, n = 1) with methyl

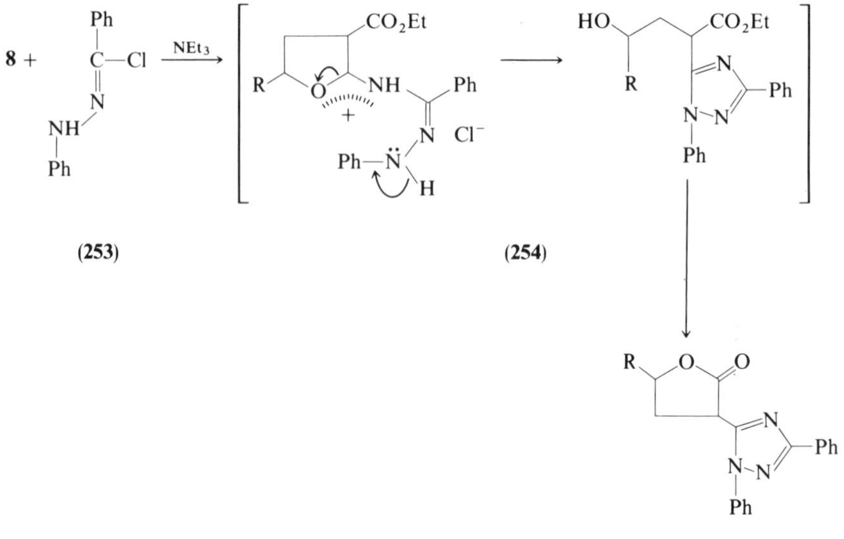

(253)   (254)

(255)

SCHEME 68

---

[141] R. Huisgen, L. Möbius, and G. Szeimies, *Chem. Ber.* **98**, 1138 (1965).
[142] H. Wamhoff and P. Sohár, *Chem. Ber.* **104**, 3510 (1971).
[143] R. Huisgen, M. Seidel, G. Wallbillich, and H. Knupfer, *Tetrahedron* **17**, 3 (1962); H. Wamhoff and M. Zahran, unpublished results.

propiolate and dimethyl acetylenedicarboxylate[144] illustrates the different paths available for the primary adduct **256**. Polar cycloaddition affords (via ring enlargement of the nonisolable condensed cyclobutene **257**)

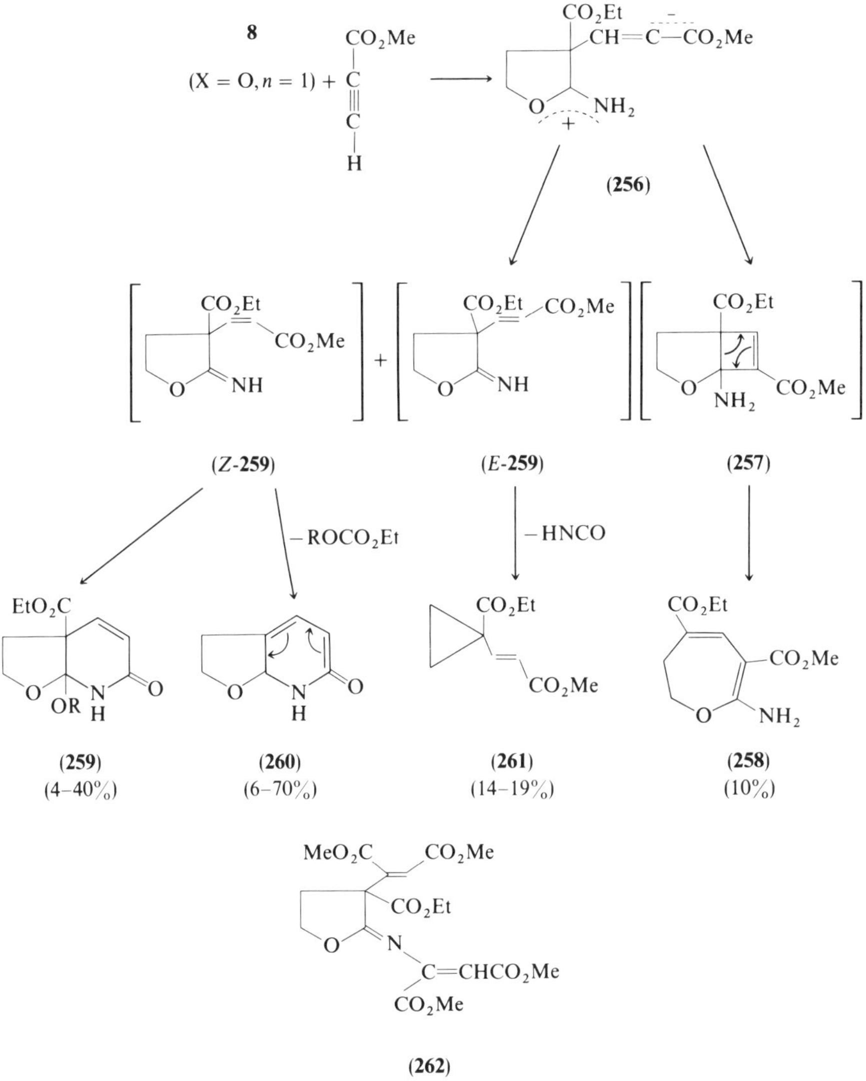

SCHEME 69

[144] H. Wamhoff and J. Hartlapp, *Chem. Ber.* **109**, 1270 (1976).

6,7-dihydroazepines **258**. The *Z* and *E* adducts *E,Z*-**259** react in different ways: either to form the furo[2,3-*b*]pyridoles **259** and **260**[145] or, after expulsion of isocyanic acid, vinylcyclopropanes **261**[144] (Scheme 69). By this method the interesting ring system in furo[2,3-*b*]pyridines[146] is constructed, for the first time starting from the furan moiety. Dimethyl acetylenedicarboxylate adds also to the 2-amino group to afford imino lactone **262**.[144]

Upon acid-catalyzed hydrolysis, the furo[2,3-*b*]pyridine **259** is cleaved to **263**, which in turn cyclizes either to another furo[3,2-*c*]pyridine (**264**) or to a pyrano[4,3-*c*]pyridine (**265**)[144] (Scheme 70). Better results in the ring enlargement reaction leading to **258** are obtained with the corresponding 2-(triphenylphosphoranylidenamino) esters (cf. Section VIII).

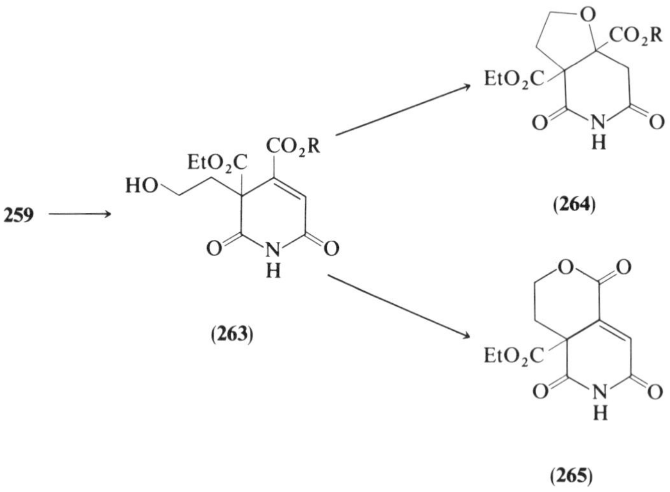

SCHEME 70

The adducts of heteroaromatic *β*-enamino esters with dimethyl acetylenedicarboxylates **266** are cyclized in a base-catalyzed reaction to yield 4-oxo-1,4-dihydro-2,3-pyridinedicarboxylic derivatives **267**[147] (Scheme 71). Ethyl 2-amino-3-thiophenecarboxylate (**26**)[21] also gives, with dimethyl acetylenedicarboxylate, a wide variety of cyclization products (**268**–**271**, E = CO_2Me)[148] (Scheme 72).

[145] E. Spinner and G. B. Yeoh, *Tetrahedron Lett.*, 5691 (1968); *J. Chem. Soc. B*, 279 (1971).
[146] J. R. Stevens, R. H. Beutel, and E. Chamberlin, *J. Am. Chem. Soc.* **64**, 1093 (1942); E. Ritchie, *Aust. J. Chem.* **9**, 244 (1956).
[147] H. Biere and W. Seelen, *Liebigs Ann. Chem.*, 1972 (1976).
[148] H. Biere, C. Herrmann, and G. A. Hoyer, *Chem. Ber.* **111**, 770 (1978).

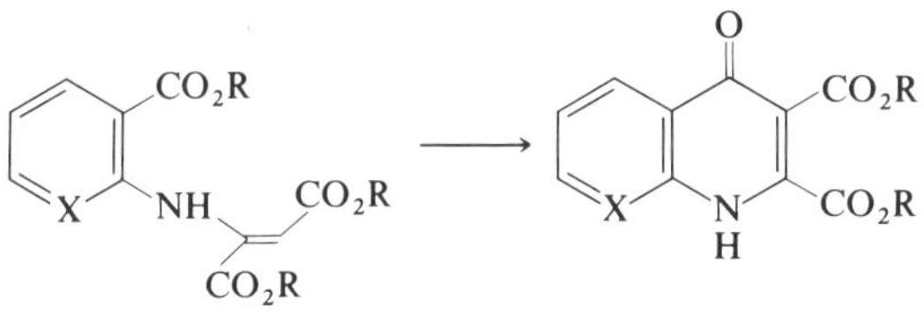

Sec. VIII.A] HETEROCYCLIC β-ENAMINO ESTERS 351

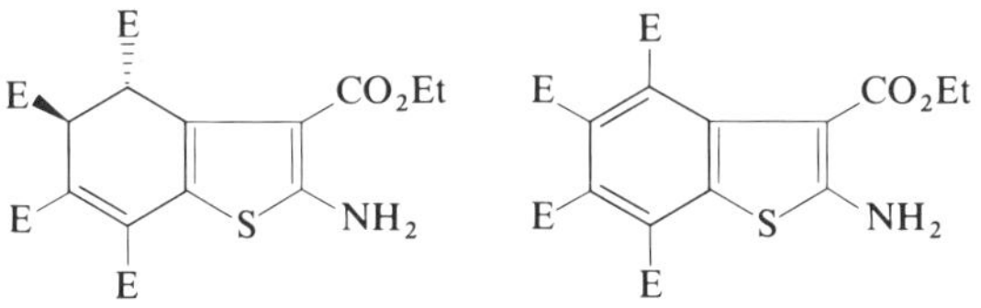

SCHEME 71

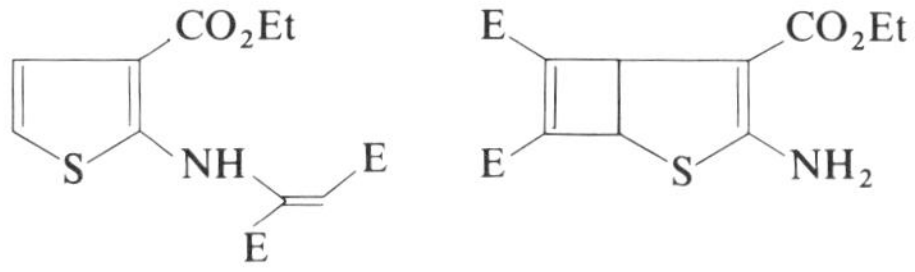

SCHEME 72

VIII. 2-(Triphenylphosphorylidenamino) Esters. The Cycloaddition–Ring Enlargement Sequence

A. Formation and Properties

Heterocyclic β-enamino esters of type **8** (X = O, S; n = 1,2) with dihalotriphenylphosphoranes[105] afford the resonance-stabilized iminophosphoranes **272** and **273** in high yields[149] (Scheme 73). As supported by spectroscopic data, the iminophosphorane group participates in the push–pull resonance of the electron distribution depicted in Section III,C and in

---

[149] H. Wamhoff, G. Haffmanns, and H. Schmidt, *Chem. Ber.* **116**, 1691 (1983).

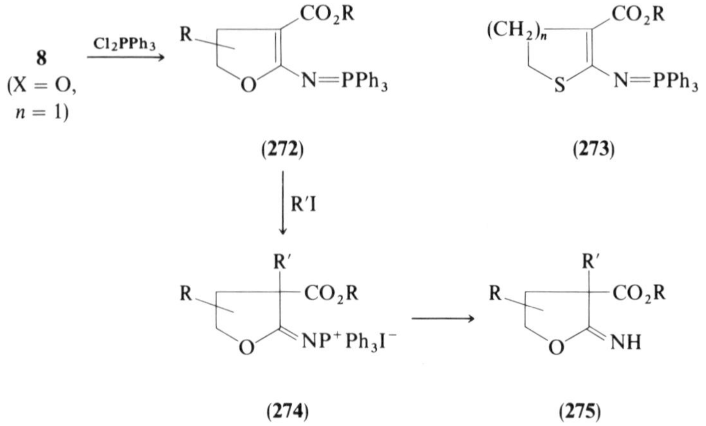

[Scheme 73 structures with compounds 272, 273, 274, 275]

SCHEME 73

canonical forms **272/273** I–III (Scheme 74). Alkyl iodides alkylate at the place with the highest electron density, i.e., at C-3, giving the phosphonium iodide **274**. Compound **274** is smoothly converted by aqueous alkali to 2-imino lactones **275**[149] (Scheme 73).

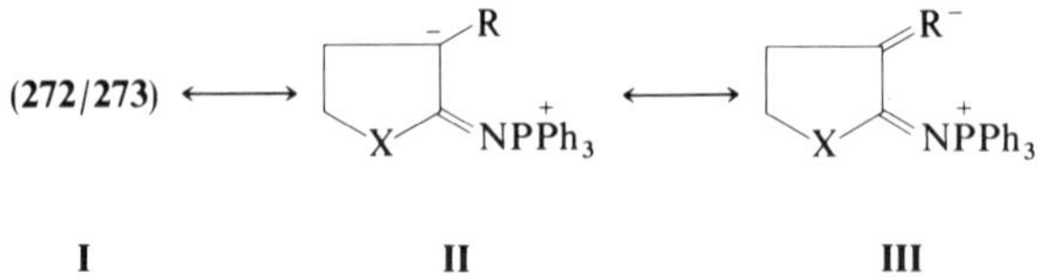

[Scheme 74 showing resonance structures I, II, III]

$R = CO_2R, CN$

SCHEME 74

B. POLAR CYCLOADDITION–RING-ENLARGEMENT REACTIONS WITH ACETYLENIC ESTERS

Unlike the unsubstituted enamino esters **8**, the iminophosphorane group in **272/273** can be considered as a protecting group with a latent functionality. Thus methyl propiolate and dialkyl acetylenedicarboxylates now give high yields of the polar cycloaddition products. An independent and detailed report concerning this cycloaddition–ring enlargement sequence will be published elsewhere[150]; here only the general principles and some of the primary results are discussed.

---

[150] H. Wamhoff, in preparation.

Sec. VIII.B] HETEROCYCLIC β-ENAMINO ESTERS 353

Under mild reaction conditions, the intermediate cyclobutene adducts **276** undergo *in situ* a $\sigma \to \pi$ ring-enlargement reaction (**277**) owing to the strong polarizing effect of both push (2-iminophosphorane group) and pull (3-ester function) groups. In a simple, one-step procedure, insertion of the acetylenic moiety takes place with 5 → 7 transformation in ring size and, in the case of the six-membered thiopyrane **278** novel types of thiocines (**279**) are smoothly obtained[149] (Scheme 75). This transformation has been achieved also by other

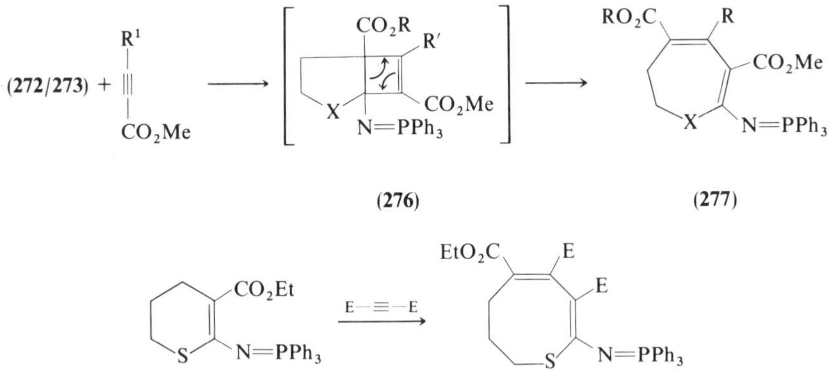

SCHEME 75

groups[151,152] with heterocyclic enamines and enamino lactones. However, in the case of the 2-(triphenylphosphoranylideneamino) esters and nitriles (**278**), derived from **32** and X_2PPh_3, stable ylides (**280**) have been obtained, most likely via adducts **279**[153] (Scheme 76).

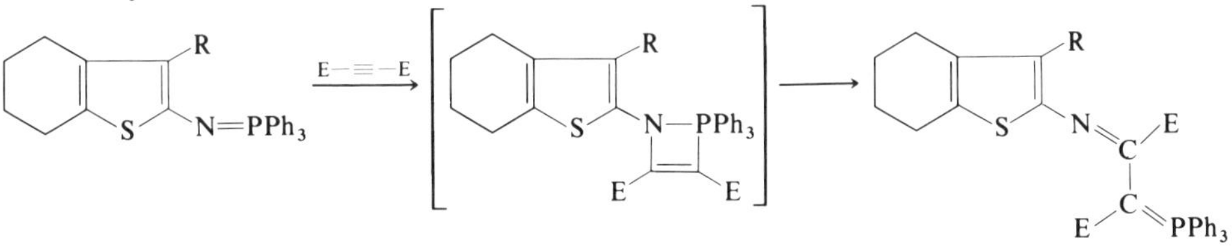

$R = CN, CO_2Et$ SCHEME 76

[151] D. N. Reinhoudt and C. G. Kouwenhoven, *Tetrahedron Lett.*, 5203 (1972); *Rec. Trav. Chim. Pays-Bas* **92**, 865 (1973); **93**, 129 (1974); D. N. Reinhoudt and C. G. Leliveld, *Tetrahedron Lett.*, 3119 (1972).
[152] D. J. Haywood and S. T. Reid, *J.C.S. Perkin I*, 2457 (1977).
[153] H. Wamhoff and G. Haffmanns, *Chem. Ber.* **117**, 585 (1984).

C. Subsequent Reactions

After the ring-enlargement reaction has been accomplished, the iminophosphorane group can be employed in a subsequent step; with phenyl isocyanate, resonance-stabilized carbodiimides (**281**) are obtained, which proved to be versatile starting materials for many heterocyclization reactions leading to novel 6:7, 6:8, and 5:6:7 combinations of heterocondensed pyrimidines (**282**)[153] (Scheme 77).

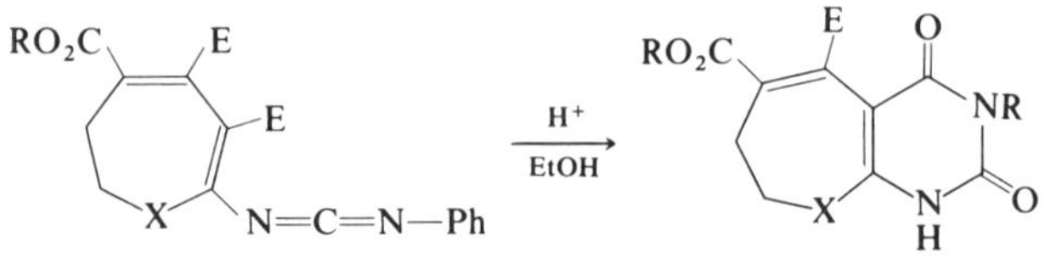

(**281**) (**282**)

Scheme 77

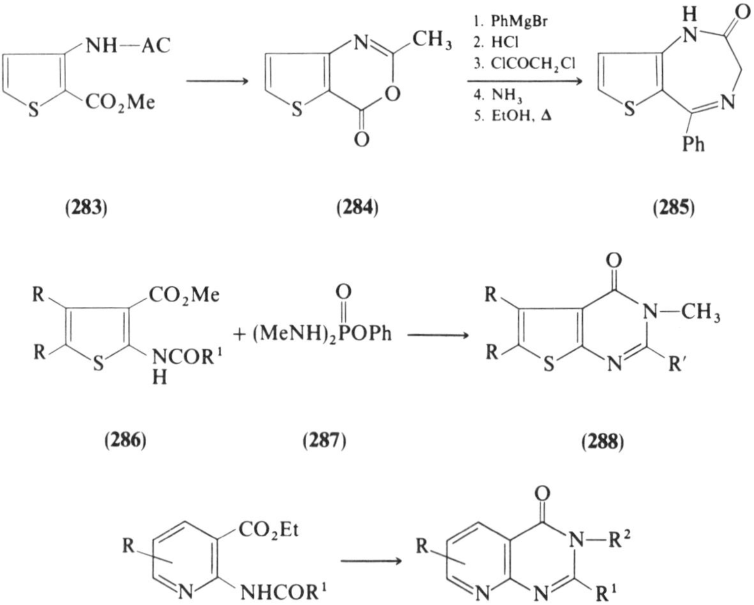

(**283**) (**284**) (**285**)

(**286**) (**287**) (**288**)

(**289**) (**290**)

Scheme 78

D. Miscellaneous Reactions Employing Phosphorus Compounds

A multistep preparation of a thieno[3,2-e]-1,4-diazepin-2(3H)-one (**285**) has been accomplished, starting from the thienooxazinone **284** obtained from methyl 3-acetylaminothiophene-3-carboxylate (**283**) and phosphorus pentachloride.[154] Similarly, 3-methylthieno[2,3-d]pyrimidin-4-(3H)-ones (**288**) were obtained in an exothermic reaction by heating methyl 2-acylamino-3-thiophenecarboxylates (**286**) with N,N'-dimethylphosphordiamidate (**287**) up to 250°C[155] (Scheme 78). Ethyl 2-acylamino-3-pyridinecarboxylate (**289**) gave, on heating with excess amine hydrochloride and phosphorus pentoxide/N,N'-dimethylcyclohexylamine, a series of pyrido[2,3-d]pyrimidine-4(3H)-ones (**290**).[156]

IX. Photochemistry

Photoisomerization of the dihydrofurans **8** (X = O, n = 1) requires triphenylene as sensitizer (E_T = 285 kJ/mol) to form 1-carbamoyl-1-ethoxycarbonylcyclopropanes (**292**)[157] (Scheme 79). Despite the bathochromic UV maximum (cf. Section III,B) caused by the enaminocarbonyl

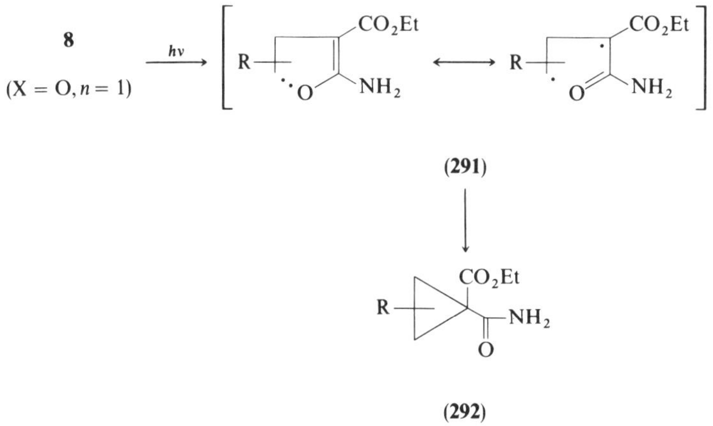

Scheme 79

---

[154] O. Hromatky and D. Binder, *Monatsh. Chem.* **104**, 1343 (1973).
[155] K. E. Nielsen and E. B. Pedersen, *Acta Chem. Scand., Ser. B* **B32**, 303 (1978).
[156] O. R. Andresen and E. B. Pedersen, *Liebigs Ann. Chem.*, 1012 (1982).
[157] H. Wamhoff, *Chem. Ber.* **105**, 748 (1972).

resonance, the enamino ester moiety is not the important site for the photochemical behavior. Instead, **8** reacts as a dihydrofuran[158] via homolysis of the C-5—O bond; rearrangement gives the stable amide group, and cyclization to cyclopropanes **292** occurs.

Upon UV irradiation, the heteroaromatic isoxazoles **35** rearrange similarly—most likely via azirines (**293**)[159]—to afford, by photocleavage of **293**, nitrilylides (**294**)[140,160] and finally the isomeric 1,3-oxazoles **295**.[157] However, ethyl 5-aminoisoxazole-4-carboxylate (**34**) isomerizes by another pathway. Similar to some thermal reactions,[161] reorganization of the nitrile function in **295** affords the vinylamine **296**[157] (Scheme 80). Similar results are known for benzisoxazoles.[162]

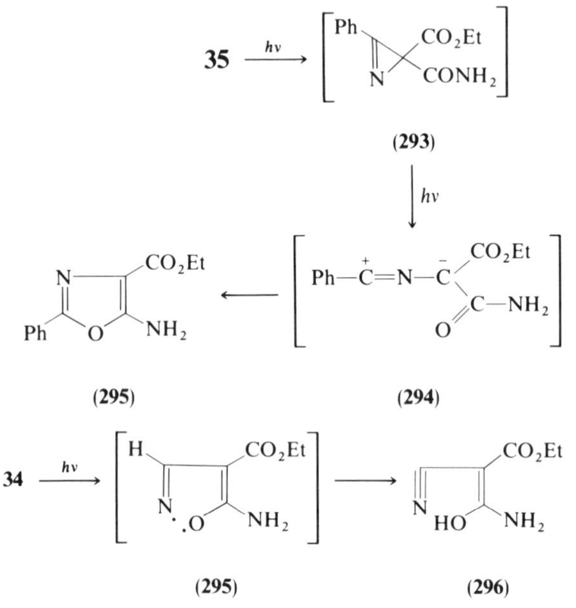

SCHEME 80

[158] Cf. D. W. Boykin and R. E. Lutz, *J. Am. Chem. Soc.* **86**, 5046 (1964); D. E. McGreer, M. G. Vinje, and R. S. McDaniel, *Can. J. Chem.* **43**, 1417 (1965); J. Wieman, N. Thoai, and F. Weisbuch, *Bull. Soc. Chim. Fr.*, 575 (1966).
[159] E. F. Ullman, and B. Singh, *J. Am. Chem. Soc.* **88**, 1844 (1966); B. Singh and E. F. Ullman, *ibid.* **89**, 6911 (1967); B. Singh, A. Zweig, and J. B. Gallivan, *ibid.* **94**, 1199 (1972).
[160] A. Padwa, M. Dharan, J. Smolanoff, and S. E. Wetmore, *Pure Appl. Chem.* **33**, 269 (1973); H. Schmid *et al., ibid.*, 339; A. Padwa, J. Smolanoff, and A. Tremper, *J. Org. Chem.* **41**, 543 (1976).
[161] A. Quilico, in "Heterocyclic Compounds" (R. H. Wiley, ed.), p. 44ff. Wiley, New York, 1962.
[162] H. Göth and H. Schmid, *Chimia* **20**, 148 (1966).

Sec. X.A]   HETEROCYCLIC β-ENAMINO ESTERS   357

Upon UV irradiation, ethyl 5-amino-1-phenyl-1,2,3-triazole-4-carboxylate (**40**) does not expel N_2 to give ketenimines or aziridines. Instead, in the presence of triphenylene as photosensitizer, **40** is smoothly isomerized, as a first example of a photo-Dimroth rearrangement,[163] to ethyl 5-anilino-1,2,3-triazole-4-carboxylate (**298**).[164] The mechanism is explained by a radical rearrangement, e.g., **297I**–**297III** (cf. Scheme 81). This isomerization type was observed also with 5-amino-1,4-diphenyl-1,2,3-triazoles.[165]

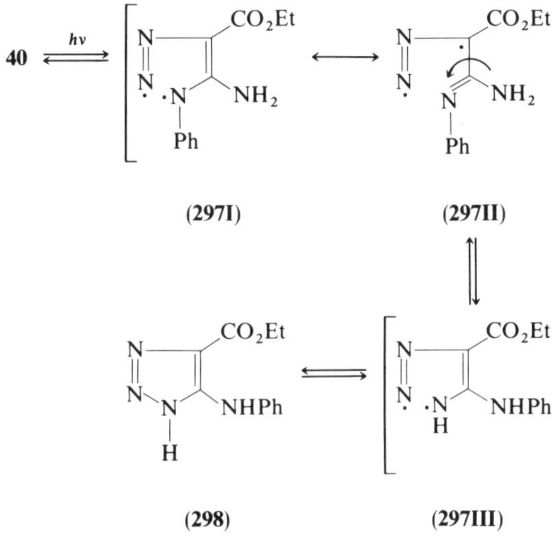

SCHEME 81

X. Heterocyclic β-Enamino Nitriles

A. Syntheses

The syntheses, properties, and chemistry of heteroaromatic β-enamino nitriles have been reviewed extensively.[166] Therefore, this section deals mainly with partially saturated heterocyclic β-enamino nitriles; in addition, some further reactions of heteroaromatic β-enamino nitriles are briefly described.

---

[163] Cf. D. J. Brown, in "Mechanisms of Molecular Migrations" (B. S. Thyagarajan, ed.), Vol. 1, p. 209. Wiley (Interscience), New York 1968.
[164] G. Szilágyi and H. Wamhoff, *Acta Chim. Acad. Sci. Hung.* **89**, 265 (1976) [*CA* **86**, 121,238t (1977)].
[165] Y. Ogata, K. Tagaki, and E. Hayashi, *Bull. Chem. Soc. Jpn.* **50**, 2505 (1977).
[166] E. C. Taylor and A. McKillop, *Adv. Org. Chem.* **7**, 1 (1970).

2-Amino-3-cyano-4,5-dihydrofurans (**299**) are prepared from oxiranes and sodium malonodinitrile,[167–169] and the corresponding dihydrothiophenes (**300**) are similarly obtained by treatment of thiirane with sodium malonodinitrile. But the procedures reported[170] had to be modified considerably; a temperature of $-20°C$ and dimethoxyethane as solvent gave better yields[169] (Scheme 82). Attempts to transfer this synthesis to *N*-tosylaziridine (**301**),

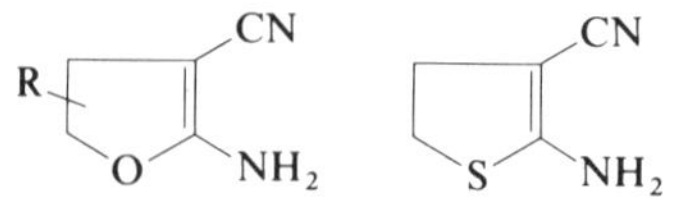

SCHEME 82

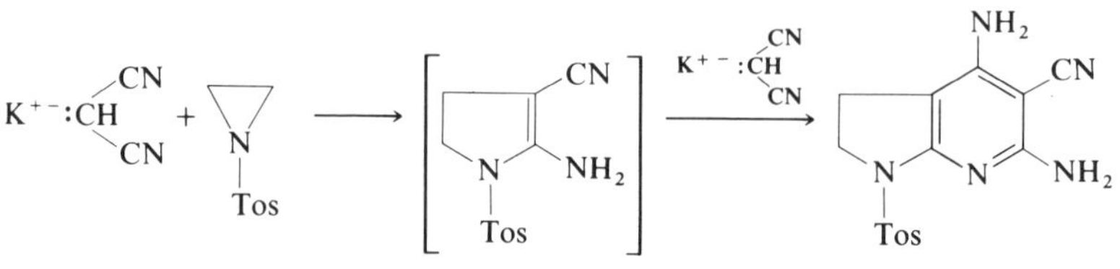

previously successfully applied in the generation of a pyrrolenamino ester,[12] lead at first to complex mixtures. But upon refluxing both components, a 2:1-adduct was isolated; this proved to be pyrrolo[2,3-*d*]pyridine **303**. The desired enamino nitrile **302** can be formulated only as an elusive intermediate. However, at 10°C in 1,2-dimethoxyethane, another 1:2-spiro adduct (**304**) is formed,[169] similar to a reaction of aziridines and malonic esters.[171] But employing $-10°C$ and the same conditions as above, the desired 1:1 adduct can be obtained in good yield.[169] Independently, workers in Japan described an analogous synthesis of ethyl 2-amino-3-cyano-4,5-dihydropyrrole-1-

---

[167] Hisamitsu Seiyaku Co. Ltd., Japanese Patent 72/05,255 (1972) (inventors: K. Yamazaki and T. Matsuda) [*CA* **76**, 140492 (1972)].
[168] S. Morgenlie, *Acta Chem. Scand.* **24**, 365 (1970).
[169] H. Wamhoff and H. A. Thiemig, *Chem. Ber.* **118** (1985), in press.
[170] Hisamitsu Pharm. Co. Ltd., *Jpn Kokai Tokkyo Koho* **74**, 13, 164 (1974) (inventors: T. Matsuda, K. Yamazaki, H. Ide, K. Noda, and K. Yamagata) [*CA* **81**, P13,375y (1975)]; K. Yamagata, Y. Tomioka, M. Yamazaki, T. Matsuda, and K. Noda, *Chem. Pharm. Bull.* **30**, 4396 (1982) [*CA* **98**, 179135w (1983)].
[171] M. Kojima, T. Kawakita, and K. Kudo, *Yakugaku Zasshi* **92**, 465 (1972) [*CA* **77**, 34,230e (1972)].

carboxylate (**302a**) and its aromatization (**302b**) by means of chloranil[171a] (Scheme 83). The 5,6-dihydro-4*H*-pyranenamino nitrile **305** is obtained in low

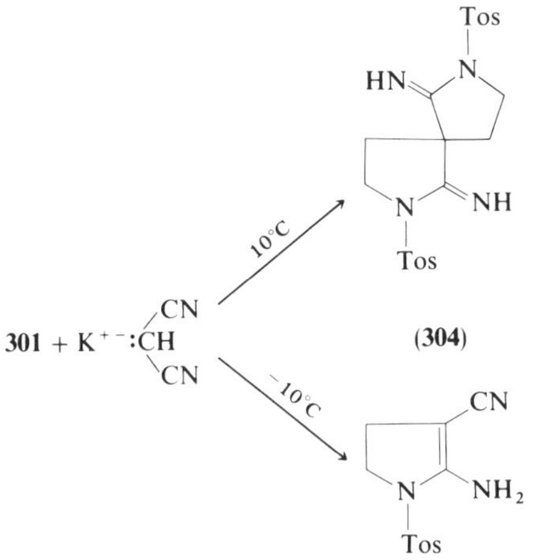

(**304**)

(**302**)

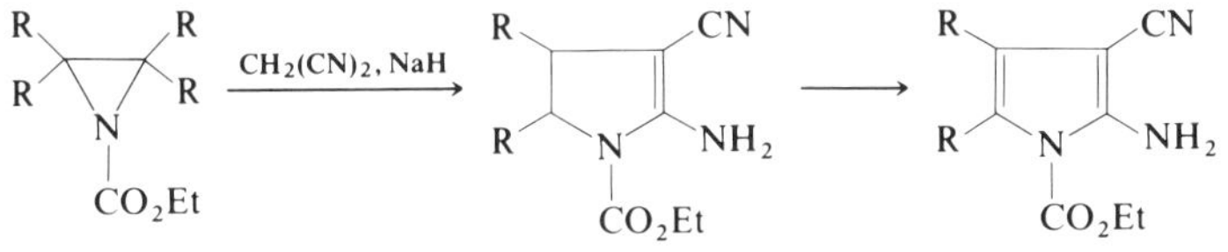

(**302a**) (**302b**)

SCHEME 83

yield from potassium malonodinitrile by fast addition of 3-bromopropanole.[168] This procedure could not be transferred to the sulfur analog (Scheme 84).

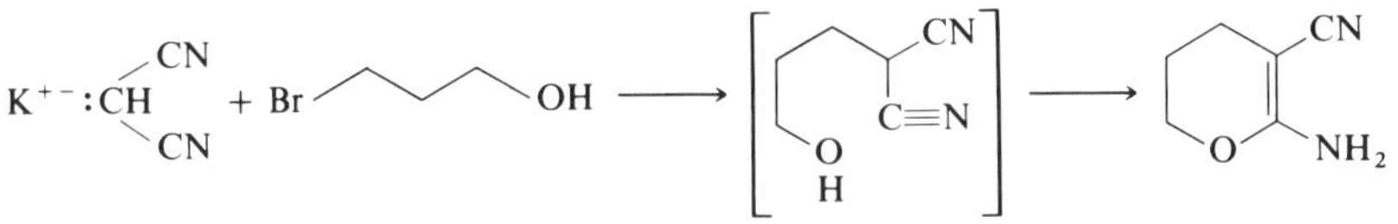

(**305**)

SCHEME 84

[171a] M. Sonoda, N. Kuriyama, Y. Tomioka, and M. Yamazaki, *Chem. Pharm. Bull.* **30**, 2357 (1982) [*CA* **97**, 162,744a (1982)].

TABLE III
Comparison of the δ ^{13}C-3 of Some β-Enamino Esters and Nitriles

Compound	n	X	Y	^{13}C-3 (ppm)
8	1	O	CO$_2$Et	73.6
71	2	S	CO$_2$Et	90.66
72	1	S	CO$_2$Et	92.1
299	1	O	CN	47.1
300	1	S	CN	65.2
302	1	N-Tos	CN	57.5
305	2	O	CN	53.9

B. Properties and Comparison with Enamino Esters

In the IR spectra, the partially saturated heterocyclic β-enamino nitriles do not exhibit the normal group frequencies that the corresponding esters display. The comparison of the ^{13}C-NMR spectra of nitriles **299, 300, 302**, and **305** with those of the esters **8, 71**, and **72** reveal for the nitriles a greater high-field shift of ~ 27 ppm (Table III). As a consequence, it is concluded that in the enamino nitriles the double bond is more polarized than that of the corresponding esters; several canonical forms can be discussed (Scheme 85). This difference in polarization (**8, 71**, and **72** versus **299, 300, 302**, and **305**) leads to several significant differences in the chemical behavior, which are briefly discussed in the following sections.

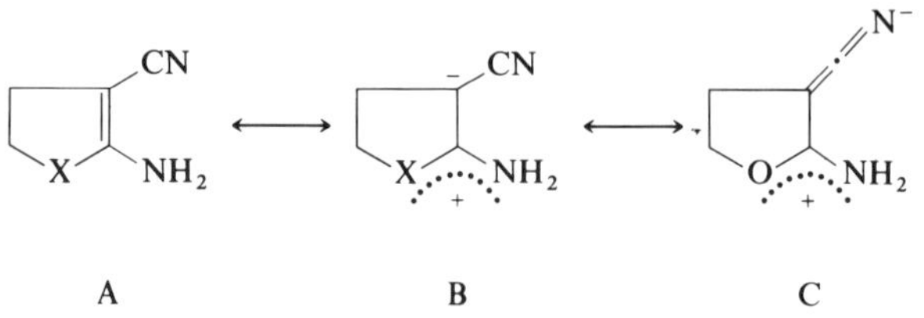

SCHEME 85

C. Heterocyclization Reactions

The enamino nitriles **299, 300**, and **302** react smoothly with various isocyanates to afford heterocondensed 4-amino-2-oxopyrimidines (**306**).[172]

---

[172] H. Wamhoff and H. A. Thiemig, *Chem. Ber.* **118** (1985), in press.

The 2:1 adduct **303** gives rise to the pyrrolo[3′,2′:5,6]pyrido[2,3-*d*]pyrimidine **307**. Similar results have been reported for imidazolenamino nitriles[173] (Scheme 86). While enamino esters, on heating with formamide, afford, in a

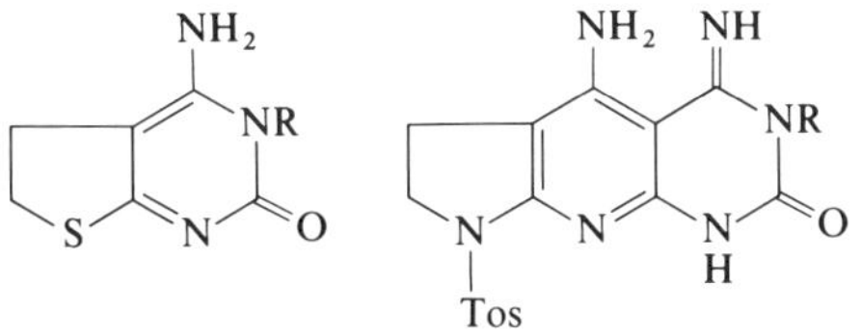

(306)   (307)

SCHEME 86

clean reaction, heterocondensed pyrimidines (cf. Section V,B), the corresponding enamino nitriles **299** give under the same conditions di- and trimerization products **308I**, **308II**,[174] and **309**, respectively. (Scheme 87). The structure of **309** was established by X-ray analysis.[174] Pyrazol- and 1,2,3-triazolenamino nitriles (**310**) are known to afford base-catalyzed dimerization products such

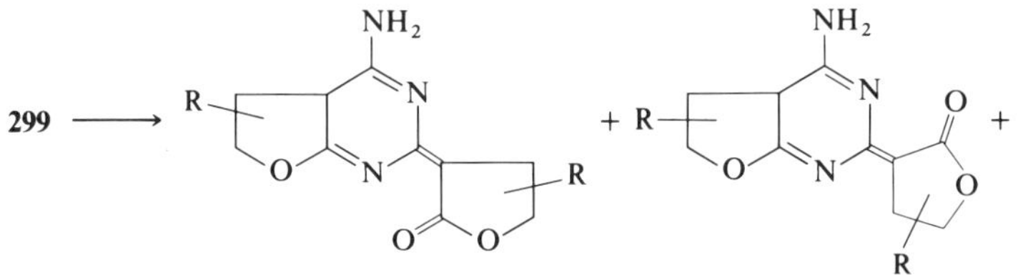

(308I)   (308II)

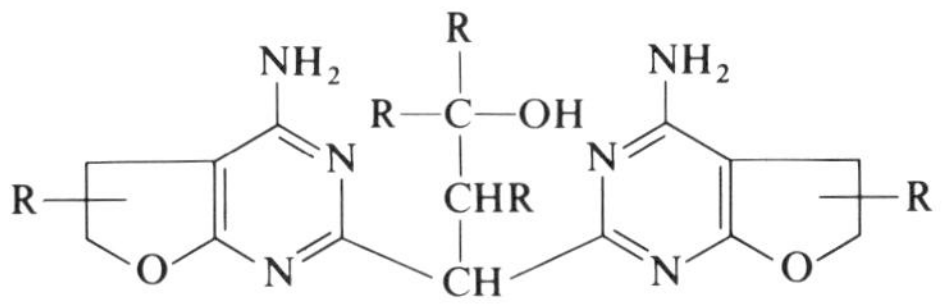

(309)

SCHEME 87

---

[173] A. F. Cook, B. A. Balder *et al.*, *J. Med. Chem.* **24**, 947 (1981).
[174] H. Wamhoff, H. A. Thiemig, E. Friedrichs, and H. Puff, *Chem. Ber.* (in press).

as **311**, which are related to **308I/308II**.[175] With ethyl cyanoacetate, **299** gives, via the condensation intermediate **312**, a novel tricyclic system (**313**)[169] (Scheme 88). Heteroaromatic β-enamino nitriles under these conditions annelate a pyridine ring to give **314** and **315**.[176]

In contrast to the broad applicability of heterocyclic β-enamino esters, the corresponding enamino nitriles with imido esters, phenylazide, or diamines, thermally as well as photochemically, give no reasonable products.

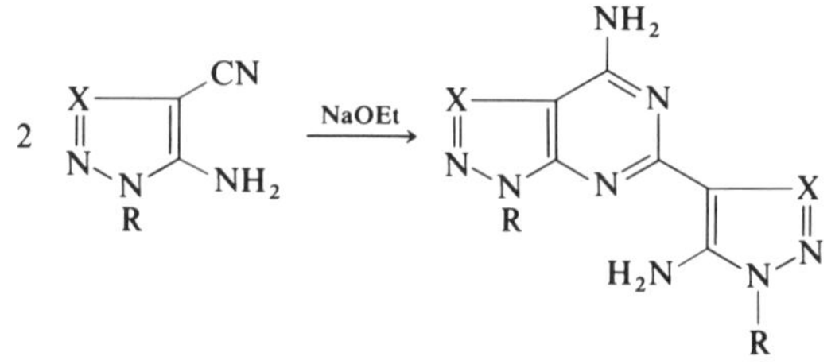

X = CH, N

(**310**)   (**311**)

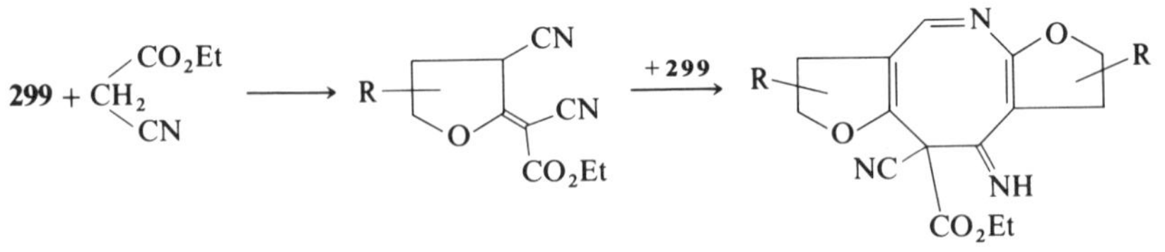

(**312**)   (**313**)

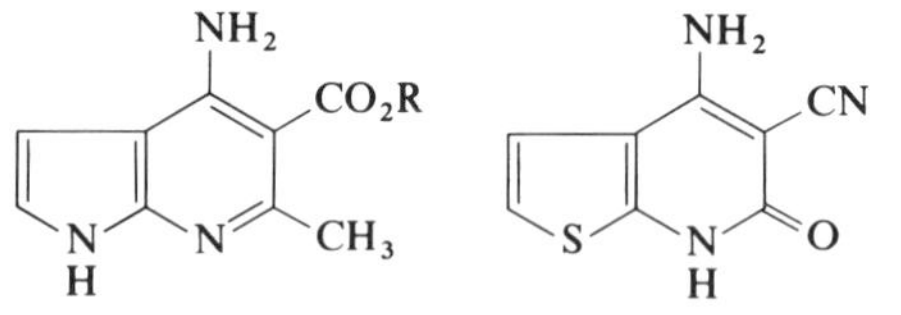

(**314**)   (**315**)

SCHEME 88

[175] D. R. Sutherland and G. Tennant, *J. Chem. Soc. C*, 706 (1971); A. Kreutzberger and K. Burgwitz, *Chem.-Ztg.* **104**, 175 (1980).
[176] W. Zimmermann, K. Eger, and H. J. Roth, *Arch. Pharm. (Weinheim Ger.)* **309**, 597 (1976); K. Gewald, H. Schäfer, and K. Sattler, *Monatsh. Chem.* **110**, 1189 (1979).

D. Dimroth Rearrangements to Thieno- and Furo[2,3-d]pyrimidophanes

Similar to the heteroannelation reactions with lactim ethers described in Section V,B, the enamino nitriles **299, 300,** and **305** on heating with lactim ethers having up to a nine-membered ring afford the linear annelated tricycles **317**.[172] However, employing medium-ring-sized lactim ethers ($m \geq 7$ and 11) the imino tricycles **317** react in a Dimroth-type reaction[163,177] to afford heterocondensed pyrimidophanes (**318**) as the thermodynamic products (Scheme 89). Such Dimroth rearrangements are influenced by water and other

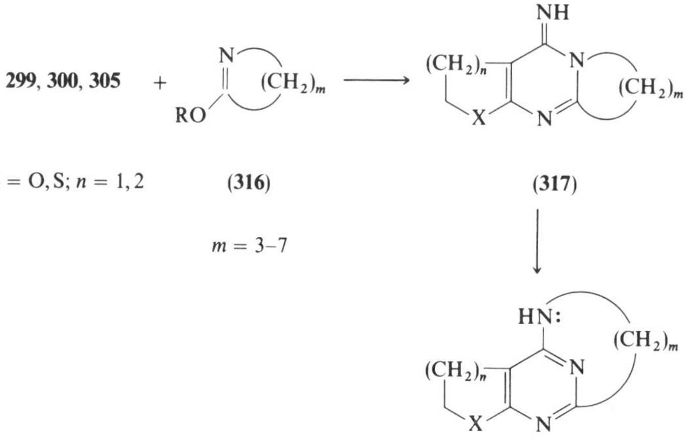

$X = O, S; n = 1, 2$ (316)

$m = 3-7$

(317)

(318)

$m = 7, 11$

SCHEME 89

nucleophiles such as alcohols and amines.[178] Thus in the presence of butanol, the rearrangement can be formulated to proceed via a macrocyclic intermediate (**317a**) to give the phanes **318** (cf. Scheme 90). Dimroth rearrangements of this type have been recently observed on open-chain, carbocyclic,

---

[177] H. C. Carrington, F. H. Curd, and D. N. Richardson, *J. Chem. Soc.*, 1858 (1955); D. J. Brown, E. Hoerger, and S. F. Mason, *ibid.*, 4035.
[178] D. J. Brown, P. W. Ford, and M. N. Paddon-Row, *J. Chem. Soc. C*, 1452 (1968).

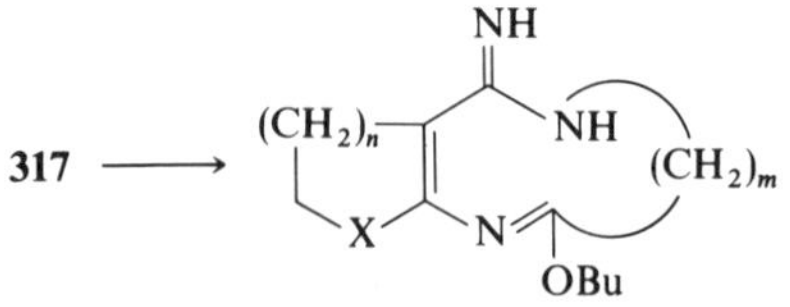

SCHEME 90

and heteroaromatic enamino nitriles with lactim ethers of ring sizes $m = 6, 7$, 9, and 11. These rearrangements (**319** → **320**) occur spontaneously or in the presence of bases[179,180] (Scheme 91).

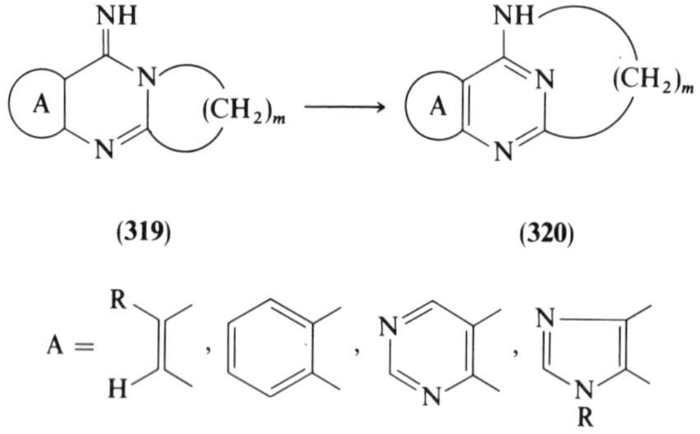

SCHEME 91

2-Benzamido-3-cyano-4,5-dihydrothiophenes (**321**) cyclize at 120°C with cyclohexylamine, morpholine, piperidine, and pyrrolidine to the corresponding 4-amino-2-phenyl-5,6-dihydrothieno[2,3-*d*]pyrimidines (**322**).[181] However, when **321** and 2 equivalents of dimethylamine hydrochloride in pyridine were refluxed for 5 hr, the 4-oxothieno[2,3-*d*]pyrimidine **323** was formed nearly quantitatively[181] (Scheme 92). Ethyl *N*-(3-cyano-4,5-dihydro-2-thienyl)oxamates (**324**) give, with cyanomethylene compounds in the presence of triethylamine, thieno[2,3-*d*]pyrimidines (**325**).[182]

---

[179] D. J. Brown and K. Ienaga, *J.C.S. Perkin I* 2182 (1975); *Aust. J. Chem.* **28**, 119 (1975).
[180] K. Ienaga and W. Pfleiderer, *Liebigs Ann. Chem.*, 1872 (1979).
[181] K. Yamagata, Y. Tomioka, M. Yamazaki, and K. Noda, *Chem. Pharm. Bull.* **31**, 401 (1983) [*CA* **99**, 88,145d (1983)].
[182] S. Hachiyama, K. Koyanagi, Y. Tomioka, and M. Yamazaki, *Chem. Pharm. Bull.* **31**, 1177 (1983) [*CA* **99**, 158,361x (1983)].

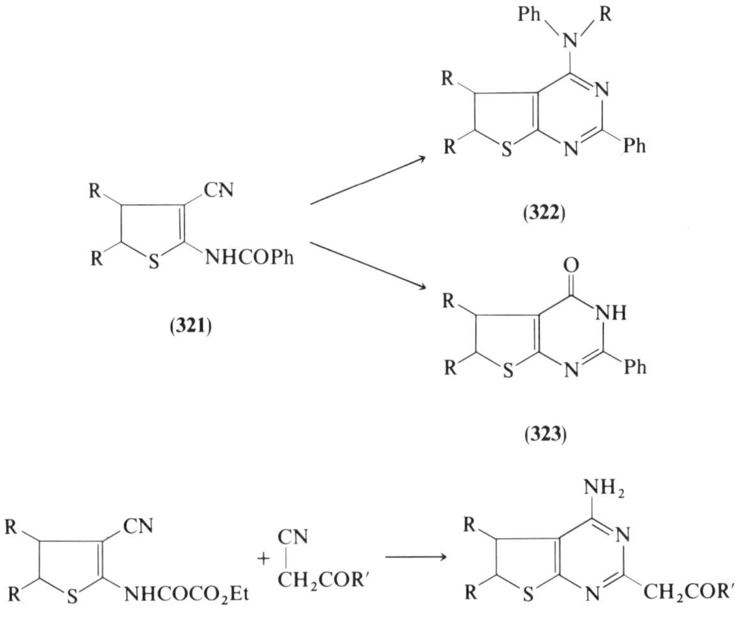

Sec. X.E] HETEROCYCLIC β-ENAMINO ESTERS 365

SCHEME 92

E. MISCELLANEOUS REACTIONS OF HETEROAROMATIC
β-ENAMINO NITRILES

2-Amino-3-cyanopyrroles (326) are versatile starting materials for the easy synthesis of 4-amino-7H-pyrrolo[2,3-d]pyrimidines (327) by treatment with triethyl orthoformate and ammonia.[183,184] The 1,2-diamino-3-cyanopyrrole 328 affords with diacetyl in acetic acid a pyrrolo[1,2-b]-as-triazine (329)[184] (Scheme 93). Malonodinitrile, sulfur, and bases afford the diaminodicyanothiophenes 330 and 331.[166] Compound 331 yields, with 2 equivalents of formamide, 4,9-diaminothieno[2,3-d:4,5-d']dipyrimidines (332)[185] (Scheme 94). Furthermore, 2-amino-3-cyanothiophenes (333) react with ethyl aminocrotonate. Cyclization of the intermediate 334 with sodium ethoxide gives

[183] E. C. Taylor and R. W. Hendess, J. Am. Chem. Soc. 86, 951 (1964); 87, 1995 (1965).
[184] K. Gewald and M. Hentschel, J. Prakt. Chem. 318, 663 (1976); with ethoxyvinyl derivatives: cf. T. Kurihara, K. Nasu, and Y. Adachi, J. Heterocycl. Chem. 20, 81 (1983).
[185] K. Gewald, M. Kleinert, B. Thiele, and M. Hentschel, J. Prakt. Chem. 314, 303 (1972).

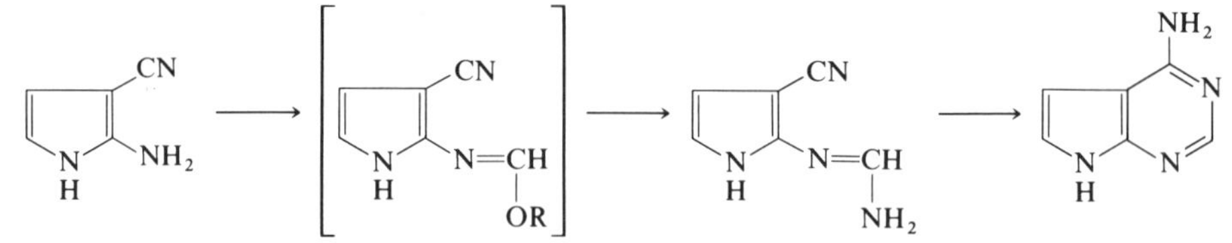

(326)

(327)

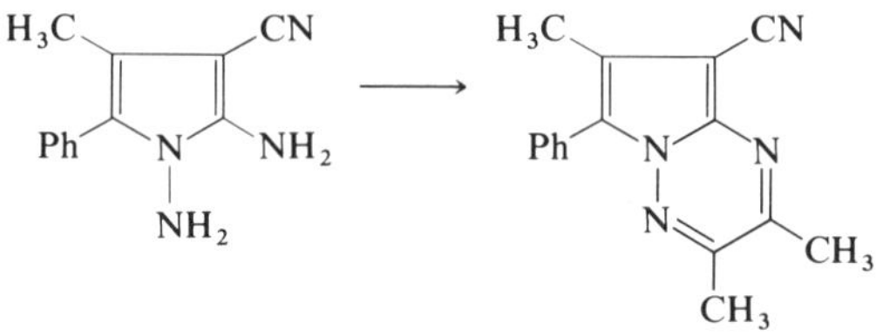

(328)

(329)

SCHEME 93

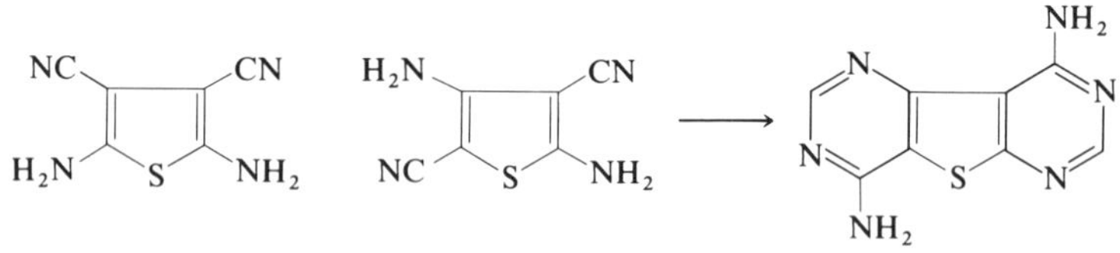

(330)

(331)

(332)

SCHEME 94

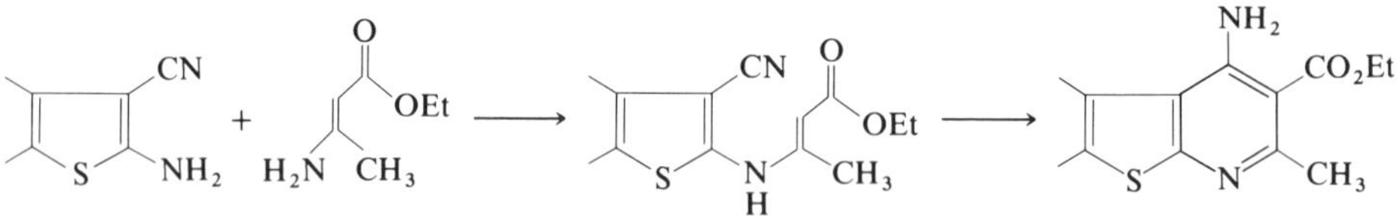

(333)

(334)

(335)

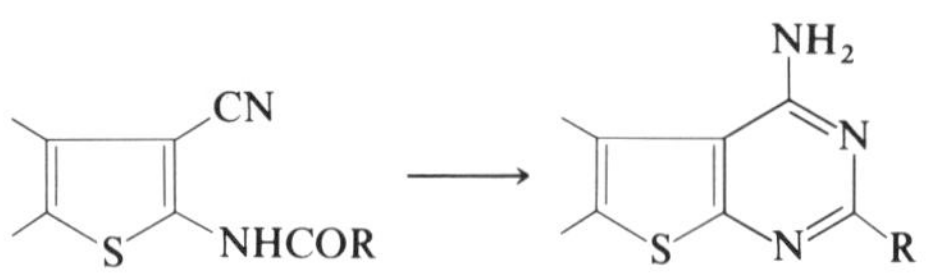

(336)

(337)

SCHEME 95

4-aminothieno[2,3-*b*]pyridines (**335**).[186] 4-Aminothieno[2,3-*d*]pyrimidines (**337**) are also formed by heating 2-acylaminothiophene-3-carbonitriles (**336**) with primary alkyl amines, P_2O_5, and cyclohexyldimethylamine[187] (Scheme 95).

XI. Conclusion and Outlook

This review on heterocyclic *β*-enamino esters and related compounds is not exhaustive. Instead, it aims to depict synthetic trends and preparative applicabilities of this rather versatile class of compounds. It is hoped that this review might stimulate a broad application of these compounds to the synthesis of novel heterocycles and especially natural products,[188] biologically active substances,[189] and drugs.[190]

One example is the investigation of the isosteric properties of benzene and thiophene moieties. Some thiophene analogues of 4-quinazolones, namely, thieno[2,3-*d*]pyrimidines, showed significant antimicrobial activity[191] as well as antiinflammatory and anticonvulsant properties (Scheme 96). Thieno analogues of the alkaloids echinorine (**338**) and echinopsine (**339**) have been obtained from thiopheneenamino ester derivatives, leading to thieno[3,2-*b*]pyridines.[192]

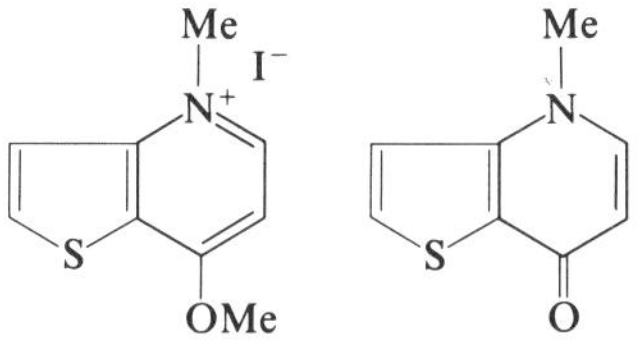

SCHEME 96

---

[186] I. Lalezari, *J. Heterocycl. Chem.* **16**, 603 (1979).
[187] K. E. Nielsen and E. B. Pedersen, *Chem. Scr.* **18**, 245 (1981) [*CA* **96**, 122739a (1982)].
[188] Cf. V. Snieckus, *Sur. Prog. Chem.* **9**, 122 (1980).
[189] Cf. A. Albert, *Intra-Sci. Chem. Rep.* **8**, 55 (1974).
[190] Cf. D. Lednicer and L. A. Mitscher, "The Organic Chemistry of Drug Synthesis." Wiley (Interscience), New York 1977.
[191] M. B. Devani, C. J. Shishoo, U. S. Pathak, S. H. Parikh, G. F. Shah, and A. C. Padhya, *J. Pharm. Sci.* **65**, 660 (1976).
[192] J. M. Barker, P. R. Huddleston, and A. W. Jones, *J. Chem. Res. Synop.*, 393 (1978); *J. Chem. Res., Miniprint*, 4701 (1978).

ACKNOWLEDGMENTS

This review would not have been possible without the skillful and untiring activity of my co-workers and collaborators: Miss Dr. A. Atta, Dr. H. Dürbeck, Prof. H. M. Elnagdi, Dr. M. Ertas, Dr. L. Farkas, E. Gierke, Dr. G. Haffmanns, Dr. J. Hartlapp, Dr. G. Hendrikx, Prof. Zhi-tang Huang, J. Keuler, Dr. G. Kunz, Prof. J. Lehmann, Dr. L. Lichtenthäler, Dr. C. Materne, J. Nagelschmitz, Dr. H. Schmidt, Prof. P. Sohár, Dr. G. Szilágyi, Dr. H. A. Thiemig, Dr. K. Wald, Dr. C. von Waldow, Dr. B. Wehling, Dr. Shi-yan Yang, and M. Zahran.

I thank the Deutsche Forschungsgemeinschaft, the Minister für Wissenschaft und Forschung des Landes Nordrhein-Westfalen, and the Fonds der Chemischen Industrie for support and Bayer AG and Hoechst Aktiengesellschaft for chemicals.

Cumulative Index of Titles

A

Acetylenecarboxylic acids and esters, reactions with N-heterocyclic compounds, **1**, 125
Acetylenecarboxylic esters, reactions with nitrogen-containing heterocycles, **23**, 263
Acetylenic esters, synthesis of heterocycles through nucleophilic additions to, **19**, 297
Acid-catalyzed polymerization of pyrroles and indoles, **2**, 287
t-Amino effect, **14**, 211
Aminochromes, **5**, 205
Anils, olefin synthesis with, **23**, 171
Annelation of a pyrimidine ring to an existing ring, **32**, 1
Annulenes, N-bridged, cyclazines and, **22**, 321
Anthracen-1,4-imines, **16**, 87
Anthranils, **8**, 277; **29**, 1
Applications of NMR spectroscopy to indole and its derivatives, **15**, 277
Applications of the Hammett equation to heterocyclic compounds, **3**, 209; **20**, 1
Arene oxides, chemistry of, **37**, 67
Aromatic azapentalenes, **22**, 183
Aromatic quinolizines, **5**, 291; **31**, 1
Aromaticity of heterocycles, **17**, 255
Aza analogs of pyrimidine and purine bases, **1**, 189
7-Azabicyclo[2.2.1]hepta-2,5-dienes, **16**, 87
1-Azabicyclo[3.1.0]hexanes and analogs with further heteroatom substitution, **27**, 1
Azapentalenes, aromatic, chemistry of, **22**, 183
Azines, reactivity with nucleophiles, **4**, 145
Azines, theoretical studies of, physicochemical properties of reactivity of, **5**, 69
Azinoazines, reactivity with nucleophiles, **4**, 145
1-Azirines, synthesis and reactions of, **13**, 45
Azodicarbonyl compounds in heterocyclic synthesis, **30**, 1

B

Barbituric acid, recent progress in chemistry of, **38**, 229
Base-catalyzed hydrogen exchange, **16**, 1
1-, 2-, and 3-Benzazepines, **17**, 45
Benzisothiazoles, **14**, 43; **38**, 105
Benzisoxazoles, **8**, 277
Benzoazines, reactivity with nucleophiles, **4**, 145
Benzo[c]cinnolines, **24**, 151
1,5-Benzodiazepines, **17**, 27
Benzo[b]furan and derivatives, recent advances in chemistry of, Part I, occurrence and synthesis, **18**, 337
Benzo[c]furans, **26**, 135
Benzofuroxans, **10**, 1; **29**, 251
2H-Benzopyrans (chrom-3-enes), **18**, 159
Benzothiazines
 1,2- and 2,1-benzothiazines and related compounds, **28**, 73
 1,4-benzothiazines and related compounds, **38**, 135
Benzo[b]thiophene chemistry, recent advances in, **11**, 177; **29**, 171
Benzo[c]thiophenes, **14**, 331
1,2,3-(Benzo)triazines, **19**, 215
Benzyne, reactions with heterocyclic compounds, **28**, 183
Biological pyrimidines, tautomerism and electronic structure of, **18**, 199
Bipyridines, **35**, 281

C

9H-Carbazoles, recent advances in, **35**, 83
Carbenes
 and nitrenes, intramolecular reactions, **28**, 231
 reactions with heterocyclic compounds, **3**, 57
Carbolines, **3**, 79
Cationic polar cycloaddition, **16**, 289 (**19**, xi)

Chemistry
  of arene oxides, **37**, 67
  of aromatic azapentalenes, **22**, 183
  of barbituric acid, recent progress in, **38**, 229
  of benzo[*b*]furan, Part I, occurrence and synthesis, **18**, 337
  of benzo[*b*]thiophenes, **11**, 177; **29**, 171
  of chrom-3-enes, **18**, 159
  of diazepines, **8**, 21
  of dibenzothiophenes, **16**, 181
  of dihydroazines, **38**, 1
  of 1,2-dioxetanes, **21**, 437
  of furans, **7**, 377
  of hydantoins, **38**, 177
  of isatin, **18**, 1
  of isoxazolidines, **21**, 207
  of lactim ethers, **12**, 185
  of mononuclear isothiazoles, **14**, 1
  of 4-oxy- and 4-keto-1,2,3,4-tetrahydroisoquinolines, **15**, 99
  of phenanthridines, **13**, 315
  of phenothiazines, **9**, 321
  of polycyclic isothiazoles, **38**, 1
  of 1-pyridines, **15**, 197
  of pyrrolizines, **37**, 1
  of tetrazoles, **21**, 323
  of 1,3,4-thiadiazoles, **9**, 165
  of thienothiophenes, **19**, 123
  of thiophenes, **1**, 1
Chrom-3-ene chemistry, advances in, **18**, 159
Claisen rearrangements, in nitrogen heterocyclic systems, **8**, 143
Complex metal hydrides, reduction of nitrogen heterocycles with, **6**, 45
Covalent hydration
  in heteroaromatic compounds, **4**, 1, 43
  in nitrogen heterocycles, **20**, 117
Current views on some physicochemical aspects of purines, **24**, 215
Cyclazines, and related N-bridged annulenes, **22**, 321
Cyclic enamines and imines, **6**, 147
Cyclic hydroxamic acids, **10**, 199
Cyclic peroxides, **8**, 165
Cyclizations under Vilsmeier conditions, **31**, 207
Cycloaddition, cationic polar, **16**, 289 (**19**, xi)
(2 + 2)-Cycloaddition and (2 + 2)-cycloreversion reactions of heterocyclic compounds, **21**, 253

D

Developments in the chemistry
  of furans (1952-1963), **7**, 377
  of Reissert compounds (1968-1978), **24**, 187
Dewar heterocycles and related compounds, **31**, 169
2,4-Dialkoxypyrimidines, Hilbert–Johnson reaction of, **8**, 115
Diazepines, chemistry of, **8**, 21
1,4-Diazepines, 2,3-dihydro-, **17**, 1
Diazirines, diaziridines, **2**, 83; **24**, 63
Diazo compounds, heterocyclic, **8**, 1
Diazomethane, reactions with heterocyclic compounds, **2**, 245
Dibenzothiophenes, chemistry of, **16**, 181
Dihydroazines, recent advances in chemistry of, **38**, 1
Dihydro-1,4-benzothiazines, and related compounds, **38**, 135
2,3-Dihydro-1,4-diazepines, **17**, 1
1,2-Dihydroisoquinolines, **14**, 279
1,2-Dioxetanes, chemistry of, **21**, 437
Diquinolylmethane and its analogs, **7**, 153
gem-Dithienylalkanes and their derivatives, **32**, 83
1,2- and 1,3-Dithiolium ions, **7**, 39; **27**, 151
1,3-Dithiol-3-thiones and 1,2-dithiol-3-ones, **31**, 63

E

Electrochemical synthesis of pyridines, **37**, 167
Electrolysis of *N*-heterocyclic compounds
  part I, **12**, 213
  part II, **36**, 235
Electronic aspects of purine tautomerism, **13**, 77
Electronic structure of biological pyrimidines, tautomerism and, **18**, 199
Electronic structure of heterocyclic sulfur compounds, **5**, 1
Electrophilic substitutions of five-membered rings, **13**, 235
β-Enamino esters, heterocyclic, as heterocyclic synthons, **38**, 299
π-Excessive heteroannulenes, medium-large and large, **23**, 55

F

Ferrocenes, heterocyclic, **13**, 1
Five-membered rings, electrophilic substitutions of, **13**, 235

Formation of anionic σ-adducts from heteroaromatic compounds, **34**, 305
Free radical substitutions of heteroaromatic compounds, **2**, 131
Furans
 development of the chemistry of (1952–1963), **7**, 377
 dibenzo-, **35**, 1
 recent advances in chemistry, Part I, **30**, 167
Furoxans, **29**, 251

G

Grignard reagents, indole, **10**, 43

H

Halogenation of heterocyclic compounds, **7**, 1
Hammett equation, applications to heterocyclic compounds, **3**, 209; **20**, 1
Hetarynes, **4**, 121
Heteroadamantane, **30**, 79
Heteroannulenes, medium-large and large π-excessive, **23**, 55
Heteroaromatic compounds
 N-aminoazonium salts, **29**, 71
 free-radical substitutions of, **2**, 131
 homolytic substitution of, **16**, 123
 nitrogen, covalent hydration in, **4**, 1, 43
 prototropic tautomerism of, **1**, 311, 339; **2**, 1, 27; Suppl. 1
 quaternization of, **22**, 71
Heteroaromatic N-imines, **17**, 213; **29**, 71
Heteroaromatic nitro compounds, ring synthesis of, **25**, 113
Heteroaromatic radicals, Part I, general properties; radicals with Group V ring heteroatoms, **25**, 205; Part II, radicals with Group VI and Groups V and VI ring heteroatoms, **27**, 31
Heteroaromatic substitution, nucleophilic, **3**, 285
Heterocycles
 aromaticity of, **17**, 255
 nomenclature of, **20**, 175
 photochemistry of, **11**, 1
 by ring closure of ortho-substituted r-anilines, **14**, 211
Heterocyclic betaine derivatives of alternant hydrocarbons, **26**, 1

Heterocyclic chemistry, literature of, **7**, 225; **25**, 303
Heterocyclic compounds
 application of Hammett equation to, **3**, 209; **20**, 1
 (2 + 2)-cycloaddition and (2 + 2)-cycloreversion reactions of, **21**, 253
 halogenation of, **7**, 1
 isotopic hydrogen labeling of, **15**, 137
 mass spectrometry of, **7**, 301
 quaternization of, **3**, 1; **22**, 71
 reactions of, with carbenes, **3**, 57
 reactions of diazomethane with, **2**, 245
N-Heterocyclic compounds
 electrolysis of, **12**, 213
 reaction of acetylenecarboxylic acids and esters with, **1**, 125; **23**, 263
Heterocyclic diazo compounds, **8**, 1
Heterocyclic ferrocenes, **13**, 1
Heterocyclic oligomers, **15**, 1
Heterocyclic pseudobases, **1**, 167; **25**, 1
Heterocyclic sulphur compounds, electronic structure of, **5**, 1
Heterocyclic synthesis
 heterocyclic β-enamino esters and, **38**, 299
 from nitrilium salts under acidic conditions, **6**, 95
 through nucleophilic additions to acetylenic esters, **19**, 279
 thioureas in, **18**, 99
Hilbert–Johnson reaction of 2,4-dialkoxypyrimidines, **8**, 115
Homolytic substitution of heteroaromatic compounds, **16**, 123
Hydantoins, chemistry of, **38**, 177
Hydrogen exchange
 base-catalyzed, **16**, 1
 one-step (labeling) methods, **15**, 137
Hydroxamic acids, cyclic, **10**, 199

I

Imidazole chemistry
 advances in, **12**, 103; **27**, 241
 2H-, **35**, 375
 4H-, **35**, 413
Indole Grignard reagents, **10**, 43
Indole(s)
 acid-catalyzed polymerization, **2**, 287
 and derivatives, application of NMR spectroscopy to, **15**, 277

Indolizine chemistry, advances in, **23**, 103
Indolones, isatogens and, **22**, 123
Indoxazenes, **8**, 277; **29**, 1
Isatin, chemistry of, **18**, 1
Isatogens and indolones, **22**, 123
Isatoic anhydrides, uses in heterocyclic synthesis, **28**, 127
Isoindoles, **10**, 113; **29**, 341
Isoquinolines
 1,2-dihydro-, **14**, 279
 4-oxy- and 4-keto-1,2,3,4-tetrahydro-, **15**, 99
Isothiazoles, **14**, 107
 recent advances in the chemistry of monocyclic, **14**, 1
 polycyclic, recent advances in chemistry of, **38**, 105
Isotopic hydrogen labeling of heterocyclic compounds, one-step methods, **15**, 137
Isoxazole chemistry, recent developments in, **2**, 365; since 1963, **25**, 147
Isoxazolidines, chemistry of, **21**, 207

L

Lactim ethers, chemistry of, **12**, 185
Literature of heterocyclic chemistry, **7**, 225; **25**, 303

M

Mass spectrometry of heterocyclic compounds, **7**, 301
Medium-large and large π-excessive heteroannulenes, **23**, 55
Meso-ionic compounds, **19**, 1
Metal catalysts, action on pyridines, **2**, 179
Monoazaindoles, **9**, 27
Monocyclic pyrroles, oxidation, of, **15**, 67
Monocyclic sulfur-containing pyrones, **8**, 219
Mononuclear heterocyclic rearrangements, **29**, 141
Mononuclear isothiazoles, recent advances in chemistry of, **14**, 1

N

Naphthalen-1,4,imines, **16**, 87
Naphthyridines, **11**, 124
 reactivity of, toward nitrogen nucleophiles, **33**, 95
 recent developments in chemistry of, **33**, 147
Nitriles and nitrilium salts, heterocyclic synthesis involving, **6**, 95
Nitrogen-bridged six-membered ring systems, **16**, 87
Nitrogen heterocycles
 Conformational equilibria in saturated six-membered rings, **36**, 1
 covalent hydration in, **20**, 117
 reactions of acetylenecarboxylic esters with, **23**, 263
 reduction of, with complex metal hydrides, **6**, 45
Nitrogen heterocyclic systems, Claisen rearrangements in, **8**, 143
Nomenclature of heterocycles, **20**, 175
Nuclear magnetic resonance spectroscopy, application to indoles, **15**, 277
Nucleophiles, reactivity of azine derivatives with, **4**, 145
Nucleophilic additions to acetylenic esters, synthesis of heterocycles through, **19**, 299
Nucleophilic heteroaromatic substitution, **3**, 285

O

Olefin synthesis with anils, **23**, 171
Oligomers, heterocyclic, **15**, 1
1,2,4-Oxadiazoles, **20**, 65
1,3,4-Oxadiazole chemistry, recent advances in, **7**, 183
1,3-Oxazine derivatives, **2**, 311; **23**, 1
Oxaziridines, **2**, 83; **24**, 63
Oxazole chemistry, advances in, **17**, 99
Oxazolone chemistry
 new developments in, **21**, 175
 recent advances in, **4**, 75
Oxidation of monocyclic pyrroles, **15**, 67
3-Oxo-2,3-dihydrobenz[d]isothiazole 1,1-dioxide (saccharin) and derivatives, **15**, 233
4-Oxy- and 4-keto-1,2,3,4-tetrahydroisoquinolines, chemistry of, **15**, 99

P

Pentazoles, **3**, 373
Peroxides, cyclic, **8**, 165 (*see also* 1,2-Dioxetanes)
Phase transfer catalysis, applications in heterocyclic chemistry, **36**, 175

Phenanthridine chemistry, recent developments in, **13**, 315
Phenanthrolines, **22**, 1
Phenothiazines, chemistry of, **9**, 321
Phenoxazines, **8**, 83
Photochemistry
  of heterocycles, **11**, 1
  of nitrogen-containing heterocycles, **30**, 239
  of oxygen- and sulfur-containing heterocycles, **33**, 1
Physicochemical aspects of purines, **6**, 1; **24**, 215
Physicochemical properties
  of azines, **5**, 69
  of pyrroles, **11**, 383
3-Piperideines, **12**, 43
Polyfluoroheteroaromatic compounds, **28**, 1
Polymerization of pyrroles and indoles, acid-catalyzed, **2**, 1
Prototropic tautomerism of heteroaromatic compounds, **1**, 311, 339; **2**, 1, 27; Suppl. 1
Pseudoazulenes, **33**, 185
Pseudobases, heterocyclic, **1**, 167; **25**, 1
Purine bases, aza analogs of, **1**, 189
Purines
  physicochemical aspects of, **6**, 1; **24**, 215
  tautomerism, electronic aspects of, **13**, 77
Pyrans, thiopyrans, and selenopyrans, **34**, 145
Pyrazine chemistry, recent advances in, **14**, 99
Pyrazole chemistry
  3H-, **34**, 1
  4H-, **34**, 53
  progress in, **6**, 347
Pyrazolopyridines, **36**, 343
Pyridazines, **9**, 211; **24**, 363
Pyridine(s)
  action of metal catalysts on, **2**, 179
  effect of substituents on substitution in, **6**, 229
  synthesis by electrochemical methods, **37**, 167
  1,2,3,6-tetrahydro-, **12**, 43
Pyridoindoles (the carboline, **3**, 79
Pyridopyrimidines, **10**, 149
Pyrido[1,2-a]pyrimidines, chemistry of, **33**, 241
Pyrimidine bases, aza analogs of, **1**, 189
Pyrimidines
  2,4-dialkoxy-, Hilbert–Johnson reaction of, **8**, 115
  tautomerism and electronic structure of biological, **18**, 199
1-Pyrindines, chemistry of, **15**, 197

Pyrones, monocyclic sulfur-containing, **8**, 219
Pyrroles
  acid-catalyzed polymerization of, **2**, 287
  2H- and 3H-, **32**, 233
  oxidation of monocyclic, **15**, 67
  physicochemical properties of, **11**, 383
Pyrrolizidine chemistry, **5**, 315; **24**, 247
Pyrrolizines, chemistry of, **37**, 1
Pyrrolodiazines, with a bridgehead nitrogen, **21**, 1
Pyrrolopyridines, **9**, 27
Pyrylium salts, syntheses, **10**, 241

Q

Quaternization
  of heteroaromatic compounds, **22**, 71
  of heterocyclic compounds, **3**, 1
Quinazolines, **1**, 253; **24**, 1
Quinolizines, aromatic, **5**, 291
Quinoxaline chemistry
  developments 1963–1975, **22**, 367
  recent advances in, **2**, 203
Quinuclidine chemistry, **11**, 473

R

Recent advances in furan chemistry, Part I, **30**, 168; Part II, **31**, 237
Reduction of nitrogen heterocycles with complex metal hydrides, **6**, 45
Reissert compounds, **9**, 1j **24**, 187
Ring closure of ortho-substituted t-anilines, for heterocycles, **14**, 211
Ring synthesis of heteroaromatic nitro compounds, **25**, 113

S

Saccharin and derivatives, **15**, 233
Selenazole chemistry, present state of, **2**, 343
Selenium–nitrogen heterocycles, **24**, 109
Selenophene chemistry, advances in, **12**, 1
Selenophenes, **30**, 127
Six-membered ring systems, nitrogen bridged, **16**, 87

Substitution(s)
  electrophilic, of five-membered rings, **13**, 235
  homolytic, of heteroaromatic compounds, **16**, 123
  nucleophilic heteroaromatic, **3**, 285
  in pyridines, effect of substituents, **6**, 229
Sulfur compounds
  electronic structure of heterocyclic, **5**, 1
  four-membered rings, **35**, 199
Sulfur transfer reagents in heterocyclic synthesis, **30**, 47
Synthesis, of tetracyclic and pentacyclic condensed thiophene systems, **32**, 127
Synthesis and reactions of 1-azirines, **13**, 45
Synthesis of heterocycles
  heterocyclic β-enamino esters as synthons in, **38**, 299
  from nitrilium salts under acidic conditions, **6**, 95
  through nucleophilic additions to acetylenic esters, **19**, 279
  thioureas in, **18**, 99

T

Tautomerism
  electronic aspects of purine, **13**, 77
  and electronic structure of biological pyrimidines, **18**, 199
  prototropic, of heteroaromatic compounds, **1**, 311, 339; **2**, 1, 27; Suppl. 1
Tellurophene and related compounds, **21**, 119
1,2,3,4-Tetrahydroisoquinolines, 4-oxy- and 4-keto-, **15**, 99
1,2,3,6-Tetrahydropyridines, **12**, 43

Tetrazole chemistry, recent advances in, **21**, 323
Theoretical studies of physicochemical properties and reactivity of azines, **5**, 69
1,2,4-Thiadiazoles, **5**, 119; **32**, 285
1,2,5-Thiadiazoles, chemistry of, **9**, 107
1,3,4-Thiadiazoles, recent advances in the chemistry of, **9**, 165
Thiathiophthenes (1,6,6aS^{IV}-trithiapentalenes), **13**, 161
1,2,3,4-Thiatriazoles, **3**, 263; **20**, 145
1,4-Thiazines and their dihydro derivatives, **24**, 293
4-Thiazolidinones, **25**, 83
Thienopyridines, **21**, 65
Thienothiophenes and related systems, chemistry of, **19**, 123
Thiochromanones and related compounds, **18**, 59
Thiocoumarins, **26**, 115
Thiophenes, chemistry of, recent advances in, **1**, 1
Thiopyrones (monocyclic sulfur-containing pyrones), **8**, 219
Thioureas in synthesis of heterocycles, **18**, 99
Three-membered rings with two heteroatoms, **2**, 83; **24**, 63
Transition organometallic compounds in heterocyclic synthesis, use of, **30**, 321
1,3,5-, 1,3,6-, 1,3,7-, and 1,3,8-Triazanaphthalenes, **10**, 149
1,2,3-Triazines, **19**, 215
1,2,3-Triazoles, **16**, 33
Δ^2-1,2,3-Triazolines, **37**, 217
Δ^3- and Δ^4-1,2,3-Triazolines, **37**, 351
Triazolopyridines, **34**, 79
1,6,5aS^{IV}-Trithiapentalenes, **13**, 161